NATIONAL 5

MATHS

WITH ANSWERS

SECOND EDITION

David Alcorn
Consultant editors:
Bob Barclay and Mike Smith

Orders: please contact Hachette UK Distribution, Hely Hutchinson Centre, Milton Road, Didcot, Oxfordshire, OX11 7HH. Telephone: +44 (0)1235 827827. Email education@hachette.co.uk. Lines are open from 9 a.m. to 5 p.m., Monday to Friday. You can also order through our website: www.hoddereducation.co.uk. If you have queries or questions that aren't about an order, you can contact us at hoddergibson@hodder.co.uk

First published in 2017 by
Hodder Gibson, an imprint of Hodder Education
An Hachette UK Company
50 Frederick Street
Edinburgh, EH2 1EX

Impression number 5 4
Year 2022

Illustrations by David Alcorn
Typeset in 11pt Times New Roman by Billy Johnson, San Francisco, California, USA
Printed and bound by CPI Group (UK) Ltd, Croydon CR0 4YY

A catalogue record for this title is available from the British Library

ISBN: 978 1 5104 2917 8

PREFACE

National 5 Maths Second Edition has been specifically written to meet the latest requirements of the SQA Mathematics Course and provides full coverage of the specifications of the **SQA Mathematics (National 5) Course**.

In preparing the text, full account has been made of the requirements for students to be able to use and apply mathematics in written examination papers and be able to solve problems both with and without a calculator.

To provide efficient, yet flexible, coverage of the specifications, the book has been split into sections.

Chapters 1 - 5 **Number**
Chapters 6 - 14 **Algebra**
Chapters 15 - 20 **Geometry**
Chapters 21 - 24 **Trigonometry**
Chapters 25 - 26 **Statistics**

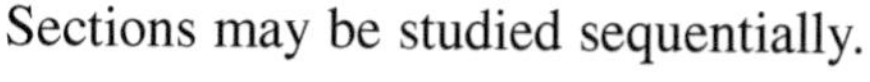

Sections may be studied sequentially.

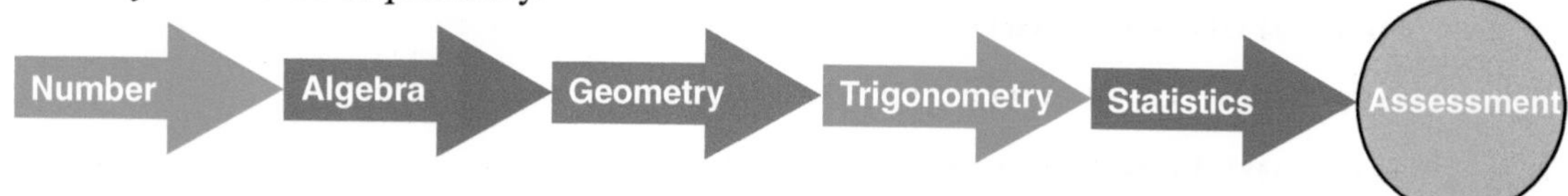

Alternatively, you may wish to study material from different chapters across all five sections.

The chapters within each unit have been organised to facilitate either approach.

You can best decide the approach to use depending on the individual needs of the students.

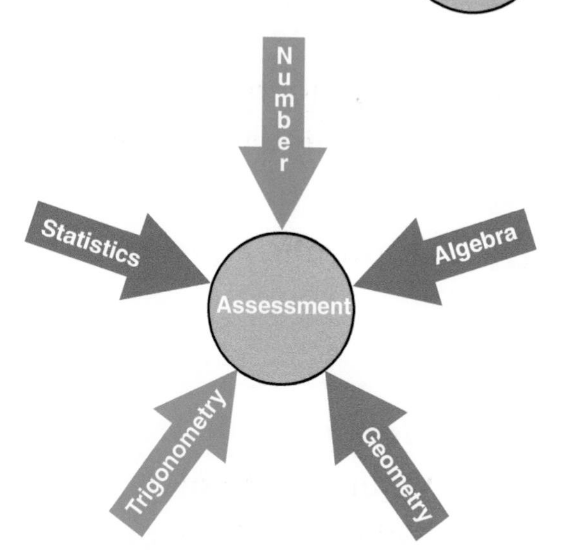

Each chapter consists of fully worked examples with explanatory notes and commentary, carefully graded questions, a summary of key points and a review exercise.
The review exercises provide the opportunity to consolidate topics introduced in the chapter and an efficient method of monitoring progress through the course.

Some chapters include ideas for investigation. These give students the opportunity to improve and practise their skills of using and applying mathematics.

Twelve revision exercises, organised to provide practice for non-calculator and calculator papers, provide opportunities to consolidate skills acquired during the course.

As final preparation for the exams, a further compilation of exam practice questions has been provided, which has been organised for non-calculator paper and calculator paper practice.

CONTENTS

Number   Chapters 1 - 5

Algebra   Chapters 6 - 14

Geometry   Chapters 15 - 20

Trigonometry Chapters 21 - 24 **Page**

Statistics Chapters 25 - 26

Revision Exercises

Exam Practice

1 Approximation

Whole numbers and decimals

The numbers 0, 1, 2, 3, 4, 5, ... can be used to count objects.
Such numbers are called **whole numbers**.

Our number system is made up of the digits 0, 1, 2, 3, 4, 5, 6, 7, 8 and 9.
The position a digit has in a number is called its **place value**.
In the number 5384 the digit 8 is worth 80, but in the number 4853 the digit 8 is worth 800.

Numbers and quantities are not always whole numbers.
The number system can be extended to include **decimal numbers**.

A **decimal point** is used to separate the whole number part from the decimal part of the number.

73.26 — This number is read as seventy-three point two six.

whole number, 73 — decimal part, 2 tenths + 6 hundredths (which is the same as 26 hundredths)

Many measurements are recorded using decimals, including money, time, distance, weight, volume, etc.

Approximation

In real-life it is not always necessary to use exact numbers. A number can be **rounded** to an **approximate** number. Numbers are rounded according to how accurately we wish to give details.
For example, the distance to the Sun can be given as 93 million miles.

Can you think of other situations where approximations might be used?

Rounding using decimal places

What is the cost of 1.75 metres of material costing £3.99 a metre?

$1.75 \times 3.99 = 6.9825$

The cost of the material is £6.9825 or 698.25p.
As you can only pay in pence, a sensible answer is £6.98,
correct to two decimal places (nearest penny).
This means that there are only two decimal places after the decimal point.

Often it is not necessary to use an exact answer. Sometimes it is impossible, or impractical, to use the exact answer.

To round a number to a given number of decimal places

When rounding a number to one, two or more decimal places:

1. Write the number using one more decimal place than asked for.
2. Look at the last decimal place and
   - if the figure is 5 or more round up,
   - if the figure is less than 5 round down.
3. When answering a problem remember to include any units and state the degree of approximation used.

Example 1

Write 2.76435 to 2 decimal places.

Look at the third decimal place. **4**
This is less than 5, so, round down.
Answer 2.76

Example 2

Write 7.104 to 2 decimal places.
7.104 = 7.10 to 2 d.p.
The zero is written down because it shows the accuracy used, 2 decimal places.

Notation: Often decimal place is shortened to d.p.

Practice Exercise 1.1

1. Write the number 3.9617 correct to
   (a) 3 decimal places, (b) 2 decimal places, (c) 1 decimal place.

2. The display on a calculator shows the result of $34 \div 7$.
   What is the result correct to two decimal places?

3. 68.847 kg The scales show Gary's weight.
   Write Gary's weight correct to one decimal place.

4. Copy and complete this table.

Number	2.367	0.964	0.965	15.2806	0.056	4.991	4.996
d.p.	1	2	2	3	2	2	2
Answer	2.4						

5. Carry out these calculations giving the answers correct to
   (a) 1 d.p. (b) 2 d.p. (c) 3 d.p.
   (i) 6.12×7.54 (ii) 89.1×0.67 (iii) 90.53×6.29
   (iv) $98.6 \div 5.78$ (v) $67.2 \div 101.45$

6. In each of these short problems decide upon the most suitable accuracy for the answer.
   Then calculate the answer. Give a reason for your degree of accuracy.
   (a) One gallon is 4.54596… litres. How many litres is 9 gallons?
   (b) What is the cost of 0.454 kg of cheese at £9.47 per kilogram?
   (c) The total length of 7 equal sticks, lying end to end, is 250 cm. How long is each stick?
   (d) A packet of 6 bandages costs £7.99. How much does one bandage cost?
   (e) Petrol costs 133.9 pence a litre. I buy 15.6 litres. How much will I have to pay?

Rounding using significant figures

Consider the calculation $600.02 \times 7500.97 = 4500732.0194$
To 1 d.p. it is 4500732.0, to 2 d.p. it is 4500732.02.
The answers to either 1 or 2 d.p. are very close to the actual answer and are almost as long.
There is little advantage in using either of these two roundings.
The point of a rounding is that it is a more convenient number to use.

Another kind of rounding uses **significant figures**.
The **most** significant figure in a number is the figure which has the greatest place value.

Consider the number 237.
The figure 2 has the greatest place value. It is worth 200.
So, 2 is the most significant figure.

In the number 0.00328, the figure 3 has the greatest place value.
So, 3 is the most significant figure.

> Noughts which are used to locate the decimal point and preserve the place value of other figures are not significant.

To round a number to a given number of significant figures

When rounding a number to one, two or more significant figures:

1. Start from the most significant figure and count the required number of figures.
2. Look at the next figure to the right of this and
   - if the figure is 5 or more round up,
   - if the figure is less than 5 round down.
3. Add noughts, as necessary, to locate the decimal point and preserve the place value.
4. When answering a problem remember to include any units and state the degree of approximation used.

Example 3

Write 4 500 732.0194 to 2 significant figures.

The figure after the first 2 significant figures **45** is 0.
This is less than 5, so, round down, leaving 45 unchanged.
Add noughts to 45 to locate the decimal point and preserve place value.
So, 4 500 732.0194 = 4 500 000 to 2 sig. fig.

Notation:
Often significant figure is shortened to sig. fig.

Example 4

Write 0.000364907 to 1 significant figure.

The figure after the first significant figure **3** is 6.
This is 5 or more, so, round up, 3 becomes 4.
So, 0.000364907 = 0.0004 to 1 sig. fig.

Notice that the noughts before the 4 locate the decimal point and preserve place value.

Choosing a suitable degree of accuracy

In some calculations it would be wrong to use the complete answer from the calculator.
The result of a calculation involving measurement should not be given to a greater degree of accuracy than the measurements used in the calculation.

Example 5

What is the area of a rectangle measuring 4.6 cm by 7.2 cm?

$4.6 \times 7.2 = 33.12$
Since the measurements used in the calculation (4.6 cm and 7.2 cm) are given to 2 significant figures the answer should be as well.
33 cm² is a more suitable answer.

Note:
To find the area of a rectangle: multiply length by breadth.

Practice Exercise 1.2

1. Write these numbers correct to one significant figure.
   (a) 17 (b) 523 (c) 350 (d) 1900 (e) 24.6
   (f) 0.083 (g) 0.086 (h) 0.00948 (i) 0.0095

2. Copy and complete this table.

Number	456 000	454 000	7 981 234	0.000567	0.093748	0.093748
sig. fig.	2	2	3	2	2	3
Answer	460 000					

3. This display shows the result of $3400 \div 7$.
   What is the result correct to two significant figures?

   485.7142857

4. Carry out these calculations giving the answers correct to
   (a) 1 sig. fig. (b) 2 sig. fig. (c) 3 sig. fig.
   (i) 672×123 (ii) 6.72×12.3 (iii) 78.2×12.8
   (iv) $7.19 \div 987.5$ (v) $124 \div 65300$

5. A rectangular field measures 18.6 m by 25.4 m.
   Calculate the area of the field, giving your answer to a suitable degree of accuracy.

6. In each of these short problems decide upon the most suitable accuracy for the answer.
   Then work out the answer, remembering to state the units.
   Give a reason for your degree of accuracy.
   (a) The area of a rectangle measuring 13.2 cm by 11.9 cm.
   (b) The area of a football pitch measuring 99 m by 62 m.
   (c) The total length of 13 tables placed end to end measures 16 m. How long is each table?
   (d) The area of carpet needed to cover a rectangular floor measuring 3.65 m by 4.35 m.

Key Points

- In real-life it is not always necessary to use exact numbers. A number can be **rounded** to an **approximate** number. Numbers are rounded according to how accurately we wish to give details. For example, the distance to the Sun can be given as 93 million miles.
- You should be able to approximate using **decimal places**.

> Write the number using one more decimal place than asked for.
> Look at the last decimal place and
> - if the figure is 5 or more round up,
> - if the figure is less than 5 round down.

- You should be able to approximate using **significant figures**.

> Start from the most significant figure and count the required number of figures.
> Look at the next figure to the right of this and
> - if the figure is 5 or more round up,
> - if the figure is less than 5 round down.
>
> Add noughts, as necessary, to preserve the place value.

- You should be able to choose a suitable degree of accuracy.

> The result of a calculation involving measurement should not be given to a greater degree of accuracy than the measurements used in the calculation.

Review Exercise 1

1. Write these numbers correct to 2 decimal places.
   (a) 28.714 (b) 6.91288 (c) 12.397 (d) 0.0418 (e) 0.00912
2. Write these numbers correct to 3 significant figures.
   (a) 2313 (b) 23.58 (c) 36.97 (d) 503.89 (e) 0.0005646
3. The display shows the result of $179 \div 7$.
   What is the result correct to:
   (a) two decimal places,
   (b) one decimal place,
   (c) one significant figure?

   25.57142857
4. Calculate 7.25×0.79
   (a) to 1 decimal place, (b) to 2 decimal places, (c) to 3 decimal places.
5. Calculate $107.9 \div 72.5$ (a) to 1 significant figure, (b) to 2 significant figures.
6. Daniel has a part-time job in a factory. He is paid £36 for each shift he works.
   Last year he worked 108 shifts.
   Calculate Daniel's total pay for the year. Give your answer to the nearest £100.
7. The floor of a lounge is a rectangle which measures 5.23 m by 3.62 m.
   The floor is to be carpeted.
   (a) Calculate the area of carpet needed.
   Give your answer to an appropriate degree of accuracy.
   (b) Explain why you chose this degree of accuracy.
8. Flour costs 79p per kilogram from the flour mill.
   Rachel bought 300 kg of flour from the mill.
   She shared the flour equally between 18 people.
   How much should each person pay?

2 Working with Surds

All real numbers are either **rational** or **irrational**.

Rational numbers

Numbers which can be written in the form $\frac{a}{b}$, where a and b are integers ($b \neq 0$) are **rational**.
Examples of rational numbers are:

$$2 \qquad -5 \qquad \frac{2}{5} \qquad 0.\dot{6} \qquad 3.47 \qquad 1\frac{3}{4}$$

$\frac{a}{b}$ is a **proper fraction** if $a < b$.

$\frac{a}{b}$ is an **improper fraction** (top heavy) if $a > b$.

All fractions can be written as decimals.

$\frac{3}{4}$ can be thought of as $3 \div 4$ and is equal to 0.75.

Some decimals have recurring digits.
These are shown by:
a single dot above a single recurring digit,
a dot above the first and last digit of a set of recurring digits.

For example: $\frac{5}{9} = 0.5555\ldots = 0.\dot{5}$ $\qquad \frac{5}{11} = 0.454545\ldots = 0.\dot{4}\dot{5}$

$$\frac{123}{999} = 0.123123123\ldots = 0.\dot{1}2\dot{3}$$

Irrational numbers

An **irrational** number **cannot** be written as a fraction.
Irrational numbers include:
square roots of non-square numbers,
cube roots of non-cube numbers.

Examples of irrational numbers are: $\sqrt{2} \quad \sqrt[3]{7} \quad \pi \quad \sqrt{13}$

Example 1

State whether each of the following are rational or irrational numbers.

$$\pi \qquad \sqrt[3]{7} \qquad \sqrt{36}$$

$\pi = 3.141592654\ldots$ irrational π is a non-recurring decimal and has no exact value.

$\sqrt[3]{7} = 1.91293118\ldots$ irrational $\sqrt[3]{7}$ is a non-recurring decimal and has no exact value.

$\sqrt{36} = 6$ rational Note: $\sqrt{36}$ means the positive square root of 36.

Practice Exercise 2.1

1. Which of these numbers are rational and which are irrational?
   (a) $\sqrt{2}$ (b) 3.14 (c) $\sqrt[3]{9}$ (d) $\sqrt{\frac{1}{4}}$ (e) $\frac{\sqrt{64}}{3}$
   (f) $\left(\frac{\sqrt{3}}{2}\right)^2$ (g) $\sqrt{\frac{1}{2}}$ (h) $\sqrt{1\frac{7}{9}}$ (i) $\sqrt{6\frac{1}{4}}$ (j) π^2

2. m and n represent two different irrational numbers.
   In each case, write down one example to show
   (a) mn is rational, (b) mn is irrational, (c) $\frac{m}{n}$ is rational, (d) $\frac{m}{n}$ is irrational.

Surds

Roots of rational numbers which **cannot** be expressed as rational numbers are called **surds**.
A surd is an irrational number.

These are examples of surds:

$\sqrt{2}$ $\sqrt{0.37}$ $\sqrt[3]{10}$ $3 + \sqrt{2}$ $\sqrt{7}$

Numbers like $\sqrt{64}$, $\sqrt{0.25}$, $\sqrt[3]{27}$ are not surds because the root of each number is rational.

($\sqrt{64} = 8$, $\sqrt{0.25} = 0.5$, $\sqrt[3]{27} = 3$.)

$\sqrt{a}$ means the positive square root of a.

Manipulating and simplifying surds

$\sqrt{ab} = \sqrt{a} \times \sqrt{b}$ $m\sqrt{a} + n\sqrt{a} = (m + n)\sqrt{a}$ $\sqrt{\frac{a}{b}} = \frac{\sqrt{a}}{\sqrt{b}}$

To simplify surds, look for factors that are square numbers.

Example 2

Simplify the following leaving the answers in surd form.

(a) $\sqrt{28}$ (b) $\sqrt{50} - \sqrt{32}$ (c) $\sqrt{48} + \sqrt{75}$ (d) $\sqrt{\frac{72}{20}}$

(a) $\sqrt{28} = \sqrt{4} \times \sqrt{7}$
$= 2\sqrt{7}$

(b) $\sqrt{50} - \sqrt{32} = \sqrt{25} \times \sqrt{2} - \sqrt{16} \times \sqrt{2}$
$= 5\sqrt{2} - 4\sqrt{2}$
$= \sqrt{2}$

(c) $\sqrt{48} + \sqrt{75} = \sqrt{16} \times \sqrt{3} + \sqrt{25} \times \sqrt{3}$
$= 4\sqrt{3} + 5\sqrt{3}$
$= 9\sqrt{3}$

(d) $\sqrt{\frac{72}{20}} = \frac{\sqrt{72}}{\sqrt{20}}$
$= \frac{\sqrt{36} \times \sqrt{2}}{\sqrt{4} \times \sqrt{5}}$
$= \frac{6\sqrt{2}}{2\sqrt{5}}$
$= \frac{3\sqrt{2}}{\sqrt{5}}$

Example 3

Remove the brackets and simplify $\sqrt{3}(\sqrt{6} + 2)$.

$\sqrt{3}(\sqrt{6} + 2) = \sqrt{3} \times \sqrt{6} + 2\sqrt{3}$
$= \sqrt{18} + 2\sqrt{3}$
$= \sqrt{9} \times \sqrt{2} + 2\sqrt{3}$
$= 3\sqrt{2} + 2\sqrt{3}$

$2\sqrt{3} = 2 \times \sqrt{3} = \sqrt{3} \times 2$

Practice Exercise 2.2

1. Which of the following are surds?

(a) $\sqrt{2}$ (b) $\sqrt{4}$ (c) $\sqrt{9}$ (d) $\sqrt{10}$ (e) $\sqrt{40}$

(f) $\sqrt{1}$ (g) $\sqrt[3]{1}$ (h) $\sqrt[3]{8}$ (i) $\sqrt[3]{9}$ (j) $\sqrt[3]{27}$

(k) $\sqrt{0.4}$ (l) $\sqrt{0.09}$ (m) $(\sqrt{3})^3$ (n) $(\sqrt{0.4})^2$ (o) $\sqrt{54}$

2. Write the following surds in their simplest form.

(a) $\sqrt{12}$ (b) $\sqrt{27}$ (c) $\sqrt{45}$ (d) $\sqrt{48}$ (e) $\sqrt{32}$

(f) $\sqrt{50}$ (g) $\sqrt{54}$ (h) $\sqrt{24}$ (i) $\sqrt{98}$ (j) $\sqrt{80}$

3. Express in the form $a\sqrt{b}$, where $a\sqrt{b}$ is in its simplest form.

(a) $\sqrt{44}$ (b) $\sqrt{75}$ (c) $\sqrt{128}$ (d) $\sqrt{72}$ (e) $\sqrt{200}$

4. Simplify the following.

(a) $\sqrt{\frac{9}{16}}$ (b) $\sqrt{\frac{49}{64}}$ (c) $\sqrt{\frac{18}{8}}$ (d) $\frac{\sqrt{28}}{\sqrt{7}}$ (e) $\frac{\sqrt{48}}{\sqrt{3}}$

5. Simplify.

(a) $\sqrt{2} + \sqrt{2}$ (b) $2\sqrt{5} - \sqrt{5}$ (c) $5\sqrt{3} + 2\sqrt{3}$

(d) $5\sqrt{2} - 3\sqrt{2}$ (e) $2\sqrt{5} + 3\sqrt{5}$ (f) $7\sqrt{3} - 3\sqrt{3} + \sqrt{3}$

(g) $\sqrt{18} + \sqrt{8}$ (h) $\sqrt{50} - \sqrt{32}$ (i) $\sqrt{45} + \sqrt{80}$

(j) $\sqrt{75} - \sqrt{12}$ (k) $\sqrt{300} - \sqrt{48}$ (l) $\sqrt{50} + \sqrt{18} - \sqrt{8}$

(m) $3\sqrt{20} + 2\sqrt{45}$ (n) $2\sqrt{48} + 3\sqrt{12}$ (o) $3\sqrt{45} - 2\sqrt{20}$

(p) $\sqrt{200} - 2\sqrt{18} + \sqrt{72}$ (q) $\sqrt{300} + \sqrt{48} - 3\sqrt{27}$ (r) $3\sqrt{18} - 2\sqrt{8} + \sqrt{2}$

6. Simplify the following.

(a) $\sqrt{3} \times \sqrt{3}$ (b) $\sqrt{3} \times 2\sqrt{3}$ (c) $2\sqrt{5} \times 3\sqrt{5}$ (d) $\sqrt{2} \times \sqrt{8}$

(e) $\sqrt{12} \times \sqrt{3}$ (f) $\sqrt{5} \times \sqrt{10}$ (g) $2\sqrt{6} \times \sqrt{3}$ (h) $2\sqrt{5} \times \sqrt{10}$

(i) $\sqrt{8} \times \sqrt{18}$ (j) $3\sqrt{2} \times 2\sqrt{3}$ (k) $4\sqrt{3} \times 2\sqrt{2}$ (l) $\sqrt{27} \times \sqrt{32}$

7. Remove the brackets and simplify the following.

(a) $\sqrt{2}(\sqrt{2} + 1)$ (b) $\sqrt{3}(\sqrt{6} - \sqrt{3})$ (c) $\sqrt{2}(\sqrt{6} + \sqrt{2})$ (d) $\sqrt{5}(\sqrt{10} - \sqrt{5})$

Rationalising denominators of fractions

When the denominator of a fraction is a surd it is usual to remove the surd from the denominator. This process is called **rationalising the denominator**.

For fractions of the form $\frac{a}{\sqrt{b}}$, multiply both the numerator (top) and the denominator (bottom) of the fraction by $\sqrt{b}$ and then simplify where possible.

$$\frac{a}{\sqrt{b}} = \frac{a\sqrt{b}}{\sqrt{b}\sqrt{b}} = \frac{a\sqrt{b}}{b}$$

Example 4

Rationalise the denominator and simplify where possible.

(a) $\frac{1}{\sqrt{2}}$ (b) $\frac{3\sqrt{2}}{\sqrt{6}}$

(a) $\frac{1}{\sqrt{2}} = \frac{1}{\sqrt{2}} \times \frac{\sqrt{2}}{\sqrt{2}} = \frac{\sqrt{2}}{2}$

(b) $\frac{3\sqrt{2}}{\sqrt{6}} = \frac{3\sqrt{2} \times \sqrt{6}}{\sqrt{6} \times \sqrt{6}} = \frac{3\sqrt{2}\sqrt{6}}{6} = \frac{3\sqrt{2}\sqrt{2}\sqrt{3}}{6} = \frac{6\sqrt{3}}{6} = \sqrt{3}$

Practice Exercise 2.3

1. Rationalise the denominator in each of the following and then simplify the fraction.

(a) $\frac{1}{\sqrt{3}}$ (b) $\frac{1}{\sqrt{5}}$ (c) $\frac{1}{\sqrt{7}}$ (d) $\frac{2}{\sqrt{2}}$ (e) $\frac{5}{\sqrt{5}}$

(f) $\frac{4}{\sqrt{2}}$ (g) $\frac{6}{\sqrt{3}}$ (h) $\frac{14}{\sqrt{7}}$ (i) $\frac{3}{\sqrt{6}}$ (j) $\frac{15}{\sqrt{5}}$

(k) $\frac{9}{\sqrt{3}}$ (l) $\frac{5}{\sqrt{15}}$ (m) $\frac{18}{\sqrt{6}}$ (n) $\frac{35}{\sqrt{5}}$ (o) $\frac{7}{\sqrt{21}}$

(p) $\frac{11}{\sqrt{22}}$ (q) $\frac{10}{\sqrt{30}}$ (r) $\frac{21}{\sqrt{14}}$ (s) $\frac{14}{\sqrt{35}}$ (t) $\frac{15}{\sqrt{10}}$

2. Express each of the following in its simplest form with a rational denominator.

(a) $\frac{6}{\sqrt{8}}$ (b) $\frac{6}{\sqrt{12}}$ (c) $\frac{6}{\sqrt{24}}$ (d) $\frac{8}{\sqrt{32}}$ (e) $\frac{9}{\sqrt{18}}$

(f) $\frac{\sqrt{3}}{\sqrt{6}}$ (g) $\frac{\sqrt{15}}{\sqrt{5}}$ (h) $\frac{\sqrt{8}}{\sqrt{2}}$ (i) $\frac{\sqrt{12}}{\sqrt{3}}$ (j) $\frac{\sqrt{18}}{\sqrt{2}}$

(k) $\frac{\sqrt{5}}{\sqrt{20}}$ (l) $\frac{\sqrt{32}}{\sqrt{2}}$ (m) $\frac{\sqrt{75}}{\sqrt{100}}$ (n) $\frac{2\sqrt{3}}{\sqrt{12}}$ (o) $\frac{3\sqrt{5}}{\sqrt{15}}$

(p) $\frac{4\sqrt{6}}{\sqrt{12}}$ (q) $\frac{2\sqrt{8}}{\sqrt{32}}$ (r) $\frac{5\sqrt{7}}{\sqrt{35}}$ (s) $\frac{\sqrt{3}\sqrt{5}}{\sqrt{30}}$ (t) $\frac{\sqrt{2}\sqrt{3}}{\sqrt{18}}$

Key Points

- All real numbers are either **rational** or **irrational**.
- **Rational numbers** can be written in the form $\frac{a}{b}$, where a and b are integers ($b \neq 0$).
  Examples of rational numbers are: 2, -5, $\frac{2}{5}$, $0.\dot{6}$, 3.47, $1\frac{3}{4}$.
- An **irrational** number **cannot** be written as a fraction.
  Irrational numbers include: square roots of non-square numbers,
  cube roots of non-cube numbers.
  Examples of irrational numbers are: $\sqrt{2}$ $\sqrt[3]{7}$ π $\sqrt{13}$
- Roots of rational numbers which **cannot** be expressed as rational numbers are called **surds**.
  A surd is an irrational number.
  These are examples of surds: $\sqrt{2}$ $\sqrt{0.37}$ $\sqrt[3]{10}$ $3+\sqrt{2}$ $\sqrt{7}$

  $\sqrt{a}$ means the positive square root of a.

  Numbers like $\sqrt{64}$, $\sqrt{0.25}$, $\sqrt[3]{27}$ are not surds because the root of each number is rational.
  ($\sqrt{64} = 8$, $\sqrt{0.25} = 0.5$, $\sqrt[3]{27} = 3$.)
- Rules for manipulating and simplifying surds:

  $$\sqrt{ab} = \sqrt{a} \times \sqrt{b} \qquad m\sqrt{a} + n\sqrt{a} = (m+n)\sqrt{a} \qquad \sqrt{\frac{a}{b}} = \frac{\sqrt{a}}{\sqrt{b}}$$

  Look for factors that are square numbers.
- To **rationalise** the denominator of a fraction of the form $\frac{a}{\sqrt{b}}$ multiply both the numerator (top) and the denominator (bottom) of the fraction by $\sqrt{b}$.

  $$\frac{a}{\sqrt{b}} = \frac{a\sqrt{b}}{\sqrt{b} \times \sqrt{b}} = \frac{a\sqrt{b}}{b}$$
- You should be able to use surds in calculations.

  To keep an answer **exact** it is necessary to keep numbers like $\sqrt{3}$ in surd form.

Review Exercise 2

1. An integer, n, is such that $n < \sqrt{250} < n + 1$.
What is the value of n?

2. Write these numbers in order of size, smallest first, and say whether they are rational or irrational.
$$5\tfrac{1}{7}, \quad \pi + 2, \quad 5.1416, \quad \sqrt{27}, \quad 2.268^2.$$

3. Which of the following are rational and which are irrational?
(a) $\frac{\sqrt{5}}{2}$ (b) $\left(\frac{\sqrt{5}}{2}\right)^2$ (c) $\sqrt{2\frac{1}{4}}$ (d) $\left(\frac{1}{3}\right)^2 + \frac{2}{3}$

4. Which of the following are surds?
(a) $\sqrt{16}$ (b) $\sqrt{3}$ (c) $\sqrt[3]{64}$ (d) $\sqrt{0.09}$ (e) $\sqrt{\frac{25}{36}}$

5. Write the following surds in their simplest form.
(a) $\sqrt{20}$ (b) $\sqrt{18}$ (c) $\sqrt{75}$ (d) $\sqrt{200}$ (e) $\sqrt{112}$

6. Simplify.
(a) $3\sqrt{2} + 5\sqrt{2}$ (b) $4\sqrt{3} + \sqrt{27}$ (c) $\sqrt{175} - \sqrt{63}$
(d) $\sqrt{5} \times \sqrt{45}$ (e) $\sqrt{6} \times \sqrt{2} \times \sqrt{24}$ (f) $\frac{\sqrt{500}}{\sqrt{20}}$

7. Write $\sqrt{147}$ in the form $a\sqrt{b}$, where a and b are prime numbers.

8. Express each of the following in its simplest form with a rational denominator.
(a) $\frac{5}{\sqrt{6}}$ (b) $\frac{8}{\sqrt{2}}$ (c) $\frac{\sqrt{5}}{\sqrt{8}}$ (d) $\frac{\sqrt{3}}{\sqrt{18}}$

9. (a) Expand and simplify $\sqrt{3}(4\sqrt{3} - 3)$
(b) Given that $\sqrt{150} = k\sqrt{s}$, find the value of k.
(c) Rationalise the denominator of $\frac{3}{\sqrt{6}}$ and simplify your answer.

10. (a) Simplify $\frac{10}{2\sqrt{3}}$.
Express your answer with a rational denominator.
(b) Given that $\sqrt{x} = 5\sqrt{7}$, find the value of x.

11. If $a = 4$ and $b = 50$, explain if $\sqrt{ab}$ represents a rational number or an irrational number.

12. Simplify the following expressions, leaving your answers, where appropriate, in surd form.
(a) $\sqrt{\frac{50}{18}}$ (b) $(2 + \sqrt{3})(3 + \sqrt{2})$ (c) $\frac{5}{2\sqrt{10}}$

A sequence of triangles, OPQ, OQR, ORS, ..., etc., is formed, as shown.
Calculate the lengths of OQ, OR and OS as surds in their simplest form.
Write, in terms of n, the hypotenuse of the nth triangle.
Investigate for different lengths of PO.

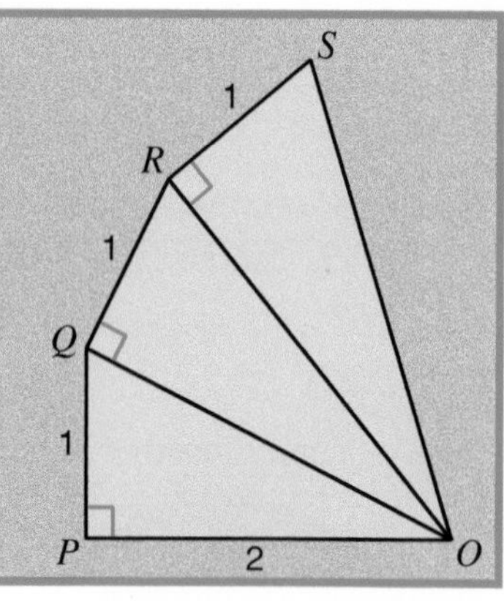

3 Using Indices

Powers

Products of the same number, like

$$3 \times 3, \quad 5 \times 5 \times 5, \quad 10 \times 10 \times 10 \times 10 \times 10,$$

can be written in a shorthand form using **powers**.

For example:

$3 \times 3 = 3^2$ This is read as '3 to the power of 2'. 3^2 has the value 9.

$5 \times 5 \times 5 = 5^3$ This is read as '5 to the power of 3'. 5^3 has the value 125.

$10 \times 10 \times 10 \times 10 \times 10 = 10^5$ This is read as '10 to the power of 5'. 10^5 has the value 100 000.

Index form

Numbers written in shorthand form like 3^2, 5^3 and 10^5 are said to be in **index form**.
This is sometimes called **power** form.

An expression of the form $a \times a \times a \times a \times a$ can be written in index form as a^5.
a^5 is read as 'a to the **power** 5'. a is the **base** of the expression. 5 is the **index** or **power**.

Multiplying and dividing numbers with powers

$3 \times 3 = 3^2$ and $5 \times 5 \times 5 = 5^3$ Both 3^2 and 5^3 are examples of numbers with powers.

These examples introduce methods for multiplying and dividing powers of the same number.

Example 1

Calculate the value of $6^5 \times 6^4$ in power form.

$6^5 = 6 \times 6 \times 6 \times 6 \times 6$ and $6^4 = 6 \times 6 \times 6 \times 6$

$$6^5 \times 6^4 = (6 \times 6 \times 6 \times 6 \times 6) \times (6 \times 6 \times 6 \times 6)$$
$$= 6 \times 6 \times 6 \times 6 \times 6 \times 6 \times 6 \times 6 \times 6$$

This gives: $6^5 \times 6^4 = 6^9$

Example 2

Calculate the value of $6^7 \div 6^4$ in power form.

$$6^7 \div 6^4 = \frac{6^7}{6^4} = \frac{6 \times 6 \times 6 \times \cancel{6} \times \cancel{6} \times \cancel{6} \times \cancel{6}}{\cancel{6} \times \cancel{6} \times \cancel{6} \times \cancel{6}}$$
$$= 6 \times 6 \times 6 = 6^3$$

This gives: $6^7 \div 6^4 = 6^3$

Can you see a quick way of working out the index (power) for each example?

Rules of indices

When **multiplying**: powers of the same base are **added**. In general: $a^m \times a^n = a^{m+n}$
When **dividing**: powers of the same base are **subtracted**. In general: $a^m \div a^n = a^{m-n}$

Raising a power to a power
$(a^m)^n = a^{mn}$

Two special results
$a^1 = a \qquad a^0 = 1$

Example 3

Simplify. Leave your answers in index form.

(a) $2^9 \times 2^4 = 2^{9+4} = 2^{13}$
Add indices

(b) $2^9 \div 2^4 = 2^{9-4} = 2^5$
Subtract indices

(c) $(4^9)^3 = 4^{9 \times 3} = 4^{27}$
Multiply indices

Multiplying and dividing algebraic expressions with powers

a^2 and b^3 are examples of **algebraic expressions with powers**.
The rules for multiplying and dividing powers of the same number can be used to simplify algebraic expressions involving powers.

Example 4

Simplify.

(a) $x^3 \times x^8 = x^{3+8} = x^{11}$
Add indices

(b) $x^8 \div x^5 = x^{8-5} = x^3$
Subtract indices

(c) $(x^3y^2)^2 = x^{3\times 2} \times y^{2\times 2} = x^6y^4$
Multiply indices

Practice Exercise 3.1

Do not use a calculator for this exercise.

1. Simplify. Leave your answers in index form.
(a) $2^3 \times 2^2$ (b) $3^5 \times 3^2$ (c) $5^6 \times 5^2$ (d) $7^3 \times 7$ (e) $9^5 \times 9^0$

2. Simplify. Leave your answers in index form.
(a) $2^3 \div 2^2$ (b) $3^5 \div 3^2$ (c) $5^6 \div 5^2$ (d) $7^3 \div 7$ (e) $9^5 \div 9^0$

3. Simplify.
(a) $y^2 \times y$ (b) $t^3 \times t^2$ (c) $g^7 \times g^3$ (d) $m \times m^3 \times m^2$
(e) $y^3 \div y$ (f) $a^4 \div a^3$ (g) $x^5 \div x^5$ (h) $g \div g^2$

4. Simplify. Leave your answers in index form.
(a) $3 \times 3^2 \times 3^3$ (b) $\frac{10 \times 10^3}{10^2}$ (c) $\frac{4^3 \times 4^3}{4}$ (d) $\frac{5^5 \times 5^2}{5^4}$
(e) $\frac{2 \times 2^5}{2^3}$ (f) $\frac{5 \times 5^2}{5^3}$ (g) $\frac{7^3 \times 7^2}{7^7}$ (h) $\frac{3^5 \times 3^0}{3^2}$

5. Simplify. Leave your answers in index form.
(a) $(2^3)^2$ (b) $(3^5)^2$ (c) $(5^2)^3$ (d) $(7^3)^3$ (e) $(9^0)^5$

6. Simplify.
(a) $(t^2)^3$ (b) $(y^3)^2$ (c) $(xy)^2$ (d) $(mn^2)^3$

7. Find x in each of the following.
(a) $2^3 \times 2^5 = 2^x$ (b) $3^5 \times 3^2 = 3^x$ (c) $7^2 \times 7^8 \times 7 = 7^x$ (d) $3^5 \div 3^2 = 3^x$
(e) $8^7 \div 8^6 = 8^x$ (f) $6^5 \div 6 = 6^x$ (g) $(2^3)^2 = 2^x$ (h) $(5^4)^5 = 5^x$
(i) $\frac{2^3 \times (2^2)^5}{2^8 \times 2^3} = 2^x$ (j) $\frac{(3^3 \times 3^2)^3}{3^7} = 3^x$ (k) $\frac{5^x \times 5^3 \times 5^4}{(5 \times 5^3)^3} = 5$ (l) $\frac{(3^x \times 3^2)^3}{(3 \times 3^5)^2} = 3^3$

Negative powers and reciprocals

Using patterns of powers
This list shows the powers of two extended to include negative indices.

$2^3 = 2 \times 2 \times 2 = 8$
$2^2 = 2 \times 2 = 4$
$2^1 = 2 = 2$
$2^0 = 1 = 1$
$2^{-1} = \frac{1}{2} = 0.5 = \frac{1}{2^1}$
$2^{-2} = \frac{1}{4} = 0.25 = \frac{1}{2^2}$
$2^{-3} = \frac{1}{8} = 0.125 = \frac{1}{2^3}$

Using the rules of indices

$\frac{1}{10} = 1 \div 10$
$\frac{1}{10} = 10^0 \div 10^1$
$\frac{1}{10} = 10^{0-1}$
$\frac{1}{10^1} = 10^{-1}$

$1 = 10^0$
$10 = 10^1$

$a^m \div a^n = a^{m-n}$

$\frac{1}{32} = 1 \div 32$
$\frac{1}{32} = 2^0 \div 2^5$
$\frac{1}{32} = 2^{0-5}$
$\frac{1}{2^5} = 2^{-5}$

$1 = 2^0$
$32 = 2^5$

All of the examples illustrate this general rule for negative indices.

$$a^{-m} = \frac{1}{a^m}$$

$\frac{1}{a^m}$ and, hence, a^{-m},

is called the **reciprocal** of a^m.

When multiplying and dividing powers with different bases each base must be dealt with separately.

Example 5

Simplify $10^{-4} \div 10^{-2}$.

$10^{-4} \div 10^{-2} = 10^{-4 - -2}$
$= 10^{-4+2}$
$= 10^{-2}$

Example 6

Simplify $3^4 \times 2^3 \times 3^{-5} \times 2^5$.

$3^4 \times 2^3 \times 3^{-5} \times 2^5 = 3^4 \times 3^{-5} \times 2^3 \times 2^5$
$= 3^{4+-5} \times 2^{3+5}$
$= 3^{-1} \times 2^8$

Example 7

Find the reciprocal of $\left(\frac{1}{6}\right)^{-2}$.

$\left(\frac{1}{6}\right)^{-2} = \frac{1}{6^{-2}} = 6^2$.

The reciprocal of $\left(\frac{1}{6}\right)^{-2}$ is 6^{-2}.

This is equivalent to $\frac{1}{36}$.

Example 8

If $2^x \div 2^5 = \frac{1}{8}$ find the value of x.

$\frac{1}{8} = \frac{1}{2^3} = 2^{-3}$

$2^{x-5} = 2^{-3}$

So, $x - 5 = -3$

This gives $x = 2$.

When multiplying or dividing expressions that include both numbers and powers:
multiply or **divide** the **numbers**, **add** or **subtract** the **powers**.
Deal with the powers of different bases **separately**.

Example 9

Simplify.

(a) $2x^2 \times 5x^7 = (2 \times 5) \times (x^2 \times x^7) = 10 \times x^{2+7} = 10x^9$
(b) $x^3y^2 \times xy^4 = (x^3 \times x) \times (y^2 \times y^4) = x^{3+1} \times y^{2+4} = x^4y^6$
(c) $(3x^4)^2 = (3 \times 3) \times (x^{4\times2}) = 9x^8$
(d) $6y^6 \div 2y^4 = (6 \div 2) \times (y^6 \div y^4) = 3 \times y^{6-4} = 3y^2$

$a^m \times a^n = a^{m+n}$
$a^m \div a^n = a^{m-n}$
$(a^m)^n = a^{m \times n}$

Practice Exercise 3.2

Do not use a calculator for this exercise.

1. Simplify. Leave your answers in index form.
   (a) $9^2 \times 9^{-2}$ (b) $2^{-3} \times 2$ (c) $5^5 \times 5^{-7}$ (d) $8^{-2} \times 8^{-3}$
   (e) $2^{-3} \div 2$ (f) $5^5 \div 5^{-7}$ (g) $11^{-2} \div 11^3$ (h) $7^{-4} \div 7^{-3}$

2. Simplify.
   (a) $3d^2 \times 2d^3$ (b) $4x^2 \times 2x^3$ (c) $2t \times t^2 \times 3t^2$ (d) $2r \times 3r^2 \times 4r^3$
   (e) $6b^3 \div b$ (f) $10m^3 \div 2m^2$ (g) $16t^3 \div 4t^2$ (h) $9h^2 \div 3h^3$
   (i) $(3a)^2$ (j) $(2h)^3$ (k) $2 \times (m^3)^2$ (l) $(2m^3)^2$

3. Simplify. Leave your answers in index form.
   (a) $8^{-3} \times 8^5$ (b) $7^2 \div 7^7$ (c) $2.5^{-2} \div 2.5^{-1}$
   (d) $4^3 \times 4^2 \times 4^{-5}$ (e) $10^{-3} \div 10^{-2}$ (f) $6^{-3} \times 6^4 \div 6^5$
   (g) $0.1^{-7} \div 0.1^5$ (h) $5^{-7} \div (5^2 \times 5^6)$ (i) $4^2 \div (4^{-1} \times 4^{-2})$
   (j) $4^{-3} \times 4^5 \times 8^5 \times 8^2$ (k) $4^{-1} \times 5^5 \times 5^{-7} \times 4^2$ (l) $2^{-5} \times 5^3 \times 2^3 \times 5^2$
   (m) $\frac{3^5 \times 3^{-2}}{3^2}$ (n) $\frac{5^{-3} \times 5^4}{5^{-2}}$ (o) $\frac{2 \times 2^{-3} \times 2^{-1}}{2^2 \times 2}$

4. Express with positive indices.
   (a) 3^{-2} (b) 2^{-3} (c) 3×3^{-2} (d) $5^{-2} \times 5^{-1}$ (e) $\frac{1}{3^{-2}}$
   (f) $\frac{5}{5^{-1}}$ (g) $2 \div 2^{-3}$ (h) $3^{-3} \div 3^{-1}$ (i) $(5^{-3})^2$ (j) $(3^{-2})^{-3}$

5. Simplify and express with positive indices.
   (a) $t^5 \times t^{-3}$ (b) $x^3 \times x \times x^{-2}$ (c) $x^7 \div x^{-3}$
   (d) $6x^3 \div 2x^{-1}$ (e) $(a^{-2})^3$ (f) $(3n^{-2})^{-2}$

Remember: $a^{-m} = \frac{1}{a^m}$

6. Simplify.

(a) $\frac{t^3}{t^2}$ (b) $\frac{g^2}{g^3}$ (c) $\frac{m^2 \times m}{m}$ (d) $\frac{y^2 \times y^3}{y^4}$

7. Write down the value of:

(a) 5^0 (b) 3^{-1} (c) 2^{-3} (d) $\frac{1}{3^2}$ (e) $\frac{1}{2^{-3}}$ (f) $\frac{3}{3^{-2}}$

8. Calculate each of the following.

(a) $10 + 10^0 + 10^{-1}$ (b) $2 + 2^{-1} + 2^{-2} + 2^{-3}$ (c) $5^0 + 5^{-1} + 5^{-2}$

(d) $3^{-1} + 2^{-1}$ (e) $5^{-2} + 2^{-2}$ (f) $5^{-2} \times 2^{-2}$

(g) $5^{-2} \div 2^{-2}$ (h) $5 \times 6^{-1} + 2 \times 5^{-1}$

9. Find x in each of the following.

(a) $3^x = \frac{1}{81}$ (b) $5^x = \frac{1}{25}$ (c) $\left(\frac{1}{4}\right)^x = 16$ (d) $\left(\frac{1}{6}\right)^x = 216$

(e) $2^x = 0.5$ (f) $5^x = 0.04$ (g) $3 \times 10^x = 0.003$ (h) $2 \times 5^x = 0.4$

10. Calculate $\frac{4^5}{4^{-2}}$ giving your answer in the form 2^n.

Scientific notation

Scientific notation is a shorthand way of writing very large and very small numbers.
Scientific notation is often called **standard form** or **standard index form.**

Scientists who study the planets and the stars work with very large numbers.

Approximate distances from the Sun to some planets are:

Earth 149 000 000 km Mars 228 000 000 km Pluto 5 898 000 000 km

A calculator displays very large numbers in **scientific notation.**
To represent large numbers in scientific notation you need to use powers of 10.

For example: 149 000 000 km = $1.49 \times 100\,000\,000$ km = 1.49×10^8 km
Therefore 149 000 000 km = 1.49×10^8 km in **scientific notation.**

Scientists who study microbiology work with numbers that are very small.
The smallest living cells are bacteria which have a diameter of about 0.000 025 cm.
Blood cells have a diameter of about 0.000 75 cm.

A calculator displays small numbers in **scientific notation.**
It does this in the same sort of way that it does for large numbers.
To represent very small numbers in scientific notation you need to use powers of 10 for numbers less than 1.

For example: 0.000 75 cm = 7.5×0.0001 cm = 7.5×10^{-4} cm
Therefore 0.000 75 cm = 7.5×10^{-4} cm in **scientific notation.**

Calculator displays

Work out 3 000 000 × 25 000 000 *on your calculator.*
Write down the display.

Most calculators will show the answer as: 

In **scientific notation** the answer should be written as 7.5×10^{13}.

Work out 0.000 007 × 0.000 9 *on your calculator.*
Write down the display.

Most calculators will show the answer as: 

In scientific notation the answer should be written as 6.3×10^{-9}.

> In **scientific notation** a number is written as: a number between 1 and 10 × a power of 10
> A **large** number has a **positive power.** E.g. 160 000 000 = 1.6×10^8
> A **small** number has a **negative power.** E.g. 0.000 000 06 = 6×10^{-8}

Example 10

Write 370 000 in scientific notation.

$370\ 000 = 3.7 \times 100\ 000$
$= 3.7 \times 10^5$

Example 11

Write 5.6×10^7 as an ordinary number.

$5.6 \times 10^7 = 5.6 \times 10\ 000\ 000$
$= 56\ 000\ 000$

Example 12

Write 0.000 73 in scientific notation.

$0.000\ 73 = 7.3 \times 0.000\ 1$
$= 7.3 \times 10^{-4}$

Example 13

Write 2.9×10^{-6} as an ordinary number.

$2.9 \times 10^{-6} = 2.9 \times 0.000\ 001$
$= 0.000\ 002\ 9$

Practice Exercise 3.3

1. Write each of these numbers in scientific notation.
   (a) 300 000 000 000 (b) 80 000 000 (c) 700 000 000 (d) 2 000 000 000
   (e) 42 000 000 (f) 21 000 000 000 (g) 3 700 000 000 (h) 630

2. Change each of these numbers to an ordinary number.
   (a) 6×10^5 (b) 2×10^3 (c) 5×10^7 (d) 9×10^8
   (e) 3.7×10^9 (f) 2.8×10^1 (g) 9.9×10^{10} (h) 7.1×10^4

3. Write these calculator displays:
   (a) in scientific notation,
   (b) as an ordinary number.

   (i) 

   (ii) 

   (iii) 

   (iv)  

4. Write each of these numbers in scientific notation.
   (a) 0.007 (b) 0.04 (c) 0.000 000 005
   (d) 0.000 8 (e) 0.000 000 002 3 (f) 0.000 000 045
   (g) 0.023 4 (h) 0.000 000 002 34 (i) 0.006 7

5. Change each of these numbers to an ordinary number:
   (a) 3.5×10^{-1} (b) 5×10^{-4} (c) 7.2×10^{-5} (d) 6.1×10^{-3}
   (e) 1.17×10^{-10} (f) 8.135×10^{-7} (g) 6.462×10^{-2} (h) 4.001×10^{-9}

6. Write these calculator displays:
   (a) in scientific notation,
   (b) as an ordinary number.

   (i) 

   (ii) 

   (iii) 

   (iv) 

7. The table shows the estimated population for some countries in 2020.

Country	Estimated population in 2020
Albania	5.3×10^6
Brazil	2.5×10^8
Fiji	9.0×10^5
Greece	1.2×10^7

   (a) (i) Which of these countries is estimated to have the largest population in 2020?
   (ii) Write the estimated population as an ordinary number.
   (b) In 2020, China is estimated to have a population of 1288 million people.
   Write 1288 million in scientific notation.

8. The smallest living cells are bacteria cells which have a diameter of 0.000 025 cm.

Here are some other very small numbers.

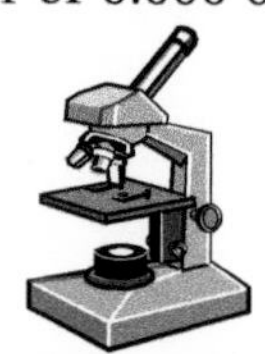

Blood cell:	diameter 0.000 75 cm.
Hydrogen atom:	diameter 0.000 000 2 mm.
Mumps virus:	diameter 0.000 225 mm.

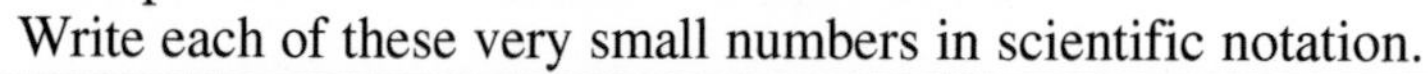

Write each of these very small numbers in scientific notation.

Calculations involving scientific notation

Calculations with large and small numbers can be done on a calculator by:

- changing the numbers to scientific notation,
- entering the numbers into the calculator using the Exp button.

If your calculator works in a different way to the example shown, refer to the instruction booklet supplied with the calculator or ask someone for help.

Example 14

The following figures refer to the population of China and the USA in 1993.

China 1.01×10^9 USA 2.32×10^8

By how much did the population of China exceed that of the USA in 1993?

1.01×10^9 is greater than 2.32×10^8.

You need to work out $1.01 \times 10^9 - 2.32 \times 10^8$

The greater the power ... the bigger the number.

To do the calculation enter the following sequence into your calculator.

1 . 0 1 Exp 9 − 2 . 3 2 Exp 8 =

Giving: $1.01 \times 10^9 - 2.32 \times 10^8 = 778\,000\,000 = 7.78 \times 10^8$

Practice Exercise 3.4

Use the Exp button on your calculator to answer these questions.

1. Give the answers to the following calculations as ordinary numbers.

   (a) $(5.25 \times 10^9) \times (5 \times 10^{-5})$
   (b) $(5.25 \times 10^9) \div (5 \times 10^{-5})$
   (c) $(8.5 \times 10^6)^2$
   (d) $(5 \times 10^{-3})^3$
   (e) $(7.2 \times 10^5) \div (2.4 \times 10^{-5})$
   (f) $(9.5 \times 10^6) \div (1.9 \times 10^{-7})^2$

2. Give the answers to the following calculations in scientific notation.

   (a) $33\,500\,000\,000 \times 2\,800\,000\,000$
   (b) $0.000\,000\,000\,2 \times 80\,000\,000\,000$
   (c) $15\,000\,000\,000\,000^2$
   (d) $0.000\,000\,000\,000\,5^3$
   (e) $48\,000\,000\,000 \div 0.000\,000\,000\,2$
   (f) $25\,000\,000\,000 \div 500\,000\,000\,000$

3. (a) In 1992 about 1 400 000 000 steel cans and about 688 000 000 aluminium cans were recycled.
   What was the total number of cans that were recycled in 1992?
   Give your answer in scientific notation.

   (b) Alpha Centauri is about 40 350 000 000 000 km from the Sun.
   Alpha Cygni is about 15 300 000 000 000 000 km from the Sun.
   How much further is it from the Sun to Alpha Cygni than from the Sun to Alpha Centauri?
   Give your answer in scientific notation.

4. Here are the diameters of some planets.

   Saturn 1.2×10^5 km Jupiter 1.42×10^5 km Pluto 2.3×10^3 km

   (a) List the planets in order of size starting with the smallest.
   (b) What is the difference between the diameters of the largest and smallest planets?
   Give your answer in scientific notation and as an ordinary number.

5. Here are the areas of some of the world's largest deserts.

The Sahara desert in North Africa	$8.6 \times 10^6\,km^2$
The Gobi desert in Mongolia and North East China	$1.166 \times 10^6\,km^2$
The Patagonian desert in Argentina	$6.73 \times 10^5\,km^2$

(a) What is the total area of the Sahara and Patagonian deserts?
(b) What is the difference in area between the Gobi and the Patagonian deserts?
Give your answer in scientific notation.

6. Calculate $5.42 \times 10^6 \times 4.65 \times 10^5$
giving your answer in scientific notation correct to 3 significant figures.

7. Calculate $1.7 \times 10^3 \div 7.6 \times 10^7$
giving your answer in scientific notation correct to 2 significant figures.

8. The area of the surface of the Earth is about 5.095×10^9 square miles.
Approximately 29.2% of this is land.
Use these figures to estimate the area of land surface on Earth.

9. James wins a lottery prize of £1.764×10^6. He pays £5.29×10^5 for a house.
What percentage of his prize did he spend on the house?
Give your answer to a suitable degree of accuracy.

10. The mass of an oxygen atom is 2.7×10^{-23} grams.
The mass of an electron at rest is approximately 30 000 times smaller than this.
Estimate the mass of an electron at rest.

11. The modern human appeared on the Earth about 3.5×10^4 years ago.
The Earth has been in existence for something like 1.3×10^5 times as long as this.
(a) Estimate the age of the Earth.

Reptiles appeared on the Earth about 2.3×10^8 years ago.
(b) How many times longer than the modern human have reptiles been alive?
Give your answer in scientific notation.

Scientific notation calculations without a calculator

In some scientific notation problems the calculations can be handled without using a calculator.

Example 15

Calculate the value of $(3 \times 10^2) + (4 \times 10^3)$.
Give your answer in scientific notation.

$3 \times 10^2 = 300$
$4 \times 10^3 = 4000$

$(3 \times 10^2) + (4 \times 10^3) = 300 + 4000$
$= 4300$
$= 4.3 \times 10^3$

When adding or subtracting numbers in scientific notation without a calculator change to ordinary numbers first.

Example 16

Calculate the value of ab where $a = 8 \times 10^3$ and $b = 4 \times 10^5$.

$ab = (8 \times 10^3) \times (4 \times 10^5)$
$= 8 \times 4 \times 10^3 \times 10^5$
$= 32 \times 10^8$
$= 3.2 \times 10 \times 10^8$
$= 3.2 \times 10^9$

When **multiplying** the powers are **added**.
$10^3 \times 10^5 = 10^{3+5} = 10^8$
$10 \times 10^8 = 10^{1+8} = 10^9$

Example 17

Calculate the value of x^2 where $x = 7 \times 10^{-8}$.

$$x^2 = (7 \times 10^{-8})^2$$
$$= 49 \times 10^{-16}$$
$$= 4.9 \times 10 \times 10^{-16}$$
$$= 4.9 \times 10^{-15}$$

Remember:
$(7 \times 10^{-8})^2 = 7^2 \times (10^{-8})^2$
$10 \times 10^{-16} = 10^{1-16} = 10^{-15}$

Example 18

Calculate the value of $(1.2 \times 10^3) \div (4 \times 10^{-8})$.

$$(1.2 \times 10^3) \div (4 \times 10^{-8}) = (1.2 \div 4) \times (10^3 \div 10^{-8})$$
$$= 0.3 \times 10^{11}$$
$$= 3 \times 10^{-1} \times 10^{11}$$
$$= 3 \times 10^{10}$$

When **dividing** the powers are **subtracted**.
$10^3 \div 10^{-8} = 10^{3--8} = 10^{11}$

Practice Exercise 3.5

Do not use a calculator. Give your answers in scientific notation.

1. For each of the following calculate the value of $p + q$.
   (a) $p = 5 \times 10^3$ and $q = 2 \times 10^2$
   (b) $p = 4 \times 10^5$ and $q = 8 \times 10^6$
   (c) $p = 3.08 \times 10^4$ and $q = 9.2 \times 10^3$
   (d) $p = 4.25 \times 10^4$ and $q = 7.5 \times 10^3$
2. For each of the following calculate the value of $p - q$.
   (a) $p = 3 \times 10^3$ and $q = 2 \times 10^2$
   (b) $p = 9.05 \times 10^5$ and $q = 5 \times 10^3$
   (c) $p = 3.05 \times 10^7$ and $q = 5 \times 10^5$
   (d) $p = 9.545 \times 10^8$ and $q = 4.5 \times 10^6$
3. For each of the following calculate the value of $p \times q$.
   (a) $p = 4 \times 10^3$ and $q = 2 \times 10^4$
   (b) $p = 2 \times 10^4$ and $q = 3 \times 10^3$
   (c) $p = 4 \times 10^5$ and $q = 6 \times 10^2$
   (d) $p = 9 \times 10^9$ and $q = 3 \times 10^5$
4. For each of the following calculate the value of $p \div q$.
   (a) $p = 6 \times 10^5$ and $q = 2 \times 10^2$
   (b) $p = 9 \times 10^5$ and $q = 3 \times 10^2$
   (c) $p = 2.5 \times 10^5$ and $q = 5 \times 10^3$
   (d) $p = 4 \times 10^8$ and $q = 2 \times 10^{-3}$
   (e) $p = 1.2 \times 10^3$ and $q = 3 \times 10^{-3}$
   (f) $p = 1.5 \times 10^{-5}$ and $q = 5 \times 10^{-3}$
5. $x = 3 \times 10^4$ and $y = 5 \times 10^{-5}$. Work out the value of each of these expressions.
   (a) xy (b) x^3 (c) x^2y (d) y^3 (e) $\frac{x}{y}$

Powers and roots

The inverse (opposite) of raising to a power is finding a **root**.
The inverse of squaring is finding the **square root**.
The inverse of cubing is finding the **cube root**.

This symbol $\sqrt{\ }$ stands for root.
$\sqrt{16}$ means the square root of 16.
$\sqrt[3]{27}$ means the cube root of 27.

The connection between powers and roots

If $a = b^n$, $\sqrt[n]{a} = \sqrt[n]{b^n} = b$

Using the rules of powers

$(b^n)^{\frac{1}{n}} = b^{n \times \frac{1}{n}} = b$

So, $(b^n)^{\frac{1}{n}}$ is the same as $\sqrt[n]{b^n}$.

So, $a^{\frac{1}{n}}$ is the same as $\sqrt[n]{a}$.

The inverse of "raising to the power n" is finding the nth root.
In general, finding the nth root of a number, a, can be written as:
$\sqrt[n]{a}$ or $a^{\frac{1}{n}}$.

Example 19

Calculate $81^{\frac{1}{4}}$.

$81^{\frac{1}{4}} = \sqrt[4]{81} = 3$

Because $3^4 = 81$.

Example 20

Calculate $25^{-\frac{1}{2}}$.

$25^{-\frac{1}{2}}$ is the **reciprocal** of $25^{\frac{1}{2}}$.

$25^{-\frac{1}{2}} = \frac{1}{25^{\frac{1}{2}}} = \frac{1}{\sqrt{25}} = \frac{1}{5} = 0.2$

Using a calculator

The powers and roots of numbers can be worked out on a calculator.
The x^y button can be used to calculate the value of a number x raised to the power of y.
Square roots can be calculated using the $\sqrt{\ }$ button.
Cube roots can be calculated using the $\sqrt[3]{\ }$ button.
The y root can be calculated using the $x^{1/y}$ button.

Example 21

Calculate $512^{-\frac{1}{9}}$.

Use this key sequence: 5 1 2 $x^{1/y}$ 9 +/– =

This gives $512^{-\frac{1}{9}} = 0.5$ *Try examples 19 and 20 using a calculator.*

Further fractional powers

This section deals with evaluating expressions of the form $a^{\frac{m}{n}}$ and $a^{-\frac{m}{n}}$.

To find the value of $a^{\frac{m}{n}}$:

1. Find the nth root of a.
2. Raise the nth root of a to the power m.

In general, using the rules of indices:

$$a^{\frac{m}{n}} = \left(a^{\frac{1}{n}}\right)^m = \left(\sqrt[n]{a}\right)^m$$

$$a^{-\frac{m}{n}} = \frac{1}{a^{\frac{m}{n}}} = \frac{1}{\left(\sqrt[n]{a}\right)^m}$$

To find the value of $a^{-\frac{m}{n}}$:

Carry out the first two steps, as before.

3. Write down the reciprocal of $a^{\frac{m}{n}}$.

Example 22

Find the value of $32^{\frac{4}{5}}$.

Find the 5th root of 32.

$32^{\frac{1}{5}} = 2$

Raise to the power of 4.

$2^4 = 16$

$32^{\frac{4}{5}} = 16$

Example 23

Find the value of $27^{-\frac{4}{3}}$.

$27^{-\frac{4}{3}}$ is the reciprocal of $27^{\frac{4}{3}}$.

So, first work out $27^{\frac{4}{3}}$.

$27^{\frac{4}{3}} = \left(27^{\frac{1}{3}}\right)^4 = 3^4 = 81$

The reciprocal of 81 is $\frac{1}{81}$.

So, $27^{-\frac{4}{3}} = \frac{1}{81}$.

Practice Exercise 3.6

Do not use a calculator for questions 1 to 9.

1. Find the value of each of the following.
   (a) $\sqrt{400}$ (b) $\sqrt[3]{27}$ (c) $\sqrt[3]{1000}$ (d) $\sqrt[4]{16}$ (e) $\sqrt[3]{64}$ (f) $\sqrt{6.25}$
2. Find the value of each of the following.
   (a) $64^{\frac{1}{2}}$ (b) $8^{\frac{1}{3}}$ (c) $81^{\frac{1}{4}}$ (d) $32^{\frac{1}{5}}$ (e) $625^{\frac{1}{4}}$ (f) $36^{0.5}$
3. Find the value of each of the following.
   (a) $100^{-\frac{1}{2}}$ (b) $49^{-0.5}$ (c) $16^{-\frac{1}{4}}$ (d) $125^{-\frac{1}{3}}$ (e) $256^{-\frac{1}{4}}$ (f) $243^{-\frac{1}{5}}$
4. Find the value of each of the following.
   (a) $1000^{\frac{2}{3}}$ (b) $9^{\frac{3}{2}}$ (c) $16^{\frac{3}{4}}$ (d) $32^{\frac{2}{5}}$ (e) $4^{\frac{5}{2}}$
   (f) $9^{2.5}$ (g) $125^{\frac{2}{3}}$ (h) $16^{\frac{5}{4}}$ (i) $243^{\frac{4}{5}}$ (j) $36^{1.5}$
5. Find the value of each of the following.
   (a) $1000^{-\frac{2}{3}}$ (b) $16^{-\frac{3}{2}}$ (c) $8^{-\frac{2}{3}}$ (d) $32^{-\frac{3}{5}}$ (e) $4^{-\frac{3}{2}}$
   (f) $100^{-\frac{5}{2}}$ (g) $25^{-\frac{3}{2}}$ (h) $16^{-\frac{3}{4}}$ (i) $128^{-\frac{5}{7}}$ (j) $125^{-\frac{2}{3}}$

6. (a) Evaluate. (i) $16^{0.5} \times 2^{-3}$ (ii) $49^{-0.5} \times 81^{0.25}$
Give your answers as fractions.
(b) Work out. (i) $(\sqrt{5})^4$ (ii) $27^{\frac{2}{3}}$

7. Find x in each of the following.
(a) $\sqrt[x]{27} = 3$ (b) $\sqrt[x]{16} = 2$ (c) $\sqrt[x]{32} = 2$ (d) $25^x = 5$ (e) $27^x = 9$
(f) $16^x = 2$ (g) $16^x = 8$ (h) $x^{\frac{3}{4}} = 64$ (i) $x^{\frac{2}{3}} = 25$ (j) $25^x = 125$

8. Calculate each of the following.
(a) $2^{-1} + \left(\frac{1}{16}\right)^{\frac{1}{2}}$ (b) $\left(\frac{1}{32}\right)^{-\frac{1}{5}} \times 8^{-\frac{2}{3}}$ (c) $27^{\frac{2}{3}} \times 3^{-1}$ (d) $49^{-\frac{1}{2}} + 16^{-\frac{3}{4}}$ (e) $9^{\frac{3}{2}} \div 8^{\frac{2}{3}}$

9. Use the rules of powers to show that:
(a) $0.25^{-\frac{7}{2}} = 16^{\frac{7}{4}}$ (b) $32^{\frac{3}{5}} \times 4^{-\frac{3}{2}} = 1$ (c) $25^{\frac{1}{2}} \times 36^{-\frac{1}{2}} = 8^{-\frac{1}{3}} + 3^{-1}$

10. Use your calculator to find the value of each of the following, writing the full display.
(a) $\sqrt{\frac{59.6}{(0.4)^2}}$ (b) $\left(\frac{3.51}{\sqrt{0.28}}\right)^3$ (c) $\left(\frac{2.96 \times 8.7}{5.4 + 13.9}\right)^4$
(d) $\sqrt[3]{\frac{47.6}{8.51 - 6.79}}$ (e) $\frac{9.3^2}{6.2 + \sqrt[3]{59.7}}$ (f) $\frac{69.7}{2.9^2} - \frac{3.7}{\sqrt{5.4}}$

Key Points

- An expression such as $3 \times 3 \times 3 \times 3 \times 3$ can be written in a shorthand way as 3^5.
  This is read as '3 to the power of 5'.
  The number 3 is the **base** of the expression. 5 is the **power** (index).
- The opposite of squaring a number is called finding the **square root**.
  To find other roots use the notation $\sqrt[n]{a}$ or $a^{\frac{1}{n}}$.
- Calculations involving powers and roots can be simplified, or evaluated, using the rules of indices.

Multiplying powers with the same base	$a^m \times a^n = a^{m+n}$	
Dividing powers with the same base	$a^m \div a^n = a^{m-n}$	
Raising a power to a power	$(a^m)^n = a^{mn}$	
Raising any number to the power zero	$a^0 = 1$	(also $a^1 = a$)
Negative powers	$a^{-m} = \frac{1}{a^m}$	a^{-m} is the **reciprocal** of a^m
Fractional powers and roots	$a^{\frac{1}{n}} = \sqrt[n]{a}$ and	$a^{\frac{m}{n}} = \left(a^{\frac{1}{n}}\right)^m = \left(\sqrt[n]{a}\right)^m$

- a^2 and b^3 are examples of **algebraic expressions with powers**.
  The rules for multiplying and dividing powers of the same number can be used to simplify algebraic expressions involving powers.

> When multiplying or dividing expressions that include both numbers and powers:
> **multiply** or **divide** the **numbers**, **add** or **subtract** the **powers**.
> Deal with the powers of different bases **separately**.

- **Powers**
  The squares and the cubes of numbers can be worked out on a calculator by using the x^y button.
  The x^y button can be used to calculate the value of a number x raised to the power of y.
- **Roots** can be calculated using the $x^{1/y}$ button.
- **Scientific notation** (or **standard form**) is a shorthand way of writing very large and very small numbers.
- In **scientific notation** a number is written as: **a number between 1 and 10 × a power of 10.**
  Large numbers (ten, or more) have a **positive** power of 10.

Review Exercise 3

Do not use a calculator for questions 1 to 7.

1. Find the value of:
   (a) 2^4 (b) 7^3 (c) 10^4 (d) $64^{\frac{1}{2}}$
   (e) $27^{\frac{1}{3}}$ (f) $2^2 \times 3^3$ (g) $2^2 \times 3 \times 5^2$ (h) $3^2 \times 10^4$

2. Simplify, leaving in index form.
   (a) $3^3 \times 3^5$ (b) $7^6 \div 7^2$ (c) $4^3 \times 4^5 \times 4$
   (d) $(2^4)^3$ (e) $2^3 \div 2^5$ (f) $(5^{-2})^2$
   (g) $(6^8)^{\frac{1}{2}}$ (h) $\left(4^{\frac{1}{3}}\right)^2$ (i) $\left(3^{1\frac{1}{2}}\right)^4$

3. Simplify, leaving in index form.
   (a) $3y^3 \times 2y^2$ (b) $8t^6 \div 4t^3$ (c) $mn \times m^2n$ (d) $a^2b \times ba^2$
   (e) $a^{-3} \div a^{-2}$ (f) $xy^2 \div xy$ (g) $m^3n \div mn^2$ (h) $6r^2s^3 \div 2rs$
   (i) $(3d^3)^2$ (j) $(3p^2q)^2$ (k) $\left(x^{\frac{5}{2}}\right)^2$ (l) $x^3 \div \frac{x^0}{x^4}$
   (m) $\frac{y \times y^3}{y^2}$ (n) $\frac{m^2 \times m^3}{m^6}$ (o) $\frac{2t^3 \times t}{t^2}$ (p) $\frac{6g^2 \times g}{2g^3}$

4. Find the value of:
   (a) 7^{-2} (b) 8^0 (c) 2^{-4} (d) $\left(\frac{1}{4}\right)^{-1}$
   (e) $2^3 \times 4^0 \times 6^{-1}$ (f) $81^{\frac{1}{2}}$ (g) $1000^{\frac{2}{3}}$ (h) $\left(\frac{1}{4}\right)^{\frac{1}{2}}$
   (i) $81^{\frac{3}{4}}$ (j) $25^{-\frac{3}{2}}$ (k) $64^{-\frac{1}{3}}$ (l) $8^{1\frac{1}{3}}$

5. (a) Write $2^{-2} \times 4$ as a single power of 2. (b) Write $2^3 \div \frac{1}{8}$ as a single power of 2.

6. Find x in each of the following.
   (a) $5^3 \times 5^2 = 5^x$ (b) $2^5 \div 2^4 = 2^x$ (c) $(7^3)^2 = 7^x$ (d) $\frac{5^5 \times 5^3}{5^4} = 5^x$

7. Work out, without using a calculator.
   (a) $(6 \times 10^3) + (5 \times 10^4)$ (b) $(6 \times 10^3) \times (5 \times 10^4)$ (c) $(6 \times 10^3) \div (5 \times 10^4)$
   Give your answers in scientific notation.

8. (a) Write 34 500 000 000 in scientific notation.
   (b) Write 0.000 000 543 in scientific notation.
   (c) Work out $\frac{7.2 \times 10^5}{6.4 \times 10^3}$. Give your answer in scientific notation.

9. Calculate $\frac{3.6 \times 10^9}{4.5 \times 10^4}$. Give your answer in scientific notation.

10. Work out $\frac{3 \times 10^4}{5 \times 10^{-5}}$, giving your answer in scientific notation.

11. $p = 1.65 \times 10^7$, $q = 4.82 \times 10^6$ and $r = 6.17 \times 10^{-2}$.
    Calculate the value of the following. Give your answers in scientific notation.
    (a) $2p + 3q$ (b) $p \div r$

12. In 2005, a company paid out a total of £1.14×10^{10} in wages to 0.63 million employees.
    Calculate the average annual wage per employee.
    Give your answer in scientific notation correct to three significant figures.

13. The volume of water on Earth is approximately 1.436×10^9 km³.
    About 94% of this is contained in the Earth's oceans.
    Use these figures to estimate the volume of water in the Earth's oceans.

14. The Earth is approximately 93 million miles from the Sun.
    Taking 1 mile as equivalent to 1.6 km, find this distance in kilometres, to 2 significant figures.
    Express your answer in scientific notation.

4 Using Percentages

The meaning of a percentage

'Per cent' means 'out of 100'.
The symbol for per cent is %.
A percentage can be written as a fraction with denominator 100.

> 10% means 10 out of 100.
> 10% can be written as $\frac{10}{100}$.
> 10% can be written as 0.1.
> 10% is read as '10 per cent'.

Finding a percentage of a quantity

Example 1

Find 20% of £56.

Step 1 Divide by 100.
£56 ÷ 100 = £0.56

Step 2 Multiply by 20.
£0.56 × 20 = £11.20

So, 20% of £56 is £11.20.

> To find 1% of a quantity divide the quantity by 100.
> To find 20% of a quantity multiply 1% of the quantity by 20.
> This is the same as the method you would use to find $\frac{20}{100}$ of a quantity.

Reverse percentage problems

Example 2

A shop sells dvds with a 20% discount.
Petra buys a dvd and pays £10.
How much does the dvd normally cost?

Discount price is normal price less 20%.
So, 80% of normal price = £10.
So, 1% of normal price = £10 ÷ 80.
= £0.125
So, normal price = £0.125 × 100.
= £12.50

Example 3

Tara gets a 5% wage rise.
Her new wage is £273 per week.
What was Tara's wage before her wage rise?

New wage = old wage + 5%.
So, 105% of old wage = £273.
1% of old wage is 273 ÷ 105 = £2.60.
Old wage = 2.6 × 100 = £260.

Practice Exercise 4.1

Do not use a calculator for questions 1 to 4.

1. A special bottle of pop contains 10% more than a normal bottle.
   The special bottle contains 660 ml.
   How much does the normal bottle contain?
2. Jim saves 15% of his monthly pension.
   Each month he saves £90.
   What is his monthly pension?
3. May gets a 20% wage rise.
   Her new wage is £360 per week.
   What was May's wage before her wage rise?
4. In a high jump event, Nick jumps 1.44 metres.
   This is 10% lower than the best height he can jump.
   What is the best height he can jump?
5. A house is valued at £350 000.
   This is a 12% increase on the value of the house a year ago.
   What was the value of the house a year ago?

6. 30 grams of a breakfast cereal provides 16.2 mg of vitamin C.
This is 24% of the recommended daily intake.
What daily intake of vitamin C is recommended?

7. Tom gets a 3% increase in his salary.
His new salary is £1668.60 per month.
What was Tom's salary before his wage rise?

8. A one-year-old car is worth £6720.
This is a decrease of 16% of its value from new.
What was the price of the new car?

9. Here is some data about the changes in the numbers of pupils in schools A and B.
School A's numbers increased by 4% to 442.
School B's numbers decreased by 6% to 423.
How many pupils were in schools A and B before the changes in numbers?

10. John sells his computer to Dan and makes a 15% profit.
Dan then sells the computer to Ron for £391.
Dan makes a 15% loss.
How much did John pay for the computer?
Explain why it is not £391.

11. Kim sells her bike to Sara.
Sara sells it to Tina for £121.50.
Both Kim and Sara make a 10% loss.
How much did Kim pay for the bike?
Explain why it is not 20% more than £121.50.

Savings

Money invested in a savings account or a bank or building society earns **interest**, which is usually paid once a year.

Banks and building societies advertise the **yearly rates** of interest payable.
For example, 6% per year.
Interest, usually calculated annually, can also be calculated for shorter periods of time.

Simple Interest

With **Simple Interest**, the interest is paid out each year and not added to your account.

The amount of Simple Interest an investment earns can be calculated using:

$$\text{Simple Interest} = \text{Amount invested} \times \text{Time in years} \times \text{Rate of interest per year}$$

Compound Interest

With **Compound Interest**, the interest earned each year is added to your account and also earns interest the following year.

For example, an investment of 5% per annum means that the amount invested earns £5 for every £100 invested for one year.

So, after the first year of the investment, every £100 invested becomes £100 + 5% of £100.

£100 + 5% of £100 = £100 + £5 = £105

So, after the second year of the investment, every £100 of the original investment becomes £105 + 5% of £105.

£105 + 5% of £105 = £105 + £5.25 = £110.25

This can also be calculated as: $100 \times (1.05)^2 = £110.25$

Explain why this works.

Example 4

Find the Simple Interest paid on £600 invested for 6 months at 8% per year.

> **Note:**
> Interest rates are given 'per year'. The length of time for which an investment is made is also given in years.
> 6 months = $\frac{6}{12}$ years.
> *Explain why.*

Simple Interest $= 600 \times \frac{6}{12} \times \frac{8}{100}$
$= 600 \times 0.5 \times 0.08$
$= £24$

The Simple Interest paid is £24.

Example 5

Find the Compound Interest paid on £600 invested for 3 years at 6% per year.

1st year	Investment	= £600
	Interest: £600 × 0.06	= £ 36
	Value of investment after one year	= £636
2nd year	Investment	= £636
	Interest: £636 × 0.06	= £ 38.16
	Value of investment after two years	= £674.16
3rd year	Investment	= £674.16
	Interest: £674.16 × 0.06	= £ 40.45
	Value of investment after three years	= £714.61

Compound Interest = Final value − Original value
= £714.61 − £600 = £114.61

This could also be calculated as follows:
$600 \times (1.06)^3 - 600 = £114.61$

> In general, when Compound Interest rates are applied:
>
> Total value of investment = Amount invested $\times \left(\frac{100 + r}{100}\right)^n$,
>
> where r is the compound rate of interest and n is the number of years.
>
> $\left(\frac{100 + r}{100}\right)$ is sometimes called the **multiplier**.
>
> Compound Interest = Amount invested $\times \left(\frac{100 + r}{100}\right)^n$ − Amount invested.

Example 6

Calculate the final value when £1400 is invested at 3% per year for 5 years.

The final value is given by Amount invested $\times \left(\frac{100 + r}{100}\right)^n$.

Substitute $r = 3$ into $\left(\frac{100 + r}{100}\right)$.

$\frac{100 + r}{100} = \frac{100 + 3}{100} = \frac{103}{100} = 1.03$ (multiplier)

> With practice, the multiplier can be written down quickly.

Final value $= £1400 \times 1.03^5$.
$= £1622.98$

The Compound Interest can be found by subtracting the amount invested.

Compound Interest = £1622.98 − £1400 = £222.98.

Practice Exercise 4.2

Do not use a calculator for questions 1 and 2.

1. Find the simple interest paid on £200 for 1 year at 5% per year.
2. Calculate the simple interest on £500 invested at 6% per year after:
   (a) 1 year, (b) 6 months.
3. Calculate the simple interest paid on an investment of £6000 at 7.5% per year after 6 months.
4. Find the simple interest on £800 invested for 9 months at 8% per year.
5. Calculate the simple interest on £10 000 invested for 3 months at 9% per year.
6. Jenny invests £200 at 10% per annum compound interest.
   What is the value of her investment after 2 years?

7. Which of the following investments earn more interest?
   (a) £200 for 3 years at 5% compound.
   (b) £300 for 2 years at 5% compound.
   Show your working.

8. (a) Find the Simple Interest if £600 is invested for 4 years at 4% per year.
   (b) Find the Compound Interest if £600 is invested for 4 years at 4% per year.
   (c) How much more interest is gained by applying a compound rate of interest?

9. £10 000 is to be invested for 3 years.
   Calculate the final value of the investment if the interest rate per annum is 6%.
   Give your answer to a suitable degree of accuracy.

10. (a) Calculate the final value when £1800 is invested at 5% per year for 4 years.
    (b) Calculate the final value when £15 000 is invested for 6 years at a rate of 3% per year.

11. (a) Calculate the total value of an investment when £8000 is invested for 8 years at 3.5% per year.
    (b) How much Compound Interest was paid on the investment?

12. (a) Alex invests £5000 for 8 years.
    Compound Interest is paid at a rate of 2% per year for the first 3 years and then at 3% for the remainder of the term of investment.
    Find the total value of Alex's investment after 8 years.
    (b) What was the total amount of Compound Interest paid over 8 years?

13.

INTEREST RATES PAYABLE	
Option 1:	2.1% per year.
Option 2:	1.5% per year for 5 years, then 3% per year.

Gail has £7500 and wants to invest her money for 9 years.
Which option will provide the best return on her investment?
You must show your working.

Appreciation and depreciation

Appreciation is when items go **up** in value over time.
Depreciation is when items go **down** in value over time.
At the end of each year the appreciation or depreciation on an item is usually described in terms of a percentage of the value of the item at the beginning of that year.
The percentage value at the end of each year is changed to a decimal to give a constant **multiplier**.
Appreciation and depreciation are worked out in a similar way to compound interest.

Example 7

The value of a vintage car appreciates by 8% each year.
What is the value of a car bought for £20 000 after 4 years?

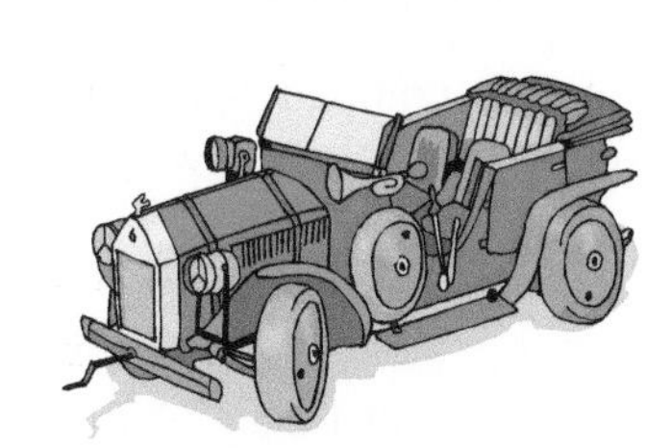

The value of the car is increasing.
So, its value at the end of each year is
100% + 8% = 108%, or 1.08 of its previous value.

Using 1.08 as a multiplier gives:
Value at end of year 1 = £20 000 × 1.08 = £21 600.
Value at end of year 2 = £21 600 × 1.08 = £23 328.
Value at end of year 3 = £23 328 × 1.08 = £25 194.
Value at end of year 4 = £25 194 × 1.08 = £27 210.
The value of the car after 4 years is £27 210.
This could be calculated as £20 000 × 1.08^4 = £27 210.

This works in the same way as Compound Interest.

Keep the exact answer in your calculator to preserve accuracy.

Example 8

A machine was originally worth £8000.
It depreciates in value by 10% each year.
What will the machine be worth at the end of 3 years?

The value of the machine is decreasing.
So, its value at the end of each year is $100\% - 10\% = 90\%$, or 0.9 of its previous value.

Using 0.9 as a multiplier gives:
Value at end of 3 years = £8000 × 0.9^3.
Value at end of 3 years = £5832.

This is similar to Compound Interest in reverse.

Practice Exercise 4.3

1. A firm owns machinery which is judged to depreciate in value by 5% each year.
   If it was valued at £2000 three years ago, what is it worth today, to the nearest £?

2. A colony of insects is increasing in numbers at the rate of 6% per day.
   There were originally 2400 insects.
   (a) Estimate the number of insects in the colony after 3 days.
   (b) How long will it take the colony of insects to double in size?

3. The value of an oil painting increases at the rate of 12% per year.
   The painting was originally valued at £30 000.
   What will the value of the painting be after 4 years?
   Give your answer to a suitable degree of accuracy.

4. A man buys a new car for £13 000.
   The car loses value at the rate of 14% per annum.
   (a) (i) What is its value after 3 years?
   (ii) Express its value after 3 years as a percentage of its original value.
   (b) Repeat (a) for a new car originally valued at £20 000.
   (c) What do you notice about your answers to (a) and (b)?

5. For accounting purposes, the value of a computer is judged to depreciate at 15% per year.
   The computer was originally bought for £1600.
   (a) Find the value of the computer after 2 years.
   (b) After how many years is the computer worth less than £800?

6. Machinery which cost £9600 when new is judged to depreciate in value by 20% in its first year and by 10% each year in future years.
   What is the estimated value of the machinery after three years?

7. The population of a town is 35 000.
   It is estimated that the population of the town will grow at a rate of 3% per year.
   (a) Estimate the population of the town after 4 years.
   (b) How long will it take for the population of the town to exceed 45 000?

8. A car insurance company estimates that a particular car will depreciate in value according to rates given in the table below.

Age of car (years)	1	2	3	4	5
Depreciation during year	40%	20%	10%	5%	5%

   Rory buys a new car for £12 500.
   Find the value of his car, according to the insurance company, at the end of each of the first 5 years.

9. The value of a flat bought for £210 000 gained in value for the first 4 years at a rate of 8% per year.
   The value of flats in the area then decreased at a rate of 10% per year over the next 3 years.
   What is the value of the flat 7 years after it was purchased?

10. A painting was bought as a long-term investment for £10 000.
During the first 3 years the value of the painting increased by 2% per year.
However, during the following 4 years, its value decreased by 3% per year.
Now it is estimated that the value of the painting will increase steadily at a rate of 1% per year.
(a) Find the value of the painting after 3 years.
(b) What was the minimum value of the painting?
(c) What will the painting be worth 10 years after it was purchased?
(d) After how many years from buying the painting will its value exceed its purchase price?

Key Points

- 'Per cent' means 'out of 100'.
  The symbol for per cent is %.
  10% can be written as $\frac{10}{100}$ or 0.1.
- Money invested in a savings account at a bank or building society earns **interest**, which is usually paid once a year.
  With **Simple Interest**, the interest is paid out each year and not added to your account.

  Simple Interest = Amount invested × Time in years × Rate of interest per year

- With **Compound Interest**, the interest earned each year is added to your account and also earns interest the following year.

  In general, when Compound Interest rates are applied:

  $$\text{Total value of investment} = \text{Amount invested} \times \left(\frac{100 + r}{100}\right)^n,$$

  where r is the compound rate of interest and n is the number of years.

  $\left(\frac{100 + r}{100}\right)$ is sometimes called the **multiplier**.

  $$\text{Compound Interest} = \text{Amount invested} \times \left(\frac{100 + r}{100}\right)^n - \text{Amount invested.}$$

- **Appreciation** is when items go **up** in value with time.
  Depreciation is when items go **down** in value over time.
  At the end of each year the appreciation or depreciation on an item is usually described in terms of a percentage of the value of the item at the beginning of that year.
  The percentage value at the end of each year is changed to a decimal to give a constant **multiplier**.
  Appreciation and depreciation are worked out in a similar way to compound interest.

Review Exercise 4

Do not use a calculator for questions 1 and 2.

1. There are 450 seats in a theatre. 60% of the seats are in the stalls.
How many seats are in the stalls?

2. A salesman earns a bonus of 3% of his weekly sales.
How much bonus does the salesman earn in a week when his sales are £1400?

3. By selling a vase for £980 an antiques dealer made a 40% profit on the amount he had paid for it.
How much did the antiques dealer pay for the vase?

4. To clear goods in a sale, a store manager reduced the price of jars of honey by 10% and sold them for £3.60 each.
What was the price of a jar of honey before the sale?

5. A shop assistant earns £17 620 after a 2.5% pay increase.
What was the shop assistant's salary before the pay rise?

6. The population of a village increased by 15% between 2007 and 2018.
The population in 2018 was 897.
What was the population in 2007?

7. How long does it take £100 to double in value at 9% Compound Interest per annum?

8. (a) Find the Simple Interest if £1000 is invested for 4 years at 3.5% per annum.
(b) Find the Compound Interest if £1000 is invested for 4 years at 3.5% per annum.

9. Sally invests £780 at a compound rate of interest of 4.6% per annum.
What will be the value of her investment after 3 years?

10. A young oak tree gains 30% more height each year.
It is 5 m tall.
How tall will it be in 3 years' time?

11. A bouncy ball is dropped from the top of a skyscraper 256 m high.
After each bounce it reaches a height 25% less than its previous height.
What is the height of the ball after 4 bounces?

12. The number of salmon in the River Dribble is expected to increase by about 15% every year up to the year 2020.
At the beginning of 2016 there were approximately 6000 salmon in the River Dribble.

(a) How many salmon are expected to be in the River Dribble at the beginning of 2017?
(b) The River Authority stated that by the year 2020 there will be over 10 000 salmon in the River Dribble.
Do you agree with this?
Show all your working.

13. (a) A water lily doubles its leaf area every day.
It covers a pond after 50 days.
How long does it take to cover half of the pond?
(b) A frog jumps from the centre of this pond to the edge.
Each jump he halves the distance between himself and the edge of the pond.
How long does he take to reach the edge?

14. These wage increases were announced by a company.
The managing director will receive 6% more and will now be earning £47 700.
The foreman will receive 4% more and will now be earning £31 200.
The skilled workers will receive 3.5% more and will now be earning £20 700.
The unskilled workers will receive 3% more and will now be earning £12 875.
The company has 1 managing director, 1 foreman, 5 skilled workers and 9 unskilled workers.
Calculate the percentage increase in the total wage bill.

15. An investor bought an item for £8000.
He estimated that the item would depreciate in value for the first 3 years at a rate of 4% per annum.
During the following 5 years he estimates that the value of the item will appreciate at 7.5% per annum.
(a) Calculate the estimated value of the item after 8 years.
(b) Would it have been better to invest the £8000 in an investment bond which guarantees 3% per year?
You must show your working.

5 Working with Fractions

Fractions

$\frac{3}{8}$ is a **fraction**.

In a fraction: the top number is called the **numerator**,
the bottom number is called the **denominator**.

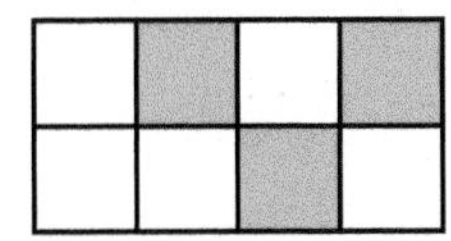

$\frac{3}{8}$ of this rectangle is shaded.

Equivalent fractions

Fractions which are equal are called **equivalent fractions**.

Each of the fractions $\frac{1}{4}$, $\frac{3}{12}$, $\frac{6}{24}$,
is the same fraction written in different ways.

These fractions are all equivalent to $\frac{1}{4}$.

To write an equivalent fraction:
Multiply the numerator and denominator by the **same** number.

For example. $\frac{1}{4} = \frac{1 \times 3}{4 \times 3} = \frac{3}{12}$

Simplifying fractions

Fractions can be **simplified** if both the numerator and denominator can be divided by the **same number**.
To write a fraction in its **simplest form** divide both the numerator and denominator by the largest number that divides into them both.
This is sometimes called **cancelling** a fraction.

Remember:
Multiplication and division are inverse (opposite) operations.
Equivalent fractions can also be made by dividing the numerator and denominator of a fraction by the same number.

Example 1

Write the fraction $\frac{25}{30}$ in its simplest form.

The largest number that divides into both the numerator and denominator of $\frac{25}{30}$ is 5.

$\frac{25}{30} = \frac{25 \div 5}{30 \div 5} = \frac{5}{6}$ $\qquad$ $\frac{25}{30} = \frac{5}{6}$ in its simplest form.

Types of fractions

Numbers like $2\frac{1}{2}$ and $1\frac{1}{4}$ are called **mixed numbers** because they are a mixture of whole numbers and fractions.
Mixed numbers can be written as **improper** or **'top heavy'** fractions.
These are fractions where the numerator is larger than the denominator.

Example 2

Write $3\frac{4}{7}$ as an improper fraction.

$3\frac{4}{7} = \frac{(3 \times 7) + 4}{7} = \frac{21 + 4}{7} = \frac{25}{7}$

Example 3

Write $\frac{32}{5}$ as a mixed number.

$32 \div 5 = 6$ remainder 2.

$\frac{32}{5} = 6\frac{2}{5}$

Finding fractions of quantities

Example 4

Find $\frac{2}{5}$ of £65.

Divide £65 into 5 equal parts.

£65 ÷ 5 = £13.

Each of these parts is $\frac{1}{5}$ of £65.

Two of these parts is $\frac{2}{5}$ of £65.

So, $\frac{2}{5}$ of £65 = 2 × £13 = £26.

Example 5

A coat costing £138 is reduced by $\frac{1}{3}$.

What is the reduced price of the coat?

Find $\frac{1}{3}$ of £138.

$\frac{1}{3}$ of £138 = £138 ÷ 3 = £46

So, reduced price = £138 − £46 = £92.

Practice Exercise 5.1

Do not use a calculator for this exercise.

1. Write three equivalent fractions for the shaded part of this rectangle. What is the simplest form of the fraction for the shaded part?

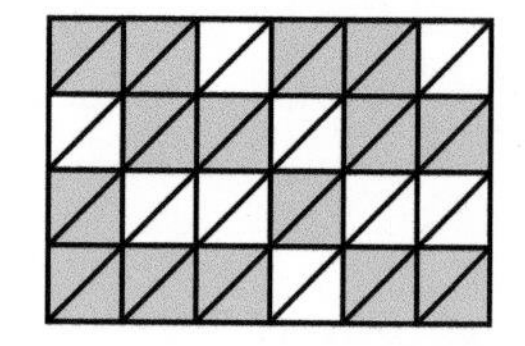

2. Write these fractions in order, smallest first. $\frac{2}{3}$ $\frac{3}{5}$ $\frac{7}{10}$ $\frac{8}{15}$
3. Which of these fractions is the largest? $\frac{4}{5}$ $\frac{13}{20}$ $\frac{7}{10}$ $\frac{3}{4}$
4. Write these fractions in their simplest form. (a) $\frac{12}{15}$ (b) $\frac{18}{27}$ (c) $\frac{50}{75}$ (d) $\frac{16}{40}$
5. Change these fractions to mixed numbers: (a) $\frac{3}{2}$ (b) $\frac{17}{8}$ (c) $\frac{15}{4}$ (d) $\frac{23}{5}$
6. Change these numbers to improper fractions: (a) $2\frac{7}{10}$ (b) $1\frac{3}{5}$ (c) $5\frac{5}{6}$
7. Write 42 as a fraction of 70. Give your answer in its simplest form.
8. A box of 50 chocolates includes 30 soft-centred chocolates. What fraction of the chocolates is soft-centred?
9. Calculate:
   (a) $\frac{1}{4}$ of 12 (b) $\frac{1}{5}$ of 20 (c) $\frac{1}{10}$ of 30 (d) $\frac{1}{6}$ of 48 (e) $\frac{2}{5}$ of 20
   (f) $\frac{3}{10}$ of 30 (g) $\frac{2}{7}$ of 42 (h) $\frac{5}{9}$ of 36 (i) $\frac{5}{6}$ of 48 (j) $\frac{3}{8}$ of 32
10. Richard has 30 marbles. He gives $\frac{1}{5}$ of them away.
    (a) How many marbles does he give away?
    (b) How many marbles has he got left?
11. Alfie collects £12.50 for charity. He gives $\frac{3}{5}$ of it to Oxfam. How much does he give to other charities?
12. In a sale all prices are reduced by $\frac{3}{10}$. What is the sale price of a microwave which was originally priced at £212?
13. A publisher offers a discount of $\frac{3}{20}$ for orders of more than 100 books. How much would a shop pay for an order of 250 books costing £15 each?
14. Lauren and Amelia share a bar of chocolate. The chocolate bar has 24 squares. Lauren eats $\frac{3}{8}$ of the bar. Amelia eats $\frac{5}{12}$ of the bar.
    (a) How many squares has Lauren eaten?
    (b) How many squares has Amelia eaten?
    (c) What fraction of the bar is left?

Adding and subtracting fractions

Calculate $1\frac{3}{4} + \frac{5}{6}$.

Change mixed numbers to improper ('top heavy') fractions. $1\frac{3}{4} = \frac{7}{4}$

The calculation then becomes $\frac{7}{4} + \frac{5}{6}$.

Find the smallest number into which both 4 and 6 will divide.
4 divides into: 4, 8, 12, 16, ...
6 divides into: 6, 12, 18, ...
So, 12 is the smallest.

Change the original fractions to equivalent fractions using the smallest number as the new denominator. $\frac{7}{4} = \frac{21}{12}$ and $\frac{5}{6} = \frac{10}{12}$

Add the new numerators. Keep the new denominator the same. $\frac{21}{12} + \frac{10}{12} = \frac{21 + 10}{12} = \frac{31}{12}$

Write the answer in its simplest form. $\frac{31}{12} = 2\frac{7}{12}$

> Fractions must have the **same denominator** before addition (or subtraction) can take place.
>
> *What happens when you use a denominator that is **not** the smallest?*

Example 6

Work out $\frac{5}{8} - \frac{7}{12}$.

24 is the smallest number into which both 8 and 12 divide.

$\frac{5}{8} = \frac{5 \times 3}{8 \times 3} = \frac{15}{24}$ and $\frac{7}{12} = \frac{7 \times 2}{12 \times 2} = \frac{14}{24}$

$\frac{5}{8} - \frac{7}{12} = \frac{15}{24} - \frac{14}{24} = \frac{1}{24}$

Example 7

Calculate $3\frac{3}{10} - 1\frac{5}{6}$.

$3\frac{3}{10} = \frac{33}{10}$ and $1\frac{5}{6} = \frac{11}{6}$

$\frac{33}{10} = \frac{99}{30}$ and $\frac{11}{6} = \frac{55}{30}$

$3\frac{3}{10} - 1\frac{5}{6} = \frac{99}{30} - \frac{55}{30} = \frac{44}{30}$

$\frac{44}{30} = \frac{22}{15} = 1\frac{7}{15}$

Remember:
- Add the numerators only.
- When the answer is an improper fraction change it into a mixed number.

Practice Exercise 5.2

Do not use a calculator for this exercise.

1. Work out:
   (a) $\frac{1}{4} + \frac{1}{8}$ (b) $\frac{1}{3} + \frac{1}{4}$ (c) $\frac{1}{2} + \frac{1}{5}$ (d) $\frac{1}{3} + \frac{1}{5}$ (e) $\frac{1}{2} + \frac{1}{7}$
2. Work out:
   (a) $\frac{1}{4} - \frac{1}{8}$ (b) $\frac{1}{3} - \frac{1}{4}$ (c) $\frac{1}{2} - \frac{1}{5}$ (d) $\frac{1}{3} - \frac{1}{5}$ (e) $\frac{1}{2} - \frac{1}{7}$
3. Work out:
   (a) $\frac{1}{2} + \frac{3}{4}$ (b) $\frac{2}{3} + \frac{5}{6}$ (c) $\frac{3}{4} + \frac{4}{5}$ (d) $\frac{5}{7} + \frac{2}{3}$ (e) $\frac{3}{8} + \frac{5}{6}$
4. Calculate:
   (a) $\frac{5}{8} - \frac{1}{2}$ (b) $\frac{13}{15} - \frac{1}{3}$ (c) $\frac{5}{6} - \frac{5}{24}$ (d) $\frac{7}{15} - \frac{2}{5}$ (e) $\frac{3}{4} - \frac{5}{12}$
5. Calculate:
   (a) $2\frac{3}{4} + 1\frac{1}{2}$ (b) $1\frac{1}{2} + 2\frac{1}{3}$ (c) $1\frac{3}{4} + 2\frac{5}{8}$ (d) $2\frac{1}{4} + 3\frac{3}{5}$ (e) $4\frac{3}{5} + 1\frac{5}{6}$
6. Calculate:
   (a) $2\frac{1}{2} - 1\frac{2}{5}$ (b) $1\frac{2}{3} - 1\frac{1}{4}$ (c) $3\frac{3}{4} - 2\frac{3}{8}$ (d) $5\frac{2}{5} - 2\frac{1}{10}$ (e) $4\frac{5}{12} - 2\frac{1}{6}$
7. Colin buys a bag of flour.
   He uses $\frac{1}{3}$ to bake a cake and $\frac{2}{5}$ to make a loaf.
   What fraction of the bag of flour is left?
8. Edward, Marc, Dee and Lin share an apple pie.
   Edward has $\frac{1}{3}$, Marc has $\frac{1}{5}$ and Dee has $\frac{1}{4}$.
   How much is left for Lin?
9. Both Lee and Mary have a packet of the same sweets.
   Mary eats $\frac{1}{3}$ of her packet.
   Lee eats $\frac{3}{4}$ of his packet.
   (a) Find the difference between the fraction Mary eats and the fraction Lee eats.

   Lee gives his remaining sweets to Mary.
   (b) What fraction of a packet does Mary now have?
10. Jon, Billy and Cathy are the only candidates in a school election.
    Jon got $\frac{7}{20}$ of the votes.
    Billy got $\frac{2}{5}$ of the votes.
    (a) What fraction of the votes did Cathy get?
    (b) Which candidate won the election?

Multiplying fractions

Calculate $\frac{3}{8} \times \frac{1}{9}$

Simplify, where possible, by cancelling. $\frac{\not{3}^{1}}{8} \times \frac{1}{\not{9}_{3}}$

Multiply the numerators.
Multiply the denominators. $\frac{1 \times 1}{8 \times 3} = \frac{1}{24}$

Write the answer in its simplest form. $\frac{3}{8} \times \frac{1}{9} = \frac{1}{24}$

To simplify:
Divide a numerator **and** a denominator by the **same number.**
In this case:
3 and 9 can be divided by 3.
$3 \div 3 = 1$ and $9 \div 3 = 3$.

Example 8

Work out $\frac{3}{8} \times 12$.

Note:
$12 = \frac{12}{1}$

$\frac{3}{8} \times \frac{12}{1}$

Simplify by cancelling.

$= \frac{3}{\not{8}_{2}} \times \frac{\not{12}^{3}}{1}$

Multiply out.

$= \frac{3 \times 3}{2 \times 1} = \frac{9}{2}$

$= 4\frac{1}{2}$

Example 9

Calculate $1\frac{1}{2} \times 1\frac{3}{5}$.

Change mixed numbers to improper fractions.

$1\frac{1}{2} \times 1\frac{3}{5} = \frac{3}{2} \times \frac{8}{5}$

Simplify by cancelling.

$= \frac{3}{\not{2}_{1}} \times \frac{\not{8}^{4}}{5}$

Multiply out.

$= \frac{3 \times 4}{1 \times 5} = \frac{12}{5} = 2\frac{2}{5}$

Practice Exercise 5.3

Do not use a calculator for this exercise.

1. Work out. Give your answers as mixed numbers.
   (a) $\frac{1}{2} \times 7$ (b) $\frac{3}{5} \times 3$ (c) $\frac{5}{8} \times 10$ (d) $\frac{3}{4} \times 12$ (e) $\frac{7}{10} \times 15$
2. Work out:
   (a) $\frac{1}{4} \times \frac{1}{5}$ (b) $\frac{3}{5} \times \frac{1}{6}$ (c) $\frac{1}{2} \times \frac{3}{4}$ (d) $\frac{1}{10} \times \frac{5}{8}$ (e) $\frac{3}{10} \times \frac{1}{6}$
3. Work out: **Hint** - Change mixed numbers to improper fractions first.
   (a) $1\frac{1}{2} \times 5$ (b) $1\frac{1}{4} \times 6$ (c) $1\frac{1}{5} \times 4$ (d) $3\frac{2}{3} \times 2$ (e) $2\frac{3}{5} \times 3$
   (f) $4\frac{1}{2} \times \frac{1}{2}$ (g) $3\frac{3}{4} \times \frac{1}{5}$ (h) $2\frac{2}{5} \times \frac{1}{3}$ (i) $2\frac{2}{3} \times \frac{1}{4}$ (j) $4\frac{3}{8} \times \frac{1}{7}$
4. Calculate:
   (a) $\frac{2}{3} \times \frac{3}{4}$ (b) $\frac{3}{4} \times \frac{2}{5}$ (c) $\frac{2}{5} \times \frac{5}{6}$ (d) $\frac{2}{5} \times \frac{5}{7}$ (e) $\frac{3}{10} \times \frac{5}{6}$
5. Calculate:
   (a) $1\frac{1}{2} \times \frac{3}{4}$ (b) $1\frac{1}{2} \times 1\frac{1}{3}$ (c) $2\frac{1}{4} \times 1\frac{2}{3}$ (d) $2\frac{3}{4} \times 1\frac{2}{5}$ (e) $1\frac{1}{2} \times 3\frac{1}{4}$
   (f) $1\frac{3}{5} \times 1\frac{1}{6}$ (g) $6\frac{3}{10} \times 2\frac{2}{9}$ (h) $3\frac{3}{4} \times 3\frac{3}{5}$ (i) $1\frac{1}{4} \times 2\frac{4}{25}$ (j) $2\frac{5}{8} \times 3\frac{1}{3}$
6. Kylie has $\frac{2}{3}$ of a litre of orange. She drinks $\frac{2}{5}$ of the orange.
   (a) What fraction of a litre does she drink?
   (b) What fraction of a litre is left?
7. Tony eats $\frac{1}{5}$ of a bag of sweets.
   He shares the remaining sweets equally among Bob, Jo and David.
   (a) What fraction of the bag of sweets does Bob get?
   (b) What is the smallest possible number of sweets in the bag?
8. In a school $\frac{8}{15}$ of the pupils are girls.
   $\frac{3}{16}$ of the girls are left-handed.
   What fraction of the pupils in the school are left-handed girls?

Dividing fractions

The method normally used when one fraction is divided by another is to change the division to a multiplication. The fractions can then be multiplied in the usual way.

Calculate $2\frac{2}{15} \div 1\frac{3}{5}$

Change mixed numbers to improper ('top heavy') fractions. $2\frac{2}{15} = \frac{32}{15}$ and $1\frac{3}{5} = \frac{8}{5}$

Change the division to a multiplication. $\frac{32}{15} \div \frac{8}{5} = \frac{32}{15} \times \frac{5}{8}$

Simplify, where possible, by cancelling. $\frac{\cancel{32}^{4}}{\cancel{15}_{3}} \times \frac{\cancel{5}^{1}}{\cancel{8}_{1}}$

Multiply the numerators. Multiply the denominators. $\frac{4 \times 1}{3 \times 1} = \frac{4}{3}$

Write the answer in its simplest form. $\frac{4}{3} = 1\frac{1}{3}$

Example 10

Work out $\frac{2}{3} \div 5$.

$\frac{2}{3} \div 5$

$= \frac{2}{3} \times \frac{1}{5}$

$= \frac{2}{15}$

Divide by 5 is the same as multiply by $\frac{1}{5}$.

Example 11

Calculate $1\frac{3}{5} \div \frac{4}{9}$.

$1\frac{3}{5} \div \frac{4}{9} = \frac{8}{5} \div \frac{4}{9}$

$= \frac{8}{5} \times \frac{9}{4}$

$= \frac{\cancel{8}^{2}}{5} \times \frac{9}{\cancel{4}_{1}}$

$= \frac{18}{5} = 3\frac{3}{5}$

Divide by $\frac{4}{9}$ is the same as multiply by $\frac{9}{4}$.

Practice Exercise 5.4

Do not use a calculator for this exercise.

1. Work out. Give your answers in their simplest form.
(a) $\frac{1}{2} \div 5$ (b) $\frac{3}{4} \div 2$ (c) $\frac{2}{3} \div 2$ (d) $\frac{2}{5} \div 4$ (e) $\frac{6}{7} \div 3$

2. Work out: **Hint** - Change mixed numbers to improper fractions first.
(a) $1\frac{1}{2} \div 2$ (b) $1\frac{2}{3} \div 5$ (c) $1\frac{5}{7} \div 3$ (d) $3\frac{1}{9} \div 7$ (e) $2\frac{2}{3} \div 4$

3. Calculate:
(a) $\frac{1}{2} \div \frac{1}{4}$ (b) $\frac{1}{5} \div \frac{1}{2}$ (c) $\frac{7}{8} \div \frac{1}{3}$ (d) $\frac{2}{3} \div \frac{1}{5}$ (e) $\frac{3}{4} \div \frac{1}{8}$

4. Calculate:
(a) $\frac{2}{3} \div \frac{4}{5}$ (b) $\frac{3}{8} \div \frac{2}{3}$ (c) $\frac{3}{5} \div \frac{3}{4}$ (d) $\frac{2}{5} \div \frac{3}{10}$ (e) $\frac{3}{8} \div \frac{9}{16}$
(f) $\frac{7}{12} \div \frac{7}{18}$ (g) $\frac{4}{9} \div \frac{2}{3}$ (h) $\frac{7}{10} \div \frac{3}{5}$ (i) $\frac{9}{20} \div \frac{3}{10}$ (j) $\frac{21}{25} \div \frac{7}{15}$

5. Calculate:
(a) $1\frac{3}{4} \div \frac{1}{8}$ (b) $1\frac{1}{2} \div \frac{3}{10}$ (c) $1\frac{3}{5} \div \frac{4}{5}$ (d) $2\frac{1}{10} \div \frac{7}{20}$ (e) $3\frac{3}{4} \div \frac{5}{8}$
(f) $1\frac{1}{4} \div 1\frac{9}{16}$ (g) $3\frac{1}{5} \div 2\frac{2}{15}$ (h) $2\frac{1}{4} \div 1\frac{4}{5}$ (i) $2\frac{1}{7} \div 1\frac{7}{8}$ (j) $1\frac{2}{5} \div 1\frac{2}{3}$

6. Neil uses $\frac{1}{2}$ of a block of paté to make 5 sandwiches.
What fraction of the block of paté does he put on each sandwich?

7. Lauren uses $\frac{2}{3}$ of a bag of flour to make 6 muffins.
What fraction of the bag of flour is used for each muffin?

8. A shelf is $40\frac{3}{4}$ cm long.
How many books of width $1\frac{3}{4}$ cm can stand on the shelf?

Problems involving fractions

Practice Exercise 5.5

Do not use a calculator for this exercise.

1. Ben spends $\frac{5}{8}$ of his pocket money.
   He has £1.20 left.
   How much pocket money did Ben get?

2. A cyclist travels from A to B in two stages.
   Stage 1 is 28 km which is $\frac{2}{7}$ of the total journey.
   How long is stage 2?

3. Robert pays £9.60 for a CD in the sale.
   How much did he save?

   Music Sale - $\frac{1}{4}$ off all CDs

4. Sara's hourly wage is increased by $\frac{1}{10}$.
   Her new hourly wage is £8.80.
   What was her original hourly wage?

5. A young tree is $\frac{3}{8}$ taller in August than it was in May.
   In August it is 132 centimetres tall.
   How tall was it in May?

Fractions on a calculator

Fraction calculations can be done quickly using the fraction button on a calculator.
On most calculators the fraction button looks like this ... $a^b\!/_c$

Example 12

Use a calculator to work out $4\frac{3}{5} \div 2\frac{1}{4}$.

This can be calculated with this calculator sequence.

This gives the answer $2\frac{2}{45}$.

Practice Exercise 5.6

You may use a calculator for this exercise.

1. Work out.
   (a) $2\frac{11}{12} + \frac{7}{8}$ (b) $2\frac{5}{6} - 1\frac{1}{4}$ (c) $7\frac{5}{12} + 5\frac{8}{9}$ (d) $1\frac{7}{20} - \frac{4}{5}$
   (e) $2\frac{5}{12} + 3\frac{2}{3}$ (f) $2\frac{2}{3} - 2\frac{5}{9}$ (g) $4\frac{7}{8} + 1\frac{11}{12}$ (h) $4\frac{3}{20} - 1\frac{7}{12}$

2. Work out.
   (a) $\frac{3}{8} \times \frac{2}{3}$ (b) $2\frac{2}{3} \times 1\frac{1}{8}$ (c) $1\frac{3}{4} \times 2\frac{2}{5}$ (d) $4\frac{1}{2} \times 2\frac{5}{6}$
   (e) $\frac{5}{6} \div \frac{7}{8}$ (f) $4\frac{1}{8} \div 2\frac{3}{4}$ (g) $7 \div 1\frac{3}{4}$ (h) $11\frac{1}{4} \div 2\frac{3}{16}$

3. Work out.
   (a) $6\frac{3}{4} - 1\frac{2}{3}$ (b) $5\frac{1}{12} \div 7\frac{5}{8}$ (c) $2\frac{2}{3} \times 2\frac{3}{4}$ (d) $3\frac{1}{6} + 1\frac{1}{4}$
   (e) $1\frac{5}{8} \times 1\frac{3}{5}$ (f) $4\frac{2}{3} - 1\frac{1}{6}$ (g) $2\frac{3}{5} \div 1\frac{3}{10}$ (h) $3\frac{1}{7} \times \frac{4}{11}$

4. Work out.
   (a) $1\frac{3}{4} \times 2\frac{5}{12} + 3\frac{5}{6}$ (b) $2\frac{1}{16} - 1\frac{3}{5} + 6\frac{1}{2}$ (c) $2\frac{1}{2} + 3\frac{7}{8} - 4\frac{1}{4}$

5. Work these out. Remember to find the value of the contents of the brackets first.
   (a) $\left(\frac{15}{28} \times \frac{7}{30}\right) + \frac{7}{8}$ (b) $\left(2\frac{2}{3} - 1\frac{3}{4}\right) \times 4$ (c) $8\frac{3}{4} \times 1\frac{3}{5} \div 4\frac{2}{3}$
   (d) $1\frac{5}{12} \div \left(3\frac{1}{5} + 1\frac{1}{3}\right)$ (e) $2\frac{1}{4} - \left(1\frac{1}{2} \times \frac{2}{5}\right)$ (f) $\left(\frac{2}{3} - \frac{1}{6}\right)^2$

Key Points

- The top number of a fraction is called the **numerator**. The bottom number of a fraction is called the **denominator**.
- Fractions which are equal are called **equivalent fractions**.
- Fractions can be **simplified** if both the numerator and denominator can be divided by the **same number**. This is sometimes called **cancelling**.
- $2\frac{1}{2}$ is an example of a **mixed number**. It is a mixture of whole numbers and fractions.
- $\frac{5}{2}$ is an **improper** (or **'top heavy'**) fraction.
- You should be able to find fractions of quantities. E.g., Find $\frac{2}{5}$ of £65.
- You should be able to add and subtract fractions. Fractions must have the **same denominator** before addition (or subtraction) can take place.
- You should be able to multiply fractions.
- You should be able to divide fractions. When one fraction is divided by another, change division to a multiplication. The fractions can then be multiplied in the usual way.
- Fraction calculations can be done quickly using the fraction button on a calculator. On most calculators the fraction button looks like this … $a^{b}/_{c}$

Review Exercise 5

Do not use a calculator for questions 1 to 6.

1. Write the fractions in their simplest form.
   (a) $\frac{24}{88}$ (b) $\frac{18}{45}$ (c) $\frac{75}{200}$ (d) $\frac{35}{56}$ (e) $\frac{11}{110}$

2. An examination in French is marked out of 80.
   (a) Jean scored $\frac{4}{5}$ of the marks. How many marks did she score?
   (b) Pete scored 35 marks. What fraction of the total did he score?

3. 36 girls and 24 boys applied to go on a rock climbing course. $\frac{2}{3}$ of the girls and $\frac{3}{4}$ of the boys went on the course. What fraction of the 60 students who applied went on the course? Write the fraction in its simplest form.

4. The price of a coat is reduced by $\frac{2}{5}$ to £48. What was the original price of the coat?

5. Work out.
   (a) $\frac{1}{2}+\frac{1}{3}+\frac{1}{4}$ (b) $2\frac{7}{10}+1\frac{3}{5}$ (c) $3\frac{11}{12}-1\frac{7}{8}+1\frac{1}{4}$
   (d) $2\frac{1}{2}+\frac{7}{10}-\frac{3}{5}$ (e) $\frac{1}{2}+\frac{1}{3}-\frac{1}{6}$ (f) $1\frac{3}{4}-\frac{4}{5}+3\frac{7}{8}$

6. Work out.
   (a) $\frac{5}{7}\times\frac{3}{4}$ (b) $2\frac{1}{6}\times\frac{9}{13}$ (c) $3\frac{3}{8}\times1\frac{1}{9}$
   (d) $\frac{3}{10}\div1\frac{1}{6}$ (e) $3\frac{3}{4}\div2\frac{2}{5}$ (f) $2\frac{1}{12}\div5\frac{5}{8}$

7. Calculate.
   (a) $2\frac{5}{7}+1\frac{4}{9}$ (b) $4\frac{3}{10}-1\frac{2}{3}$ (c) $2\frac{2}{3}\times3\frac{3}{4}$ (d) $3\frac{3}{5}\div2\frac{1}{10}$

8. Work out.

(a) $\left(3\frac{1}{7} \times 8\frac{3}{4}\right) - 2\frac{1}{3}$ (b) $12\frac{1}{5} - \left(2\frac{2}{9} \times 4\frac{1}{2}\right)$

(c) $2\frac{3}{4} - \left(2\frac{1}{2} \times \frac{2}{5}\right)$ (d) $2\frac{1}{10} - 1\frac{4}{5} + 6\frac{3}{4}$

9. Use your calculator to find these amounts to the nearest penny.

(a) $\frac{5}{6}$ of £5 (b) $\frac{2}{7}$ of £10.30 (c) $\frac{3}{8}$ of £1.05

(d) $\frac{1}{8}$ of £11.49 (e) $\frac{7}{8}$ of £23

10. On a farm there were 560 animals.
$\frac{5}{8}$ of them were sheep. $\frac{1}{10}$ of them were pigs.
There were 2 ponies, 4 dogs, 3 cats, 5 rabbits and the rest were cows.

(a) How many cows were there?

(b) What fraction of the total number of animals were cows?

11. (a) Copy and complete this number pattern to the line starting with $\frac{9}{10}$.

$$\frac{2}{3} - \frac{1}{2} = \frac{2 \times 2 - 3 \times 1}{3 \times 2} = \frac{1}{6}$$

$$\frac{3}{4} - \frac{2}{3} = \frac{3 \times 3 - 4 \times 2}{4 \times 3} = \frac{1}{12}$$

$$\frac{4}{5} - \frac{3}{4} = \frac{4 \times 4 - 5 \times 3}{5 \times 4} = \frac{1}{20}$$

...

(b) Use the pattern to find the value of $\frac{19}{20} - \frac{18}{19}$.

12. Five people in a group of 50 people are left-handed.
There are 20 females in the group.
A person is picked at random from the group.
Claire draws the following tree diagram.

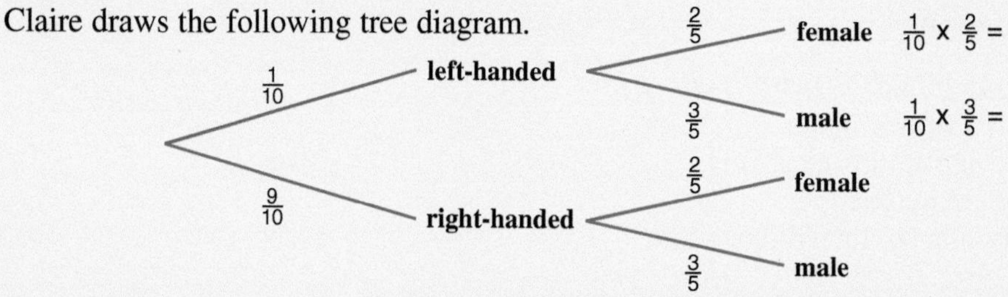

To find the probability of each outcome, multiply the probabilities along the branches of the tree diagram.
For example, the probability of a left-handed female being picked is given by:

$$\frac{1}{10} \times \frac{2}{5} = \frac{1}{25}.$$

Copy and complete the tree diagram above to show the probability of each outcome.

13. Three-fifths of the people at a party are boys.
Three-quarters of the boys are wearing fancy dress.
What fraction of the people at the party are boys wearing fancy dress?

14. Tom spends his wages as follows: $\frac{3}{20}$ on tax, $\frac{1}{4}$ on rent and $\frac{1}{5}$ on fares.
He has £120 left.
How much were his wages?

15. Is the value of $\frac{a+c}{b+d}$ always between the fractions $\frac{a}{b}$ and $\frac{c}{d}$?
Explain your answer.

16. Use your calculator to evaluate:

(a) $7\frac{5}{8} + 6\frac{7}{12}$ (b) $6\frac{3}{4} - \left(2\frac{4}{5} \div \frac{7}{10}\right)$ (c) $4\frac{3}{7}\left(3\frac{1}{2} \times 5\frac{1}{4}\right)$

6 Algebraic Expressions

Expressions and terms

Consider this situation:

$2n$ students start a typing course. 3 of the students leave the course.
How many students remain on the course?

$2n - 3$ students remain.
$2n - 3$ is an **algebraic expression**, or simply an **expression**.

A term includes the sign, + or −.
$2n$ has the same value as $+2n$.

An expression is just an answer made up of letters and numbers.
$+2n$ and -3 are **terms** of the expression.

Simplifying expressions

Adding and subtracting terms

You can add and subtract terms with the same letter.
This is sometimes called **simplifying an expression**.

$x + x = 2x$ $\quad 6x - 2x = 4x$
$4x - 4x = 0$ $\quad 5x - 6x = -x$
$3x + 5 + x - 1 = 4x + 4$
$x^2 + 2x^2 = 3x^2$ $\quad x^2 + x$ cannot be simplified.

A simpler way to write $1x$ is just x.
$-1x$ can be written as $-x$.
$0x$ is the same as 0.
Just as with ordinary numbers, you can add terms in any order.
$x - 2x + 5x = x + 5x - 2x = 4x$

Multiplying terms

When multiplying expressions that include both numbers and powers:

- multiply the numbers
- add the powers $x^m \times x^n = x^{m+n}$

$3 \times x = 3x$ $\quad x \times 4 = 4x$ $\quad x \times x = x^2$
$2x \times x = 2x^2$ $\quad x \times 5x = 5x^2$ $\quad 2x \times 3x = 6x^2$
$2x \times 3x^2 = (2 \times 3) \times (x \times x^2) = 6x^3$

Practice Exercise 6.1

1. Simplify these expressions.
   (a) $n + n + n$ (b) $2y + 3y$ (c) $5g + g + 4g$ (d) $5y - y$
   (e) $6m - m + 3m$ (f) $-2x - 3x + x$ (g) $2a - 5a - 12a + a$

2. Simplify where possible.
   (a) $5x + 3x + 4$ (b) $3 + 2x - x$ (c) $2x + 4 - x$
   (d) $-x + 3 + 4x$ (e) $5 - 9x + 4x$ (f) $2x - 5x - 7$
   (g) $3x + 2 + 4x + 2$ (h) $2x - 5 - 3x + 4$ (i) $7 - 2x + 3x - 5$

3. Simplify.
   (a) $3x^2 - x^2$ (b) $2x^2 + 3x + 4x$ (c) $5x - 2x^2 + 4x^2$
   (d) $3x^2 + 2x - x^2 - 5x$ (e) $3x^2 - 4x^2 - 7 + 3$ (f) $-3 + 2x^2 + 6 - x^2$

4. Simplify.
   (a) $3 \times (-x)$ (b) $x \times (-5)$
   (c) $(-2) \times (-x)$ (d) $3 \times (-2x)$
   (e) $x \times (-x)$ (f) $2x \times (-x)$
   (g) $(-5x) \times 2x$ (h) $(-2x) \times (-5x)$

Remember:
$2 \times (-x) = -2x$
$(-2) \times (-x) = 2x$

Brackets

Some expressions contain brackets. $2(x + 3)$ means $2 \times (x + 3)$.

You can multiply out brackets in an expression either by using a diagram or by expanding.

To multiply out $2(x + 3)$ using the **diagram method**:

$2(x + 3)$ means $2 \times (x + 3)$.
This can be shown using a rectangle.
The areas of the two parts are $2x$ and 6.
The total area is $2x + 6$.
$2(x + 3) = 2x + 6$

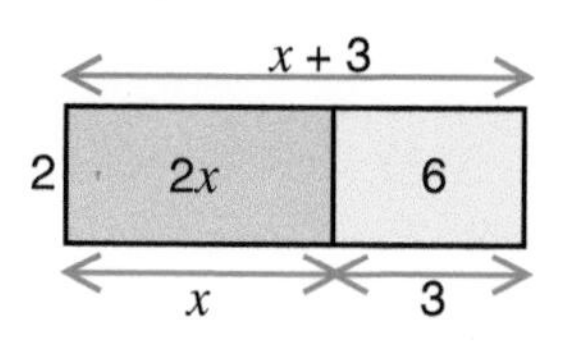

To multiply out $2(x + 3)$ by **expanding**:

$$2(x + 3) = 2 \times x + 2 \times 3$$
$$= 2x + 6$$

Example 1

Multiply out $2x(x + 3)$.

Diagram method

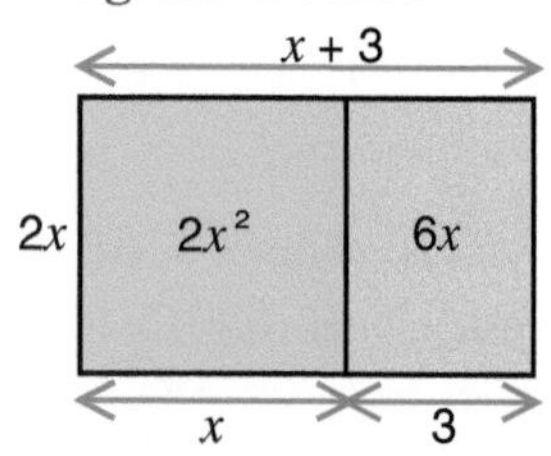

The areas of the two parts are $2x^2$ and $6x$.
$2x(x + 3) = 2x^2 + 6x$

Expanding

$$2x(x + 3) = 2x \times x + 2x \times 3$$
$$= 2x^2 + 6x$$

Practice Exercise 6.2

1. Use the diagrams to multiply out the brackets.

(a)

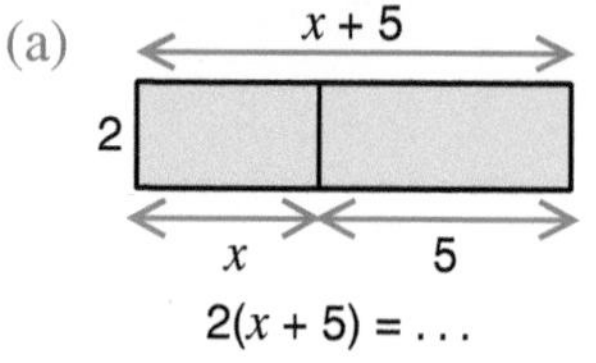

$2(x + 5) = \ldots$

(b)

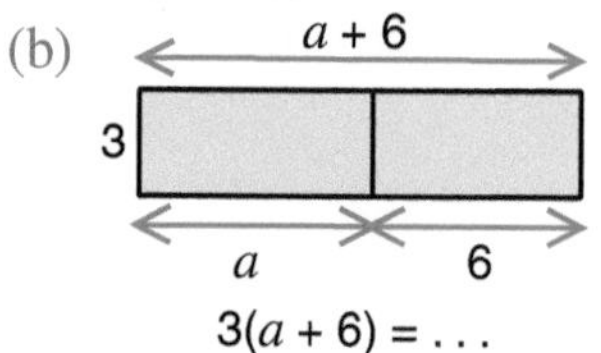

$3(a + 6) = \ldots$

(c)

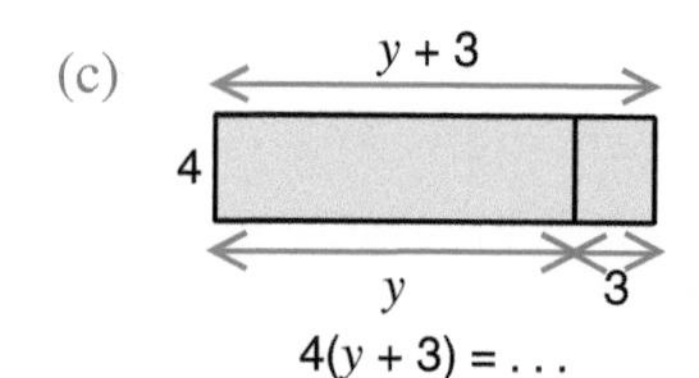

$4(y + 3) = \ldots$

2. Draw your own diagrams to multiply out these brackets.
   (a) $3(x + 2)$ (b) $2(y + 5)$ (c) $2(2x + 1)$ (d) $3(p + q)$

3. Use the diagrams to multiply out the brackets.

(a)

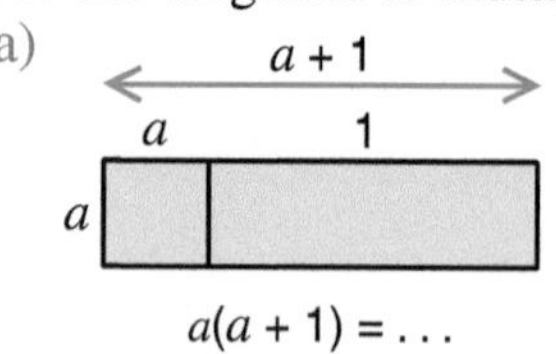

$a(a + 1) = \ldots$

(b)

$d(2 + d) = \ldots$

(c)

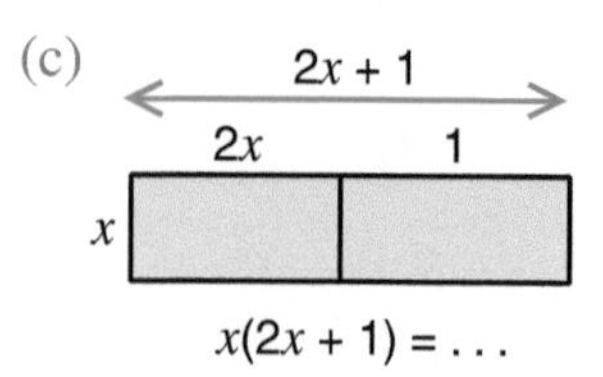

$x(2x + 1) = \ldots$

4. Multiply out the brackets by expanding.
   (a) $2(x + 4)$ (b) $3(t - 2)$ (c) $2(3a + b)$ (d) $5(3 - 2d)$
   (e) $x(x + 3)$ (f) $t(t - 3)$ (g) $g(2g + 3)$ (h) $m(2 - 3m)$

5. Multiply out the brackets by expanding.
   (a) $2p(p + 3)$ (b) $3d(2 - 3d)$ (c) $m(m - n)$ (d) $5x(x + 2y)$

To simplify an expression involving brackets:

- Remove the brackets.
- Simplify by collecting like terms together.

$3(t + 4) + 2$ — Remove the brackets
$= 3t + 12 + 2$ — Simplify
$= 3t + 14$

6. Multiply out the brackets and simplify.

(a) $2(x + 1) + 3$ (b) $3(a + 2) + 5$ (c) $6(w - 4) + 7$
(d) $4 + 2(p + 3)$ (e) $3 + 3(q - 1)$ (f) $1 + 3(2 - t)$
(g) $4(z + 2) + z$ (h) $5(t + 3) + 3t$ (i) $3(c - 2) - c$
(j) $2a + 3(a - 3)$ (k) $y + 2(y - 5)$ (l) $5x + 3(2 - x)$
(m) $4(2a + 5) + 3$ (n) $-2x + 4(3x - 3)$ (o) $3(p - 5) - p + 4$
(p) $3a + 2(a + b)$ (q) $3(x + y) - 2y$ (r) $2(p - q) - 3q$
(s) $2x + x(3 - x)$ (t) $a(a - 3) + a$ (u) $y(2 - y) + y^2$

7. Remove the brackets and simplify.

(a) $2(x + 1) + 3(x + 2)$ (b) $3(a + 1) + 2(a + 5)$ (c) $4(y + 2) + 5(y + 3)$
(d) $2(3a + 1) + 3(a + 1)$ (e) $3(2t + 5) + 5(4t + 3)$ (f) $3(z + 5) + 2(z - 1)$
(g) $7(q - 2) + 5(q + 6)$ (h) $5(x + 3) + 6(x - 3)$ (i) $8(2e - 1) + 4(e - 2)$
(j) $2(5d + 4) + 2(d - 1)$ (k) $m(m - 2) + m(2m - 1)$ (l) $a(3a + 2) + 2a(a - 3)$

8. Multiply out the brackets and simplify.

(a) $-3(x + 2)$ (b) $-3(x - 2)$
(c) $-2(y - 5)$ (d) $-2(3 - x)$
(e) $-3(5 - y)$ (f) $-4(1 + a)$
(g) $5 - 2(a + 1)$ (h) $5d - 3(d - 2)$
(i) $4b - 2(3 + b)$ (j) $-3(2p + 3)$
(k) $5m - 2(3 + 2m)$ (l) $2(3d - 1) - d + 3$
(m) $-a(a - 2)$ (n) $2d - d(1 + d)$ (o) $x^2 - x(1 - x) + x$
(p) $-3g(2g + 3)$ (q) $t^2 - 2t(3 - 3t)$ (r) $2m - 2m(m - 3)$

Remember:
$(-2) \times (+3) = -6$
$(-2) \times (-3) = +6$
so, $-2(x + 3) = -2x - 6$
and $-2(x - 3) = -2x + 6$

9. Expand the brackets and simplify.

(a) $2(a + 1) - (3 - a)$ (b) $3(y - 1) - 2(y + 1)$ (c) $2(1 + m) - 3(1 - m)$
(d) $5(x - 2) - 2(x - 3)$ (e) $2(5 + d) - 3(3 + d)$ (f) $4(t - 1) - 3(t + 2)$
(g) $3(2m + 1) - 2(m - 3)$ (h) $4(2x - 3) - 3(3x + 2)$ (i) $2(4 + 3a) - 3(4a - 5)$

More brackets

The diagram method can be extended to multiply out $(x + 2)(x + 3)$.
The areas of the four parts are: x^2, $3x$, $2x$ and 6.
$(x + 2)(x + 3) = x^2 + 3x + 2x + 6$
Collect like terms and simplify (i.e. $3x + 2x = 5x$)
$= x^2 + 5x + 6$

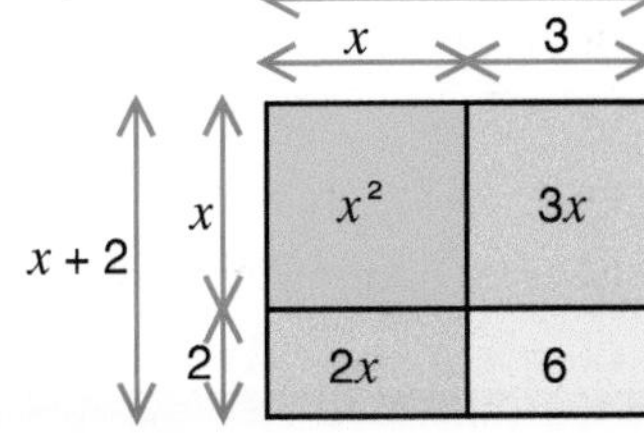

Example 2

Expand $(2x + 3)(x + 4)$.

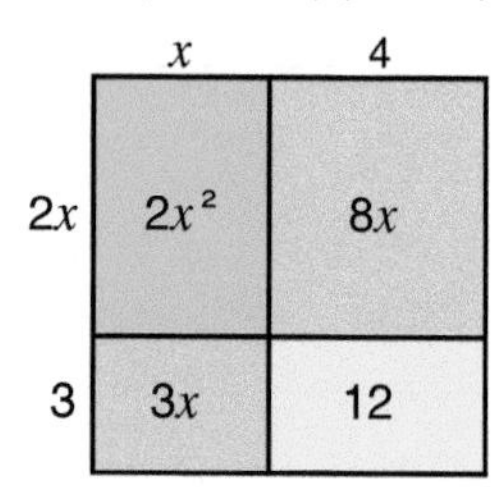

$$(2x + 3)(x + 4) = 2x^2 + 8x + 3x + 12$$
$$= 2x^2 + 11x + 12$$

Example 3

Expand $(2x - 5)^2$.

	$2x$	-5
$2x$	$4x^2$	$-10x$
-5	$-10x$	25

The diagram method works with negative numbers.

$$(2x - 5)^2 = (2x - 5)(2x - 5)$$
$$= 4x^2 - 10x - 10x + 25$$
$$= 4x^2 - 20x + 25$$

Example 4

Expand $(x + 3)(x - 5)$.

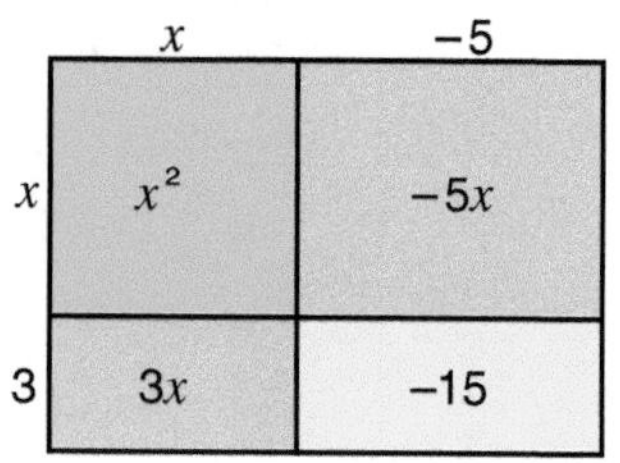

$$(x + 3)(x - 5) = x^2 - 5x + 3x - 15$$
$$= x^2 - 2x - 15$$

Example 5

Expand $(x - 1)(2x + 3)$.

As you become more confident you may not need a diagram to expand the brackets.

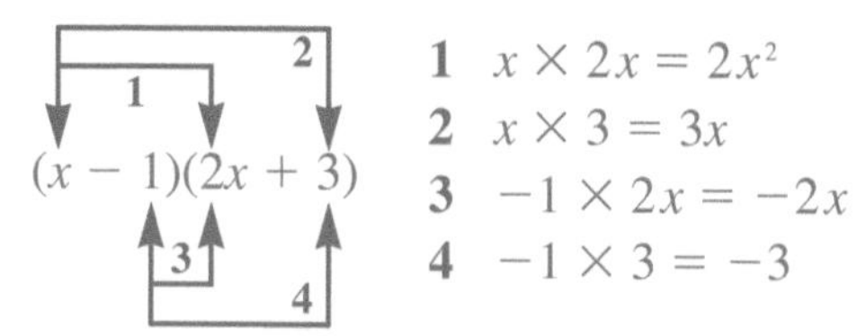

$$(x - 1)(2x + 3) = 2x^2 + 3x - 2x - 3$$
$$= 2x^2 + x - 3$$

Practice Exercise 6.3

Questions 1 to 6. Use diagrams to multiply out the brackets.

1. $(x + 3)(x + 4)$ **2.** $(x + 1)(x + 5)$ **3.** $(x - 5)(x + 2)$

4. $(2x + 1)(x - 2)$ **5.** $(3x - 2)(x - 6)$ **6.** $(2x + 1)(3x + 2)$

Questions 7 to 24. Expand the following brackets.
Only draw a diagram if necessary.

7. $(x + 8)(x - 2)$ **8.** $(x + 5)(x - 2)$ **9.** $(x - 1)(x + 3)$

10. $(x - 3)(x - 2)$ **11.** $(x - 4)(x - 1)$ **12.** $(x - 7)(x + 2)$

13. $(2x + 1)(x + 3)$ **14.** $(x + 5)(3x - 2)$ **15.** $(5x - 3)(x - 1)$

16. $(2x + 3)(2x - 1)$ **17.** $(3x - 1)(2x + 5)$ **18.** $(4x - 2)(3x + 5)$

19. $(x + 3)(x - 3)$ **20.** $(x + 5)(x - 5)$ **21.** $(x + 7)(x - 7)$

22. $(x + 5)^2$ **23.** $(x - 7)^2$ **24.** $(2x - 3)^2$

25. Expand the following brackets and simplify where possible.

(a) $(x + 3)(x^2 + 2x + 4)$

(b) $(x - 2)(2x^2 + x - 3)$

(c) $(2x - 1)(3x^2 - x + 4)$

(d) $(2 - x^2)(3x^2 - 2x + 5)$

Factorising

Factorising is the opposite operation to removing brackets.
For example: to remove brackets $2(x + 5) = 2x + 10$

To factorise $3x + 6$ we can see that $3x$ and 6 have a **common factor** of 3 so,

$$3x + 6 = 3(x + 2)$$

A **common factor** can also be a **letter**.
Both y^2 and $5y$ can be divided by y.
To factorise $y^2 - 5y$ we take y as the common factor,
so, $y^2 - 5y = y(y - 5)$

In some instances the terms will have both a number and a letter as common factors and **both** must be taken out.

Common factors:
The **factors** of a number are all the numbers that will divide exactly into the number.
Factors of 6 are 1, 2, 3 and 6.
A **common factor** is a factor which will divide into two or more terms.

You can check that you have factorised an expression correctly by multiplying out the brackets.

Example 6

Factorise $4x - 6$.

Each term has a factor of 2.
So, the common factor is 2.
$4x - 6 = 2(2x - 3)$

Example 7

Factorise $2x^2 - 6x$.

Each term has a factor of 2.
Each term has a factor of x.
So, the common factor is $2x$.
$2x^2 - 6x = 2x(x - 3)$

Example 8

Factorise completely $2ab^2 - 6b^3$.

$2ab^2 - 6b^3 = 2b^2(a - 3b)$

Practice Exercise 6.4

1. Copy and complete.
   (a) $2x + 2y = 2(\ldots + \ldots)$ (b) $3a - 6b = 3(\ldots - \ldots)$ (c) $6m + 8n = 2(\ldots + \ldots)$
   (d) $x^2 - 2x = x(\quad)$ (e) $ab + a = a(\quad)$ (f) $2x - xy = x(\quad)$
   (g) $2ab - 4a = 2a(\quad)$ (h) $4x^2 + 6x = 2x(\quad)$ (i) $dg - dg^2 = dg(\quad)$
2. Factorise.
   (a) $2a + 2b$ (b) $5x - 5y$ (c) $3d + 6e$ (d) $4m - 2n$
   (e) $6a + 9b$ (f) $6a - 8b$ (g) $8t + 12$ (h) $5a - 10$
   (i) $4d - 2$ (j) $3 - 9g$ (k) $5 - 20m$ (l) $4k + 4$
3. Factorise.
   (a) $xy - xz$ (b) $fg + gh$ (c) $ab - 2b$ (d) $3q + pq$
   (e) $a + ab$ (f) $gh - g$ (g) $a^2 + 3a$ (h) $5t - t^2$
   (i) $d - d^2$ (j) $m^2 + m$ (k) $5r^2 - 3r$ (l) $3x^2 + 2x$
4. Factorise completely.
   (a) $3y + 6 - 9x$ (b) $t^3 - t^2 + t$ (c) $2d^2 + 4d$ (d) $3m - 6mn$
   (e) $2fg + 4g^2$ (f) $4pq - 8q$ (g) $6y - 15y^2$ (h) $6x^2 + 4xy$
   (i) $6n^2 - 2n$ (j) $4ab + 6b$ (k) $\frac{1}{2}a - \frac{1}{2}a^2$ (l) $20x + 4xy$
   (m) $a^3 + a^5 + a^2$ (n) $2\pi r + \pi r^2$ (o) $20a^2b + 12ab^2$ (p) $3pq - 9p^2q$

Factorising quadratic expressions

$x^2 + 8x + 15$, $x^2 - 4$ and $x^2 + 7x$ are examples of **quadratic expressions**.

You will need to be able to factorise quadratic expressions in order to solve quadratic equations.

The general form of a quadratic expression is $ax^2 + bx + c$, where a cannot be equal to 0.

Difference of two squares

In the expression $x^2 - 4$,
$x^2 = x \times x$ and $4 = 2^2 = 2 \times 2$.

$x^2 - 4 = (x + 2)(x - 2)$
This result is called the **difference of two squares**.
In general: $a^2 - b^2 = (a + b)(a - b)$

Checking your work:
To check your work, expand the brackets and simplify, where necessary.
The result should be the same as the original expression.

Example 9

Factorise $x^2 - 100$.

$x^2 - 100 = x^2 - 10^2$
$= (x + 10)(x - 10)$

Example 10

Factorise $25 - x^2$.

$25 - x^2 = 5^2 - x^2$
$= (5 + x)(5 - x)$

Example 11

Factorise $s^2 - t^2$.

$s^2 - t^2 = (s + t)(s - t)$

Example 12

Factorise $4x^2 - 9y^2$.

$4x^2 - 9y^2$
$= (2x + 3y)(2x - 3y)$

Quadratics of the form $x^2 + bx + c$

The expression $x^2 + 8x + 15$ can be factorised.

From experience, we know that the answer is likely to be of the form:

$x^2 + 8x + 15 = (x + ?)(x + ?)$

where the question marks represent numbers.

Replacing the question marks with letters, p and q, we get:

$x^2 + 8x + 15 = (x + p)(x + q)$

Multiply the brackets out, using either the diagram method or by expanding, and compare the results with the original expression.

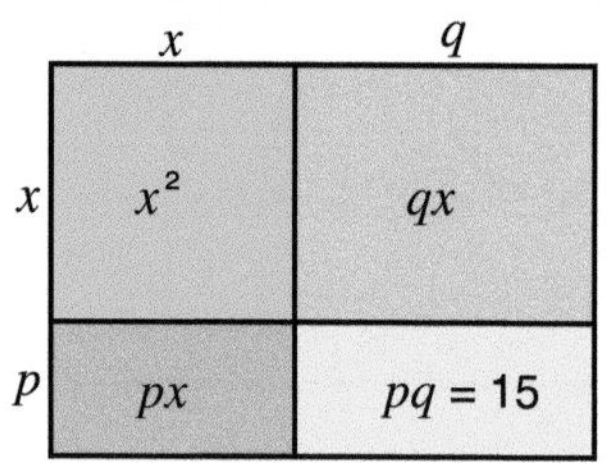

$px + qx = x(p + q) = 8x$
and $p \times q = 15$

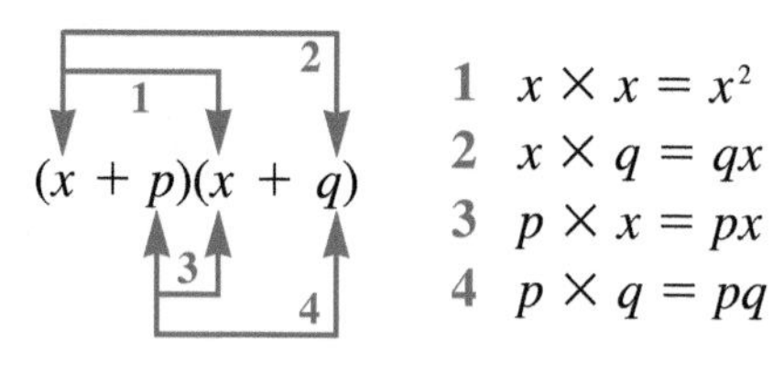

$(x + p)(x + q) = x^2 + qx + px + pq$
$= x^2 + (q + p)x + pq$
$= x^2 + 8x + 15$

Two numbers are required which: when multiplied give $+15$, **and** when added give $+8$.

$+5$ and $+3$ satisfy **both** conditions.

$x^2 + 8x + 15 = (x + 5)(x + 3)$

Example 13

Factorise $x^2 + 6x + 9$.

$x^2 + 6x + 9 = (x + 3)(x + 3)$
$= (x + 3)^2$

Because $3 \times 3 = 9$
and $3 + 3 = 6$.

Example 14

Factorise $x^2 - 8x + 12$.

$x^2 - 8x + 12 = (x - 6)(x - 2)$

Because $-6 \times -2 = 12$
and $-6 + -2 = -8$.

Example 15

Factorise $x^2 + 2x - 15$.

$x^2 + 2x - 15 = (x + 5)(x - 3)$

Because $+5 \times -3 = -15$
and $+5 + -3 = +2$.

Example 16

Factorise $x^2 - 2x - 8$.

$x^2 - 2x - 8 = (x + 2)(x - 4)$

Because $+2 \times -4 = -8$
and $+2 + -4 = -2$.

Practice Exercise 6.5

1. Factorise these expressions.
   (a) $x^2 + 5x$ (b) $x^2 - 7x$ (c) $y^2 - 6y$ (d) $2y^2 - 12y$
   (e) $5t - t^2$ (f) $8y + y^2$ (g) $x^2 - 20x$ (h) $3x^2 - 60x$

2. Factorise.
   (a) $x^2 - 9$ (b) $x^2 - 81$ (c) $y^2 - 25$ (d) $y^2 - 1$
   (e) $x^2 - 64$ (f) $100 - x^2$ (g) $36 - x^2$ (h) $x^2 - a^2$
   (i) $100x^2 - 9$ (j) $25 - 16x^2$ (k) $25y^2 - 49$ (l) $9x^2 - 25y^2$

3. Copy and complete the following.
   (a) $x^2 + 6x + 5 = (x + 5)(x + \ldots)$
   (b) $x^2 + 9x + 14 = (x + 7)(x + \ldots)$
   (c) $x^2 + 6x + 8 = (x + \ldots)(x + 4)$
   (d) $x^2 + 9x + 18 = (x + \ldots)(x + 6)$
   (e) $x^2 - 6x + 5 = (x - 5)(x - \ldots)$
   (f) $x^2 - 7x + 10 = (x - 5)(x - \ldots)$
   (g) $x^2 - 7x + 12 = (x - \ldots)(x - 3)$
   (h) $x^2 + 3x - 4 = (x + 4)(x - \ldots)$
   (i) $x^2 + 5x - 14 = (x + 7)(x - \ldots)$
   (j) $x^2 - 4x - 5 = (x - 5)(x + \ldots)$

   Multiply the brackets out mentally to check your answers.

4. Factorise.
   (a) $x^2 + 3x + 2$ (b) $x^2 + 8x + 7$ (c) $x^2 + 8x + 15$ (d) $x^2 + 8x + 12$
   (e) $x^2 + 12x + 11$ (f) $x^2 + 9x + 20$ (g) $x^2 + 10x + 24$ (h) $x^2 + 13x + 36$
   (i) $x^2 + 15x + 14$ (j) $x^2 + 10x + 16$

5. Factorise.
   (a) $x^2 - 6x + 9$ (b) $x^2 - 6x + 8$ (c) $x^2 - 11x + 10$ (d) $x^2 - 16x + 15$
   (e) $x^2 - 8x + 15$ (f) $x^2 - 10x + 16$ (g) $x^2 - 12x + 20$ (h) $x^2 - 11x + 24$
   (i) $x^2 - 13x + 12$ (j) $x^2 - 8x + 12$

6. Factorise.
   (a) $x^2 - x - 6$ (b) $x^2 - 5x - 6$ (c) $x^2 + 2x - 24$ (d) $x^2 + 5x - 24$
   (e) $x^2 - 2x - 15$ (f) $x^2 + 3x - 18$ (g) $x^2 - 3x - 40$ (h) $x^2 - 4x - 12$
   (i) $x^2 + 3x - 10$ (j) $x^2 - x - 20$

7. Factorise.
   (a) $x^2 - 4x + 4$ (b) $x^2 + 11x + 30$ (c) $x^2 + 2x - 8$ (d) $x^2 - 4x - 21$
   (e) $x^2 + x - 20$ (f) $x^2 + 7x + 12$ (g) $x^2 + 8x + 16$ (h) $x^2 - 2x + 1$
   (i) $x^2 - 49$ (j) $t^2 + 12t$ (k) $x^2 - 9x + 14$ (l) $x^2 - 7x + 6$
   (m) $x^2 + 11x + 18$ (n) $x^2 + 11x + 24$ (o) $x^2 + 19x + 18$ (p) $x^2 - y^2$
   (q) $x^2 + x - 6$ (r) $4x^2 - 36$ (s) $y^2 - 10y + 25$ (t) $x^2 - 12x + 36$

Quadratics of the form $ax^2 + bx + c$

The expression $2x^2 + 11x + 15$ can be factorised.

First look at the term in x^2.
The coefficient of x^2 is 2, so, the brackets start:
$(2x \ldots \ldots)(x \ldots \ldots)$

Next, look for two numbers which when multiplied give $+15$.
The possibilities are: 1 and 15 or 3 and 5.

Use these pairs of numbers to complete the brackets.
Multiply out pairs of brackets until the correct factorisation of $2x^2 + 11x + 15$ is found.

$$(2x + 1)(x + 15) = 2x^2 + 30x + x + 15 = 2x^2 + 31x + 15$$
$$(2x + 15)(x + 1) = 2x^2 + 2x + 15x + 15 = 2x^2 + 17x + 15$$
$$(2x + 3)(x + 5) = 2x^2 + 10x + 3x + 15 = 2x^2 + 13x + 15$$
$$(2x + 5)(x + 3) = 2x^2 + 6x + 5x + 15 = 2x^2 + 11x + 15$$

$$2x^2 + 11x + 15 = (2x + 5)(x + 3)$$

Rather than expanding brackets fully it is only necessary to look at two terms. $(2x + 5)(x + 3)$
Explain why.

Example 17

Factorise $3x^2 + 5x + 2$.
$3x^2 + 5x + 2 = (3x + 2)(x + 1)$

Multiply the brackets out mentally to check the answers.

Example 18

Factorise $4x^2 + 19x - 5$.
$4x^2 + 19x - 5 = (4x - 1)(x + 5)$
The brackets contain different signs.
Explain why.

Example 19

Factorise $3a^2 + 7a + 2$.
$3a^2 + 7a + 2 = (3a + 1)(a + 2)$

Example 20

Factorise $4x^2 - 16x + 15$.
$4x^2 - 16x + 15 = (2x - 5)(2x - 3)$
Each bracket contains a minus sign.
Explain why.

Practice Exercise 6.6

1. Copy and complete the following.
   (a) $2x^2 + 12x + 18 = (2x + 6)(\dots\ \dots\ \dots)$ (b) $2y^2 - 9y - 18 = (\dots\ \dots\ \dots)(y - 6)$
   (c) $2a^2 + 5a - 18 = (2a + \dots)(a - \dots)$ (d) $2m^2 - 20m + 18 = (2m\ \dots\ \dots)(m\ \dots\ \dots)$
   (e) $4x^2 - 13x + 3 = (4x\ \dots\ \dots)(x\ \dots\ \dots)$ (f) $4d^2 + 4d - 3 = (2d - 1)(\dots\ \dots\ \dots)$
   Multiply the brackets out mentally to check your answers.

2. Factorise.
   (a) $2x^2 + 7x + 3$ (b) $2x^2 + 11x + 12$ (c) $2a^2 + 25a + 12$
   (d) $4x^2 + 4x + 1$ (e) $6y^2 + 7y + 2$ (f) $6x^2 + 17x + 12$
   (g) $11m^2 + 12m + 1$ (h) $3x^2 + 17x + 10$ (i) $5k^2 + 12k + 4$

3. Factorise.
   (a) $2a^2 - 5a + 3$ (b) $2y^2 - 9y + 7$ (c) $4x^2 - 4x + 1$
   (d) $3x^2 - 5x + 2$ (e) $6d^2 - 11d + 3$ (f) $2x^2 - 3x + 1$
   (g) $2x^2 - 11x + 15$ (h) $9y^2 - 12y + 4$ (i) $4t^2 - 12t + 9$

4. Factorise.
   (a) $2x^2 + x - 6$ (b) $2a^2 - a - 3$ (c) $3x^2 - 14x - 5$
   (d) $3t^2 + 14t - 5$ (e) $2y^2 - 5y - 7$ (f) $4y^2 + 7y - 2$
   (g) $9x^2 + 3x - 2$ (h) $10x^2 + 27x - 9$ (i) $5m^2 - 2m - 3$

Further factorising

When factorising, work logically.

Does the expression have a common factor?

Is the expression a difference of two squares?

Will the expression factorise into two brackets?

Example 21

Factorise $2x^2 - 14x$.

$2x^2 - 14x = 2x(x - 7)$

Example 22

Factorise $25 - 4y^2$.

$25 - 4y^2 = 5^2 - (2y)^2$
$= (5 + 2y)(5 - 2y)$

Example 23

Factorise $2x^2 - 18$.

$2x^2 - 18 = 2(x^2 - 9)$
$= 2(x + 3)(x - 3)$

Example 24

Factorise $2x^2 - 6x - 8$.

$2x^2 - 6x - 8 = 2(x^2 - 3x - 4)$
$= 2(x + 1)(x - 4)$

Practice Exercise 6.7

1. Factorise.
   (a) $3x + 12y$ (b) $t^2 - 16$ (c) $x^2 + 4x + 3$
   (d) $y^2 - y$ (e) $2d^2 - 6$ (f) $p^2 - q^2$
   (g) $a^2 - 2a$ (h) $x^2 - 2x + 1$ (i) $2y^2 - 8y$
   (j) $a^2 - 6a + 9$ (k) $3m^2 - 12$ (l) $v^2 - v - 6$
   (m) $ax^2 - ay^2$ (n) $-8 - 2x$ (o) $4 - 4k + k^2$
   (p) $18 - 2x^2$ (q) $50 - 2x^2$ (r) $12x^2 - 27y^2$
   (s) $6a^2 - 3a$ (t) $x^2 - 15x + 56$ (u) $x^2 + 2x - 15$

2. Factorise.
   (a) $3a^2 - 2a - 8$ (b) $3x^2 + 31x + 10$ (c) $64y^2 - 49$
   (d) $3x^2 + 11x + 10$ (e) $7x^2 - 19x + 10$ (f) $6x^2 - 25x - 25$
   (g) $5m^2 - 18m + 9$ (h) $8x^2 + 2x - 15$ (i) $13x - 20 - 2x^2$
   (j) $4y^2 + 22y + 24$ (k) $6x^2 + 3x - 18$ (l) $28n - 10 + 6n^2$
   (m) $4x^2 + 22x + 24$ (n) $3x^2 - 12x + 12$ (o) $x^3 - x$

Completing the square

16, x^2, $9y^2$, $(m-3)^2$ are all examples of **perfect squares**.

What must be added to $x^2 + 8x$ to make it a perfect square?
A diagram can be used to show the process.

	x	8
x	x^2	$8x$

→

	x	4
x	x^2	$4x$
4	$4x$	16

The number to be added is 16.

$x^2 + 8x + 16 = (x+4)^2$

16 came from (half of 8) squared.

The process of adding a constant term to a quadratic expression to make it a perfect square is called **completing the square**.

Quadratics can be rearranged into a **square term** and a **constant value** (number).
This is sometimes called **completed square form** and can be used to solve quadratic equations.

> To write the quadratic $x^2 + bx + c$ in completed square form:
> write the **square term** as $(x + \frac{1}{2}$ the coefficient of $x)^2$ and then adjust for the **constant value**.

Example 25

Write each of these quadratics in the form $(x+a)^2 + b$.

(a) $x^2 + 10x + 7$

$\frac{1}{2}$ of 10 is 5.

$$(x+5)^2 = x^2 + 10x + 25$$
$$x^2 + 10x + 7 = x^2 + 10x + 25 - 18$$
$$= (x+5)^2 - 18$$

The minimum value of $(x+5)^2 - 18$ occurs when $(x+5)^2 = 0$.

So, minimum value of $x^2 + 10x + 7$ is -18.

(b) $x^2 - 8x - 5$

$\frac{1}{2}$ of -8 is -4.

$$(x-4)^2 = x^2 - 8x + 16$$
$$x^2 - 8x - 5 = x^2 - 8x + 16 - 21$$
$$= (x-4)^2 - 21$$

The minimum value of $(x-4)^2 - 21$ occurs when $(x-4)^2 = 0$.

So, minimum value of $x^2 - 8x - 5$ is -21.

Practice Exercise 6.8

1. What must be added to each expression to make a perfect square?
   (a) $x^2 + 2x$ (b) $x^2 + 4x$ (c) $x^2 - 6x$ (d) $y^2 + 8y$
   (e) $p^2 - 2p$ (f) $a^2 - 3a$ (g) $m^2 + 10m$ (h) $x^2 - 5x$

2. Complete these squares.
   (a) $x^2 + 6x + \ldots = (x + \ldots)^2$ (b) $a^2 - 4a + \ldots = (a - \ldots)^2$
   (c) $b^2 + 2b + \ldots = (b + \ldots)^2$ (d) $m^2 - 8m + \ldots = (m - \ldots)^2$
   (e) $n^2 - n + \ldots = (n - \ldots)^2$ (f) $x^2 + 5x + \ldots = (x + \ldots)^2$

3. Change these quadratics into the form $(x+a)^2 + b$.
   (a) $x^2 + 6x + 20$ (b) $x^2 + 6x + 5$ (c) $x^2 + 10x - 4$
   (d) $x^2 - 4x + 5$ (e) $x^2 - 4x + 2$ (f) $x^2 - 4x - 4$
   (g) $x^2 - 6x + 4$ (h) $x^2 - 8x$ (i) $x^2 + 12x$

4. These quadratics can be written in the form $(x+a)^2 + b$.
   State the values of a and b.
   (a) $x^2 + 6x + 15$ (b) $x^2 + 10x + 5$ (c) $x^2 - 6x - 5$
   (d) $x^2 - 8x + 4$ (e) $x^2 + 12x + 4$ (f) $x^2 + 6x + 9$

5. The expression $x^2 - 4x - 1$ can be written in the form $(x + a)^2 + b$.
   (a) Find the values of a and b.
   (b) What is the minimum value of the expression $x^2 - 4x - 1$?

6. $x^2 - 6x + a = (x + b)^2 - 1$.
   Find the values of a and b.

7. What is the minimum value of the expression $x^2 - 10x + 26$?

Key Points

- $n - 3$ is an **algebraic expression**, or simply an **expression**.
  An expression is just an answer made up of letters and numbers.
  $+n$ and -3 are **terms** of the expression.

> A term includes the sign, + or −.
> n has the same value as $+n$.

- You can add and subtract terms with the same letter.
  This is sometimes called **simplifying an expression**.
  E.g. $2d + 3d = 5d$ *and* $3x + 2 - x + 4 = 2x + 6$
- You should be able to **multiply out brackets**, either by using a diagram or by expanding.
  Multiply out the bracket $3(4a + 5)$.

Diagram method

$3(4a + 5) = 12a + 15$

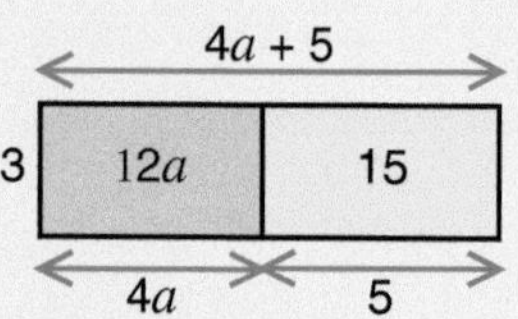

Expanding

$3(4a + 5) = 3 \times 4a + 3 \times 5$
$= 12a + 15$

Multiply out $(x + 2)(x + 3)$.

Diagram method

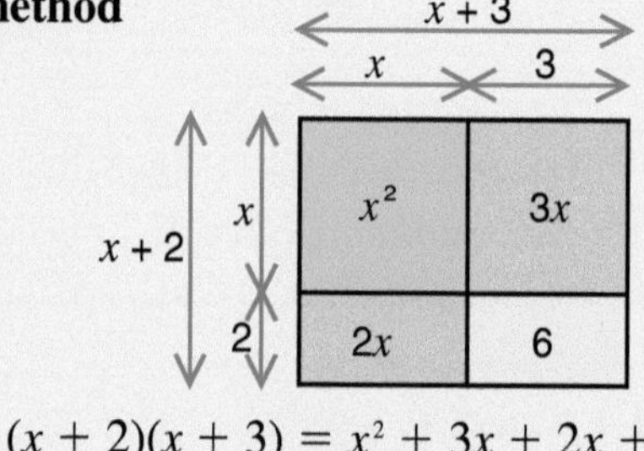

$(x + 2)(x + 3) = x^2 + 3x + 2x + 6$
$= x^2 + 5x + 6$

Expanding

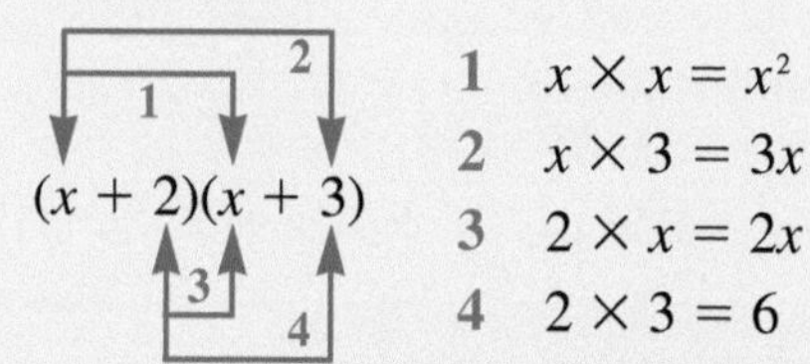

$(x + 2)(x + 3) = x^2 + 3x + 2x + 6$
$= x^2 + 5x + 6$

- **Factorising** is the opposite operation to removing brackets.
  When factorising, work logically.
  Does the expression have a **common factor**?
  Is the expression a difference of **two squares**?
  Will the expression factorise into **two brackets**?
- A **common factor** is a factor which divides into two, or more, numbers (or terms).
  For example: $x^2 + 7x = x(x + 7)$
- **Difference of two squares**: $a^2 - b^2 = (a + b)(a - b)$
- Quadratic expressions, such as $x^2 + 8x + 20$, can be written in the form $(x + a)^2 + b$, where a and b are integers.
  In **completed square form**, $(x + a)^2 + b$:
  the value of a is half the coefficient of x,
  the value of b is found by subtracting the value of a^2 from the constant term of the original expression.
  For example: $x^2 + 8x + 20 = (x + 4)^2 + 4$

Review Exercise 6

1. Multiply out and simplify where possible.
   (a) $y(y - 4)$ (b) $4(3y + 1) - 5y$

2. Multiply out.
   (a) $3(x - 2y)$ (b) $2x(x + 3)$ (c) $x(x^2 - 3x)$

3. Multiply out the brackets and simplify.
   (a) $x^2 - x(1 - x)$ (b) $4 - 3(x + 1)$

4. (a) Factorise. (i) $6x - 15$ (ii) $y^2 + 7y$
   (b) Multiply out and simplify. $3(y + 2) - 2(y - 3)$

5. Simplify.
   (a) $x(x - 4) + 3(x - 2)$ (b) $x(2x + 3) - 4(3x - 1)$
   (c) $x(x^2 + 1) - x^2(x + 1)$ (d) $2x - 3 - (x - 4)$

6. Expand and simplify $(x - 5)(x + 2)$.

7. (a) Expand $(2x + 1)(x + 4)$.
   (b) Factorise completely $8x^2 - 6x$.

8. Factorise.
   (a) $2xy^2 + 4xy$ (b) $m^2 - 9m + 8$

9. Expand the following.
   (a) $(x - 7)(x - 3)$ (b) $(x + 4)^2$ (c) $(3x + 5)(2x - 1)$
   (d) $(4x - 3)(4x + 3)$ (e) $(2x - 1)^2$ (f) $(x + 1)(2x^2 - 3x - 4)$

10. Factorise.
    (a) $3x^2 - 6$ (b) $9x^2 - 25$ (c) $x^3 - x$
    (d) $x^2 + 11x + 10$ (e) $x^2 - 2x + 1$ (f) $x^2 + 8x - 33$
    (g) $2x^2 + 9x + 7$ (h) $2x^2 - 3x - 5$ (i) $4x^2 - 16$

11. Add a number to complete the square in these expressions.
    Then write the complete expression as a square.
    (a) $x^2 + 16x$ (b) $x^2 - 24x$

12. Find a common factor of $3x^2 + x - 2$ and $3x^2 - 5x + 2$.

13. The expression $x^2 - 8x + 20$ can be written in the form $(x - p)^2 + q$.
    Calculate the values of p and q.

14. The area of a rectangle is:
    $(2x^2 + 5x - 3)$ cm².
    The length of one side is:
    $(2x - 1)$ cm.
    What is the length of the other side?

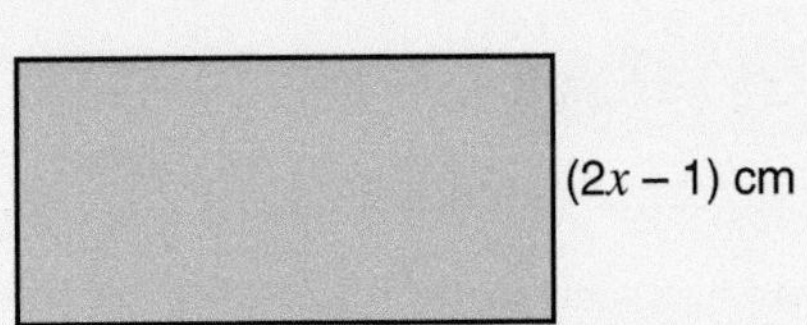

15. Find the value of $101^2 - 99^2$ without using a calculator.
    Using the difference of two squares.
    $101^2 - 99^2 = (101 + 99)(101 - 99) = 200 \times 2 = 400$

    Without using a calculator, find the value of the following.
    (a) $6.36^2 - 3.64^2$ (b) $7.5^2 - 2.5^2$ (c) $8.987^2 - 1.013^2$

16. Simplify $(x + y)^2 + x(x - 2y) - 9y^2$ and factorise your answer.

7 Algebraic Fractions

Algebraic fractions

Algebraic fractions have a numerator and a denominator (just as an ordinary fraction) but at least one of them is an expression involving an unknown.

E.g. $\frac{1}{x}$ $\frac{x}{2}$ $\frac{2}{x-5}$ $\frac{x+1}{x-6}$ $\frac{3}{x+7}$ $\frac{x^2+2x+1}{x^2-1}$

Algebraic fractions can be simplified, added, subtracted, multiplied, divided and used in equations in the same way as numerical fractions.

Simplifying algebraic fractions

Fractions can be simplified if the numerator and the denominator have a common factor.
In its **simplest form**, the numerator and denominator of an algebraic fraction have no common factor other than 1.

The numerator and denominator of $\frac{15}{25}$ have a highest common factor of 5.

$\frac{15}{25} = \frac{15 \div 5}{25 \div 5} = \frac{3}{5}$

3 and 5 have no common factors, other than 1.

$\frac{15}{25} = \frac{3}{5}$ in its simplest form.

Algebraic fractions work in a similar way.

> To write an algebraic fraction in its simplest form:
> - factorise the numerator and denominator of the fraction,
> - divide the numerator and denominator by their highest common factor.
>
> This is sometimes called **cancelling** a fraction.

Example 1

Write the following in their simplest form.

(a) $\frac{3x-6}{3} = \frac{3(x-2)}{3} = x-2$

(b) $\frac{3x+9}{4x^2+12x} = \frac{3(x+3)}{4x(x+3)} = \frac{3}{4x}$

(c) $\frac{x^2y+3xy^2}{xy-2x^2y^2} = \frac{xy(x+3y)}{xy(1-2xy)} = \frac{x+3y}{1-2xy}$

(d) $\frac{x^2+5x+6}{x^2-2x-8} = \frac{(x+2)(x+3)}{(x+2)(x-4)} = \frac{x+3}{x-4}$

(e) $\frac{x^2+2x+1}{x^2-1} = \frac{(x+1)^2}{(x+1)(x-1)} = \frac{x+1}{x-1}$

(f) $\frac{9-3y}{y-3} = \frac{3(3-y)}{y-3} = \frac{-3(y-3)}{y-3} = -3$

Practice Exercise 7.1

1. Simplify these algebraic fractions.

(a) $\frac{4d+6}{2}$ (b) $\frac{9x+6}{3}$ (c) $\frac{8a+10b}{2}$ (d) $\frac{15m-10n}{5}$

(e) $\frac{8x-4y}{2}$ (f) $\frac{ax+bx}{x}$ (g) $\frac{x^2-x}{x}$ (h) $\frac{2x^2-4}{2}$

(i) $\frac{12}{3x-9}$ (j) $\frac{5x+10}{15}$ (k) $\frac{2x+4y+6z}{4x-6y+2z}$ (l) $\frac{-2x-4}{-6x-4}$

(m) $\frac{2x}{6x-4}$ (n) $\frac{3x}{6x^2-3}$ (o) $\frac{3x}{6x^2-3x}$ (p) $\frac{2x-1}{6x^2-3x}$

(q) $\frac{3m-6}{2m-4}$ (r) $\frac{m^2+3m}{3m+9}$ (s) $\frac{x^2-3x}{x^2+2x}$ (t) $\frac{5x-10}{6-3x}$

2. Match the algebraic fractions with the values **A** to **E**.

(a) $\frac{3x+6}{4x+8}$ (b) $\frac{5x-15}{2x-6}$ (c) $\frac{8-4y}{6-3y}$ (d) $\frac{8x+10}{4x+5}$

(e) $\frac{-2x-4}{-4x-8}$ **A** 2 **B** $\frac{4}{3}$ **C** $2\frac{1}{2}$ **D** $\frac{1}{2}$ **E** $\frac{3}{4}$

3. Simplify these algebraic fractions.

(a) $\frac{x^2y+5xy^2}{3xy+4x^2y^2}$ (b) $\frac{x^2+xy+2xz}{2x^2-3xy+xz}$ (c) $\frac{a^2b^2-3ab}{ab^2+a^2b}$ (d) $\frac{ax+ab+ay}{a^2b-2ay}$

4. Simplify these algebraic fractions.

(a) $\frac{x^2+3x}{x^2+4x+3}$ (b) $\frac{x^2+3x+2}{x^2+4x+3}$ (c) $\frac{x^2-2x}{x^2+x-6}$

(d) $\frac{x^2-x-20}{x^2+7x+12}$ (e) $\frac{x^2-4x+4}{x^2-5x+6}$ (f) $\frac{x^2-5x}{x^2+4x}$

(g) $\frac{x^2-1}{x+1}$ (h) $\frac{x-2}{x^2-4}$ (i) $\frac{x^2-2x-3}{2x-6}$

(j) $\frac{2x^2-2x-12}{2x^2-18}$ (k) $\frac{2x^2-7x-15}{2x^2-5x-12}$ (l) $\frac{6x^2-13x-5}{9x^2-1}$

Arithmetic of algebraic fractions

The same methods used for adding, subtracting, multiplying and dividing numeric fractions can be applied to algebraic fractions.

Multiplication

Numeric

$\frac{8}{9}\times\frac{3}{14}=\frac{\cancel{2}\times4}{3\times\cancel{3}}\times\frac{\cancel{3}}{\cancel{2}\times7}$

Divide by 2 and 3.

$=\frac{4}{21}$

Algebraic

$\frac{2x}{5}\times\frac{10}{7x}=\frac{2\times\cancel{x}}{\cancel{5}}\times\frac{2\times\cancel{5}}{7\times\cancel{x}}$

Divide by x and 5.

$=\frac{4}{7}$

Division

Numeric

$\frac{3}{7}\div\frac{9}{14}=\frac{3}{7}\times\frac{14}{9}$

$=\frac{\cancel{3}}{\cancel{7}}\times\frac{2\times\cancel{7}}{3\times\cancel{3}}$

Divide by 3 and 7.

$=\frac{2}{3}$

Algebraic

$\frac{3x}{8}\div\frac{x}{2}=\frac{3x}{8}\times\frac{2}{x}$

$=\frac{3\times\cancel{x}}{\cancel{2}\times4}\times\frac{\cancel{2}}{\cancel{x}}$

Divide by x and 2.

$=\frac{3}{4}$

Addition

Numeric

$\frac{4}{11}+\frac{2}{5}=\frac{4\times5+2\times11}{11\times5}$

$=\frac{42}{55}$

Algebraic

$\frac{x}{3}+\frac{2x}{5}=\frac{x\times5+2x\times3}{3\times5}$

$=\frac{5x+6x}{15}=\frac{11x}{15}$

Subtraction

Numeric

$\frac{4}{7}-\frac{2}{5}=\frac{4\times5-2\times7}{7\times5}$

$=\frac{6}{35}$

Algebraic

$\frac{12}{x}-\frac{5}{x}=\frac{12\times x-5\times x}{x\times x}=\frac{12x-5x}{x^2}$

$=\frac{7x}{x^2}=\frac{7}{x}$

Multiplying algebraic fractions

Calculate $\frac{2x}{5} \times \frac{10}{7x}$

Simplify, where possible, by cancelling $\frac{2 \times x}{5} \times \frac{10}{7 \times x}$

> To simplify:
> Divide a numerator and a denominator by the same number (or letter).

In this case: 10 and 5 can both be divided by 5.

$10 \div 5 = 2$ and $5 \div 5 = 1$

Also, x appears in a numerator (top) and a denominator (bottom).

$x \div x = 1$ (in term $2x$) and $x \div x = 1$ (in term $7x$)

$$\frac{2 \times \not{x}}{{}_1\not{5}} \times \frac{\not{10}^2}{7 \times \not{x}}$$

Multiply the numerators.
Multiply the denominators. $\frac{2 \times 2}{1 \times 7} = \frac{4}{7}$

So, $\frac{2x}{5} \times \frac{10}{7x} = \frac{4}{7}$

Example 2

Work out $\frac{2}{3a} \times \frac{a}{4}$

Method 1

$\frac{2}{3a} \times \frac{a}{4}$

Simplify by cancelling.

$2 \div 2 = 1$ and $4 \div 2 = 2$

Also, a appears top and bottom.

$$\frac{{}^1\not{2}}{3 \times \not{a}} \times \frac{\not{a}}{\not{4}_2}$$

Multiply out.

$\frac{1}{3 \times 2} = \frac{1}{6}$

Work out $\frac{2}{3a} \times \frac{a}{4}$

Method 2

Multiply the numerators.
Multiply the denominators. $\frac{2 \times a}{3a \times 4} = \frac{2a}{12a}$

Simplify by cancelling.

2 and 12 can both be divided by 2.

$2 \div 2 = 1$ and $12 \div 2 = 6$

a (numerator) cancels with a (denominator).

$\frac{2a}{12a} = \frac{1}{6}$

Use the method you find easiest.

You must give your answer in its simplest form.

Example 3

Work out $\frac{2a}{9} \times \frac{3b}{5}$

Simplify by cancelling.

$$= \frac{2 \times a}{{}_3\not{9}} \times \frac{{}^1\not{3} \times b}{5}$$

Multiply out.

$$= \frac{2 \times a \times b}{3 \times 5}$$

$$= \frac{2ab}{15}$$

Example 4

Work out $\frac{5}{6x} \times \frac{y}{2x}$

There is no cancelling.

Multiply out.

$$= \frac{5 \times y}{6x \times 2x}$$

$$= \frac{5y}{12x^2}$$

Dividing algebraic fractions

The method normally used when one fraction is divided by another is to change the division to a multiplication. The fractions can then be multiplied in the usual way.

Calculate $\frac{3x}{10} \div \frac{2x}{15}$

Change division to multiplication. $\frac{3x}{10} \times \frac{15}{2x}$

Simplify, where possible, by cancelling.

10 and 15 can both be divided by 5.

$10 \div 5 = 2$ and $15 \div 5 = 3$

Also, x appears in a numerator (top) and a denominator (bottom).

$x \div x = 1$ (in term $3x$) and $x \div x = 1$ (in term $2x$).

$\frac{3 \times \cancel{x}}{_2\cancel{10}} \times \frac{\cancel{15}^3}{2 \times \cancel{x}}$

Multiply the numerators.
Multiply the denominators. $\frac{3 \times 3}{2 \times 2} = \frac{9}{4} = 2\frac{1}{4}$

So, $\frac{3x}{10} \div \frac{2x}{15} = 2\frac{1}{4}$

Example 5

Work out $\frac{4y}{3} \div \frac{y}{x}$

Divide by $\frac{y}{x}$ is the same as multiply by $\frac{x}{y}$.

$\frac{4y}{3} \div \frac{y}{x} = \frac{4y}{3} \times \frac{x}{y}$

Simplify by cancelling.
y appears top and bottom.

$= \frac{4 \times \cancel{y}}{3} \times \frac{x}{\cancel{y}}$

Multiply out.

$\frac{4 \times x}{3} = \frac{4x}{3}$

So, $\frac{4y}{3} \div \frac{y}{x} = \frac{4x}{3}$

Do not change an improper (top heavy) fraction to a mixed number if a letter appears in the numerator or denominator of your answer.

Practice Exercise 7.2

1. Work out.
   (a) $\frac{4x}{5} \times \frac{10}{7x}$ (b) $\frac{3x}{8} \times \frac{2x}{3}$ (c) $\frac{2}{3x} \times \frac{9x}{8}$ (d) $\frac{5x}{6} \times \frac{9}{10x}$
2. Work out.
   (a) $\frac{5x}{6} \div \frac{7x}{8}$ (b) $\frac{7}{12x} \div \frac{14}{15x}$ (c) $\frac{5x}{6} \div \frac{10x}{9}$ (d) $\frac{27}{8x} \div \frac{9}{16x}$
3. Work out.
   (a) $\frac{2x}{3y} \times \frac{y}{5}$ (b) $\frac{4}{5x} \times \frac{10x}{3y}$ (c) $\frac{8x}{9y} \times \frac{3y}{4}$ (d) $\frac{5x}{2y} \times \frac{4x}{3}$
4. Work out.
   (a) $\frac{3x}{5} \div \frac{2x}{y}$ (b) $\frac{3}{4x} \div \frac{9y}{8x}$ (c) $\frac{2x}{5y} \div \frac{8}{15y}$ (d) $\frac{5y}{8x} \div \frac{9y}{4x}$
5. Work out.
   (a) $\frac{5x}{4y} \times \frac{8x}{15}$ (b) $\frac{30y}{7x} \div \frac{8}{21x}$ (c) $\frac{35x}{8y} \div \frac{5x}{12y^2}$ (d) $\frac{10y^2}{7x} \times \frac{14x^2}{5y}$

Adding and subtracting algebraic fractions

Calculate $\frac{x}{3} + \frac{2x}{5}$

Change the original fractions to equivalent fractions with the same denominator.

> Fractions must have the same denominator before addition (or subtraction) can take place.
> *What happens when you use a denominator that is **not** the smallest?*

With algebraic fractions, it is often quicker, and easier, to multiply the original denominators.

$$3 \times 5 = 15$$

$$\frac{x}{3} = \frac{x \times 5}{3 \times 5} = \frac{5x}{15} \text{ and } \frac{2x}{5} = \frac{2x \times 3}{5 \times 3} = \frac{6x}{15}$$

$$\frac{x}{3} + \frac{2x}{5} = \frac{5x}{15} + \frac{6x}{15}$$

Add the new numerators.
Keep the new denominators the same. $\frac{5x}{15} + \frac{6x}{15} = \frac{11x}{15}$

So, $\frac{x}{3} + \frac{2x}{5} = \frac{11x}{15}$

> Always check that your answer is in its simplest form.

Example 6

Work out $\frac{x}{2} - \frac{x}{7}$

$$2 \times 7 = 14$$

$$\frac{x}{2} = \frac{x \times 7}{2 \times 7} = \frac{7x}{14}$$

$$\frac{x}{7} = \frac{x \times 2}{7 \times 2} = \frac{2x}{14}$$

$$\frac{x}{2} - \frac{x}{7} = \frac{7x}{14} - \frac{2x}{14}$$

$$= \frac{5x}{14}$$

So, $\frac{x}{2} - \frac{x}{7} = \frac{5x}{14}$

Example 7

Work out $\frac{x}{y} + \frac{3}{x}$

$$y \times x = xy$$

$$\frac{x}{y} = \frac{x \times x}{y \times x} = \frac{x^2}{xy}$$

$$\frac{3}{x} = \frac{3 \times y}{x \times y} = \frac{3y}{xy}$$

$$\frac{x}{y} + \frac{3}{x} = \frac{x^2}{xy} + \frac{3y}{xy}$$

$$= \frac{x^2 + 3y}{xy}$$

So, $\frac{x}{y} + \frac{3}{x} = \frac{x^2 + 3y}{xy}$

Practice Exercise 7.3

1. Work out.
   (a) $\frac{x}{4} + \frac{x}{3}$ (b) $\frac{x}{2} + \frac{x}{5}$ (c) $\frac{2x}{3} + \frac{3x}{8}$ (d) $\frac{4x}{7} + \frac{3x}{2}$
2. Show that $\frac{2x}{3} + \frac{4x}{9} = \frac{10x}{9}$.
3. Work out.
   (a) $\frac{7}{x} + \frac{x}{5}$ (b) $\frac{7}{x} + \frac{5}{x}$ (c) $\frac{2x}{7} + \frac{1}{5x}$ (d) $\frac{3}{2x} + \frac{5}{6x}$
4. Work out.
   (a) $\frac{x}{3} - \frac{x}{4}$ (b) $\frac{2x}{5} - \frac{x}{6}$ (c) $\frac{x}{2} - \frac{3x}{10}$ (d) $\frac{4x}{5} - \frac{3x}{8}$
5. Show that $\frac{3x}{4} - \frac{5x}{12} = \frac{x}{3}$.
6. Work out.
   (a) $\frac{3}{x} - \frac{x}{4}$ (b) $\frac{6}{x} - \frac{5}{2x}$ (c) $\frac{3x}{4} - \frac{2}{5x}$ (d) $\frac{4}{5x} - \frac{5}{12x}$

Further addition and subtraction of algebraic fractions

The numerator and/or the denominator of a fraction can involve an expression.

For example, work out

$$\frac{(x+2)}{3} + \frac{(x-1)}{4}$$

Change the original fractions to equivalent fractions.

$3 \times 4 = 12$

$$\frac{(x+2)}{3} = \frac{(x+2) \times 4}{3 \times 4} = \frac{4(x+2)}{12} \quad \text{and} \quad \frac{(x-1)}{4} = \frac{(x-1) \times 3}{4 \times 3} = \frac{3(x-1)}{12}$$

$$\frac{(x+2)}{3} + \frac{(x-1)}{4} = \frac{4(x+2)}{12} + \frac{3(x-1)}{12}$$

$$= \frac{4(x+2) + 3(x-1)}{12}$$

Multiply out the brackets and simplify.

$$= \frac{4x + 8 + 3x - 3}{12}$$

$$\frac{(x+2)}{3} + \frac{(x-1)}{4} = \frac{7x+5}{12}$$

Example 8

Work out $\frac{1}{(x-3)} - \frac{3}{(x+4)}$

$(x-3) \times (x+4) = (x-3)(x+4)$

$$\frac{1}{(x-3)} = \frac{1 \times (x+4)}{(x-3)(x+4)} = \frac{(x+4)}{(x-3)(x+4)}$$

$$\frac{3}{(x+4)} = \frac{3 \times (x-3)}{(x+4)(x-3)} = \frac{3(x-3)}{(x+4)(x-3)}$$

$$\frac{1}{(x-3)} - \frac{3}{(x+4)} = \frac{(x+4)}{(x-3)(x+4)} - \frac{3(x-3)}{(x+4)(x-3)}$$

$$= \frac{(x+4) - 3(x-3)}{(x-3)(x+4)}$$

Multiply out the brackets in the numerator and simplify.

$$= \frac{x + 4 - 3x + 9}{(x-3)(x+4)}$$

$$= \frac{-2x + 13}{(x-3)(x+4)}$$

$$\frac{1}{(x-3)} - \frac{3}{(x+4)} = \frac{13 - 2x}{(x-3)(x+4)}$$

Example 9

Work out $\frac{5}{x} + \frac{2}{(x+3)}$

$x \times (x+3) = x(x+3)$

$$\frac{5}{x} = \frac{5 \times (x+3)}{x(x+3)} = \frac{5(x+3)}{x(x+3)}$$

$$\frac{2}{(x+3)} = \frac{2 \times x}{(x+3) \times x} = \frac{2x}{x(x+3)}$$

$$\frac{5}{x} + \frac{2}{(x+3)} = \frac{5(x+3)}{x(x+3)} + \frac{2x}{x(x+3)}$$

$$= \frac{5(x+3) + 2x}{x(x+3)}$$

$$= \frac{5x + 15 + 2x}{x(x+3)}$$

$$\frac{5}{x} + \frac{2}{(x+3)} = \frac{7x + 15}{x(x+3)}$$

The denominator is usually left in factor form.

Practice Exercise 7.4

1. Work out.

   (a) $\frac{(x+1)}{2} + \frac{(x+2)}{3}$ (b) $\frac{(x+3)}{4} + \frac{(x-1)}{3}$ (c) $\frac{(x-2)}{5} + \frac{(x+3)}{2}$

2. Work out.

   (a) $\frac{(x+2)}{3} - \frac{(x+1)}{2}$ (b) $\frac{(x+2)}{2} - \frac{(x-1)}{5}$ (c) $\frac{(x-3)}{4} - \frac{(x-1)}{3}$

3. Work out.

   (a) $\frac{2}{(x+3)} + \frac{3}{(x-2)}$ (b) $\frac{3}{(x+4)} - \frac{2}{(x-1)}$ (c) $\frac{5}{(x-3)} - \frac{3}{(x+1)}$

   (d) $\frac{1}{(x-5)} + \frac{3}{(x+4)}$ (e) $\frac{2}{(x-1)} - \frac{1}{(x+1)}$ (f) $\frac{3}{(x+4)} + \frac{4}{(x-1)}$

4. Work out.

   (a) $\frac{2}{x} + \frac{3}{(x+1)}$ (b) $\frac{4}{(x-2)} - \frac{1}{x}$ (c) $\frac{3}{x} + \frac{2}{(x-3)}$

5. Express as a single fraction.

   (a) $\frac{4}{x} - \frac{2}{x+1}$ (b) $\frac{2}{y+2} + \frac{1}{y}$ (c) $\frac{3}{p} - \frac{1}{(p-2)}$

Key Points

- **Algebraic fractions** have a numerator and a denominator (just as an ordinary fraction) but at least one of them is an expression involving an unknown.
- To write an algebraic fraction in its **simplest form**:
  - factorise the numerator and denominator of the fraction,
  - divide the numerator and denominator by their highest common factor.
- The same methods used for adding, subtracting, multiplying and dividing numeric fractions can be applied to algebraic fractions.

Review Exercise 7

1. (a) Factorise: (i) $x^2 - 4$ (ii) $3x^2 + 2x - 8$

   (b) Hence, simplify as fully as possible $\frac{x^3 - 4x}{3x^2 + 2x - 8}$

2. Simplify fully the following expression. $\frac{6x - 18}{x^2 - 5x + 6}$

3. (a) Factorise (i) $x^2 - 9$, (ii) $2x^2 - 5x - 3$.

   (b) Hence, simplify $\frac{x^2 - 9}{2x^2 - 5x - 3}$.

4. Simplify fully the following expressions.

   (a) $\frac{a^2 + ad}{a^2 - d^2}$ (b) $\frac{p^2 - 4}{p^2 + 5p + 6}$ (c) $\frac{c^2 + 3c + 2}{c^2 + 5c + 4}$

5. Work out.

   (a) $\frac{8a}{15} \times \frac{5}{4a}$ (b) $\frac{3d}{10f} \times \frac{2f}{9}$ (c) $\frac{6p}{5q} \times \frac{25s}{18p}$ (d) $\frac{3}{8b} \div \frac{15}{16b}$ (e) $\frac{2g}{5} \div \frac{4h}{15}$ (f) $\frac{3p}{7} \div \frac{2p}{14q}$

6. Work out.

   (a) $\frac{3x}{8} + \frac{x}{6}$ (b) $\frac{10x}{3} - \frac{2x}{9}$ (c) $\frac{4x}{15} + \frac{5x}{6}$

7. Express as a single fraction.

   (a) $\frac{(x-2)}{4} + \frac{(x+3)}{2}$ (b) $\frac{2}{(x-3)} - \frac{1}{(x+5)}$ (c) $\frac{5}{x} + \frac{3}{(x+1)}$

8 Gradient of a Straight Line Graph

Gradient of a straight line graph

The gradient of a straight line graph is found by drawing a right-angled triangle.

$$\text{Gradient} = \frac{\text{distance up}}{\text{distance along}}$$

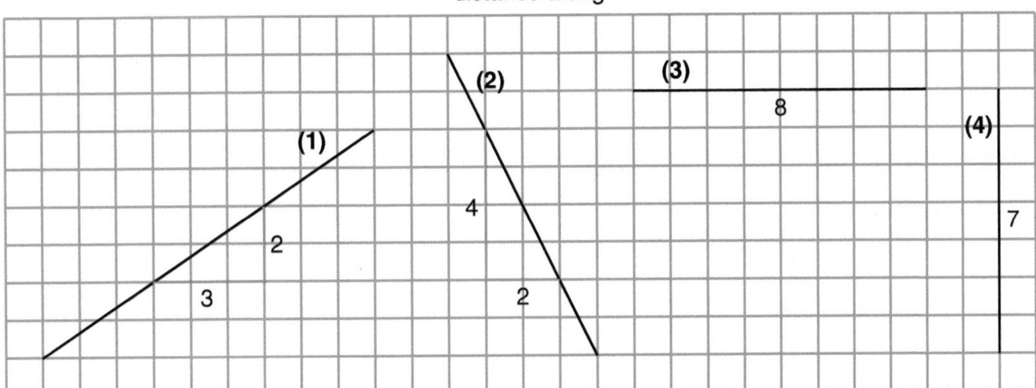

Line (**1**) has **positive gradient**.
As you go from left to right along the line it slopes **up**.

$$\text{Gradient} = \frac{\text{distance up}}{\text{distance along}} = \frac{2}{3}$$

Line (**1**) has a gradient $\frac{2}{3}$.

Line (**2**) has **negative gradient**.
As you go from left to right along the line it slopes **down**.

$$\text{Gradient} = \frac{\text{distance up}}{\text{distance along}} = -\frac{4}{2} = -2$$

Line (**2**) has a gradient -2.

Line (**3**) has **zero gradient**.
The line is **horizontal**.

$$\text{Gradient} = \frac{\text{distance up}}{\text{distance along}} = \frac{0}{8} = 0$$

Line (**4**) has **undefined gradient**.
The line is **vertical**.

$$\text{Gradient} = \frac{\text{distance up}}{\text{distance along}} = \frac{7}{0} = \text{undefined}$$

If you are finding the gradient of lines drawn on a grid, choose points on the line so that distances up and along are whole numbers.

Lines which are **parallel** have the **same gradient**.

Practice Exercise 8.1

1.

(a) Find the gradients of the lines labelled (**1**) to (**14**).
(b) Which line is parallel to line (**1**)?
(c) Which line is parallel to line (**8**)?

To find the gradient of a line drawn on a grid, you must:
- read the scale on the vertical axis to find the distance up,
- read the scale on the horizontal axis to find the distance along.

2. Find the gradients of the lines labelled (**1**) to (**6**).

(a)

(b)

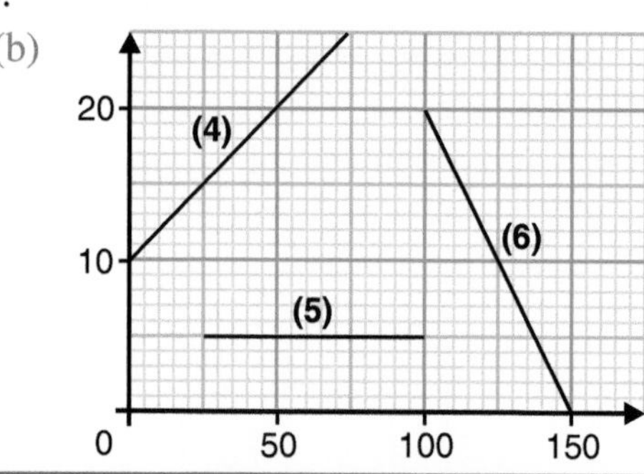

To find the gradient of a straight line given two points

Gradient of $AB = \dfrac{\text{increase in } y}{\text{increase in } x}$

$= \dfrac{y\text{-coordinate of } B - y\text{-coordinate of } A}{x\text{-coordinate of } B - x\text{-coordinate of } A}$

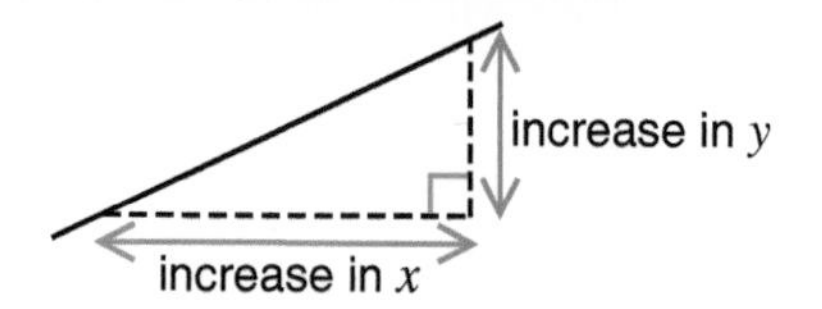

In general, the gradient, m, of any straight line passing through points $A(x_1, y_1)$ and $B(x_2, y_2)$ is given by

$$m_{AB} = \frac{y_2 - y_1}{x_2 - x_1}$$

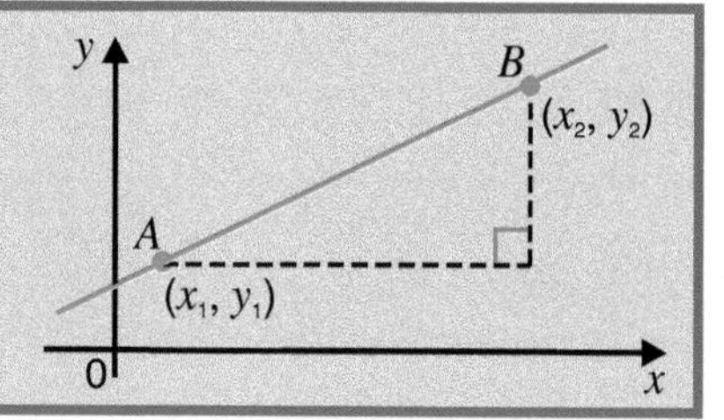

Example 1

If A is (8, 5) and B is (12, 15), find the gradient of AB.

$m_{AB} = \dfrac{y_2 - y_1}{x_2 - x_1}$

$m_{AB} = \dfrac{15 - 5}{12 - 8} = \dfrac{10}{4} = 2.5$

The gradient of AB is 2.5.

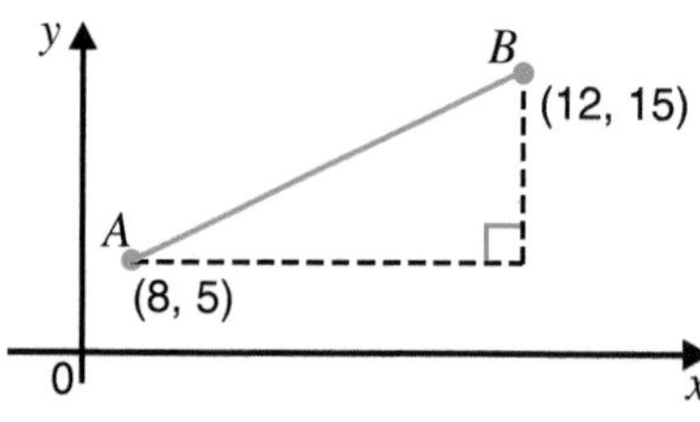

Practice Exercise 8.2

1. If A is (1, 4) and B is (6, 10), find the gradient of the straight line which passes through points A and B.

2. Find the gradients of the straight lines which pass through these pairs of points.
   (a) A(2, 4) and B(5, 6) (b) C(1, 11) and D(5, 3)
   (c) E(−5, −11) and F(−2, 1) (d) G(4, 3) and H(4, 6)
   (e) P(5, −2) and Q(7, 8) (f) R(3, 7) and S(10, 7)

3. (a) A straight line with gradient zero passes through A(5, 4) and B(7, x). What is the value of x?
   (b) A straight line with undefined gradient passes through P(y, −1) and (−4, 5). What is the value of y?

4. Line AB is parallel to line CD.
   (a) If A is (2, 4) and B is (5, 13), find m_{AB}.
   (b) C is the point (5, 2) and D has coordinates $(a, 14)$.
   Find the value of a.

5. $A(1, 1)$, $B(5, b)$ and $m_{AB} = 2$.
   Find the value of b.

6. $P(1, 8)$, $Q(a, 4)$ and $m_{PQ} = -4$.
   Find the value of a.

7. A line with gradient $\frac{1}{3}$ passes through the points $(-1, 0)$ and $(a, 4)$.
   Find the value of a.

8. A line with gradient -1.5 passes through the points $(-5, 8)$ and $(-1, b)$.
   Find the value of b.

9. $ABCD$ is a quadrilateral.
   $A(3, 2)$, $B(1, 5)$, $C(7, 5)$, $D(9, 2)$.
   (a) Find the gradients of AD and BC.
   (b) Find the gradients of AB and CD.
   (c) Explain why $ABCD$ is a parallelogram.

Key Points

- The **gradient** of a line can be found by drawing a right-angled triangle.

$$\text{Gradient} = \frac{\text{distance up}}{\text{distance along}}$$

distance up

distance along

Gradient can be positive, negative, zero or undefined.

- In general, the gradient, m, of a straight line passing through points $A(x_1, y_1)$ and $B(x_2, y_2)$ is given by
$$m_{AB} = \frac{y_2 - y_1}{x_2 - x_1}$$

Review Exercise 8

1. Find the gradient of the line AB.

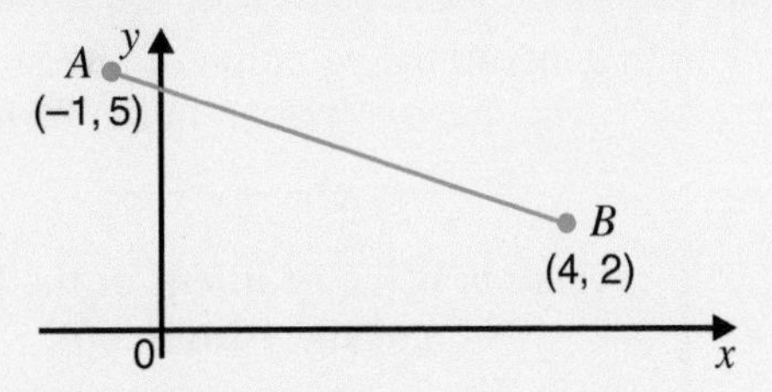

2. If A is $(-3, 1)$ and B is (5, 3), find the gradient of AB.

3. Find the gradient of the line which passes through points $P(-2, 3)$ and $Q(3, -7)$.

4. A is (3, 0), B is (4, 1), C is (1, 3) and D is (0, 2).
   (a) Find the gradient of AB, BC, AD and DC.
   (b) Show that $ABCD$ is a parallelogram.

5. A straight line with gradient 4 passes through $(-1, -2)$ and $(3, a)$.
   Find the value of a.

6. A straight line with gradient -3 passes through $(c, -4)$ and $(-3, 8)$.
   Find the value of c.

9 Straight Line Graphs

$y = 3x - 2$ and $y = -2x + 5$ are examples of a **linear function**.
The graph of a linear function is a **straight line**.

To draw a linear graph:
- find at least two corresponding values of x and y,
- plot the points,
- join the points with a straight line.

x	-1	0	1	2
$y = 3x - 2$	-5	-2	1	4

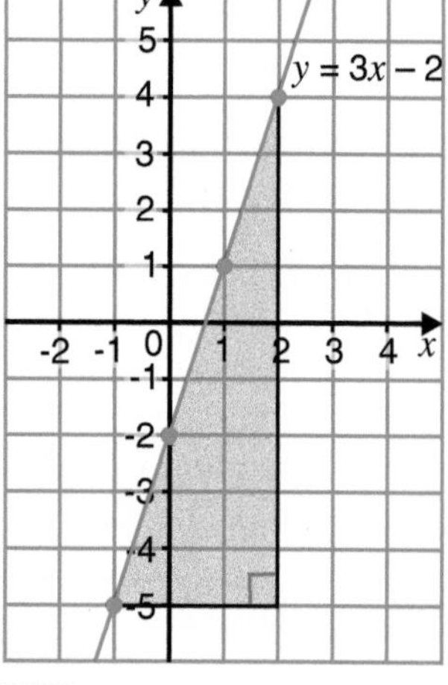

To find the gradient of $y = 3x - 2$, draw a triangle.

$$m = \frac{9}{3} = 3$$

The line $y = 3x - 2$ has **gradient 3** and it crosses the y axis at $(0, -2)$.
We say that the ***y*-intercept** is -2.

x	-1	0	1	2	3
$y = 2x + 5$	7	5	3	1	-1

To find the gradient of the line $y = -2x + 5$, draw a triangle.
The gradient is negative.

$$m = -\frac{8}{4} = -2$$

The line $y = -2x + 5$ has **gradient** -2 and its ***y*-intercept** is 5.
$y = -2x + 5$ can be written as $y = 5 - 2x$.

In general, the equation of any straight line can be written in the form

$$y = mx + c$$

where m is the **gradient** of the line and c is the ***y*-intercept**.

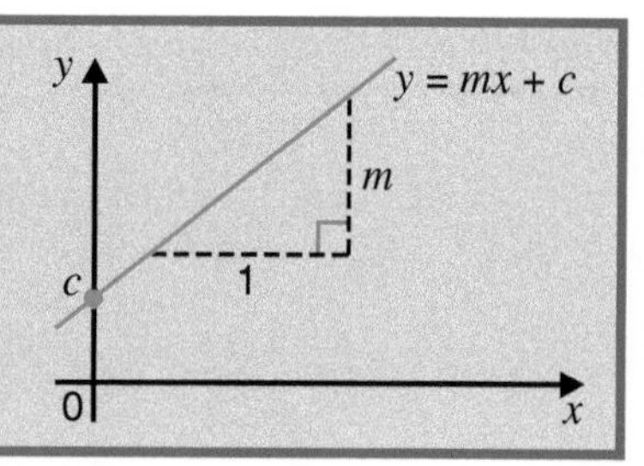

Special graphs

This diagram shows the graphs: $x = 4$ $y = 1$
$x = -2$ $y = -5$

The graph of $x = 4$ is a **vertical** line.
All points on the line have x coordinate 4.
The gradient of a vertical line is **undefined**.

The graph of $y = 1$ is a **horizontal** line.
All points on the line have y coordinate 1.
The gradient of a horizontal line is **zero**.

$x = 0$ is the y axis.
$y = 0$ is the x axis.

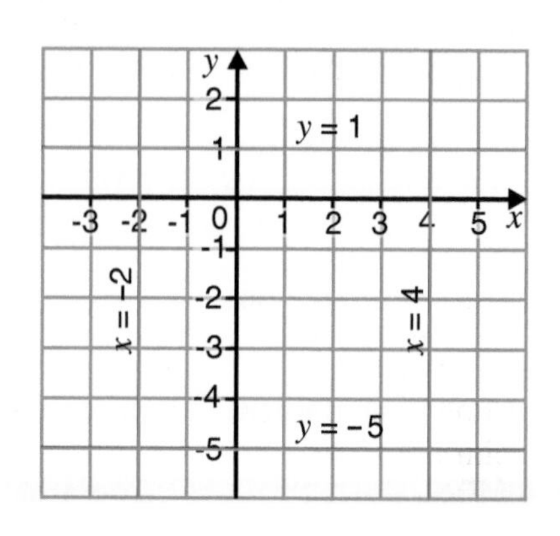

Example 1

Write down the gradient and y-intercept for each of the following graphs.

(a) $y = 3x + 5$ (b) $y = 4x - 1$ (c) $y = 6 - x$

(a) $y = 3x + 5$
Gradient $= 3$,
y-intercept $= 5$.

(b) $y = 4x - 1$
Gradient $= 4$,
y-intercept $= -1$.

(c) $y = 6 - x$
Gradient $= -1$,
y-intercept $= 6$.

Example 2

Write down the equation of the straight line which has gradient -7 and cuts the y axis at the point $(0, 4)$.

The general form for the equation of a straight line is $y = mx + c$.
The gradient, $m = -7$, and the y-intercept, $c = 4$.
Substitute these values into the general equation.
The equation of the line is $y = -7x + 4$.
This can be written as $y = 4 - 7x$.

Example 3

Find the equation of the line shown on this graph.

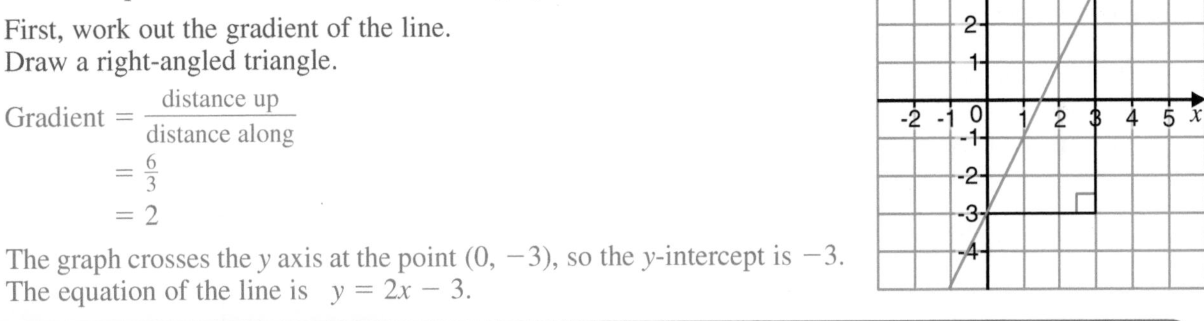

First, work out the gradient of the line.
Draw a right-angled triangle.

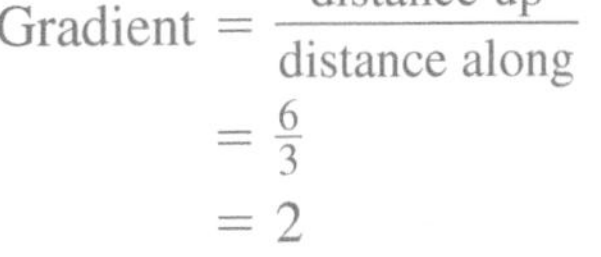

$$\text{Gradient} = \frac{\text{distance up}}{\text{distance along}} = \frac{6}{3} = 2$$

The graph crosses the y axis at the point $(0, -3)$, so the y-intercept is -3.
The equation of the line is $y = 2x - 3$.

Practice Exercise 9.1

1. (a) Draw these graphs **on the same diagram**:
   (i) $y = x + 2$ (ii) $y = x + 1$ (iii) $y = x$ (iv) $y = x - 1$
   Draw and label the x axis from 0 to 3 and the y axis from -1 to 5.
   (b) What do they all have in common? What is different?

2. (a) Write down the gradient and y-intercept of $y = 3x - 1$.
   (b) Draw the graph of $y = 3x - 1$ to check your answer.

3. Which of the following graphs are parallel?

   $y = 3x$ $y = x + 2$ $y = 2x + 3$ $y = 3x + 2$

4. Copy and complete this table.

Graph	gradient	y-intercept
$y = 3x + 5$	3	
$y = 2x - 3$		
$y = 4 - 2x$		4
$y = \frac{1}{2}x + 3$		
$y = 2x$		
$y = 3$		

5. (a) Write down the equation of the straight line which has gradient 5 and crosses the y axis at the point $(0, -4)$.
   (b) Write down the equation of the straight line which has gradient $-\frac{1}{2}$ and cuts the y axis at the point $(0, 6)$.

6. (a) Write down the equations of the lines labelled on this graph.
(b) A line, with a gradient of 3, passes through the origin. What is the equation of the line?

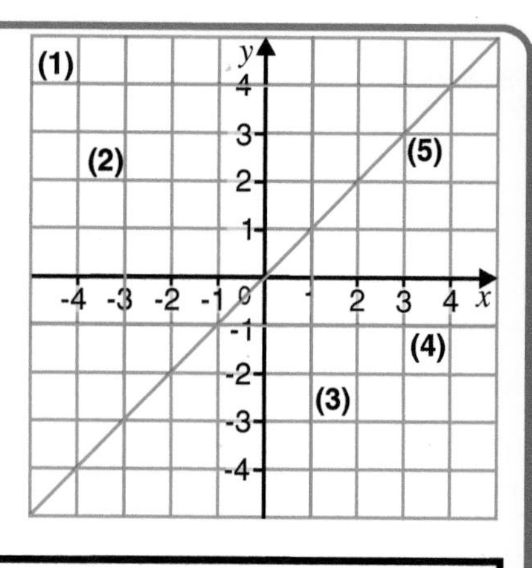

7. Match the following equations to their graphs.

(1) $y = x - 6$

(2) $y = 6 - x$

(3) $y = 2x + 1$

(4) $y = 2x - 1$

A

B

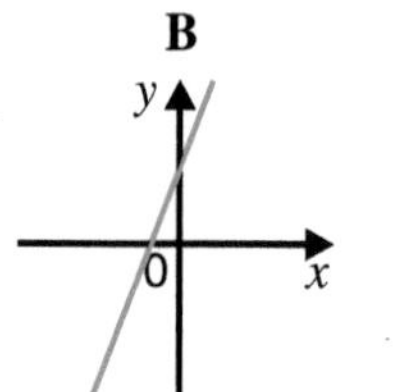

C

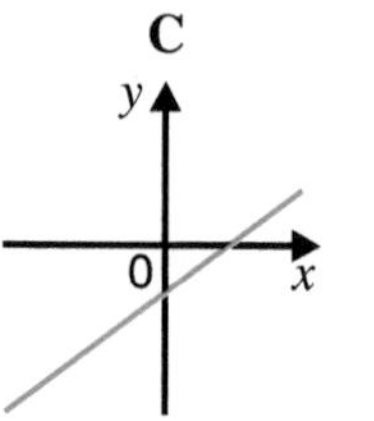

D

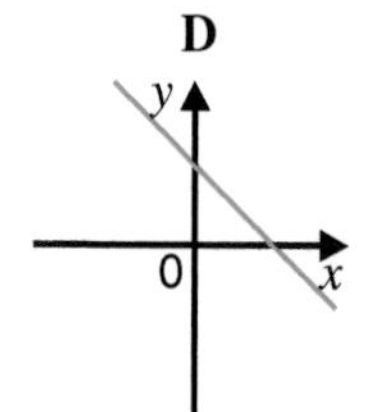

8. Find the equations of the lines shown on the following graphs.

(a)

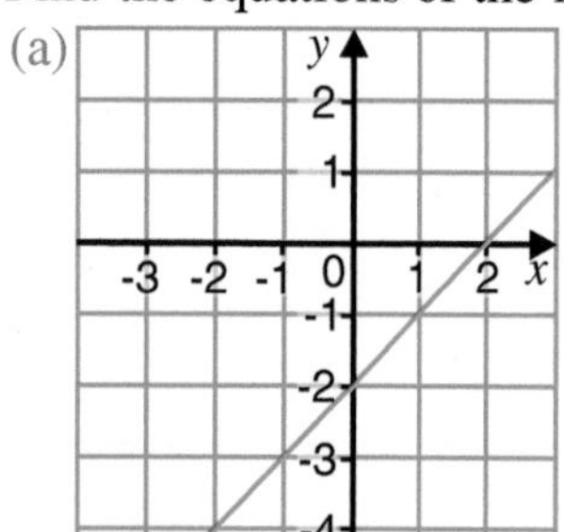

(b)

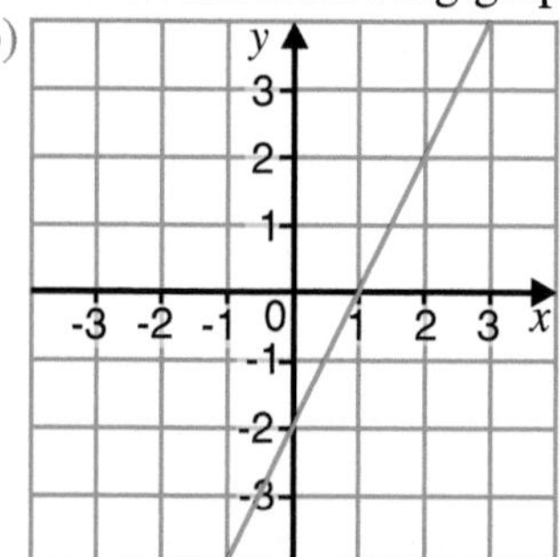

(c)

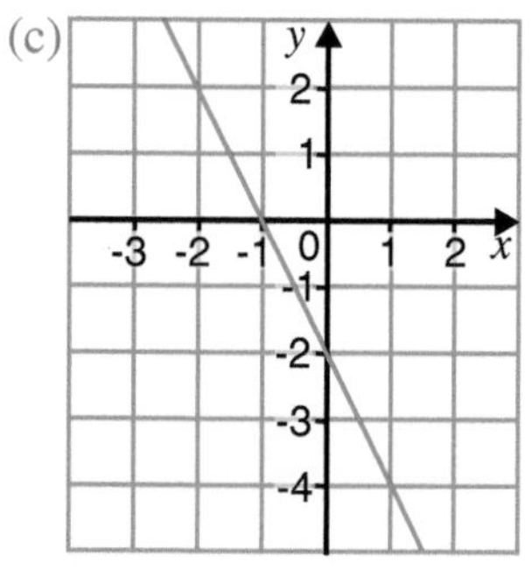

9. State the coordinates where the line $y = 4 - \frac{1}{2}x$ crosses the y axis.

Alternative formula for the equation of a straight line

Find the equation of the straight line which has gradient 3 and passes through the point (2, 8).
The diagram shows the line passing through the point (2, 8) and another point on the line (x, y).

Gradient of line is 3, so:

$$\frac{y-8}{x-2} = 3$$

Multiply both sides by $(x - 2)$.

$y - 8 = 3(x - 2)$

$y - 8 = 3x - 6$

Add 8 to both sides.

$y = 3x + 2$

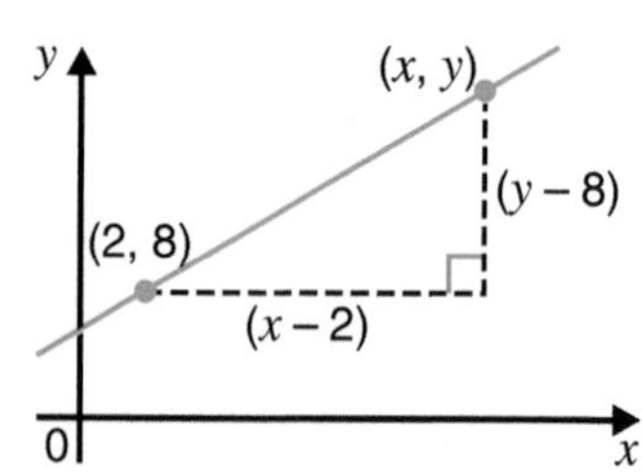

The equation of the straight line which has gradient 3 and passes through the point (2, 8) is $y = 3x + 2$.

In general, the equation of any straight line which has gradient m and passes through the point (a, b) is given by

$$y - b = m(x - a)$$

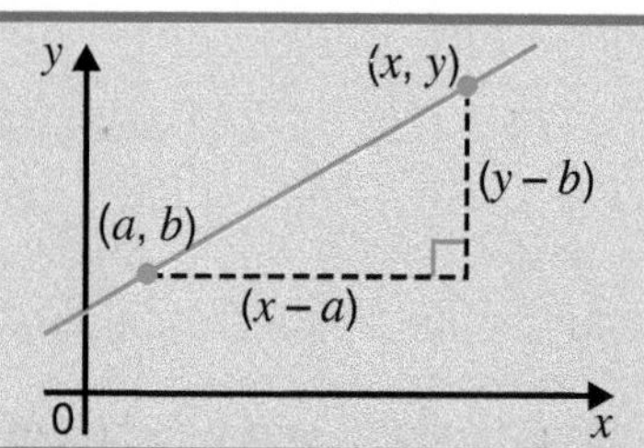

Example 4

A line passes through the points $A(-1, 9)$ and $B(2, 3)$.
Find the equation of the line.

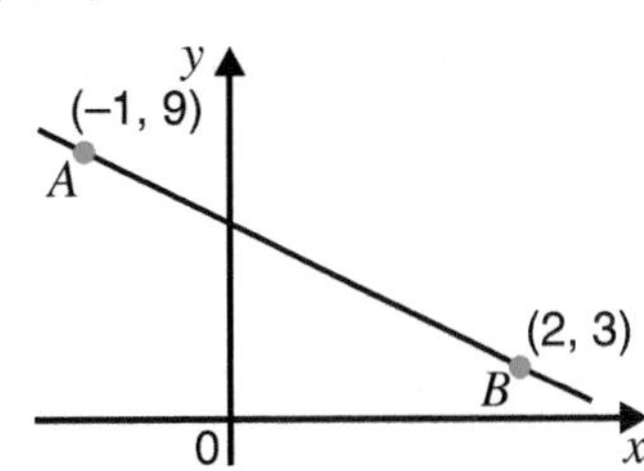

Gradient of line AB is given by

$$m_{AB} = \frac{\text{increase in } y}{\text{increase in } x}$$

$$m_{AB} = \frac{9-3}{-1-2} = \frac{6}{-3} = -2$$

Substitute $m = -2$ and $(a, b) = (2, 3)$
into $y - b = m(x - a)$

$$y - 3 = -2(x - 2)$$
$$y - 3 = -2x + 4$$

Add 3 to both sides.

$$y = -2x + 7$$

You can check your answer using the other point $A(-1, 9)$.
Substitute $x = -1$ into $y = -2x + 7$

Practice Exercise 9.2

1. Use the equation $y - b = m(x - a)$ to find the equations of these straight lines.
   (a) Gradient 1, passes through point (2, 7).
   (b) Gradient 2, passes through point (3, −5).
   (c) Gradient −5, passes through point (−4, 2).
   (d) Gradient −3, passes through point (−4, −8).

2. A line passes through the points $A(1, 3)$ and $B(3, 7)$.
   Find the equation of the line.

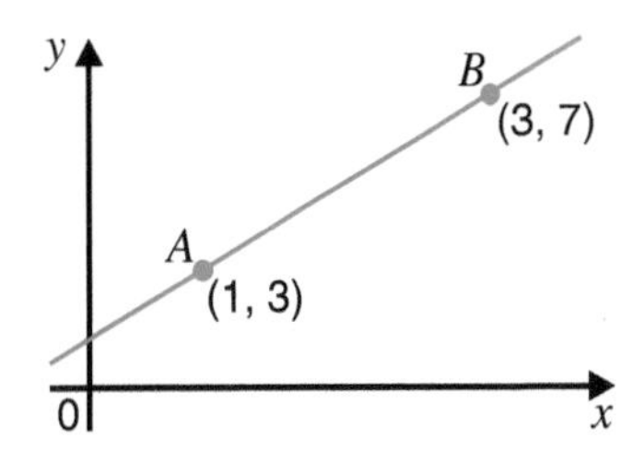

3. A line crosses the y axis at the point $(0, -2)$ and has gradient 3.5.
   Show that the equation of the line can be written as $2y = 7x - 4$.

4. The diagram shows three points:
   $A(2, 6)$, $B(5, 3)$ and $C(1, 1)$.
   (a) Find the equation of the line joining points A and B.
   (b) Find the equation of the line joining points A and C.
   (c) Show that the equation of the line joining points B and C is $y = \frac{1}{2}x + \frac{1}{2}$.

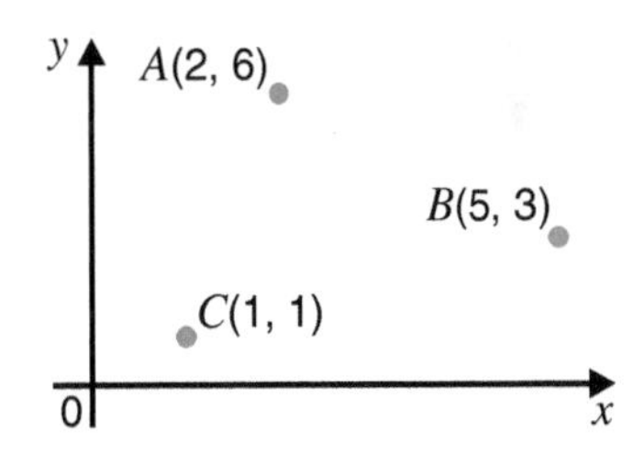

5. Find the equation of the straight line which passes through $P(-2, 3)$ and $Q(3, -7)$.

Rearranging equations

The general equation for a straight line graph is $y = mx + c$.
When an equation is in this form the gradient and y-intercept are given by the values of m and c.

The equation for a straight line can also be written in the form $px + qy = r$.
To find the gradient and y-intercept of this line we must first **rearrange** the equation.

For example, the graph of a straight line is given as $2y - 6x = 5$.

To rearrange: $2y - 6x = 5$

Add $6x$ to both sides.

$$2y = 6x + 5$$

Divide both sides by 2.

$$y = 3x + 2.5$$

Hence, the line has gradient 3 and y-intercept 2.5.

Example 5

The graph of a straight line is given by the equation $4y - 3x = 8$.

(a) Write this equation in the form $y = mx + c$.

(b) Write down the gradient and the y-intercept of the line.

(a) $4y - 3x = 8$

Add $3x$ to both sides.

$$4y = 3x + 8$$

Divide both sides by 4.

$$y = \frac{3}{4}x + 2$$

(b) The line has gradient $\frac{3}{4}$ and y-intercept 2.

Example 6

The equation of a straight line is $6x + 3y = 2$.

Write down the equation of another line which is parallel to this line.

Write the equation in the form $y = mx + c$.

$6x + 3y = 2$

Subtract $6x$ from both sides.

$$3y = -6x + 2$$

Divide both sides by 3.

$$y = -2x + \frac{2}{3}$$

The gradient of the line is -2.

To write an equation of a parallel line, keep the same gradient and change the value of the y-intercept.

For example: $y = -2x + 5$

Practice Exercise 9.3

1. The graph of a straight line is given by the equation $2y - 3x = 6$. Write this equation in the form $y = mx + c$.

2. Write the equations of the following lines in the form $y = mx + c$.
   (a) $2y + x = 4$ (b) $5y + 4x = 20$ (c) $4 - 3y = 2x$ (d) $2x - 7y = 14$

3. Find the gradients of these lines.
   (a) $2y = x$ (b) $x - y = 0$ (c) $2x - y = 0$
   (d) $3y + x = 0$ (e) $4y - 3x = 0$ (f) $2x + 5y = 0$

4. These lines cross the y axis at the point $(0, a)$. Find the value of a for each line.
   (a) $y = x$ (b) $2y = x + 1$ (c) $3y + 6 = x$
   (d) $2y - 5 = x$ (e) $4y - 3x = 8$ (f) $x + 5y = 2$

5. Write down an equation which is parallel to each of these lines.
   (a) $y = x$ (b) $y = 2x - 3$ (c) $2y = x - 4$

6. The equation of a straight line is $2y + 3x = 4$. Write down the equation of another line which is parallel to this line.

7. The diagram shows a sketch of the line $2y = x + 4$.
   (a) Find the coordinates of points A and B.
   (b) What is the gradient of the line?

   Copy the diagram.
   (c) (i) Draw the sketch of another line that has the same gradient as $2y = x + 4$.
   (ii) What is the equation of the line you have drawn?

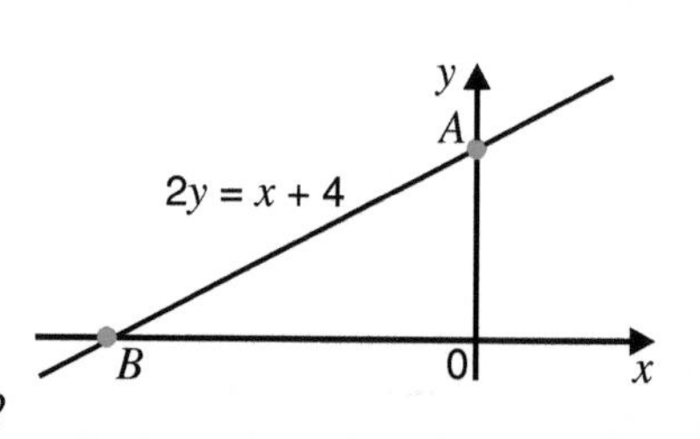

8. The diagram shows a sketch of the line $2y = 6 - x$.
(a) Find the coordinates of points P and Q.
(b) What is the gradient of the line?

Copy the diagram.
(c) (i) Draw the sketch of another line that has the same gradient as $2y = 6 - x$.
(ii) What is the equation of the line you have drawn?

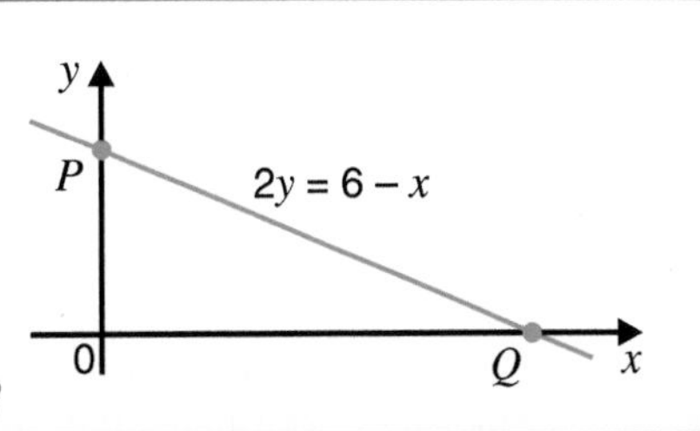

Function notation

Function notation is an alternative way of expressing the relationship between two variables.

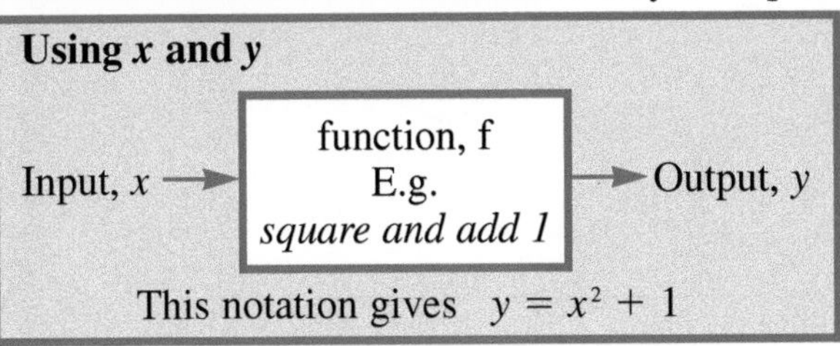

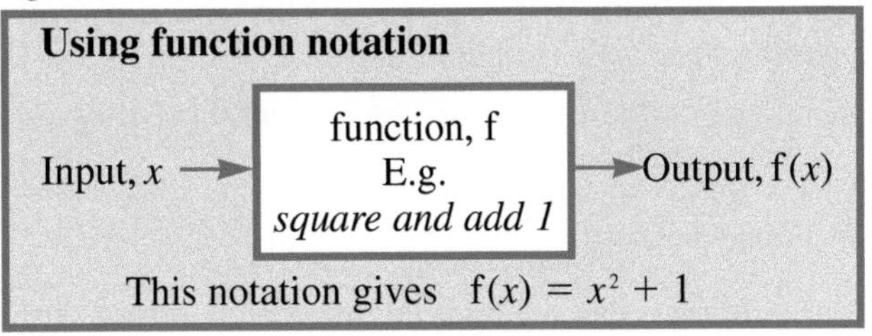

$f(x)$ means 'a function of x'.
In general, $y = f(x)$.
For example:
The equation $y = x^5$ can be expressed in function notation as $f(x) = x^5$, where $y = f(x)$.
Letters other than f can be used.

$f(-1)$ means 'the value of $f(x)$ when $x = -1$'.
For example:
If $f(x) = x^2 + 1$, $f(-1) = (-1)^2 + 1 = 2$
$f(0) = 0^2 + 1 = 1$
$f(5) = 5^2 + 1 = 26$
and so on.

Practice Exercise 9.4

1. These sketch graphs represent the functions

$f(x) = 2$ $\quad f(x) = 2x$ $\quad f(x) = x + 2$
$f(x) = 2x + 2$ $\quad f(x) = 2 - x$ $\quad f(x) = 2 - 2x$

Identify each graph.

A
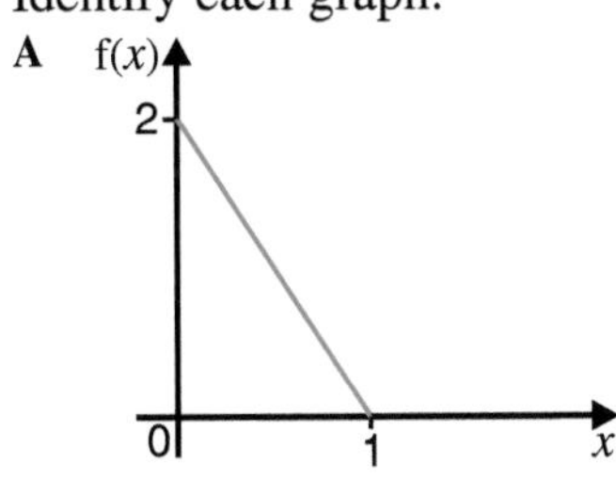

B
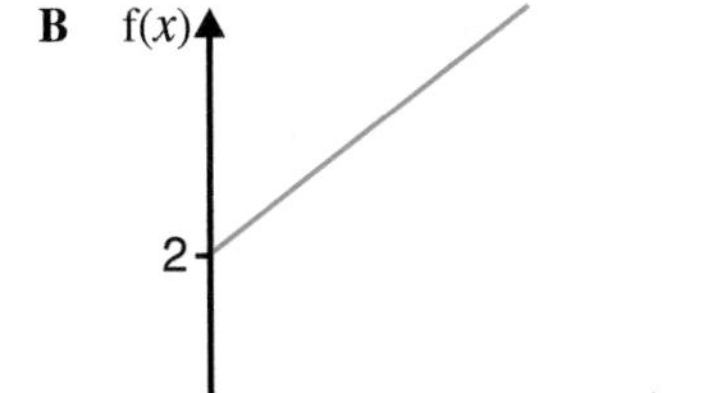

C
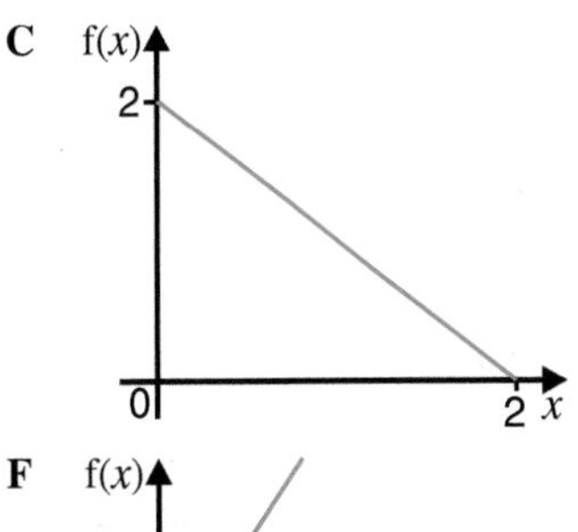

D
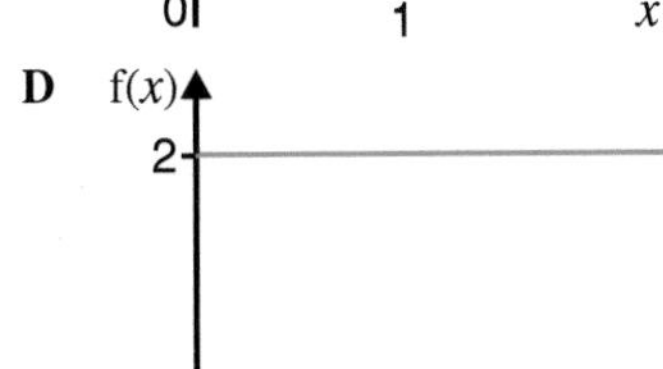

E
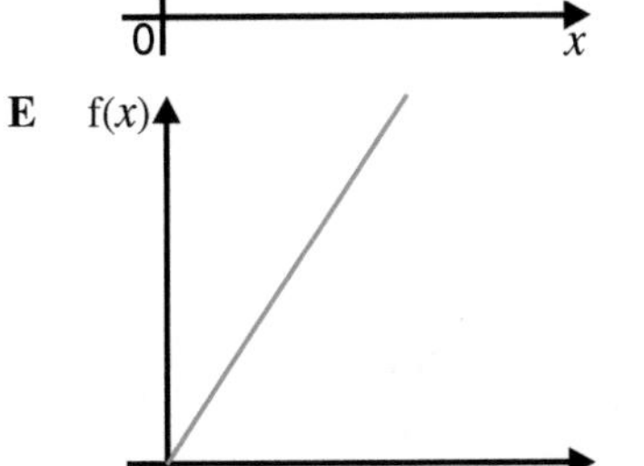

F
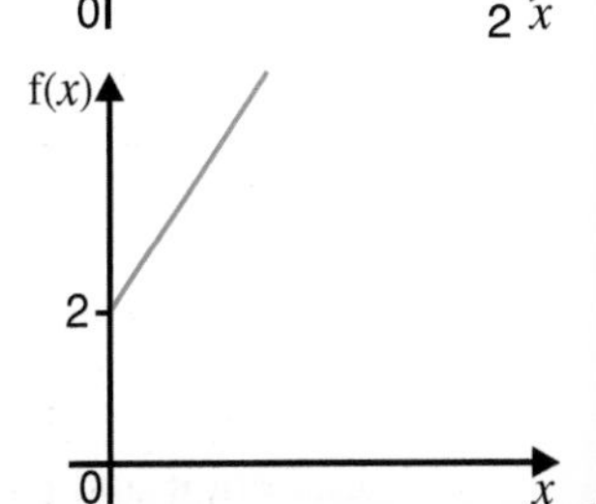

2. Consider the function $f(x) = x + 5$.
   (a) Find the value of $f(0)$.
   (b) Use your answer to part (a) to write down the coordinates of the point where the graph of $f(x)$ crosses the y axis.
   (c) For what value of x does $f(x) = 0$?
   (d) Write down the coordinates of the point where the graph of $f(x)$ crosses the x axis.

3. $f(x) = \frac{1}{2}x + 6$.
   (a) Find the y-intercept for the function $f(x)$.
   (b) Find the coordinates where the graph of $f(x)$ crosses the x axis.

Key Points

- The equation of the graph of a straight line is of the form $y = mx + c$, where m is the gradient and c is the y-intercept.
  The **gradient** of a line can be found by drawing a right-angled triangle.

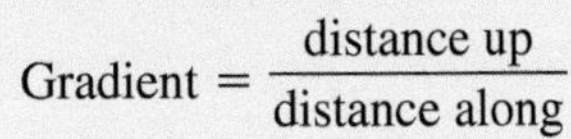

  $$\text{Gradient} = \frac{\text{distance up}}{\text{distance along}}$$

  Gradients can be positive, negative, zero or undefined.

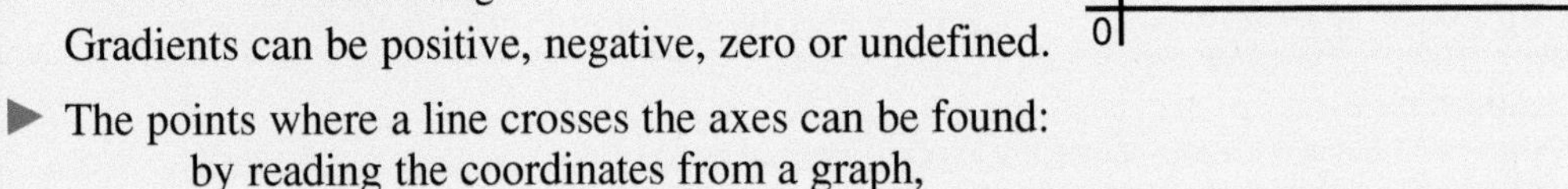

- The points where a line crosses the axes can be found:
  by reading the coordinates from a graph,
  by substituting $x = 0$ and $y = 0$ into the equation of the line.
- The formula $y - b = m(x - a)$ can be used to find the equation of a straight line which has gradient m and passes through the point (a, b).
- Equations of the form $px + qy = r$ can be rearranged to the form $y = mx + c$.
- **Function notation** is a way of expressing a relationship between two variables.
  For example

  Input, x → function, f E.g. *cube* → Output, $f(x)$

  $f(x)$ means 'a function of x'.
  In the example above, $f(x) = x^3$ is equivalent to the equation $y = x^3$ where $y = f(x)$.

Review Exercise 9

1. State the gradients of these lines, and also give the coordinates of the points where the lines cut the y axis.
   (a) $y = 4x - 1$ (b) $y = 2 - 5x$ (c) $y = 2$ (d) $y = 3 - x$
   (e) $3y = x + 3$ (f) $y = \frac{1}{2}x + 7$ (g) $x = 4$ (h) $y = 0$

2. The line $y = mx + 5$ passes through a fixed point, A, whatever the value of m. Find the coordinates of A.

3. The line $y + 4x = 8$ meets the x axis at A.
   (a) Find the coordinates of A.
   (b) Find the equation of the line with gradient 4 passing through A.

4. Find the equations of these lines.
   (a) Gradient -1 and passing through the point $(1, 7)$.
   (b) Gradient 2 and passing through the point $(2, -5)$.
   (c) Gradient $\frac{1}{3}$ and passing through the point $(-1, 0)$.

5. The diagram shows a sketch of the line $y = 3x - 12$.
   (a) Find the coordinates of the points P and Q.
   (b) The line $3y = x + 6$ passes through the point $A(9, q)$.
   What is the value of q?

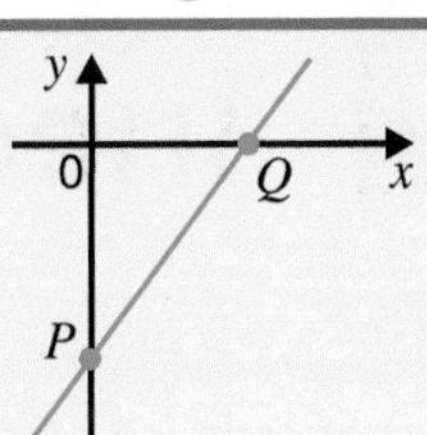

6. The table shows the values of x and y recorded in an experiment.

x	0	1	2	3	4
y	3	7	11	15	19

The results are plotted and a straight line is drawn through the points.
Find the equation of the line.

7. A line passes through the points $A(-1, 5)$ and $B(4, 2)$.
   Find its equation.

8. Draw the x axis from 0 to 10 and the y axis from 0 to 8 using equal scales on both axes.
   (a) Plot points $A(6, 3)$ and $B(10, 5)$.
   Draw the line OAB (where O is the origin).
   What is the equation of this line?
   (b) Plot point $C(9, 7)$ and join BC.
   Find the point D such that $ABCD$ is a rectangle.
   Join AD and DC.
   What are the coordinates of D?
   (c) What is the equation of the line BD?

9. What is the equation of this line?

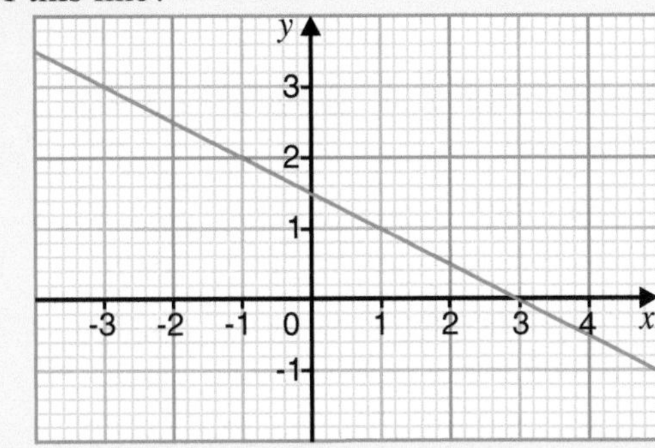

10. $f(x) = x^3 - 4x$.
    Find the values of $f(-3)$, $f(-2)$, $f(-1)$, $f(0)$, $f(1)$, $f(2)$ and $f(3)$.

11. $f(x) = 3^x$.
    (a) Find the value of $f(4)$.
    (b) Find the value of x if $f(x) = \frac{1}{3}$.

12. The diameter and circumference of different-sized circular objects were measured with the following results:

Diameter, d, in cm	2.6	5.4	8.2	13.2	15.8
Circumference, c, in cm	7.8	17.5	26.2	41.0	49.5

Plot the values of c (vertical axis) against d (horizontal axis) and show that they lie approximately on a straight line.
Draw a line of best fit through these points. This line should pass through the origin.
Find the gradient of this line.
What is the equation connecting c and d?

10 Equations *and* Inequalities

Solving equations by working backwards

I think of a number and then subtract 3.
The answer is 5.
What is the number I thought of?

Imagine that x is the number I thought of.
The steps of the problem can be shown in a diagram.

x → subtract 3 → **Answer 5**

Now work backwards, doing the opposite calculation.

8 ← add 3 ← **5**

The number I thought of is 8.

Remember:

Forwards	Backwards
add	subtract
subtract	add
multiply	divide
divide	multiply

Example 1

I think of a number, multiply it by 3 and add 4.
The answer is 19.
What is my number?

x → multiply by 3 → add 4 → **19**

5 ← divide by 3 ← 15 ← subtract 4 ← **19**

The number I thought of is 5.

Practice Exercise 10.1

Solve these equations by working backwards.

1. I think of a number and then multiply it by 2. The answer is 10.
   What is my number?
2. Jan thinks of a number and then subtracts 5. Her answer is 9.
   What is her number?
3. Lou thinks of a number. He multiplies it by 2 and then subtracts 5. The answer is 7.
   What is his number?
4. I think of a number, subtract 5 and then multiply by 2. The answer is 12.
   What is my number?
5. I think of a number, add 4 then multiply by 3. The answer is 24.
   What is my number?
6. Steve thinks of a number. He multiplies it by 5 and then adds 2. The answer is 17.
   What is his number?
7. I think of a number, multiply it by 3 and then subtract 5. The answer is 7.
   What is my number?
8. Solve this puzzle.

   Begin with x. Double it and then add 3. The result is equal to 17. What is the value of x?

9. Kathryn thinks of a number. She adds 3 and then doubles the result.
   (a) What number does Kathryn start with to get an answer of 10?
   (b) Kathryn starts with x. What is her answer in terms of x?
10. Sarah thinks of a number. She subtracts 2 and multiplies by 3.
    (a) What number does Sarah start with to get an answer of 21?
    (b) Sarah starts with x. What is her answer in terms of x?

The balance method

Mathematicians always try to find efficient methods of working.
Here is a method used to solve equations that works a bit like a balance.

These scales are balanced.

You can add the same amount to both sides and they still balance.

You can subtract the same amount from both sides and they still balance.

You can double (or halve) the amount on both sides and they still balance.

Equations work in the same way.
If you do the same to both sides of an equation, it is still true.

Example 2

Solve $d - 13 = 5$.

$d - 13 = 5$

Add 13 to both sides.

$d = 18$

Example 3

Solve $x + 7 = 16$.

$x + 7 = 16$

Subtract 7 from both sides.

$x = 9$

Example 4

Solve $4a = 20$.

$4a = 20$

Divide both sides by 4.

$a = 5$

Example 5

Solve $4n + 5 = 17$.

$4n + 5 = 17$

Subtract 5 from both sides.

$4n = 12$

Divide both sides by 4.

$n = 3$

The aim is to find out what number the letter stands for, by ending up with **one letter** on one side of the equation and a **number** on the other side.

Look at the examples carefully.
The steps taken to solve the equations are explained.

Doing the same to both sides of an equation means:

- **Adding** the **same number** to both sides.
- **Subtracting** the **same number** from both sides.
- **Dividing** both sides by the **same number**.
- **Multiplying** both sides by the **same number**.

Practice Exercise 10.2

1. Use the balance method to solve these equations.

(a) $y + 4 = 7$ (b) $n - 7 = 9$ (c) $9 + x = 11$

(d) $y - 12 = 7$ (e) $14 + b = 21$ (f) $x - 9 = 20$

(g) $7 + m = 11$ (h) $k - 2 = 3$ (i) $5 + y = 12$

2. Use the balance method to solve these equations.

(a) $3c = 12$ (b) $5a = 20$ (c) $4f = 12$

(d) $8q = 56$ (e) $7x = 35$ (f) $9y = 9$

3. Solve these equations.

(a) $2p + 1 = 9$ (b) $4t - 1 = 11$ (c) $3h - 7 = 14$

(d) $3 + 4b = 11$ (e) $5d - 8 = 42$ (f) $2x + 3 = 15$

(g) $2 + 3c = 17$ (h) $3n - 1 = 8$ (i) $4x + 3 = 11$

More equations

All the equations you have solved so far have had whole number solutions, but the solutions to equations can include negative numbers and fractions.

Example 6

Solve $-4a = 20$.

$$-4a = 20$$

Divide both sides by -4.

$$a = -5$$

Example 7

Solve $6m - 1 = 2$.

$$6m - 1 = 2$$

Add 1 to both sides.

$$6m = 3$$

Divide both sides by 6.

$$m = \frac{1}{2}$$

Practice Exercise 10.3

1. Solve these equations.

(a) $4k = 2$ (b) $2a = -6$ (c) $-3d = 12$

(d) $-8n = 4$ (e) $t + 3 = -2$ (f) $n - 3 = -2$

(g) $2m + 1 = 4$ (h) $3x - 2 = 5$ (i) $2y + 5 = 4$

2. Solve these equations.

(a) $5x = -10$ (b) $2y + 7 = 1$ (c) $4t + 10 = 2$

(d) $5 - a = 7$ (e) $2 - d = 5$ (f) $3 - 2g = 9$

(g) $4t = 2$ (h) $2x = 15$ (i) $5d = 7$

(j) $4a - 5 = 1$ (k) $3 + 5g = 4$ (l) $2b - 5 = 4$

3. Solve.

(a) $x - 1 = -3$ (b) $3 + 2n = 2$ (c) $2 - a = 3$

(d) $4 - 3y = 13$ (e) $2x - 1 = -3$ (f) $3 - 5d = 18$

(g) $4x + 1 = -5$ (h) $-2 - 3x = 10$ (i) $2 - 4x = 8$

(j) $5 - 4n = -1$ (k) $z + 2 = -7$ (l) $-2 = 5m + 13$

(m) $-3 = 17 - 5n$ (n) $-6p - 1 = 8$ (o) $-12 - 3y = 15$

Equations with brackets

Equations can include brackets. Before using the balance method any brackets must be removed by multiplying out. This is called **expanding**.

Remember: $2(x + 3)$ means $2 \times (x + 3)$

$2(x + 3) = 2 \times x + 2 \times 3 = 2x + 6$

Once the brackets have been removed the balance method can be used as before.

Example 8

Solve $3(x + 2) = 12$.

$3(x + 2) = 12$

Expand the brackets.

$3x + 6 = 12$

$3x = 6$

$x = 2$

Example 9

Solve $5(3y - 7) + 15 = 25$.

$5(3y - 7) + 15 = 25$

Expand the brackets.

$15y - 35 + 15 = 25$

$15y = 45$

$y = 3$

Practice Exercise 10.4

1. Solve.
   (a) $2(x + 3) = 12$ (b) $4(a + 1) = 12$ (c) $5(t + 4) = 30$
   (d) $3(p - 2) = 9$ (e) $6(c - 2) = 24$ (f) $2(x - 1) = 4$
   (g) $6(d - 3) = 36$ (h) $7(2 + e) = 49$ (i) $5(f + 2) = 30$

2. Solve.
   (a) $3(2w + 1) = 15$ (b) $2(4s + 5) = 34$ (c) $8(3t - 5) = 32$
   (d) $5(3y + 2) = 25$ (e) $4(7 - 2x) = 4$ (f) $5(3y - 10) = 25$

3. Solve these equations. The solution will not always be a whole number.
   (a) $3(p + 2) = 3$ (b) $2(3 - d) = 10$ (c) $2(1 - 3g) = 14$
   (d) $2(x - 5) = 7$ (e) $5(y + 1) = 7$ (f) $2(1 + 3t) = 5$
   (g) $2(2t - 1) = 5$ (h) $3(2a - 3) = 6$ (i) $5(m - 2) = 3$

4. Solve these equations.
   (a) $2(x + 3) - 5 = 9$ (b) $3(a - 1) + 2 = 5$ (c) $5(3 - 2m) - 7 = 13$
   (d) $5 + 2(y - 3) = 4$ (e) $2w + 3(1 + w) = -12$ (f) $e + 5(2e - 4) = 2$
   (g) $3a - 7(1 - a) = 8$ (h) $2 - 3(t + 5) = 20$ (i) $5x - 2(3 - x) = 1$

5. Solve these equations.
   (a) $x(x + 2) = x^2 + 6$ (b) $2x(x - 3) = x(2x - 1) - 10$
   (c) $x(x - 2) = (x + 2)(x - 3)$ (d) $(x + 1)^2 = (x - 2)(x + 3)$
   (e) $(x + 1)(x - 2) = (x + 3)^2$ (f) $x(2x + 5) - (x - 2) = 2x(x + 5)$

Equations with letters on both sides

In some questions letters appear on both sides of the equation.

Example 10

Solve $3x + 1 = x + 7$.

$3x + 1 = x + 7$

Subtract 1 from both sides.

$3x = x + 6$

Subtract x from both sides.

$2x = 6$

Divide both sides by 2.

$x = 3$

Example 11

Solve $4(3 + 2x) = 5(x + 2)$.

$4(3 + 2x) = 5(x + 2)$

$12 + 8x = 5x + 10$

$8x = 5x - 2$

$3x = -2$

$x = -\frac{2}{3}$

Practice Exercise 10.5

1. Solve the following equations.

   (a) $3x = 20 - x$ (b) $5q = 12 - q$ (c) $2t = 15 - 3t$
   (d) $5e - 9 = 2e$ (e) $3g - 8 = g$ (f) $y + 3 = 5 - y$
   (g) $4x + 1 = x + 7$ (h) $7k + 3 = 3k + 7$ (i) $3a - 1 = a + 7$
   (j) $3p - 1 = 2p + 5$ (k) $6m - 1 = m + 9$ (l) $3d - 5 = 5 + d$
   (m) $2y + 1 = y + 6$ (n) $3 + 5u = 2u + 12$ (o) $4q + 3 = q + 3$

2. Solve.

   (a) $3d = 32 - d$ (b) $3q = 12 - q$ (c) $3c + 2 = 10 - c$
   (d) $4t + 2 = 17 - t$ (e) $4w + 1 = 13 - 2w$ (f) $2e - 3 = 12 - 3e$
   (g) $2g + 5 = 25 - 2g$ (h) $2z - 6 = 14 - 3z$ (i) $5m + 2 = 20 + 2m$
   (j) $5a - 4 = 3a + 6$ (k) $3 + 4x = 15 + x$ (l) $6y - 11 = y + 4$

3. Solve these equations.
   The solution will not always be a whole number.

   (a) $3m + 8 = m$ (b) $2 - 4t = 12 + t$ (c) $5p - 3 = 3p - 7$
   (d) $5x - 7 = 3x$ (e) $3 + 5a = a + 5$ (f) $2b + 7 = 11 - 3b$
   (g) $4 - 4y = y$ (h) $7 + 3d = 10 - d$ (i) $f - 6 = 3f + 1$

4. Solve.

   (a) $2(x + 3) - 5 = 9$ (b) $2(a - 1) + a = 3$ (c) $4(3 - 2m) - 7 = 2m$
   (d) $3(a + 4) = 2 + a$ (e) $3(y - 5) = y - 4$ (f) $4(n + 2) = 2n + 5$
   (g) $4d + 3 = 2(d - 3)$ (h) $7k + 2 = 5(k - 4)$ (i) $2(4t + 5) = t - 18$
   (j) $5q - 2(q + 1) = 4$ (k) $x = 8 - 2(x + 3)$ (l) $4 - 3(a - 2) = a$

5. Solve.

   (a) $2(3h - 4) = 3(h + 1) - 5$ (b) $2(3 - 2x) = 2(6 - x)$
   (c) $2(3w - 1) + 4w = 28$ (d) $2(y + 4) + 3(2y - 5) = 5$
   (e) $3(2v + 3) = 5 - 4(3 - v)$ (f) $5c - 2(4c - 9) = 5 + 5(2 - c)$
   (g) $5(x + 2) + 2(2x - 1) = 7(x - 4)$ (h) $3(x - 4) = 5(2x - 3) - 2(3x - 5)$

Equations with fractions

Solve $\frac{3}{4}x = \frac{2}{5}$.

With equations like this, it is easier to get rid of the fractions first.
To do this, multiply both sides of the equation by the **least common multiple** of the denominators of the fractions.

The multiples of 4 are: 4, 8, 12, 16, **20**, ...
The multiples of 5 are: 5, 10, 15, **20**, ...

The least common multiple of 4 and 5 is 20.
So, the first step is to multiply both sides of the equation by 20.

$$\frac{3}{4}x \times 20 = \frac{2}{5} \times 20$$

This is the same as: $x \times \frac{3}{4} \times 20 = \frac{2}{5} \times 20$

$$15x = 8$$

Divide both sides by 15.

$$x = \frac{8}{15}$$

Remember:
$\frac{3}{4} \times 20$ is the same as $\frac{3}{4}$ of 20.
To find $\frac{3}{4}$ of 20:
$20 \div 4 = 5$ gives $\frac{1}{4}$ of 20.
$5 \times 3 = 15$ gives $\frac{3}{4}$ of 20.
So, $\frac{3}{4} \times 20 = 15$.

Example 12

Solve $\frac{x}{2} + \frac{2x}{3} = 7$.

$$\frac{x}{2} + \frac{2x}{3} = 7$$

Multiply both sides by 6.

$$6 \times \frac{x}{2} + 6 \times \frac{2x}{3} = 6 \times 7$$
$$3x + 4x = 42$$
$$7x = 42$$
$$x = 6$$

Example 13

Solve $\frac{x-1}{3} = \frac{x+1}{4}$.

$$\frac{x-1}{3} = \frac{x+1}{4}$$

Multiply both sides by 12.

$$4(x-1) = 3(x+1)$$
$$4x - 4 = 3x + 3$$
$$4x = 3x + 7$$
$$x = 7$$

Practice Exercise 10.6

1. Solve these equations.
   (a) $\frac{2}{3}x = 4$ (b) $\frac{2d}{5} = -4$ (c) $\frac{3}{4}w = 6$ (d) $\frac{5}{6}n = 20$ (e) $\frac{a}{8} = \frac{3}{4}$
   (f) $\frac{3}{5} = \frac{m}{10}$ (g) $\frac{4}{5}a = \frac{3}{8}$ (h) $\frac{3}{4}p = \frac{4}{7}$ (i) $\frac{3t}{4} = \frac{1}{3}$ (j) $\frac{2}{3}b = \frac{5}{9}$
2. Solve.
   (a) $\frac{h+1}{4} = 3$ (b) $\frac{2x-1}{3} = 5$ (c) $\frac{3a+4}{5} = -1$
   (d) $\frac{7-d}{4} = \frac{5}{2}$ (e) $\frac{2a+1}{2} = \frac{3}{5}$ (f) $\frac{2-3h}{3} = -\frac{5}{6}$
   (g) $\frac{x+2}{5} = \frac{3-x}{4}$ (h) $\frac{a-1}{2} = \frac{a+1}{3}$ (i) $\frac{2x-1}{6} = \frac{2-x}{3}$
3. Solve.
   (a) $\frac{x}{2} + \frac{x}{4} = 1$ (b) $\frac{x}{2} - \frac{x}{3} = 2$ (c) $\frac{x}{4} + \frac{3x}{8} = -1$
   (d) $\frac{2x}{3} - \frac{x}{6} = -2$ (e) $\frac{x+1}{2} + \frac{x-1}{3} = 1$ (f) $\frac{x+2}{3} - \frac{x+1}{4} = 2$
   (g) $\frac{11-x}{4} = 2 - x$ (h) $\frac{x+2}{2} + \frac{x-1}{5} = \frac{1}{10}$ (i) $\frac{2x-3}{6} + \frac{x+2}{3} = \frac{5}{2}$

Using equations to solve problems

So far, you have been given equations and asked to solve them.
The next step is to **form an equation** first using the information given in a problem.
The equation can then be solved in the usual way.

Example 14

The triangle has sides of length: x cm, $2x$ cm and 7 cm.

(a) Write an expression, in terms of x, for the perimeter of the triangle.
Give your answer in its simplest form.

(b) The triangle has a perimeter of 19 cm.
By forming an equation find the value of x.

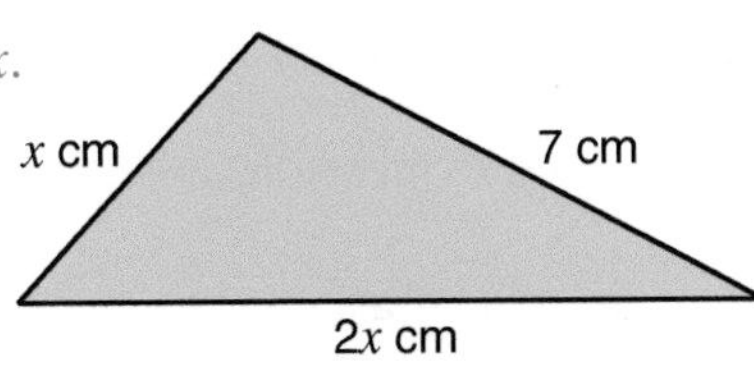

(a) The perimeter of the triangle is: $x + 2x + 7$ cm
In its simplest form, the perimeter is: $(3x + 7)$ cm

(b) The perimeter of the triangle is 19 cm, so, $3x + 7 = 19$

$$3x = 12$$
$$x = 4$$

Practice Exercise 10.7

1. The weights of three boxes are shown.

k kilograms $2k$ kilograms $3k$ kilograms

   (a) Write an expression, in terms of k, for the total weight of the boxes.
   (b) The boxes weigh 15 kilograms altogether.
   By forming an equation find the weight of the lightest box.

2. Dominic is 7 years younger than Marcie.
   (a) Dominic is n years old.
   Write an expression, in terms of n, for Marcie's age.
   (b) The sum of their ages is 43 years.
   By forming an equation find the ages of Dominic and Marcie.

3. The diagram shows the lengths of three rods.

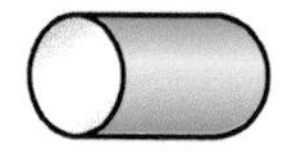

$(y - 5)$ centimetres y centimetres $(2y + 3)$ centimetres

   (a) Write an expression, in terms of y, for the total length of the rods.
   (b) The total length of the rods is 30 centimetres.
   What is the length of the longest rod?

4. Grace is given a weekly allowance of £p.
   Aimee is given £4 a week **more** than Grace.
   Lydia is given £3 a week **less** than Grace.
   (a) Write an expression, in terms of p, for the amount given to
      (i) Aimee,
      (ii) Lydia,
      (iii) all three girls.
   (b) The three girls are given a total of £25 a week altogether.
   By forming an equation find the weekly allowance given to each girl.

5. The cost of a pencil is x pence.
   The cost of a pen is 10 pence more than a pencil.
   (a) Write an expression, in terms of x, for the cost of a pen.
   (b) Write an expression, in terms of x, for the total cost of a pencil and two pens.
   (c) The total cost of a pencil and two pens is 65 pence.
   Form an equation in x and solve it to find the cost of a pencil.

6. A chocolate biscuit costs x pence.
   A cream biscuit costs 4 pence less than a chocolate biscuit.
   Jamie pays 77 pence for 3 chocolate biscuits and 2 cream biscuits.
   By forming an equation find the cost of a cream biscuit.

7. (a) Write down a simplified expression, in terms of x, for the perimeter of the triangle.

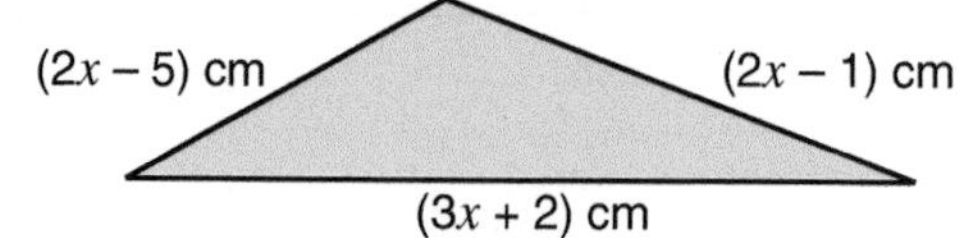

   (b) The perimeter is 59 cm.
   Write down an equation and solve it to find the value of x.
   (c) Use your answer to find the length of each side of the triangle.

8. The length of a rectangle is x cm. The width of the rectangle is 4 cm less than the length.
   (a) Write an expression, in terms of x, for the width of the rectangle.
   (b) Write an expression for the perimeter of the rectangle in terms of x.
   (c) The perimeter of the rectangle is 20 cm.
   By forming an equation find the value of x.

9. Geoffrey knows that the sum of the angles of this shape add up to 540°.
   (a) Write down an equation in x.
   (b) Use your equation to find the size of the largest angle.

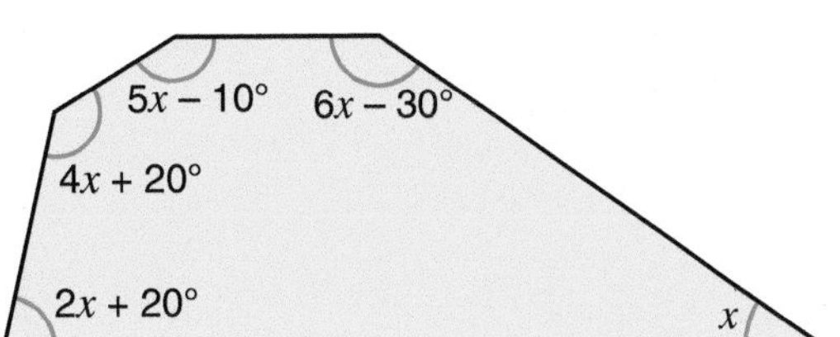

10. A drink costs x pence.
    A cake costs 7 pence more than a drink.
    The total cost of two drinks and a cake is 97 pence.
    Form an equation in x and solve it to find the cost of a cake.

11. The areas of these shapes are equal.

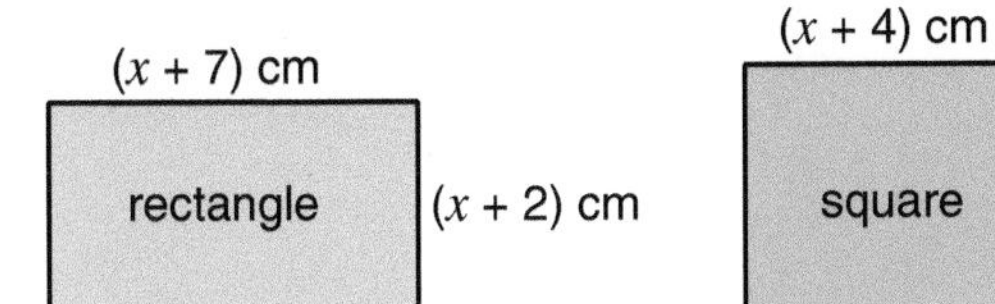

   (a) Form an equation and find the value of x.
   (b) Which shape has the greater perimeter?
   You must show your working.

12. Triangle ABC has sides of length $(x + 10)$ cm, $(x + 7)$ cm and 12 cm.
    Form an equation and find the value of x.

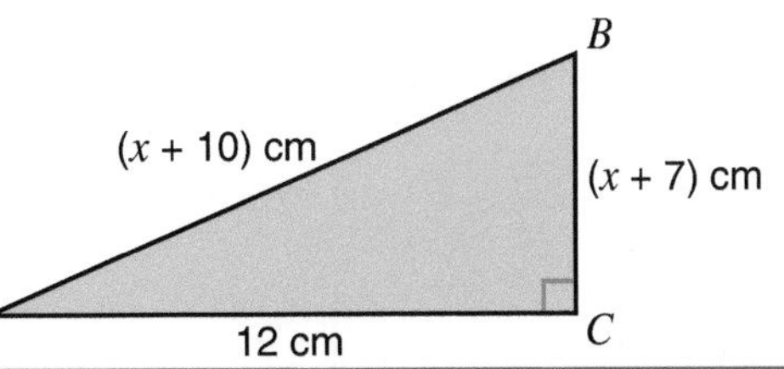

Inequalities

An **inequality** is a mathematical statement, such as $x > 1$, $a \leqslant 2$ or $-3 \leqslant n < 2$.
In the following, x is an integer.

Sign	Meaning	Example	Possible values of x
$<$	is less than	$x < 4$	3, 2, 1, 0, −1, −2, −3, ...
$\leqslant$	is less than or equal to	$x \leqslant 4$	4, 3, 2, 1, 0, −1, −2, −3, ...
$>$	is greater than	$x > 6$	7, 8, 9, 10, ...
$\geqslant$	is greater than or equal to	$x \geqslant 2$	2, 3, 4, 5, ...

An **integer** is a positive or negative whole number or zero.

Explain the difference between the meanings of the signs $<$ and $\leqslant$.
Explain the difference between the meanings of the signs $>$ and $\geqslant$.

Solving inequalities

Solve means to find the values of x which make the inequality true.
The aim is to end up with **one letter** on one side of the inequality and a **number** on the other side of the inequality.

Solving inequalities is similar to solving equations.

Example 15

Solve the inequality $5x - 3 < 27$.

$5x - 3 < 27$

Add 3 to both sides.

$5x < 30$

Divide both sides by 5.

$x < 6$

Example 16

Solve the inequality $3x - 1 > x + 7$.

$3x - 1 > x + 7$

Add 1 to both sides.

$3x > x + 8$

Subtract x from both sides.

$2x > 8$

Divide both sides by 2.

$x > 4$

Example 17

Solve the inequality $7a \geqslant a + 9$.

$7a \geqslant a + 9$

Subtract a from both sides.

$6a \geqslant 9$

Divide both sides by 6.

$a \geqslant 1.5$

This means that the inequality is true for all values of a which are equal to 1.5, or greater.

Substitute $a = 1.5$, $a = 2$ *and* $a = 1$ *into the original inequality.*

What do you notice?

Practice Exercise 10.8

1. Solve each of the following inequalities.
   (a) $3n > 6$ (b) $2x < -4$ (c) $a + 1 < 5$ (d) $a - 3 < 1$
   (e) $2d - 5 \leqslant 1$ (f) $t + 2 < -1$ (g) $5 + 2g > 1$ (h) $4 + 3y \geqslant 4$
2. Solve the following inequalities. Show your working clearly.
   (a) $a + 3 < 7$ (b) $5 + x \geqslant 3$ (c) $y + 2 < -1$ (d) $3c > 15$
   (e) $2d < -6$ (f) $b - 3 \geqslant -2$ (g) $-2 + b \leqslant -1$ (h) $2c + 5 \leqslant 11$
   (i) $3d - 4 > 8$ (j) $4 + 3f < -2$ (k) $8g - 1 \leqslant 3$ (l) $5h < h + 8$
   (m) $3x < x - 6$ (n) $6j \geqslant 2j + 10$ (o) $7k > 3k - 16$ (p) $6m - 7 \leqslant m$
3. Solve the following inequalities.
   (a) $4x - 1 \geqslant 2x + 9$ (b) $3p - 2 > 6 + 2p$ (c) $3b + 5 \geqslant 10b - 3$
   (d) $4m + 2 > 2m - 11$ (e) $7n - 3 \leqslant 13 - n$ (f) $2t - 10 > t + 3$
   (g) $7a - 2 < 4a + 5$ (h) $5 + 5x \leqslant 2 - 4x$ (i) $8 + 2x > 3(4 - x)$

Multiplying (or dividing) an inequality by a negative number

Activity

$-2 < 3$

Multiply both sides by -1.

$-2 \times (-1) = 2$ and $3 \times (-1) = -3$

$2 > -3$

To keep the statement true we have to reverse the inequality sign.

Multiply both sides of these inequalities by -1.

1. $3 > 2$ 2. $3 > -2$ 3. $-3 < -1$
4. $5 \geqslant 4$ 5. $-4 \leqslant 5$ 6. $-4 \geqslant -5$

The same rules for equations can be applied to inequalities, with one exception: When you **multiply** (or **divide**) both sides of an inequality by a negative number the inequality is reversed.

Example 18

Solve $-3x < 6$.

$-3x < 6$

Divide both sides by -3.

Because we are dividing by a negative number the inequality is reversed.

$x > -2$

Example 19

Solve $3a - 2 \geqslant 5a - 9$.

$3a - 2 \geqslant 5a - 9$

Subtract $5a$ from both sides.

$-2a - 2 \geqslant -9$

Add 2 to both sides.

$-2a \geqslant -7$

Divide both sides by -2.

$a \leqslant 3.5$

Practice Exercise 10.9

Solve the following inequalities. Show your working clearly.

1. $-4a > 8$
2. $-5b \leqslant -15$
3. $-3c \geqslant 12$
4. $3 - 2d < 5$
5. $14 - 3e \leqslant 4e$
6. $-5f > 4f - 9$
7. $4g < 7g + 12$
8. $5 - 3h \leqslant h - 3$
9. $-5 - j \geqslant 12j - 18$
10. $3 - 5k < 2(3 + 2k)$
11. $3(m - 2) > 5m$
12. $3(2n - 1) < 8n + 5$
13. $3p \geqslant 5 - 6p$
14. $2(q - 3) < 5 + 7q$
15. $n - 5 > 2(n - 7)$

Key Points

- The solution of an equation is the value of the unknown letter that fits the equation.
- You should be able to use the **balance method** to solve equations.
  If you do the same to both sides of an equation, it is still true.
  Doing the same to both sides means:
  - **Adding** the **same number** to both sides.
  - **Subtracting** the **same number** from both sides.
  - **Dividing** both sides by the **same number**.
  - **Multiplying** both sides by the **same number**.
- You should be able to solve equations:
  - with unknowns on both sides of the equals signs,
  - which include brackets, which involve fractions.
- **Inequalities** can be described using words or numbers and symbols.

Sign	Meaning
$<$	is less than
$\leqslant$	is less than or equal to

Sign	Meaning
$>$	is greater than
$\geqslant$	is greater than or equal to

- **Solving inequalities** means finding the values of x which make the inequality true.

The same rules for equations can be applied to inequalities, with one exception:
When you **multiply** (or **divide**) both sides of an inequality by a negative number the inequality is reversed. For example, if $-3x < 6$ then $x > -2$.

Review Exercise 10

1. Georgina makes up this number game.

   Think of an odd number.
   Subtract 5.
   Halve the result.
   Tell me your answer.

   (a) Naomi thinks of the number 11.
   What is her answer?
   (b) Jacob says his answer is 9.
   What number did he start with?
   (c) Alfie says his answer is 4.5.
   Georgina tells him he has made a mistake. Explain how she knows this.

2. Solve.
   (a) $\frac{x}{4} = 3$ (b) $5y = 20$ (c) $3y + 2 = 11$ (d) $2y + 5 = 2 - y$

3. (a) Hilda is twice as old as Evie. Their ages add up to 51 years.
   How old is Hilda?
   (b) Colin is 2 years older than John. Their ages add up to 36 years.
   How old is Colin?

4. A bag contains x yellow balls, $(2x + 1)$ red balls and $(3x + 2)$ blue balls.
   (a) Write an expression, in terms of x, for the total number of balls in the bag.
   (b) The bag contains 45 balls. How many yellow balls are in the bag?

5. The diagram shows two cans of oil.
   The cans hold a total of 3 litres of oil.
   By forming an equation find the amount of oil in the larger can.

6. A drink costs x pence.
   A packet of crisps costs 15 pence **less** than a drink.
   (a) Write an expression, in terms of x, for the cost of a packet of crisps.
   (b) A drink and two packets of crisps cost 96 pence.
   By forming an equation find the cost of a drink.

7. The areas of these rectangles are equal.

(2x – 1) cm
3 cm
A

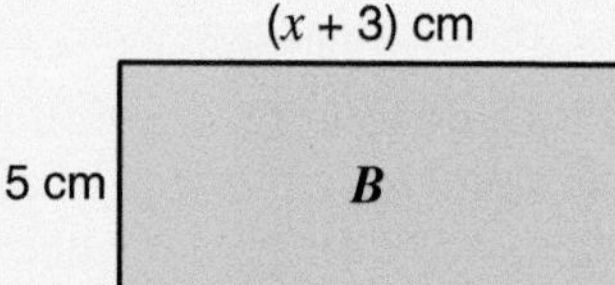

   By forming and solving an equation in x, find the area of A.

8. Robert has x computer games.
   Sarah has 5 more games than Robert.
   Tamzin has twice as many computer games as Sarah.
   (a) Write down an expression, in x, for the number of computer games that Tamzin has.

   Robert, Sarah and Tamzin have a total of 39 computer games.
   (b) Form an equation and solve it to find how many computer games Sarah has.

9.

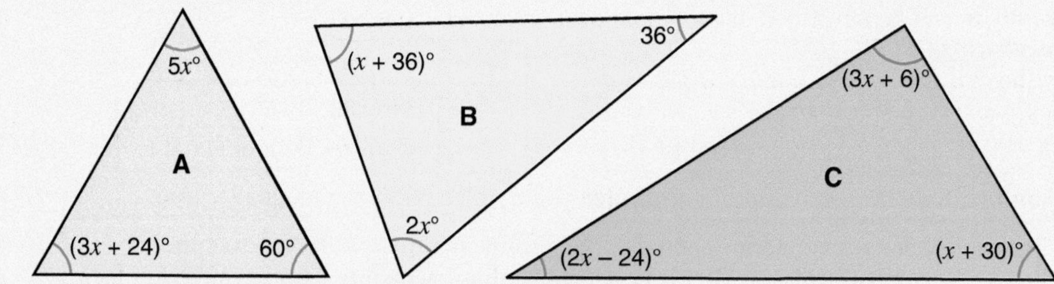

   For each triangle, write down an equation involving x.
   Solve each equation and find the sizes of the angles in each triangle.
   Also, say what type of triangle each one is.

10. A trader sells sweatshirts and hoodies.
    He sells x sweatshirts and y hoodies.
    Write mathematical statements for the following.
    (a) The trader expects to sell over 200 articles altogether.
    (b) The trader expects to sell more than five times as many hoodies as sweatshirts.
    (c) The trader expects to sell at least 70 sweatshirts.

11. Solve.
    (a) $3(a - 5) = 6a$ (b) $5(x + 2) = 14$ (c) $x + 8 = 3(2 - x)$

12. (a) Write down an expression, in terms of x, for the perimeter of the rectangle.

(b) The perimeter of the rectangle is equal to the perimeter of the square. Form an equation and find the value of x.

(c) What is the perimeter of the rectangle, in centimetres?

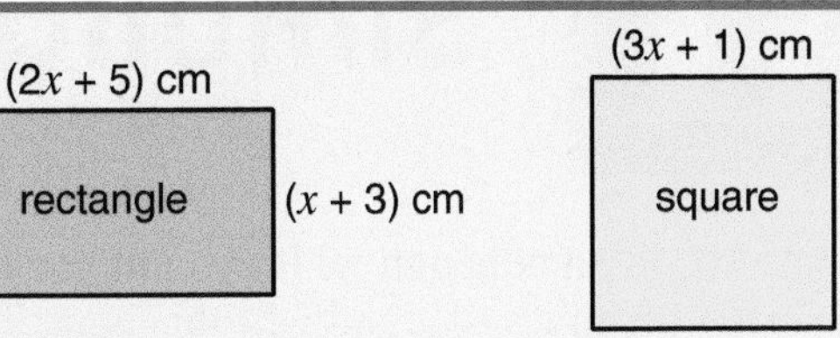

13. Solve these equations.

(a) $\frac{x}{2} + 4 = -3$ (b) $\frac{y-2}{5} = 4$ (c) $\frac{2x+1}{3} = \frac{5x+1}{7}$

14. Two painters estimate the price they will charge for painting a building.

The first painter, Mr Archer, says he will charge £400, and in addition, £5 for each hour the job takes.

The second painter, Mr Barton, says he will charge £13 for each hour the job takes.

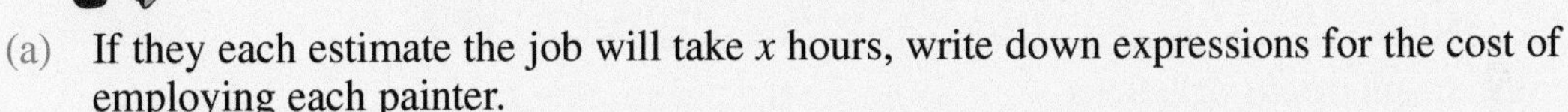

(a) If they each estimate the job will take x hours, write down expressions for the cost of employing each painter.

(b) The customer decides that Mr Barton will do the job for less money than Mr Archer. Write down a statement expressing this.

(c) Simplify your statement to find a range of values of x, for which it is cheaper to employ Mr Barton.

(d) Which of the painters will it be cheaper to employ if the job takes 60 hours?

15. Solve these inequalities.

(a) $6(x - 7) \leqslant 6$ (b) $x - 1 > 2x + 5$ (c) $5 - x \leqslant 6 - 3x$

(d) $5(x + 1) \leqslant x + 8$ (e) $8(x - 4) < 5(x - 7)$ (f) $2(x + 3) > 3(2 - x)$

16. Rachel and Jack are care workers.
At the beginning of each week they are given a list of new patients to visit.
Rachel, a senior care worker, is given twice as many new patients to visit than Jack, who has recently qualified.
By the end of Monday, Rachel and Jack have each visited 6 new patients.
Jack was given x new patients to visit at the beginning of the week.

(a) Write an expression for the number of new patients that Rachel was given at the beginning of the week.

(b) Write expressions for the numbers of new patients that Rachel and Jack have still to visit after finishing work for the day on Monday.

(c) Rachel commented that, when she starts work on Tuesday, she now has three times as many new patients to visit than Jack.
Form an equation in x and solve it to find the number of new patients given to Rachel and Jack at the beginning of the week.

A coach firm has to transport 300 guests to a wedding.
The company has 5 coaches and 7 minibuses.
A coach can carry a maximum of 50 passengers and a minibus can carry a maximum of 20 passengers.
There are 10 drivers available and they are qualified to drive either type of vehicle.
How many coaches and how many minibuses should the manager allocate to transport the guests to the wedding so as to use the least number of drivers?

11 Simultaneous Equations

$x + y = 10$ is an equation with two unknown quantities x and y.
Many pairs of values of x and y fit this equation.

For example.
$x = 1$ and $y = 9$, $x = 4$ and $y = 6$, $x = 2.9$ and $y = 7.1$, $x = 1.005$ and $y = 8.995$, ...
$x - y = 2$ is another equation with the **same** two unknown quantities x and y.
Again, many pairs of values of x and y fit this equation.

For example.
$x = 4$ and $y = 2$, $x = 7$ and $y = 5$, $x = 2.9$ and $y = 0.9$, $x = -1$ and $y = -3$, ...

There is only **one** pair of values of x and y which fit **both** of these equations ($x = 6$ and $y = 4$).
Pairs of equations like $x + y = 10$ and $x - y = 2$ are called **simultaneous equations**.

To solve simultaneous equations you need to find values which fit **both** equations simultaneously.
Simultaneous equations can be solved using different methods.

Using graphs to solve simultaneous equations

Consider the simultaneous equations $x + 2y = 5$ and $x - 2y = 1$.

Draw the graphs of $x + 2y = 5$ and $x - 2y = 1$.

For $x + 2y = 5$:
When $x = 1$, $y = 2$.
This gives the point (1, 2).

When $x = 5$, $y = 0$.
This gives the point (5, 0).

To draw the graph of $x + 2y = 5$ draw a line through the points (1, 2) and (5, 0).

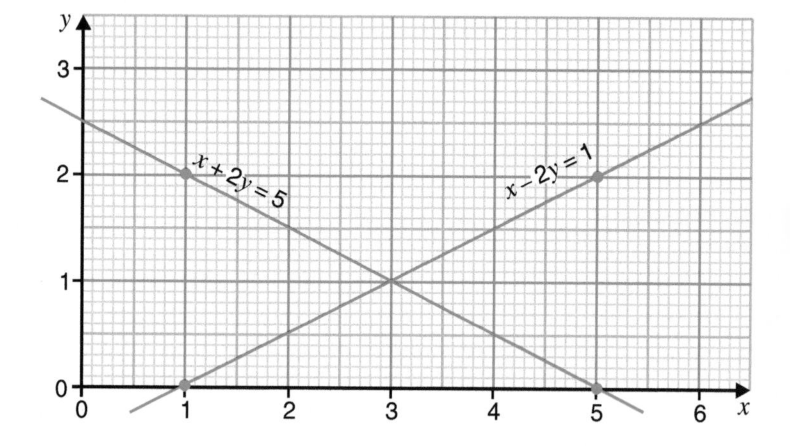

For $x - 2y = 1$:
When $x = 1$, $y = 0$.
This gives the point (1, 0).

When $x = 5$, $y = 2$.
This gives the point (5, 2).

To draw the graph of $x - 2y = 1$ draw a line through the points (1, 0) and (5, 2).

The values of x and y at the point where the lines cross give the solution to the simultaneous equations.

The lines cross at the point (3, 1).

This gives the solution $x = 3$ and $y = 1$.

To solve a pair of simultaneous equations, plot the graph of each of the equations on the same diagram.
The coordinates of the point where the two lines cross:
- fit **both equations** simultaneously,
- give the **graphical solution** of the equations.

Example 1

Use a graphical method to solve this pair of simultaneous equations: $5x + 2y = 20$
$y = 2x + 1$

Find the points that fit the equations $5x + 2y = 20$ and $y = 2x + 1$.

For $5x + 2y = 20$:

When $x = 0$, $y = 10$.
This gives the point (0, 10).

When $y = 0$, $x = 4$.
This gives the point (4, 0).

Draw a line through the points (0, 10) and (4, 0).

For $y = 2x + 1$:

When $x = 0$, $y = 1$.
This gives the point (0, 1).

When $x = 4$, $y = 9$.
This gives the point (4, 9).

Draw a line through the points (0, 1) and (4, 9).

The lines cross at the point (2, 5).
This gives the solution $x = 2$ and $y = 5$.

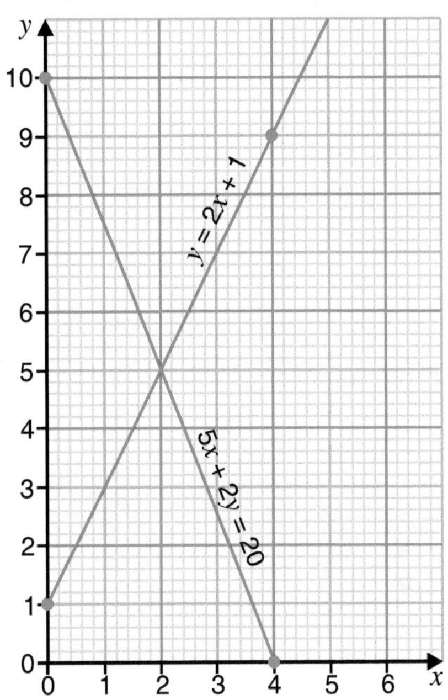

Check:
You can check a graphical solution by substituting the values of x and y into the original equations.

When $x = 2$ and $y = 5$

$5x + 2y = 5 \times 2 + 2 \times 5 = 20$

$y = 2x + 1 = 2 \times 2 + 1 = 5$

Practice Exercise 11.1

Use a graphical method to solve each of these pairs of simultaneous equations.
For each question use the sizes of axes given.

For example, $0 \leqslant x \leqslant 8$ means draw and label the x axis from 0 to 8 inclusive.

1. $x + y = 6$
   $y = x - 2$
   Axes $0 \leqslant x \leqslant 8$, $-3 \leqslant y \leqslant 7$

2. $x + y = 8$
   $y - x = 2$
   Axes $0 \leqslant x \leqslant 10$, $-3 \leqslant y \leqslant 10$

3. $x + 2y = 8$
   $2x + y = 7$
   Axes $0 \leqslant x \leqslant 10$, $0 \leqslant y \leqslant 8$

4. $3x + 2y = 12$
   $y = x + 1$
   Axes $0 \leqslant x \leqslant 5$, $0 \leqslant y \leqslant 8$

5. $x + 3y = 6$
   $y = 2x - 5$
   Axes $0 \leqslant x \leqslant 10$, $-6 \leqslant y \leqslant 4$

6. $3x + 4y = 24$
   $2y = x + 2$
   Axes $-4 \leqslant x \leqslant 10$, $-2 \leqslant y \leqslant 8$

7. $2x + y = -1$
   $x + 2y = 4$
   Axes $-3 \leqslant x \leqslant 5$, $-2 \leqslant y \leqslant 4$

8. $x + 2y = 6$
   $x - 2y = 4$
   Axes $0 \leqslant x \leqslant 6$, $-2 \leqslant y \leqslant 3$

Simultaneous equations with no solution

Some pairs of simultaneous equations do not have a solution.

Example 2

Show that this pair of simultaneous equations do not have a solution.

$y - 2x = 4$

$2y = 4x - 1$

Method 1

Draw the graph of each equation.

$y - 2x = 4$

When $x = 0$, $y = 4$.

When $y = 0$, $x = -2$.

Plot and draw a line through the points $(0, 4)$ and $(-2, 0)$.

$2y = 4x - 1$

When $x = 0$, $y = -0.5$.

When $x = 2$, $y = 3.5$.

Plot and draw a line through the points $(0, -0.5)$ and $(2, 3.5)$.

The two lines are **parallel**.

This means they never cross and there are no values of x and y which fit both equations.

So, the simultaneous equations have no solution.

Method 2

Rearrange each equation to the form $y = mx + c$.

$y - 2x = 4$

Add $2x$ to both sides.

$y = 2x + 4$

The graph of this equation has a gradient (m) of 2 and a y-intercept (c) of 4.

$2y = 4x - 1$

Divide both sides by 2.

$y = 2x - 0.5$

The graph of this equation has a gradient (m) of 2 and a y-intercept (c) of -0.5.

Both lines have the same gradient (2) and different y-intercepts which shows that the lines are parallel.

Practice Exercise 11.2

1. Draw graphs to show that each of these pairs of simultaneous equations have no solution.

   (a) $x + y = 6$, $y = 2 - x$ (b) $y - 4x = 8$, $y = 4x + 2$ (c) $3x + 4y = 12$, $8y = 24 - 6x$ (d) $5y - 2x = 10$, $5y = 2x + 20$

2. By rearranging each of these pairs of simultaneous equations to the form $y = mx + c$ show that they do not have a solution.

   (a) $2x + y = 6$, $y = 3 - 2x$ (b) $2y - 4x = 7$, $y - 2 = 2x$ (c) $5x - 2y = 8$, $4y = 10x + 7$ (d) $4y + 12x = 5$, $1 - 2y = 6x$

3. Two of these pairs of simultaneous equations have no solution.

   (a) $5x + y = 6$, $y - 5x = 2$ (b) $5x = 8 + y$, $y - 5x = 2$ (c) $2y - 10x = 5$, $4y + 20x = 5$ (d) $y + 1 = -5x$, $2y + 10x = 5$

   Use an appropriate method to find which ones.
   Use a graphical method to solve the other two.

Using the elimination method to solve simultaneous equations

The graphical method of solving simultaneous equations can be quite time consuming.
Sometimes, due to the equations involved, the coordinates of the points where the lines intersect can be difficult to read accurately.
For these reasons, other methods of solving simultaneous equations are often used.

Consider again the simultaneous equations $x + 2y = 5$ and $x - 2y = 1$.
Both equations have the same number of x's and the same number of y's.
If the two equations are added together the y's will be **eliminated** as shown.

$$\begin{aligned} x + 2y &= 5 \\ x - 2y &= 1 \\ \text{Adding gives} \quad 2x \quad &= 6 \\ \text{So,} \quad x \quad &= 3 \end{aligned}$$

Remember: $+2y + -2y = 2y - 2y = 0$

By **substituting** the value of this letter (x) into one of the original equations we can find the value of the other letter (y).

$$\begin{aligned} x + 2y &= 5 \\ 3 + 2y &= 5 \\ 2y &= 2 \\ y &= 1 \end{aligned}$$

Remember: If you do the same to both sides of an equation it is still true.

This gives the solution $x = 3$ and $y = 1$.

Example 3

Use the elimination method to solve this pair of simultaneous equations: $2x - y = 1$
$3x + y = 9$

Each equation has the **same number** of y's but the **signs** are **different**.
To eliminate the y's the equations must be **added**.

$5x = 10$
$x = 2$

Substitute $x = 2$ into $3x + y = 9$.
$3 \times 2 + y = 9$
$6 + y = 9$
$y = 3$

The solution is $x = 2$ and $y = 3$.

Check:
Substitute $x = 2$ and $y = 3$ into $2x - y = 1$.
$2 \times 2 - 3 = 1$
$4 - 3 = 1$
$1 = 1$
The equation is true, so, the solution
$x = 2$ and $y = 3$ is correct.

Example 4

Use the elimination method to solve this pair of simultaneous equations: $2x + 3y = 9$
$2x + y = 7$

Each equation has the **same number** of x's and the **signs** are the **same**.
To eliminate the x's one equation must be **subtracted** from the other.

Subtract $2x + y = 7$ from $2x + 3y = 9$.
$2y = 2$
$y = 1$

Substitute $y = 1$ into $2x + y = 7$.
$2x + 1 = 7$
$2x = 6$
$x = 3$

The solution is $x = 3$ and $y = 1$.

Check:
Substitute $x = 3$ and $y = 1$ into $2x + 3y = 9$.
Do this and make sure the solution is correct.

Practice Exercise 11.3

Use the elimination method to solve each of these pairs of simultaneous equations.

1. $3x - y = 1$, $x + y = 3$
2. $2x - y = 2$, $x + y = 7$
3. $4x + y = 9$, $2x - y = 3$
4. $-x + 2y = 13$, $x + y = 8$
5. $2x + y = 7$, $x + y = 4$
6. $3x + y = 9$, $2x + y = 7$
7. $2x + y = 12$, $x + y = 7$
8. $x + 5y = 14$, $x + 2y = 8$
9. $x + 2y = 13$, $x + 4y = 21$
10. $x + 4y = 11$, $x + y = 5$
11. $2x + 5y = 13$, $2x + y = 9$
12. $5x + 3y = 26$, $2x + 3y = 14$
13. $5x + 4y = 22$, $5x + y = 13$
14. $2x - y = 10$, $3x + y = 10$
15. $5x - 2y = 13$, $3x + 2y = 3$
16. $x + 5y = 14$, $-x + 2y = 7$
17. $2x + 3y = 8$, $2x + y = -4$
18. $2x + y = 4$, $4x - y = 11$
19. $3x + 4y = -8$, $x + 4y = 4$
20. $3x + 2y = 6$, $x - 2y = 6$
21. $x - 3y = 8$, $x + 2y = -7$
22. $2x - 2y = 9$, $4x - 2y = 16$
23. $3x - y = 5$, $3x + y = 4$
24. $5x - 3y = 5$, $5x + y = -5$

Further use of the elimination method

Look at this pair of simultaneous equations:

$$5x + 2y = 11$$
$$3x - 4y = 4$$

A useful technique is to use capital letters to label the equations.

$5x + 2y = 11$ Equation A

$3x - 4y = 4$ Equation B

These equations do not have the same number of x's or the same number of y's.

To make the number of y's the same we can multiply equation A by 2.

A × 2 gives $10x + 4y = 22$ Equation C

B × 1 gives $3x - 4y = 4$ Equation D

C + D gives $13x = 26$

$x = 2$

The number of y's in equations C and D is the **same** but the **signs** are **different**.
To eliminate the y's the equations must be **added**.

Substitute $x = 2$ into $5x + 2y = 11$.

$$5 \times 2 + 2y = 11$$
$$10 + 2y = 11$$
$$2y = 1$$
$$y = 0.5$$

In this example eliminating the y's rather than the x's is less likely to produce an error.
Try to solve the equations by eliminating the x's.

The solution is $x = 2$ and $y = 0.5$.

Check:
Substitute $x = 2$ and $y = 0.5$ into $3x - 4y = 4$.

$$3 \times 2 - 4 \times 0.5 = 4$$
$$6 - 2 = 4$$
$$4 = 4$$

The equation is true, so the solution $x = 2$ and $y = 0.5$ is correct.

Example 5

Solve this pair of simultaneous equations:

$$3x + 7y = -2$$
$$4x + 9 = -3y$$

> Both equations must be in the form $px + qy = r$ before the elimination method can be used.
> You may have to **rearrange** the equations you are given.
> $4x + 9 = -3y$ can be rearranged as $4x + 3y = -9$.

Rearrange and label the equations.

$3x + 7y = -2$ A

$4x + 3y = -9$ B

These equations do not have the same number of x's or the same number of y's.
So, the multiplying method can be used.

Method 1

Eliminating the x's.

A × 4 gives $12x + 28y = -8$ C

B × 3 gives $12x + 9y = -27$ D

C − D gives
$$19y = -8 - -27$$
$$19y = -8 + 27$$
$$19y = 19$$
$$y = 1$$

Substitute $y = 1$ into $3x + 7y = -2$.
$$3x + 7 \times 1 = -2$$
$$3x + 7 = -2$$
$$3x = -9$$
$$x = -3$$

The solution is $x = -3$ and $y = 1$.

Method 2

Eliminating the y's.

A × 3 gives $9x + 21y = -6$ C

B × 7 gives $28x + 21y = -63$ D

D − C gives
$$19x = -63 - -6$$
$$19x = -63 + 6$$
$$19x = -57$$
$$x = -3$$

Substitute $x = -3$ into $3x + 7y = -2$.
$$3 \times -3 + 7y = -2$$
$$-9 + 7y = -2$$
$$7y = 7$$
$$y = 1$$

Check the solution by substituting $x = -3$ *and* $y = 1$ *into* $4x + 9 = -3y$.

Practice Exercise 11.4

Solve each of these pairs of simultaneous equations.

1. $3x + 2y = 8$, $2x - y = 3$
2. $x + y = 5$, $5x - 3y = 1$
3. $2x + 3y = 9$, $x + 4y = 7$
4. $x + 3y = 10$, $2x + 5y = 18$
5. $5x + 2y = 8$, $2x - y = 5$
6. $3x + y = 9$, $x - 2y = 10$
7. $3x - 4y = 10$, $x + 2y = 5$
8. $x + 6y = 0$, $3x - 2y = -10$
9. $2x + 3y = 11$, $3x + y = 13$
10. $2x + y = 10$, $-x + 2y = 9$
11. $2x + 3y = 9$, $4x - y = 4$
12. $2x + 3y = 8$, $3x + 2y = 7$
13. $3x + 4y = 23$, $2x + 5y = 20$
14. $2x - 3y = 8$, $x - 5y = 11$
15. $3x + 4y = 5$, $-2x + 5y = 12$
16. $3x - 2y = 4$, $x + 4y = 6$
17. $-3x + 2y = 5$, $4x + 3y = -1$
18. $3x + 4y = 6$, $3y = 7 - x$
19. $5x + 3y = 16$, $2y = 13 - x$
20. $5x - 4y = 24$, $2x = y + 9$
21. $2x + 3y = 14$, $8x - 5y = 5$
22. $4x - 7y = 15$, $5x - 12 = 2y$
23. $8x + 3y = 2$, $5x = 1 - 2y$
24. $9x = 4y - 20$, $5x = 6y - 13$

Using the substitution method to solve simultaneous equations

For some pairs of simultaneous equations a method using **substitution** is sometimes more convenient.

Example 6

Solve this pair of simultaneous equations: $5x + y = 9$
$y = 4x$

$5x + y = 9$ Equation A
$y = 4x$ Equation B

Substitute $y = 4x$ into Equation A

$5x + 4x = 9$
$9x = 9$
$x = 1$

Substitute $x = 1$ into $y = 4x$.

$y = 4 \times 1$
$y = 4$

The solution is $x = 1$ and $y = 4$.

Check the solution by substituting $x = 1$ *and* $y = 4$ *into* $5x + y = 9$.

Practice Exercise 11.5

Use the substitution method to solve these pairs of simultaneous equations.

1. $2x + y = 10$, $y = 3x$
2. $3x - y = 9$, $y = 2x$
3. $x + 5y = 18$, $x = 4y$
4. $x + 2y = 15$, $y = 2x$
5. $2x + y = 17$, $y = 6x + 1$
6. $3x + 2y = 4$, $x = y - 2$
7. $5x + 6y = 34$, $y = x + 2$
8. $5x - 2y = 23$, $x = y + 1$
9. $5x - y = 12$, $y = 32 - 6x$
10. $x + 5y = 13$, $x = 3y + 9$
11. $5x - 3y = 26$, $y = 2x + 14$
12. $x + 4y = 32$, $x = 2y - 4$

Solving problems using simultaneous equations

Example 7

Billy buys 5 first class stamps and 3 second class stamps at a cost of £2.13.
Jane buys 3 first class stamps and 5 second class stamps at a cost of £1.95.
Calculate the cost of a first class stamp and the cost of a second class stamp.

Let x pence be the cost of a first class stamp, and let y pence be the cost of a second class stamp.

Billy's purchase of the stamps gives this equation. $5x + 3y = 213$ Equation A
Jane's purchase of the stamps gives this equation. $3x + 5y = 195$ Equation B

This gives a pair of simultaneous equations which can be solved using the elimination method.

	$5x + 3y = 213$	A
	$3x + 5y = 195$	B
A × 5 gives	$25x + 15y = 1065$	C
B × 3 gives	$9x + 15y = 585$	D
C − D gives	$16x = 480$	
	$x = 30$	

Substitute $x = 30$ into $5x + 3y = 213$.

$5 \times 30 + 3y = 213$
$3y = 63$
$y = 21$

So, the cost of a first class stamp is 30 pence and the cost of a second class stamp is 21 pence.

Check the solution by substituting the values for x and y into the original problem.

Practice Exercise 11.6

1. Pencils cost x pence each and pens cost y pence each.
   Pam buys 6 pencils and 3 pens for 93 pence.
   Ray buys 2 pencils and 5 pens for 91 pence.
   (a) Write down two equations connecting x and y.
   (b) By solving these simultaneous equations find the cost of a pencil and the cost of a pen.

2. Apples are x pence per kg.
   Oranges are y pence each.
   5 kg of apples and 30 oranges cost £9.00.
   10 kg of apples and 15 oranges cost £12.60.
   (a) Write down two equations connecting x and y.
   (b) By solving these simultaneous equations find the cost of a kilogram of apples and the cost of an orange.

3. Standard eggs cost x pence per dozen.
   Small eggs cost y pence per dozen.
   10 dozen standard eggs and 5 dozen small eggs cost £13.60.
   5 dozen standard eggs and 8 dozen small eggs cost £11.31.
   By forming two simultaneous equations find the values of x and y.

4. A group of children and adults went on a coach trip to a theme park.
   Ticket prices for the theme park were £20 for adults and £15 for children.
   Ticket prices for the coach were £10 for adults and £6 for children.
   The total cost of the tickets for the theme park was £560.
   The total cost of the coach tickets was £232.
   How many children and adults went on the trip?

5. Jenny types at x words per minute.
   Stuart types at y words per minute.
   When Jenny and Stuart both type for 1 minute they type a total of 170 words.
   When Jenny types for 5 minutes and Stuart types for 3 minutes they type a total of 710 words.
   Calculate x and y.

6. At a café, John buys 3 coffees and 2 teas for £4.30 and Susan buys 2 coffees and 3 teas for £4.20.
   Calculate the price of a coffee and the price of a tea.

7. Standard coaches hold x passengers and first class coaches hold y passengers.
   A train with 5 standard coaches and 2 first class coaches carries a total of 1040 passengers.
   A train with 7 standard coaches and 3 first class coaches carries a total of 1480 passengers.
   By forming two simultaneous equations find the values of x and y.

8. Nick writes a number in each of these boxes: $x = \square$ $y = \square$
   When he adds 5 to x the result is 2 times y.
   When he adds 17 to y the result is 2 times x.
   Find the two numbers.

9. In a game you score p points if you win and q points if you lose.
   Matt plays. He wins 2 games and loses 3 games. He scores 4 points.
   Kath plays. She wins 3 games and loses 2 games. She scores 11 points.
   By forming two simultaneous equations find the values of p and q.

10. A local pop band hold two gigs.
    Tickets for the gigs cost £x and CDs cost £y.
    At the first gig they sell 92 tickets and 8 CDs for a total of £1186.
    At the second gig they sell 104 tickets and 16 CDs for a total of £1372.
    Find the cost of a ticket and the cost of a CD.

Key Points

- A pair of **simultaneous equations** are linked equations with the same unknown letters in each equation.
- To solve a pair of simultaneous equations find values for the unknown letters that fit **both** equations.
- Simultaneous equations can be solved either **graphically** or **algebraically**.
- Solving simultaneous equations **graphically** involves:
  drawing the graphs of both equations, finding the point where the graphs cross.
  When the graphs of the equations are parallel, the equations have no solution.
- Solving simultaneous equations **algebraically** involves using either:
  the **elimination** method or the **substitution** method.

Review Exercise 11

1. Use a graphical method to solve each of these simultaneous equations.
   For each question use the size of axes given.
   (a) $x + y = 10$
   $y = 2x + 1$
   Axes $0 \leqslant x \leqslant 11, \quad 0 \leqslant y \leqslant 11$
   (b) $5x + 6y = 30$
   $2y = x - 2$
   Axes $0 \leqslant x \leqslant 8, \quad -3 \leqslant y \leqslant 8$

2. This graph shows the line $y + 2x = -4$.
   Copy the graph.

   By drawing another line, use the graph to solve the simultaneous equations:
   $y + 2x = -4$
   $y - 3x = 1$

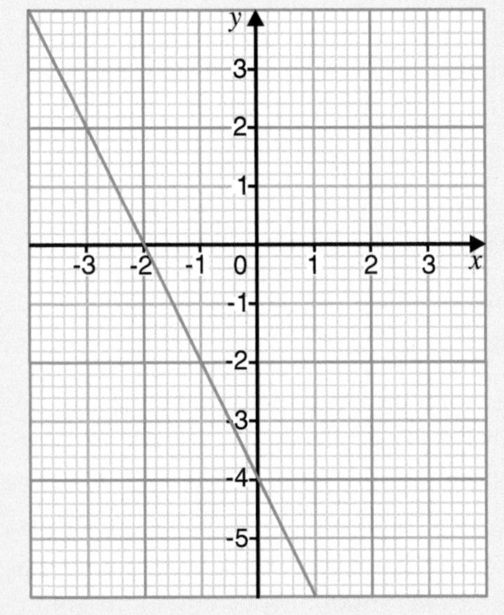

3. (a) Show that the simultaneous equations $y = 2x - 2$ and $2y - 4x = 3$ have no solution.
   (b) Show that the simultaneous equations $4y = x + 1$ and $8y - 2x = 3$ have no solution.
   (c) The simultaneous equations $y = 3x + 2$ and $y = ax + b$ have no solution.
   What can you say about the values of a and b?
   (d) The simultaneous equations $y = 3x + 2$ and $py + qx = r$ have no solution.
   Find some possible values for p, q and r.

4. Use an algebraic method to solve these simultaneous equations: $2x + y = 8$
   $x + y = 5$

5. Solve each of these pairs of simultaneous equations.
   (a) $2x + 3y = 11$
   $2x + y = 5$
   (b) $4x + y = 6$
   $2x - y = 6$
   (c) $x - y = -3$
   $x + 4y = 7$
   (d) $x + 2y = 9$
   $5x - 2y = 3$

6. By substituting $x = 3y$ into $4x + 3y = 30$, find values of x and y that are solutions to the simultaneous equations $x = 3y$ and $4x + 3y = 30$.

7. Peaches cost x pence each and oranges cost y pence each.
4 peaches and one orange cost 58p.
6 peaches and 2 oranges cost 92p.
(a) Write down two equations connecting x and y.
(b) Solve these simultaneous equations to find the values of x and y.

8. In the diagram line BD is parallel to line CE.

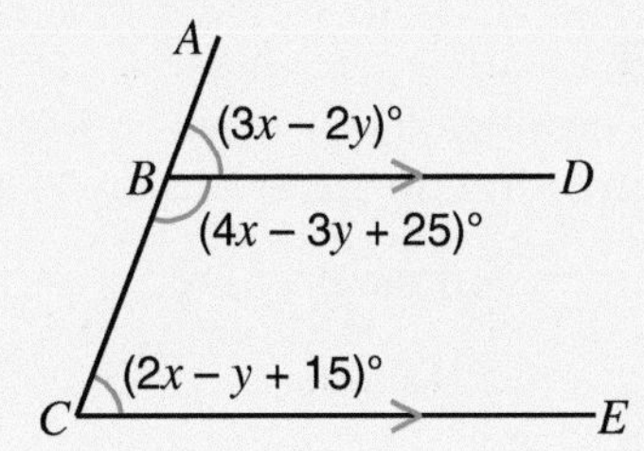

(a) Write down two equations connecting x and y.
Simplify your equations, where possible.
(b) Solve your equations simultaneously to find the values of x and y.
(c) Find the sizes of $\angle ABD$, $\angle CBD$ and $\angle BCE$.

9. Patrick buys a bunch of 10 roses and 5 carnations.
He pays £3.50.
At the same stall, Kate buys a bunch of 8 roses and 10 carnations.
She pays £4.
(a) Find the cost of 1 rose and 1 carnation.
(b) What would you expect to pay for 7 roses and 6 carnations?

10. The diagram shows an equilateral triangle.

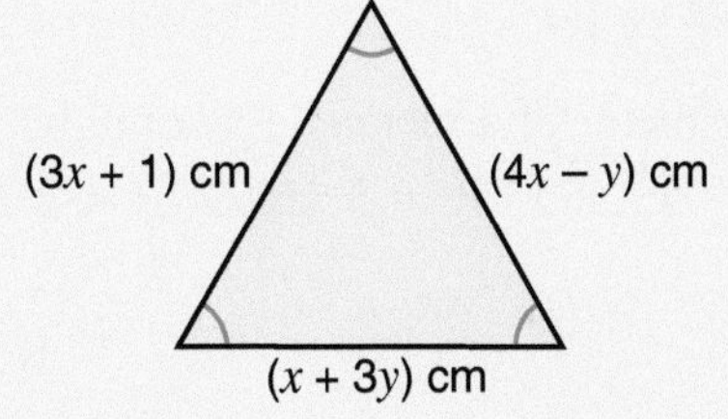

(a) Write down two equations connecting x and y.
Simplify your equations, where possible.
(b) Solve your equations simultaneously to find the values of x and y.
(c) Using your answer to part (b), find the perimeter of the triangle.

11. 16 people entered a mountain bike competition in which pairs of riders raced against each other.
Tim raced against each person in the competition, winning x races and losing y races.
(a) Write down an equation in x and y to illustrate this information.
(b) Tim scored 5 points for each race he won and 2 points for each race he lost.
He scored a total of 57 points.
Write down another equation in x and y to illustrate this information.
(c) How many races did Tim win?

12. A haulage company earns £12.50 for every mile a lorry carries a load.
It loses £5.50 for every mile a lorry travels without a load.
In one week, a lorry covered 1850 miles and earned the company £22 225.
(a) If x is the number of miles the lorry travelled carrying a load and y is the number of miles the lorry travelled empty, write an equation connecting x and y.
(b) Write an equation to show the amount of money earned from the lorry in a week, and show that it can be written as $25x - 11y = 44\,450$.
(c) Find the number of miles the lorry travelled carrying a load during the week.

12 Working with Formulae

Expressions and formulae

Most people at some time make use of **formulae** to carry out routine calculations.
A **formula** represents a rule written using numbers, letters and mathematical signs.

Example 1

A hedge is l metres long. A fence is 50 metres longer than the hedge.
Write an **expression**, in terms of l, for the length of the fence.

The fence is $(l + 50)$ metres long.

An **expression** is just an answer using letters and numbers.
A **formula** is an algebraic rule. It always has an equals sign.

Example 2

Boxes of matches each contain 48 matches.
Write down a **formula** for the number of matches, m, in n boxes.

$m = 48 \times n$ This could be written as $m = 48n$.

Using formulae

The formula for the perimeter of a rectangle is $P = 2L + 2W$.

By **substituting** values for the length, L, and the width, W, you can calculate the value of P.

For example.
To find the perimeter of a rectangle 5 cm in length and 3 cm in width,
substitute $L = 5$ and $W = 3$ into $P = 2L + 2W$.

$$
\begin{aligned}
P &= 2 \times 5 + 2 \times 3 \\
&= 10 + 6 \\
&= 16
\end{aligned}
$$

The perimeter of the rectangle is 16 cm.

Example 3

Here is a formula for the area of a rectangle.

Area = length × width

Use the formula to find the area of a rectangle 8 cm in length and 3 cm in width.

$$
\begin{aligned}
\text{Area} &= \text{length} \times \text{width} \\
&= 8 \times 3 \\
&= 24\,\text{cm}^2
\end{aligned}
$$

Example 4

$G = 4t - 1$.

Find the value of G when $t = \frac{1}{2}$.

$$
\begin{aligned}
G &= 4t - 1 \\
&= 4 \times \tfrac{1}{2} - 1 \\
&= 2 - 1 \\
&= 1
\end{aligned}
$$

Example 5

$H = 3(4x - y)$

Find the value of H when $x = 5$ and $y = 7$.

$$
\begin{aligned}
H &= 3(4x - y) \\
&= 3(4 \times 5 - 7) \\
&= 3(20 - 7) \\
&= 3(13) \\
&= 39
\end{aligned}
$$

Example 6

$W = x^2 + 2$

Find the value of W when $x = 3$.

$$
\begin{aligned}
W &= x^2 + 2 \\
&= 3 \times 3 + 2 \\
&= 9 + 2 \\
&= 11
\end{aligned}
$$

Practice Exercise 12.1

1. Write a formula for the perimeter, P, for each of these shapes in terms of the letters given.

(a) g g g g

(b) 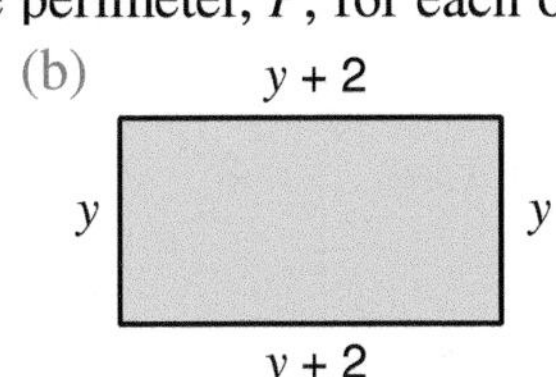

(c) 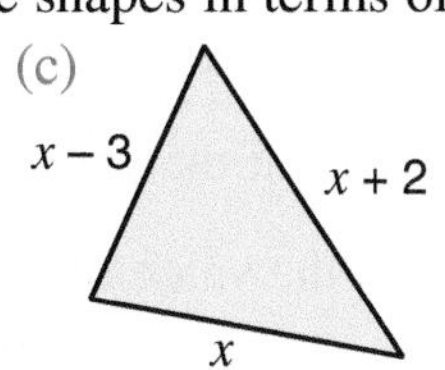

(d) 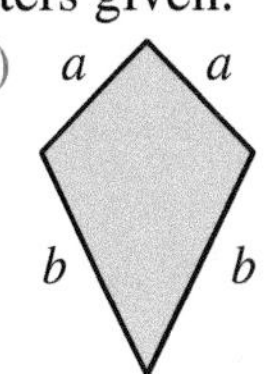

2. A caravan costs £25 per day to hire.
Write a formula for the cost, C, in £s, to hire the caravan for d days.

3. The cost of hiring a ladder is given by:

£12 per day, plus a delivery charge of £8

(a) Sam hired a ladder for 6 days.
How much did he pay?
(b) Fred hired a ladder for x days.
Write down a formula for the total cost, £C, in terms of x.

4. The number of points scored by a soccer team can be worked out using this formula.

Points scored = 3 × games won + games drawn

A team has won 5 games and drawn 2 games.
How many points have they scored?

5. $T = 5a - 3$. Find the value of T when $a = 20$.

6. $M = 4n + 1$. (a) Work out the value of M when $n = -2$.
(b) Work out the value of n when $M = 19$.

7. $H = 3g - 5$. (a) Find the value of H when $g = 2.5$.
(b) Find the value of g when $H = 13$.

8. $F = 5(v + 6)$. What is the value of F when $v = 9$?

9. $V = 2(7 + 2x)$. What is the value of V when $x = -3$?

10. $P = 3(5 - 2d)$. What is the value of P when $d = 0.5$?

11. The distance, d metres, travelled by a lawn mower in t minutes is given as: $d = 24t$.
Find the distance travelled by the lawn mower in 4 minutes.

12. Convert 30° Centigrade to Fahrenheit using the formula: $F = C \times 1.8 + 32$

13. $T = 45W + 30$ is used to calculate the time in minutes needed to cook a joint of beef weighing W kilograms.
How many minutes are needed to cook a joint of beef weighing 2.4 kg?

14. Convert 77 degrees Fahrenheit to Centigrade using the formula: $C = (F - 32) \div 1.8$

15. The voltage, V volts, in a circuit with resistance, R ohms, and current, I amps, is given by the formula: $V = IR$.
Find the voltage in a circuit when $I = 12$ and $R = 20$.

16. A simple formula for the motion of a car is $F = ma + R$.
Find F when $m = 500$, $a = 0.2$ and $R = 4000$.

17. The cost, £C, of n units of gas is calculated using the formula $C = 0.08n + 8.5$.
Calculate the cost of 458 units of gas.

18. The formula $v = u + at$ gives the speed v of a particle, t seconds after it starts with speed u. Calculate v when $u = 7.8$, $a = -10$ and $t = \frac{3}{4}$.

19. $S = a^2 - 5$. Find the value of S when $a = -3$.

20. $R = p^2 + 2p$. Find the value of R when (a) $p = 3$, (b) $p = -3$.

21. $K = m^2 - 5m$. Work out the value of K when $m = -4$.

22. $S = 2a^2$. Find the value of S when (a) $a = 3$, (b) $a = -3$.

23. $S = (2a)^2$. Find the value of S when (a) $a = 3$, (b) $a = -3$.

24. $T = \sqrt{\frac{a}{b}}$ Work out the value of T when $a = 9$ and $b = 16$.

25. $L = \sqrt{m^2 + n^2}$. Work out the value of L when
(a) $m = 6$ and $n = 8$, (b) $m = 0.3$ and $n = 0.4$.

26. The formula $F = \frac{mv^2}{r}$ describes the motion of a cyclist rounding a corner.
Find F when $m = 80$, $v = 6$ and $r = 20$.

27. Use the formula $R = \left(\frac{t}{s}\right)^2$ to calculate the value of R when $t = -0.6$ and $s = \frac{2}{5}$.

28. Use the formula $v = \sqrt{u^2 + 2as}$ to calculate the value of v when
(a) $u = 2.4$, $a = 3.2$, $s = 5.25$,
(b) $u = 9.1$, $a = -4.7$, $s = 3.04$.
Give your answers correct to one decimal place.

Rearranging formulae

Sometimes it is easier to use a formula if you **rearrange** it first.

The formula $k = \frac{8m}{5}$ can be used to change distances in miles to distances in kilometres.

Rearrange this to give a formula which can be used to change distances in kilometres into distances in miles.

$k = \frac{8m}{5}$

Multiply both sides by 5.

$5k = 8m$

Divide both sides by 8.

$\frac{5k}{8} = m$

Here is a reminder of some operations and their inverses.

Operation	Inverse operation
Addition $+a$	Subtraction $-a$
Subtraction $-a$	Addition $+a$
Multiplication $\times a$	Division $\div a$
Division $\div a$	Multiplication $\times a$

We say we have **rearranged the formula** $k = \frac{8m}{5}$ as $m = \frac{5k}{8}$ to make m the **subject** of the formula.

Example 7

Make x the subject of $y = 2x + 8$.

$y = 2x + 8$

Subtract 8 from both sides.

$y - 8 = 2x$

Divide both sides by 2.

$\frac{1}{2}y - 4 = x$

Writing the subject first

$x = \frac{1}{2}y - 4$

Example 8

Make x the subject of the formula
$ax + b = cx + d$.

$ax + b = cx + d$

Subtract cx and b from both sides.

$ax - cx = d - b$

Factorise $ax - cx$, x is a common factor.

$x(a - c) = d - b$

Divide both sides by $a - c$.

$x = \frac{d - b}{a - c}$

Practice Exercise 12.2

1. Make m the subject of these formulae.
   (a) $a = m + 5$ (b) $a = x + m$ (c) $a = m - 2$ (d) $a = m - b$
2. Make x the subject of these formulae.
   (a) $y = 4x$ (b) $y = ax$ (c) $y = \frac{x}{2}$
   (d) $y = \frac{x}{a}$ (e) $y = \frac{3x}{5}$
3. Make p the subject of these formulae.
   (a) $y = 2p + 6$ (b) $t = 5p + q$ (c) $m = 3p - 2$ (d) $r = 4p - q$
4. The cost, £C, of hiring a car for n days is given by $C = 35 + 24n$.
   Make n the subject of the formula.
5. $V = IR$. Rearrange the formula to give R in terms of V and I.
6. Make x the subject of these formulae.
   (a) $3(x - a) = x$ (b) $2(a + x) = 3(b - x)$
7. Make a the subject of each of these formulae.
   (a) $3a - x = a + 2x$ (b) $a - b = ax$ (c) $a - 2 = ax + b$
8. (a) Make p the subject of the formula $q(2 - p) = r(p + 1)$.
   (b) $x = \frac{a + b}{a - b}$. Express b in terms of a and x.

Rearranging formulae involving powers and roots

Sometimes formulae involve powers, roots or both powers and roots.

Example 9

$A = 3r^2$

Make r the subject of the formula.

$A = 3r^2$

Divide both sides by 3.

$\frac{A}{3} = r^2$

Take the square root of both sides.

$\pm\sqrt{\frac{A}{3}} = r$

So, $r = \pm\sqrt{\frac{A}{3}}$

Operation	Inverse operation
Squaring a^2	Square rooting $\sqrt{a}$
Square rooting $\sqrt{a}$	Squaring a^2
Cubing a^3	Cube rooting $\sqrt[3]{a}$
Cube rooting $\sqrt[3]{a}$	Cubing a^3

Example 10

$a\sqrt{x} - b = c$

Make x the subject of the formula.

$a\sqrt{x} - b = c$

Add b to both sides.

$a\sqrt{x} = b + c$

Divide both sides by a.

$\sqrt{x} = \frac{b + c}{a}$

Square both sides.

$x = \frac{(b + c)^2}{a^2}$

Example 11

Make v the subject of the formula $E = \frac{1}{2}mv^2$.

v is positive.

$E = \frac{1}{2}mv^2$

Multiply both sides by 2.

$2E = mv^2$

Divide both sides by m.

$\frac{2E}{m} = v^2$

Take the square root of both sides.

$v = \pm\sqrt{\frac{2E}{m}}$

Practice Exercise 12.3

1. Make c the subject of these formulae.
   (a) $y = c^2$ (b) $y = \sqrt{c}$ (c) $y = dc^2$ (d) $y = \frac{\sqrt{c}}{3}$
   (e) $y = c^2 + x$ (f) $y = x + \sqrt{c}$ (g) $y = x + \frac{c^2}{d}$ (h) $y = \frac{\sqrt{c}}{a} - x$
2. Make a the subject of these formulae.
   (a) $b = a + c^2$ (b) $a^2 = b$ (c) $p = ma + d$ (d) $ma^2 = F$
3. Make t the subject of these formulae.
   (a) $3t^2 = x$ (b) $at^2 = V$ (c) $2t^2 + a = b$
   (d) $a - t^2 = b$ (e) $a + bt^2 = c$ (f) $a - bt^2 = c$
4. Make x the subject of each of these formulae.
   (a) $\sqrt{x + 3} = a$ (b) $\frac{\sqrt{x + 2}}{3} = a$ (c) $\frac{1}{4}\sqrt{2x - a} = b$
5. Make a the subject of each of these formulae.
   (a) $3a - x = a + 2x$ (b) $a - b = ax$ (c) $a - 2 = ax + b$
   (d) $a + 2 = x(3+a)$ (e) $y = \frac{a - 3}{5 - a}$ (f) $x(a - 1) = b(a + 2)$
   (g) $y(a - 1) = 3(2 - a)$ (h) $\sqrt{\frac{a + x}{a - x}} = 2$ (i) $\sqrt{\frac{a}{x - a}} = 2x$

Solving problems by rearranging formulae

Example 12

A cuboid has length 8 cm and breadth 5 cm.
The volume of the cuboid is 140 cm^3.
Calculate the height of the cuboid.

The formula for the volume of a cuboid is $V = lbh$

Divide both sides of $V = lbh$ by lb.

So, $h = \frac{V}{lb}$

Substitute $V = 140$, $l = 8$ and $b = 5$ in $h = \frac{V}{lb}$.

$$h = \frac{140}{8 \times 5} = 3.5$$

The height of the cuboid is 3.5 cm.

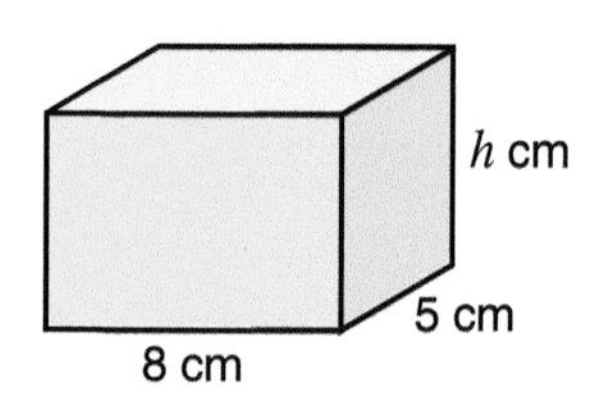

Practice Exercise 12.4

1. The perimeter of a square is $P = 4d$.
   (a) Rearrange the formula to give d in terms of P.
   (b) Find d when $P = 2.8$ cm.
2. The area of a rectangle is $A = lb$.
   (a) Rearrange the formula to give l in terms of A and b.
   (b) Find l when $A = 27\text{ cm}^2$ and $b = 4.5$ cm.
3. The speed of a car is $S = \frac{D}{T}$.
   (a) (i) Change the subject to D.
   (ii) Find D when $S = 48$ km/h and $T = 2$ hours.
   (b) (i) Change the subject to T.
   (ii) Find T when $S = 36$ km/h and $D = 90$ km.
4. The perimeter of a rectangle is $P = 2(l + b)$.
   (a) Change the subject to b.
   (b) Find b when $P = 18$ cm and $l = 4.8$ cm.

5. $y = mx + c$
   (a) Rearrange the formula to give x in terms of y, m and c.
   (b) Calculate x when $y = 0.6$, $m = -0.4$ and $c = 1.8$.

6. $A = \frac{bh}{2}$
   (a) Rearrange the formula to give b in terms of A and h.
   (b) Calculate b when $A = 9.6$ and $h = 3$.

7. $p^2 = q^2 + r^2$
   (a) Rearrange the formula to give q in terms of p and r.
   (b) Calculate q when $p = 7.3$ and $r = 2.7$.

8. The surface area of a sphere is given by $A = 4\pi r^2$.
   (a) Rearrange $A = 4\pi r^2$ to make r the subject of the formula.
   (b) Find the value of r when $A = 350\,\text{cm}^2$.
   Give your answer to a suitable degree of accuracy.

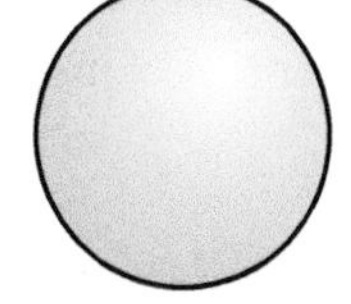

9. The volume of a cone is given by $V = \frac{1}{3}\pi r^2 h$.
   (a) Make h the subject of the formula $V = \frac{1}{3}\pi r^2 h$.
   (b) Make r the subject of the formula $V = \frac{1}{3}\pi r^2 h$.

Key Points

- An **expression** is just an answer using letters and numbers.
  A **formula** is an algebraic rule. It always has an equals sign.
- You should be able to **substitute** values into expressions and formulae.
- You should be able to **rearrange** a formula to make another letter (variable) the subject.
- You should be able to solve problems by rearranging formulae.

Review Exercise 12

1. The cost of printing business cards is:

   £5 plus 15 pence a card.

   (a) What is the total cost of printing 80 cards?
   (b) Write a formula for the total cost, £C, of printing n cards.
   (c) Fred pays £77 for some business cards to be printed.
   How many cards did he have printed?

2. The grid shows the numbers from 1 to 50.
   A **T** shape has been drawn on the grid.
   It is called $\mathbf{T}_{23}$ because the lowest number is 23.

1	2	3	4	5	6	7	8	9	10
11	12	13	14	15	16	17	18	19	20
21	22	23	24	25	26	27	28	29	30
31	32	33	34	35	36	37	38	39	40
41	42	43	44	45	46	47	48	49	50

   Calculate the sum of the numbers in:

   (a) $\mathbf{T}_{16}$ (b) $\mathbf{T}_{28}$ (c) $\mathbf{T}_2$

   (d) The diagram on the right shows $\mathbf{T}_n$.
   Copy and complete the **T** shape in terms of n.

   (e) Write a formula for the sum of the numbers, $\mathbf{S}_n$, in terms of n, for shape $\mathbf{T}_n$.
   Write your answer in its simplest form.

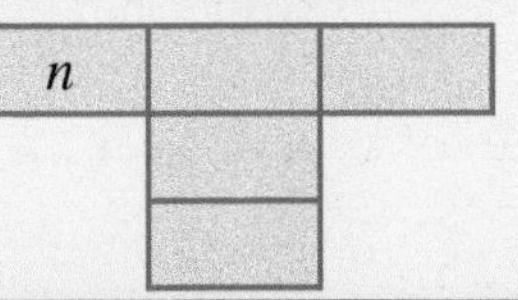

3. Match the pairs:

$y = x + 3$	$x = 3y$	$y = 3x + 1$	$x = y - 3$	$y = \frac{1}{3}x$
$x = \frac{1}{3}y$	$y = 3x$	$x = \frac{1}{3}(y - 1)$	$y = 3x - 1$	$x = \frac{1}{3}y + \frac{1}{3}$

4. The cost, £C, of hiring a car for n days is given by $C = 35 + 24n$.
   (a) Find the cost of hiring a car for 3 days.
   (b) A customer paid £251 to hire a car.
   Make n the subject of the formula and use your new formula to find for how many days the customer had hired the car.

5. Make t the subject of the formula $v = 5t + u$.

6. You are given the formula $a = \frac{1}{3}(b + c)d$.
   (a) Work out the value of a when $b = 5$, $c = -20$ and $d = 0.5$.
   (b) Express d in terms of a, b and c.

7. $F = 1.8C + 32$ changes temperatures in °C to °F.
   Rearrange the formula to give C in terms of F.

8. A formula is given as $s = t - av^2$.
   Calculate the value of s when $t = -6$, $a = 50$ and $v = 0.2$.

9 If $3a = \sqrt{x} + 2$, find x in terms of a.

10. If $V = \frac{1}{6}x^2h$,
    (a) find the value of V when $x = 3$ and $h = 4$,
    (b) find the value of h when $V = 50$ and $x = 4$,
    (c) find x in terms of V and h, where x is positive.

11. The formula for finding the curved surface area of a cylinder is $A = 2\pi rh$.
    (a) Rearrange $A = 2\pi rh$ to make h the subject of the formula.
    (b) Calculate h when $A = 72\,\text{cm}^2$ and $r = 4\,\text{cm}$.

    The formula for finding the volume of a cylinder is given by $V = \pi r^2h$.
    (c) Rearrange $V = \pi r^2h$ to make r the subject of the formula.
    (d) Calculate r when $V = 120\,\text{cm}^3$ and $h = 9.6\,\text{cm}$.

The traffic capacity of a main road may be given by the

formula $n = \dfrac{3600\,vl}{d}$

where n = the number of vehicles per hour,
v = the average speed, in m/s,
l = the number of traffic lanes,
d = the distance between vehicles in metres.

(a) If the road has 2 lanes, the traffic speed is 20 m/s, and there is 50 metres between vehicles, how many vehicles per hour can go by?

(b) If the road was widened to make 3 lanes and the traffic speed was increased to 30 m/s, but the gap between vehicles was increased to 100 metres, how many extra vehicles per hour could then go by?

(c) Write three formulae to make v, l and d the subjects.

(d) Investigate the effect of adjusting the average speed, the number of traffic lanes and the distance between vehicles.

13 Graphs of Quadratic Functions

Quadratic functions

Look at these coordinates: $(-3, 9)$, $(-2, 4)$, $(-1, 1)$, $(0, 0)$, $(1, 1)$, $(2, 4)$, $(3, 9)$.
Can you see any number patterns?

The diagram shows the coordinates plotted on a **graph**.
The points all lie on a **smooth curve**.

A **rule** connects the x coordinate with the y coordinate.
All points on the line obey the rule $y = x^2$.

$y = x^2$ is an example of a **quadratic function**.

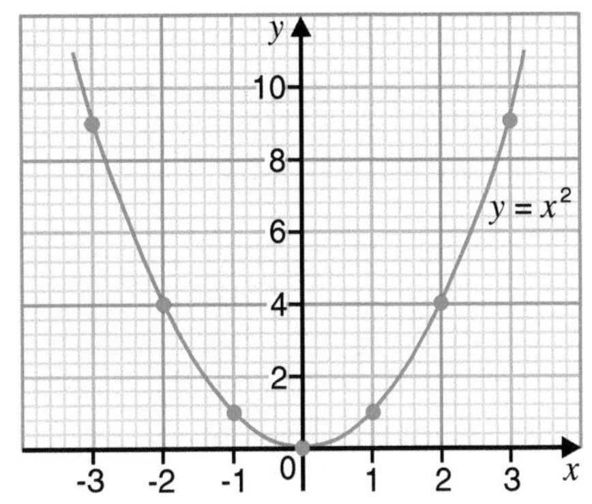

The graph of a **quadratic function** is always a smooth curve and is called a **parabola**.
The general equation of a quadratic function is $y = ax^2 + bx + c$, where a cannot be equal to zero.

The graph of a quadratic function is symmetrical and has a **maximum** or a **minimum** value.

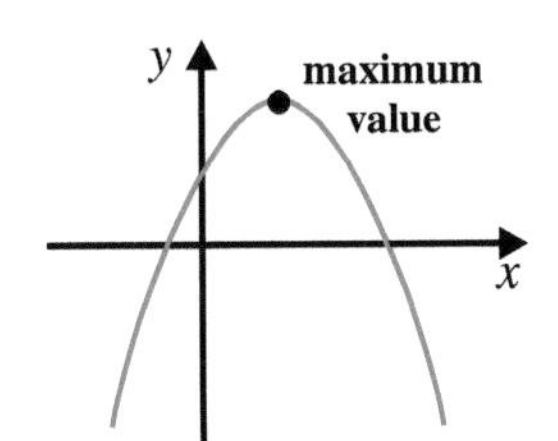

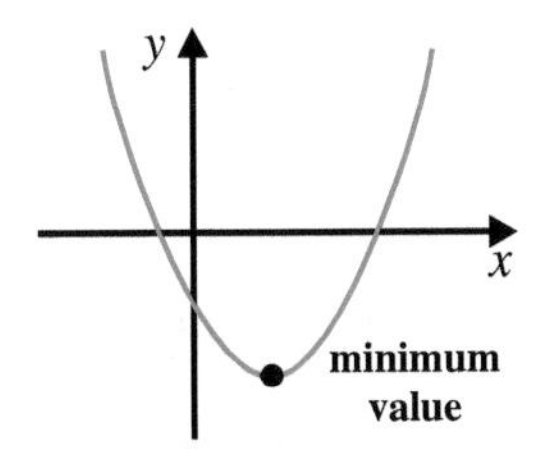

Drawing a graph of a quadratic function

To draw a quadratic graph:
- Make a table of values connecting x and y.
- Plot the points.
- Join the points with a smooth curve.

Example 1

Draw the graph of $y = x^2 - 3$ for values of x from -3 to 3.

First make a table of values for $y = x^2 - 3$.

x	-3	-2	-1	0	1	2	3
y	6	1	-2	-3	-2	1	6

Plot these points.
The curve which passes through these points is the graph of the equation $y = x^2 - 3$.

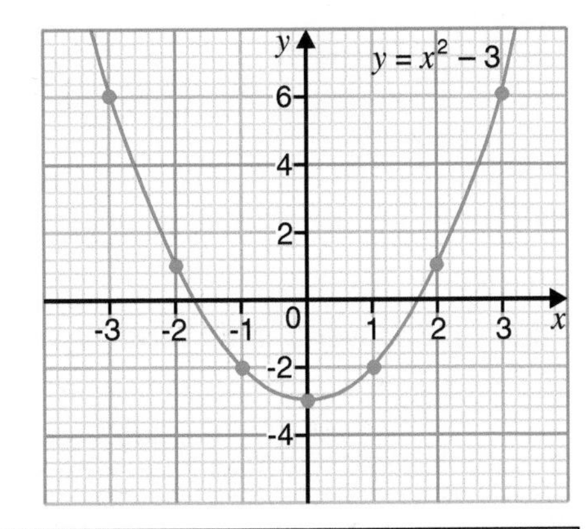

- Quadratic graphs are always symmetrical.
- Join plotted points using smooth curves and not a series of straight lines.

Practice Exercise 13.1

1. (a) Copy and complete this table of values for $y = x^2 - 2$.

x	-3	-2	-1	0	1	2	3
y		2			-1		7

   (b) Draw the graph of $y = x^2 - 2$.
   Label the x axis from -3 to 3 and the y axis from -3 to 8.
   (c) Use your graph to find the value of y when $x = -1.5$.

2. (a) Draw the graph of $y = x^2 + 1$ for values of x from -2 to 4.
   (b) Use your graph to find the value of y when $x = 2.5$.
   (c) Use your graph to find the values of x when $y = 4$.
   (d) Write down the coordinates of the point at which the graph has a minimum value.

3. (a) Copy and complete this table of values for $y = 6 - x^2$.

x	-3	-2	-1	0	1	2	3
y		2		6			-3

   (b) Draw the graph of $y = 6 - x^2$ for values of x from -3 to 3.
   (c) Write down the coordinates of the points where the graph of $y = 6 - x^2$ crosses the x axis.
   (d) Find the coordinates of the point at which the graph has a maximum value.

4. Draw the graph of $y = 2x^2$ for values of x from -2 to 2.

5. Draw the graph of $y = x^2 - x + 2$ for values of x from -2 to 3.

Transforming graphs

Transformations, such as **translations** and **stretches**, can be used to change the position and size of the graph of $y = x^2$.
The equation of the transformed (new) graph is related to the equation of the original graph.

Original graph $y = x^2$ → **Transformation** → **New graph** $y = ax^2 + bx + c$

By understanding the effects of a transformation, or combination of transformations, it is possible to:

- determine the equation of a quadratic function from its graph,
- sketch the graph for a given quadratic function.

Translating a graph

The graph of $y = x^2$ is transformed to the graph of $y = x^2 + 2$ by a **translation**.

All points on the original graph are moved the **same distance** in the **same direction** without twisting or turning.

The translation has vector $\begin{pmatrix} 0 \\ 2 \end{pmatrix}$.

This means that the curve moves 2 units vertically up.

Original graph: $y = x^2$

Transformation: translation with vector $\begin{pmatrix} 0 \\ 2 \end{pmatrix}$.

New graph: $y = x^2 + 2$

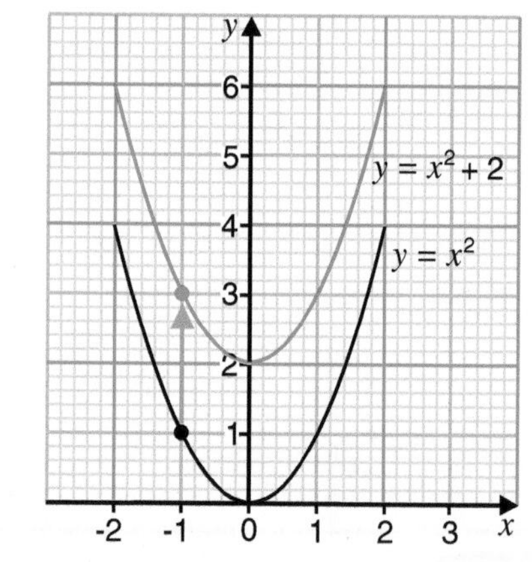

Vectors can be used to describe movement.
The **top number** describes the **horizontal movement**. Positive = right, negative = left.
The **bottom number** describes the **vertical movement**. Positive = up, negative = down.

Draw the following graphs: $y = x^2$ *for* $-2 \leq x \leq 2$,
$y = x^2 - 3$ *for* $-2 \leq x \leq 2$.
Describe how the graph of $y = x^2$ *is transformed to the graph of* $y = x^2 - 3$.

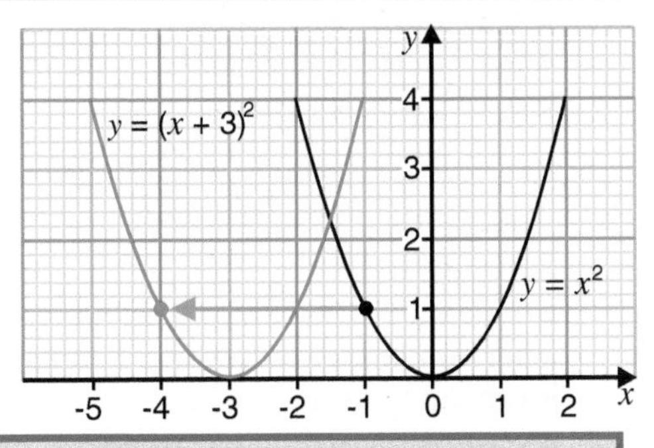

The graph of $y = x^2$ is transformed to the graph of $y = (x + 3)^2$ by a **translation**.

The translation has vector $\begin{pmatrix} -3 \\ 0 \end{pmatrix}$.

This means that the curve moves 3 units horizontally to the left.

Original graph: $y = x^2$
Transformation: translation with vector $\begin{pmatrix} -3 \\ 0 \end{pmatrix}$.
New graph: $y = (x + 3)^2$

Draw the following graphs: $y = x^2$ *for* $-2 \leq x \leq 2$,
$y = (x - 2)^2$ *for* $0 \leq x \leq 4$.
Describe how the graph of $y = x^2$ *is transformed to the graph of* $y = (x - 2)^2$.

Stretching a graph

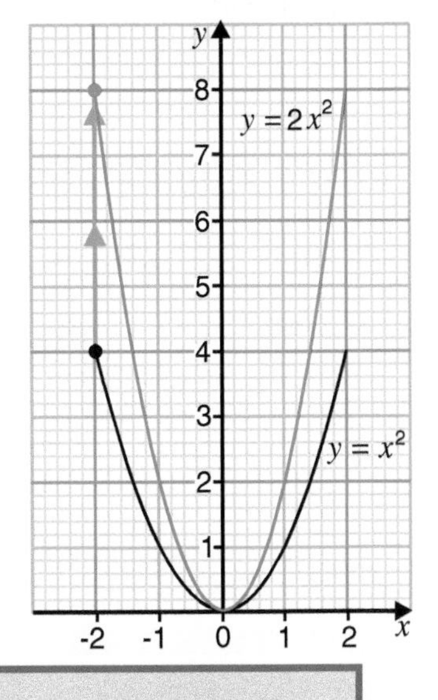

The graph of $y = x^2$ is transformed to the graph of $y = 2x^2$ by a **stretch**.
The stretch is **from** the x axis and is **parallel** to the y axis.
The stretch has scale factor 2.
This means that y coordinates on the graph of $y = x^2$ are doubled to obtain the corresponding y coordinates on the graph of $y = 2x^2$.

Original graph: $y = x^2$
Transformation: stretch from x axis, parallel to y axis, scale factor 2.
New graph: $y = 2x^2$

Draw the following graphs: $y = x^2$ *for* $-2 \leq x \leq 2$,
$y = 3x^2$ *for* $-2 \leq x \leq 2$.
Describe how the graph of $y = x^2$ *is transformed to the graph of* $y = 3x^2$.

Original graph	New graph	Transformation
$y = x^2$	$y = x^2 + a$	**translation** (vertically). If a is **positive**, curve moves a units **up**. If a is **negative**, curve moves a units **down**.
$y = x^2$	$y = (x + a)^2$	**translation** (horizontally). If a is **positive**, curve moves a units **left**. If a is **negative**, curve moves a units **right**.
$y = x^2$	$y = ax^2$	**stretch**, from the x axis, parallel to the y axis, scale factor a. The y coordinates on the graph of $y = f(x)$ are multiplied by a.

Practice Exercise 13.2

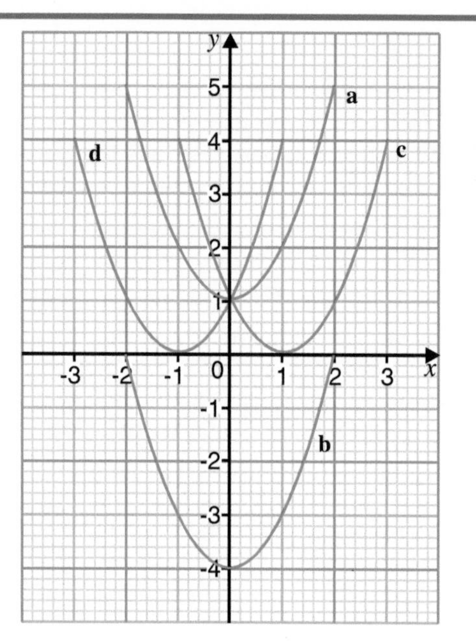

1. In this diagram, each of the graphs labelled **a**, **b**, **c** and **d** is a transformation of the graph $y = x^2$.

   What are the equations of graphs **a**, **b**, **c** and **d**?

2. On separate diagrams sketch the graphs of:
   (a) $y = x^2$ and $y = -x^2$
   (b) $y = x^2$ and $y = 2 - x^2$
   (c) $y = x^2$ and $y = x^2 - 2$
   (d) $y = x^2$ and $y = 3x^2$

3. The graph of $y = x^2$ for $-2 \leqslant x \leqslant 2$ is transformed by:
   (a) a horizontal translation of 3 units,
   (b) a horizontal translation of -5 units,
   (c) a vertical translation of -2 units,
   (d) a vertical translation of 6 units,
   (e) a stretch from the x axis, parallel to the y axis, with scale factor 2.
   In each case draw the original graph and its transformation.
   Write down the equation of each transformed graph.

4. In this diagram, each of the graphs **a**, **b** and **c** is a stretch of the graph of $y = x^2$.
   What are the equations of graphs **a**, **b** and **c**?

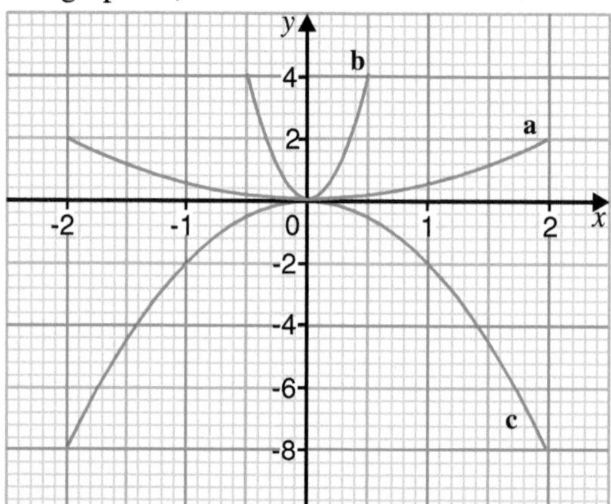

Turning points

The general form of a quadratic function is $y = ax^2 + bx + c \quad (a \neq 0)$.
Every quadratic function has a **turning point**.

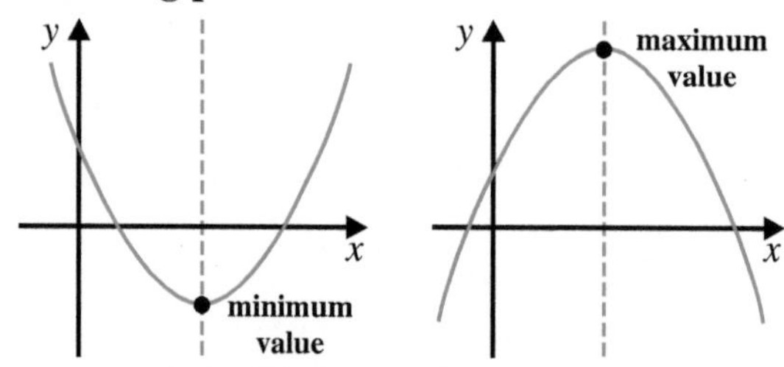

If $a > 0$, the turning point gives the **minimum value** of the function.

If $a < 0$, the turning point gives the **maximum value** of the function.

The graph is symmetrical about the vertical line which passes through the turning point.

Graphs of the form $y = k(x + p)^2 + q$

In general, the vertex (turning point) of the graph of $y = k(x + p)^2 + q$ is at the point $(-p, q)$.
The line of symmetry of the graph has equation $x = -p$.

> When $k = 1$, the graph of $y = k(x + p)^2 + q$ has a minimum value.
> When $k = -1$, the graph of $y = k(x + p)^2 + q$ has a maximum value.

Graphs of the form $y = (x - d)(x - e)$

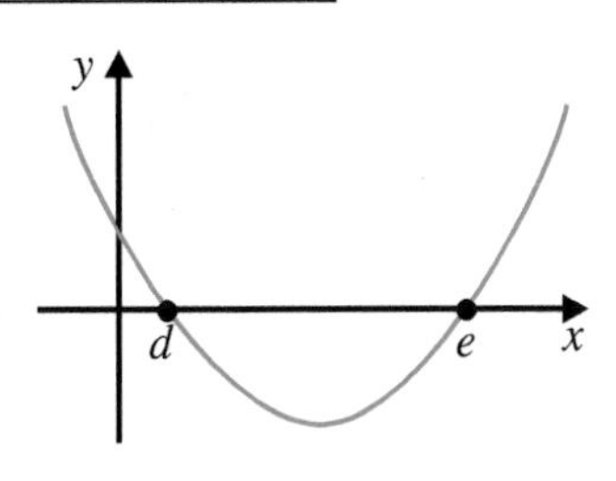

The graph of $y = (x - d)(x - e)$ crosses the x axis.
The x axis has equation $y = 0$.
So, at the point where the graph crosses the x axis $(x - d)(x - e) = 0$.
The points where the graph crosses the x axis give the **roots of the quadratic equation** $(x - d)(x - e) = 0$.
The roots of the equation are $x = d$ and $x = e$.

Example 2

(a) Write the quadratic equation $y = x^2 + 2x - 8$ in the form $y = (x + p)^2 + q$.

(b) Write down the coordinates of the turning point of $y = x^2 + 2x - 8$.

(c) Sketch the graph of $y = x^2 + 2x - 8$ by applying two transformations to the graph of $y = x^2$.

(a) $y = x^2 + 2x - 8$.

> To write the quadratic $x^2 + 2x - 8$ in completed square form:
> Write the **square term** as $(x - \frac{1}{2}$ the coefficient of $x)^2$ and then adjust for the **constant value**.

$y = x^2 + 2x - 8 = (x + 1)^2 - 9$.

(b) The turning point of the graph of $y = (x + p)^2 + q$ is at $(-p, q)$.
Turning point of $y = (x + 1)^2 - 9$ is at $(-1, -9)$.

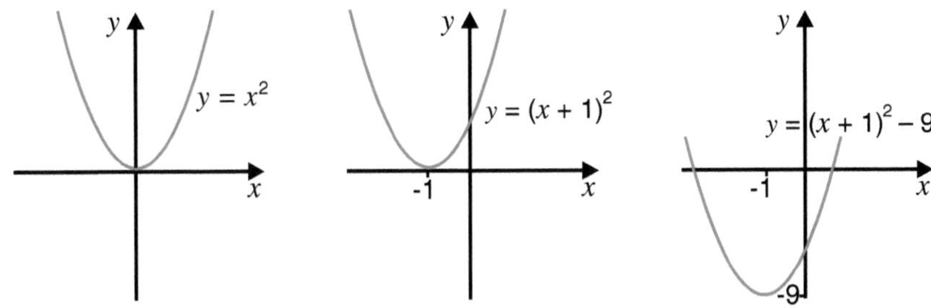

To get $y = (x + 1)^2$ translate the graph of $y = x^2$ by 1 unit to the left.
To get $y = (x + 1)^2 - 9$ translate the graph of $y = (x + 1)^2$ by 9 units down.
The translations can be applied in any order.

Example 3

This is a sketch of the quadratic graph given by $y = x^2 + bx + c$.
Find the values of b and c.

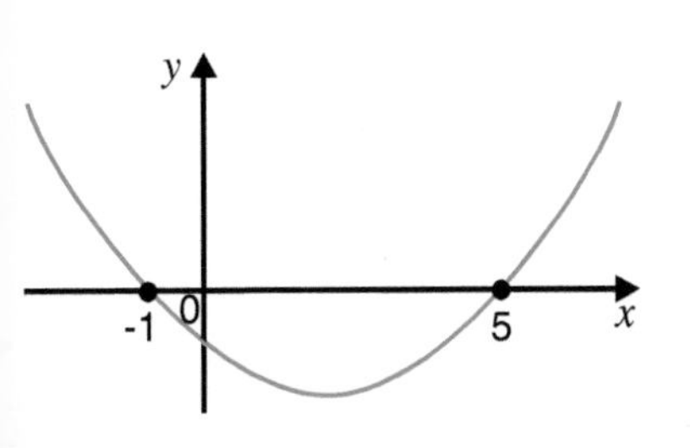

The roots of the equation are -1 and 5.
So, $(x + 1)$ and $(x - 5)$ are factors of $x^2 + bx + c$.

$y = (x + 1)(x - 5)$

> $x = -1$, so, $(x + 1) = 0$
> $x = 5$, so, $(x - 5) = 0$

Expand the brackets.

$y = x^2 - 5x + x - 5$

$y = x^2 - 4x - 5$

Compare this with $y = x^2 + bx + c$.

$b = -4$ and $c = -5$

> Compare the coefficients
> $bx = -4x$ so, $b = -4$

Practice Exercise 13.3

1. Write each quadratic equation in the form $y = (x + p)^2 + q$ and find the coordinates of the turning points.
   (a) $y = x^2 + 4x - 1$ (b) $y = x^2 + 6x + 5$ (c) $y = x^2 - 2x + 3$

2. By writing $y = x^2 - 2x + 2$ in the form $y = (x + p)^2 + q$ find:
   (a) the minimum value of $y = x^2 - 2x + 2$,
   (b) the equation of the line of symmetry of $y = x^2 - 2x + 2$.

3. 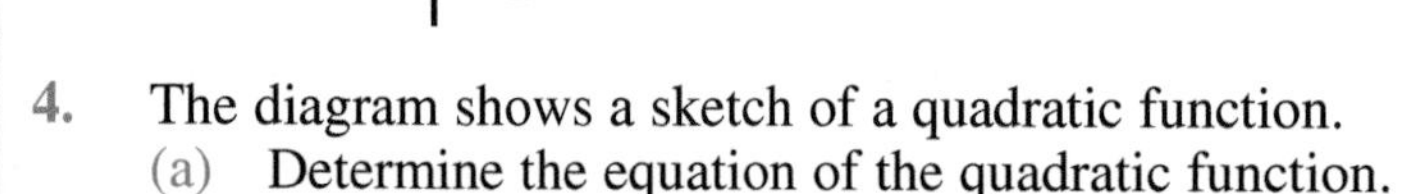

   This is a sketch of the quadratic graph given by $y = x^2 + bx + c$. Find the values of b and c.

4. The diagram shows a sketch of a quadratic function.
   (a) Determine the equation of the quadratic function.
   (b) What is the equation of the line of symmetry for this function?
   (c) Find the minimum value of the function.

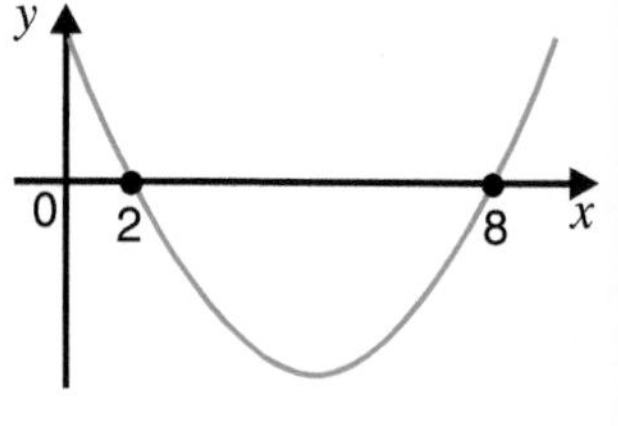

5. A quadratic function is given by $y = x^2 - x - 6$.
   (a) By factorising $x^2 - x - 6$, determine the coordinates of the points where the graph crosses the x axis.
   (b) Sketch the graph of $y = x^2 - x - 6$.

6. (a) Write the quadratic equation $y = x^2 - 4x - 1$ in the form $y = (x + p)^2 + q$.
   (b) State the two transformations that must be made to the graph of $y = x^2$ to get the graph of $y = x^2 - 4x - 1$.

7. (a) Sketch the graph of $y = -(x + 4)^2 + 2$.
   (b) Show how the graph of $y = -(x + 4)^2 + 2$ can be obtained by applying transformations to the graph of $y = x^2$.

8. The following parabolas are of the form $y = (x + p)^2 + q$.
   Write down the equation of each parabola and its axis of symmetry.

(a)
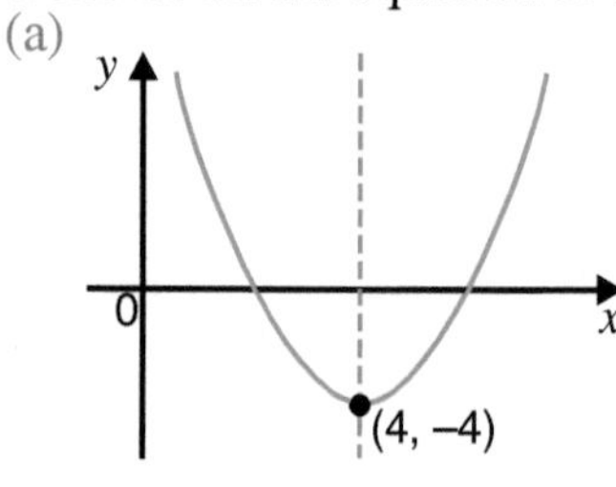

(b)
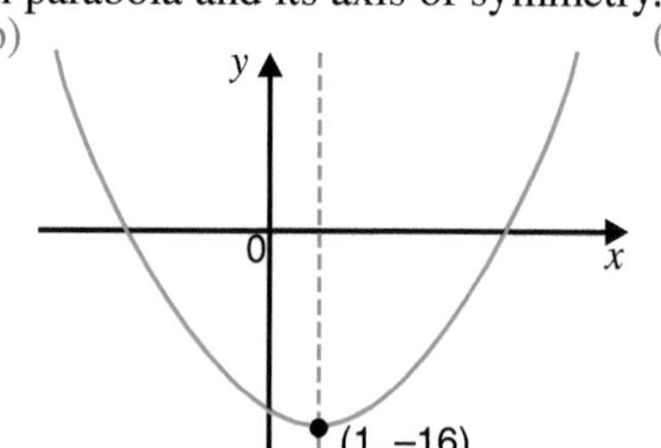

(c)
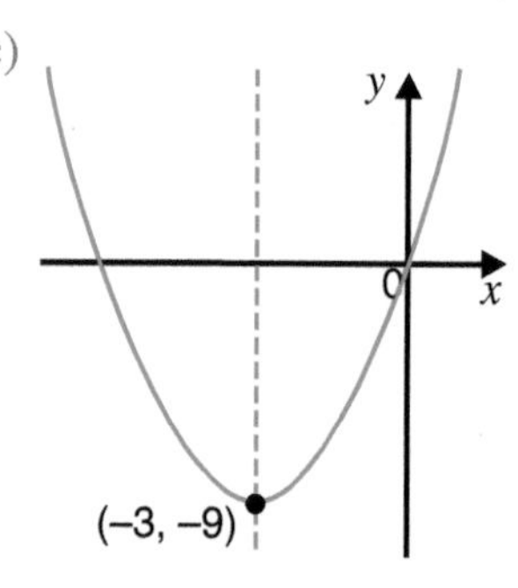

9. The following parabolas are of the form $y = q - (x + p)^2$.
   Write down the equation of each parabola and its axis of symmetry.

(a)
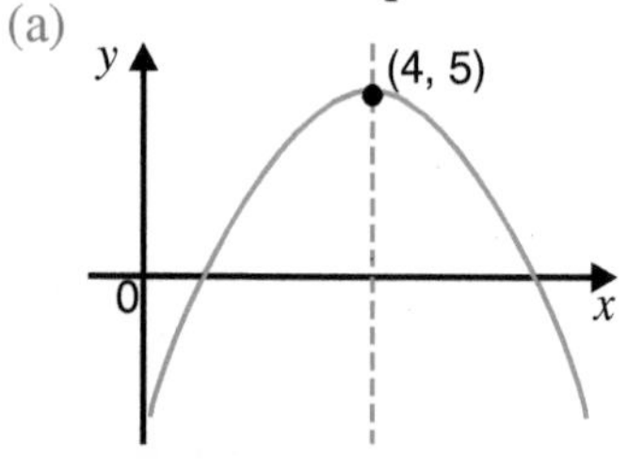

(b)
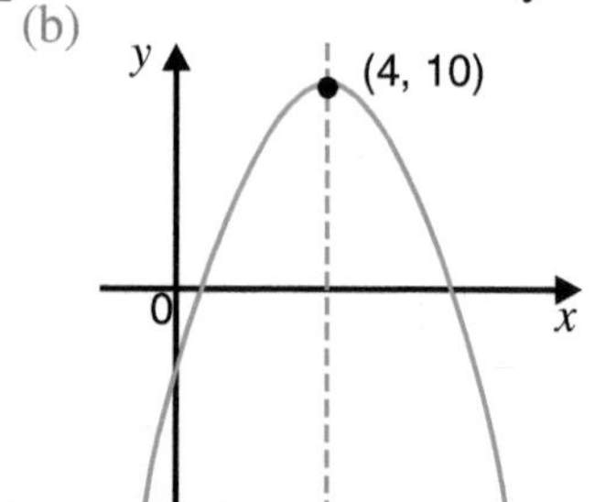

(c)
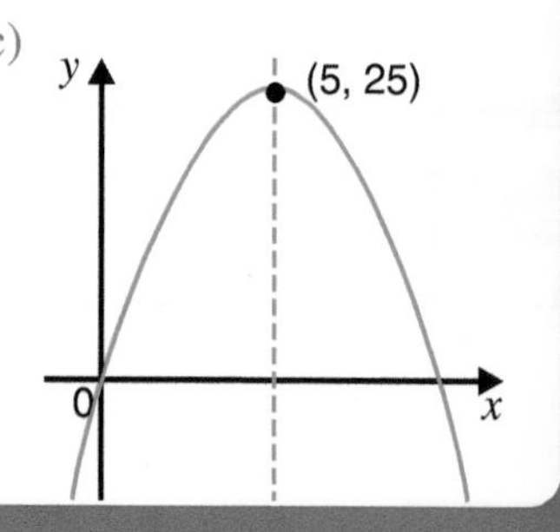

Key Points

- The graph of a **quadratic function** is always a smooth curve and is called a **parabola**.
- The general form of a **quadratic function** is $y = ax^2 + bx + c$, where a cannot be zero.

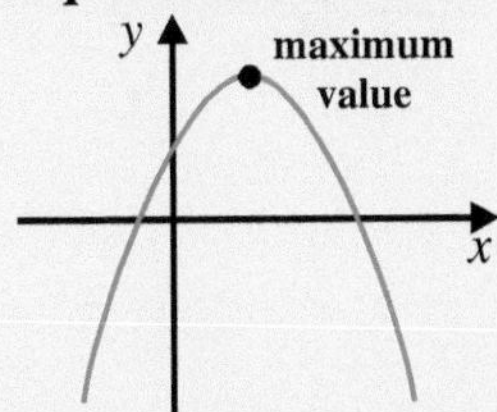

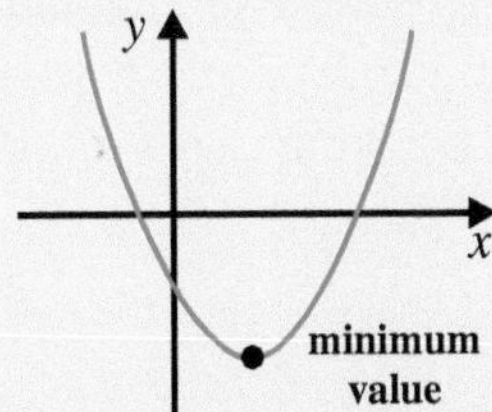

The graph of a quadratic function is symmetrical and has a **maximum** or **minimum** value.

- **Transformations**, such as **translations** and **stretches**, can be used to change the position and size of a graph.
  The equation of the transformed (new) graph is related to the equation of the original graph.
  In general

Original graph	New graph	Transformation
$y = x^2$	$y = x^2 + a$	**translation** (vertically). If a is **positive**, curve moves a units **up**. If a is **negative**, curve moves a units **down**.
$y = x^2$	$y = (x + a)^2$	**translation** (horizontally). If a is **positive**, curve moves a units **left**. If a is **negative**, curve moves a units **right**.
$y = x^2$	$y = ax^2$	**stretch**, from the x axis, parallel to the y axis, scale factor a. The y coordinates on the graph of $y = f(x)$ are multiplied by a.

- You should be able to determine the equation of a quadratic function from its graph.
- You should be able to sketch the graph of a quadratic function.
- Every quadratic function has a **turning point**.
  For the quadratic function $y = ax^2 + bx + c$ $(a \neq 0)$.
  If $a > 0$, the turning point gives the **minimum value** of the function.
  If $a < 0$, the turning point gives the **maximum value** of the function.
- For graphs of the form $y = (x + p)^2 + q$, the turning point is at the point $(-p, q)$.
  The line of symmetry of the graph has equation $x = -p$.
- Graphs of the form $y = (x - d)(x - e)$.

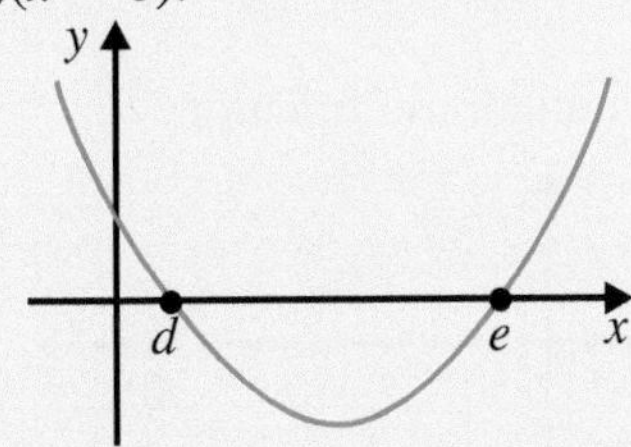

The points where the graph crosses the x axis give the **roots of the quadratic equation**
$(x - d)(x - e) = 0$.
The roots of the equation are $x = d$ and $x = e$.

Review Exercise 13

1. On separate diagrams, sketch the graphs of:
   (a) $y = x^2$ and $y = (x + 2)^2$.
   (b) $y = x^2$ and $y = 4x^2$.
   (c) $y = x^2$ and $y = (x - 5)^2$.

2. (a) Write the quadratic equation $y = x^2 - 8x + 2$ in the form $y = (x + p)^2 + q$.
(b) Using your answer to part (a), describe the two transformations that must be applied to the graph of $y = x^2$ to obtain the graph of $y = x^2 - 8x + 2$.

3. Match the following quadratic equations to their graphs.

$y = x^2 + 4$ $\quad y = x^2 - 4$ $\quad y = 4 - x^2$ $\quad y = 4x^2$ $\quad y = x^2 - 4x + 4$

A

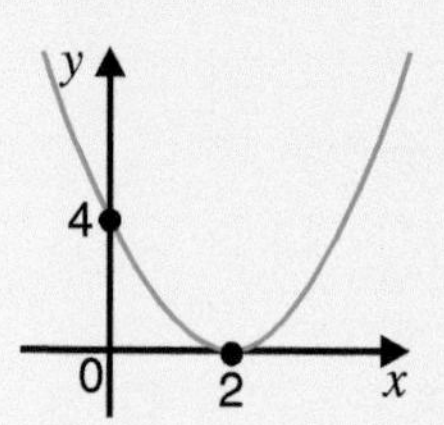

B

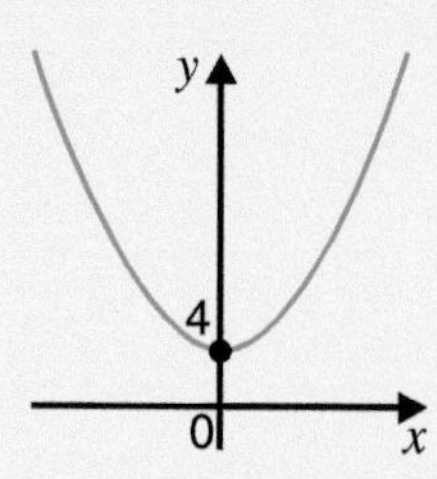

C

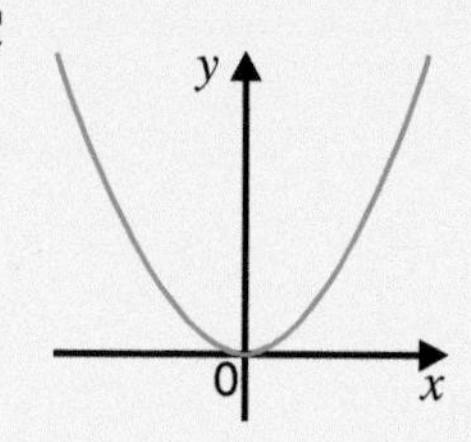

D

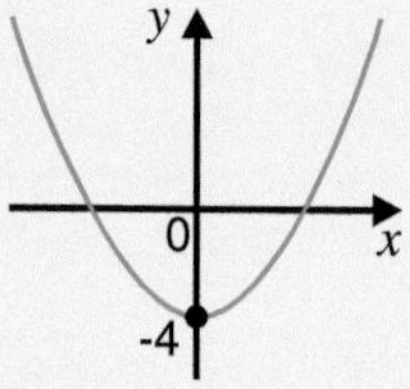

E

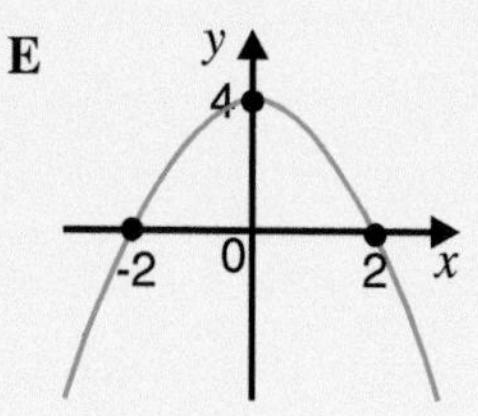

4. The diagram shows a graph of the form $y = x^2 + bx + c$.
Find the values of b and c.

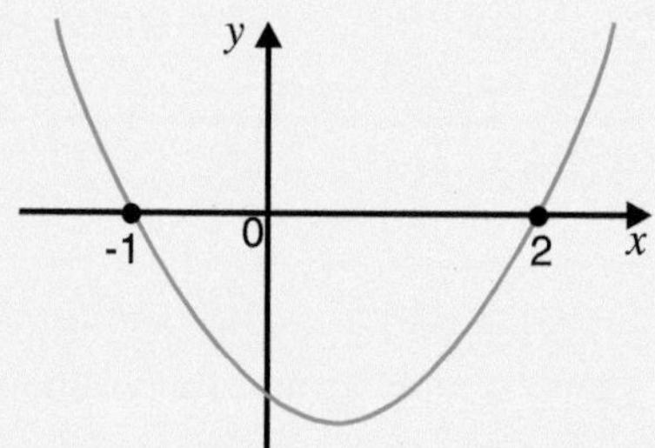

5. A quadratic function is defined by $y = (x + 3)(x - 4)$.
(a) Write down the coordinates of the points where the graph crosses the x axis.
(b) Write down the coordinates of the point where the graph crosses the y axis.
(c) Write down the equation of the line of symmetry of the graph.
(d) Find the coordinates of the turning point of the graph.
(e) Sketch the graph of $y = (x + 3)(x - 4)$.

6. The diagram shows the graph of a quadratic function.

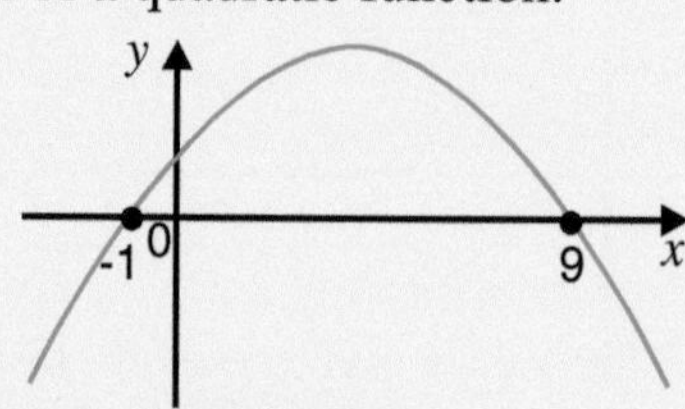

(a) Write down the roots of the quadratic function.
(b) Write down the equation of the line of symmetry of the function.
(c) Find the coordinates of the turning point.
(d) Find the coordinates of the point where the graph crosses the y axis.

7. A parabola is of the form $y = (x + p)^2 + q$.
The line of symmetry of the parabola is $x = -1$.
The minimum value of the quadratic function is 4.
Write down the equation of the parabola.

14 Quadratic Equations

Using graphs to solve quadratic equations

The diagram shows the graph of $y = x^2 - 4$.

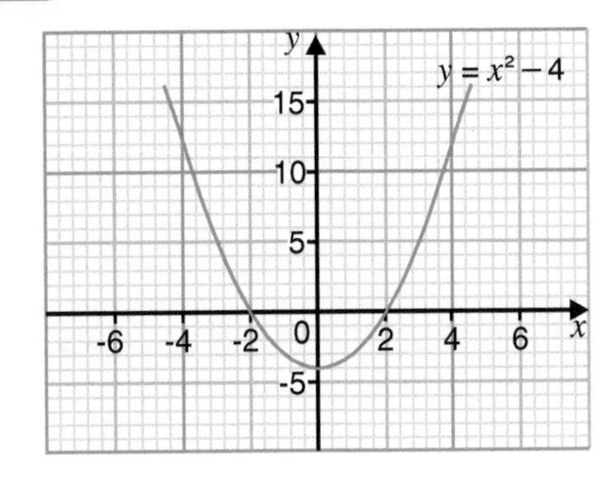

The values of x where the graphs of quadratic functions cross (or touch) the x axis give the **solutions to quadratic equations**.

At the point where the graph $y = x^2 - 4$ crosses the x axis the value of $y = 0$.

$$x^2 - 4 = 0$$

The solutions of this quadratic equation can be read from the graph: $x = -2$ and $x = 2$.

Example 1

(a) Draw the graph of $y = x^2 + 2x - 3$ for values of x from -4 to 2.

(b) Use your graph to find the solutions of the equation $x^2 + 2x - 3 = 0$.

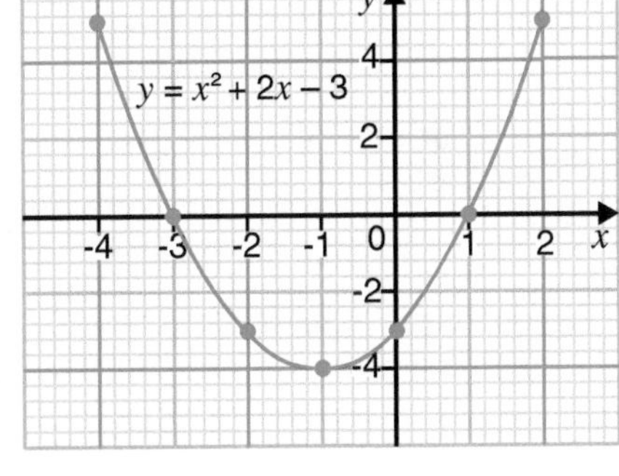

(a) First make a table of values for $y = x^2 + 2x - 3$.

x	-4	-3	-2	-1	0	1	2
y	5	0	-3	-4	-3	0	5

Plot these points.
The curve which passes through these points is the graph of the equation $y = x^2 + 2x - 3$.

(b) To solve this equation, read the values of x where the graph of $y = x^2 + 2x - 3$ crosses the x axis. $x = -3$ and $x = 1$.

Practice Exercise 14.1

1. The graph of $y = x^2 + 3x - 2$ is shown.

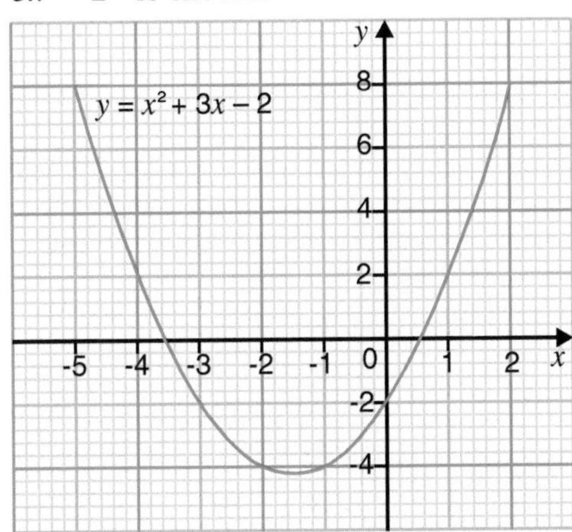

Use the graph to solve the equation $x^2 + 3x - 2 = 0$.
Give your answers correct to 1 d.p.

2. (a) Draw the graph of $y = x^2 + x$ for values of x from -3 to 2.
   (b) Use your graph to solve the equation $x^2 + x = 0$.
   (c) Find the coordinates of the point at which the graph has a minimum value.

3. (a) Copy and complete this table of values for $y = x^2 - 2x + 1$.

x	-2	-1	0	1	2	3
y						

(b) Draw the graph of $y = x^2 - 2x + 1$.

(c) Use your graph to solve the equation $x^2 - 2x + 1 = 0$.

4. (a) Draw the graph of $y = 10 - x^2$ for values of x from -4 to 4.

(b) Use your graph to solve the equation $10 - x^2 = 0$.

(c) Find the coordinates of the point at which the graph has a maximum value.

5. Draw suitable graphs to solve the following equations.

(a) $x^2 - 10 = 0$ (b) $5 - x^2 = 0$ (c) $y = x^2 - 3x + 2$ (d) $12 - 2x^2 = 0$

Solving quadratic equations by factorising

The diagram shows the graph of $y = x^2 + 2x - 3$.

The solutions of the quadratic equation $x^2 + 2x - 3 = 0$ can be read from the graph.

The solutions are: $x = -3$ and $x = 1$.

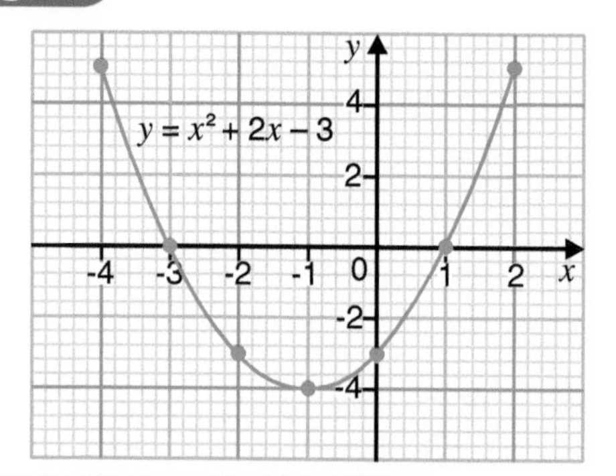

The quadratic expression can be **factorised**.

$$x^2 + 2x - 3 = (x - 1)(x + 3)$$

So, the quadratic equation $x^2 + 2x - 3 = 0$

can be written as $(x - 1)(x + 3) = 0$.

Either $x - 1 = 0$ or $x + 3 = 0$

$x = 1$ or $x = -3$

If the product of two numbers is **zero**, at least one of the numbers must be zero.

Example 2

Find x if:

(a) $x(x - 2) = 0$

Either $x = 0$ or $x - 2 = 0$.

Because one of them must be zero.

$x = 0$ or $x = 2$

(b) $(x - 3)(x + 2) = 0$

Either $x - 3 = 0$ or $x + 2 = 0$

Because one of them must be zero.

$x = 3$ or $x = -2$

Example 3

Solve $x^2 - 5x + 6 = 0$

$x^2 - 5x + 6 = 0$

Factorise the quadratic expression first.

$(x - 2)(x - 3) = 0$

Either $x - 2 = 0$ or $x - 3 = 0$.

$x = 2$ or $x = 3$

To check solutions:

Does $(2)^2 - 5 \times (2) + 6 = 0$

and $(3)^2 - 5 \times (3) + 6 = 0$?

Example 4

Solve these equations.

(a) $x^2 - 9 = 0$

$(x - 3)(x + 3) = 0$

Either $x - 3 = 0$ or $x + 3 = 0$.

$x = 3$ or $x = -3$

(b) $y^2 + 7y = 0$

$y(y + 7) = 0$

Either $y = 0$ or $y + 7 = 0$.

$y = 0$ or $y = -7$

(c) $a^2 + 4a - 5 = 0$

$(a - 1)(a + 5) = 0$

Either $a - 1 = 0$ or $a + 5 = 0$.

$a = 1$ or $a = -5$

Practice Exercise 14.2

1. Solve these equations.
   (a) $(x - 2)(x - 3) = 0$ (b) $(x + 4)(x + 6) = 0$ (c) $(x - 3)(x + 1) = 0$
   (d) $(x - 5)(x + 2) = 0$ (e) $x(x - 4) = 0$ (f) $3x(x + 2) = 0$

2. Solve.
   (a) $x^2 - 3x + 2 = 0$ (b) $y^2 + 7y + 12 = 0$ (c) $m^2 - 2m - 8 = 0$
   (d) $a^2 + a - 12 = 0$ (e) $n^2 - 5n - 36 = 0$ (f) $z^2 - 9z + 18 = 0$
   (g) $k^2 + 8k + 15 = 0$ (h) $c^2 + 15c + 56 = 0$ (i) $b^2 + b - 20 = 0$
   (j) $v^2 - 7v - 60 = 0$ (k) $w^2 + 8w - 48 = 0$ (l) $p^2 - p - 72 = 0$

3. Solve.
   (a) $x^2 - 5x = 0$ (b) $y^2 + y = 0$ (c) $p^2 + 3p = 0$
   (d) $4a - a^2 = 0$ (e) $t^2 - 6t = 0$ (f) $g^2 - 4g = 0$

4. Solve.
   (a) $x^2 - 4 = 0$ (b) $y^2 - 144 = 0$ (c) $9 - a^2 = 0$
   (d) $d^2 - 16 = 0$ (e) $x^2 - 100 = 0$ (f) $36 - x^2 = 0$
   (g) $4x^2 - 25 = 0$ (h) $81 - 16x^2 = 0$ (i) $144x^2 - 25 = 0$

5. Solve these equations.
   (a) $x^2 + 6x = 0$ (b) $x^2 + 5x + 4 = 0$ (c) $x^2 - 64 = 0$
   (d) $x^2 - 4x + 3 = 0$ (e) $5x^2 - 10x = 0$ (f) $x^2 - x - 6 = 0$
   (g) $x^2 + 2x - 15 = 0$ (h) $2x^2 - 8 = 0$ (i) $x^2 - 15x + 56 = 0$

6. Solve these equations.
   (a) $2x^2 + 7x + 5 = 0$ (b) $2x^2 - 11x + 5 = 0$ (c) $3x^2 - 17x - 28 = 0$
   (d) $6y^2 + 7y + 2 = 0$ (e) $3x^2 + 7x - 6 = 0$ (f) $3z^2 - 5z - 2 = 0$
   (g) $5m^2 - 8m + 3 = 0$ (h) $6a^2 + 3a - 63 = 0$ (i) $4y^2 - 3y - 10 = 0$

7. Rearrange these equations and then solve them.
   (a) $y^2 = 4y + 5$ (b) $x^2 = x$ (c) $x^2 = 8x - 16$
   (d) $x^2 = 2x + 15$ (e) $n^2 - 10n = 24$ (f) $7 = 8m - m^2$
   (g) $a(a - 5) = 24$ (h) $x^2 - 5x + 6 = 3 - x$ (i) $x = 6x^2 - 1$
   (j) $15 - 7m = 2m^2$ (k) $6a^2 = 2 - a$ (l) $8x^2 = 3 - 2x$

Using the quadratic formula

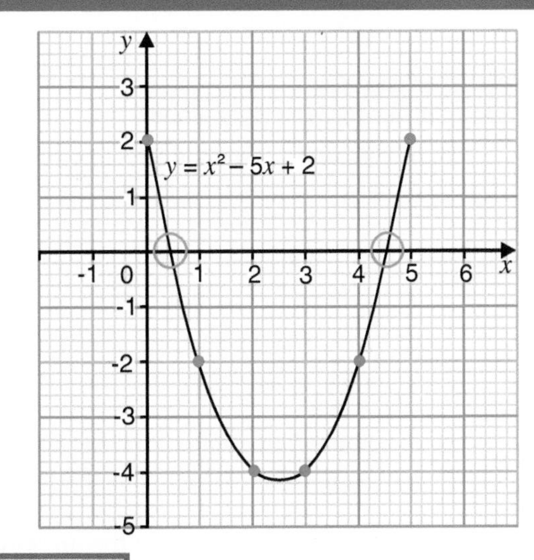

Some quadratic expressions do not factorise.
Consider the equation $x^2 - 5x + 2 = 0$.
It is not possible to find two numbers which:
when multiplied give $+2$ **and** when added give -5.
We could draw the graph of $y = x^2 - 5x + 2$
and use it to solve the equation $x^2 - 5x + 2 = 0$.
One solution lies between 0 and 1.
Another solution lies between 4 and 5.
Graphical solutions have limited accuracy and are time consuming. Explain why.

The solutions to a quadratic equation can be found using the **quadratic formula**.

If $ax^2 + bx + c = 0$ and $a \neq 0$ then $x = \dfrac{-b \pm \sqrt{b^2 - 4ac}}{2a}$

The quadratic formula can be used at any time, even if the quadratic expression factorises.

Solve $x^2 - 5x + 2 = 0$, giving the answers correct to three significant figures.

Using $x = \dfrac{-b \pm \sqrt{b^2 - 4ac}}{2a}$

Substitute $a = 1$, $b = -5$ and $c = 2$.

$$x = \frac{-(-5) \pm \sqrt{(-5)^2 - 4(1)(2)}}{2(1)}$$

$$x = \frac{5 \pm \sqrt{25 - 8}}{2}$$

$$x = \frac{5 \pm \sqrt{17}}{2}$$

Either $x = \dfrac{5 + \sqrt{17}}{2}$ or $x = \dfrac{5 - \sqrt{17}}{2}$

$x = 4.5615\ldots$ or $x = 0.43844\ldots$

$x = 0.438$ or 4.56, correct to 3 sig. figs.

Check the answers by substituting the values of x into the original equation.

To use the quadratic formula to solve $x^2 + 2x - 3$, substitute $a = 1$, $b = 2$ and $c = -3$.

Example 5

Solve $6x = 3 - 5x^2$, giving the answers correct to two decimal places.

Rearrange $6x = 3 - 5x^2$ to get $5x^2 + 6x - 3 = 0$.

Using $x = \dfrac{-b \pm \sqrt{b^2 - 4ac}}{2a}$

Substitute $a = 5$, $b = 6$ and $c = -3$.

$$x = \frac{-6 \pm \sqrt{6^2 - 4(5)(-3)}}{2(5)}$$

$$x = \frac{-6 \pm \sqrt{36 + 60}}{10}$$

$$x = \frac{-6 \pm \sqrt{96}}{10}$$

Either $x = \dfrac{-6 + \sqrt{96}}{10}$ or $x = \dfrac{-6 - \sqrt{96}}{10}$

$x = 0.3797\ldots$ or $x = -1.5797\ldots$

$x = -1.58$ or 0.38, correct to 2 d.p.

Practice Exercise 14.3

1. Use the quadratic formula to solve these equations.
   (a) $x^2 + 4x + 3 = 0$ (b) $x^2 - 3x + 2 = 0$ (c) $x^2 + 2x - 3 = 0$

2. Solve these equations, giving the answers correct to 2 decimal places.
   (a) $x^2 + 4x + 2 = 0$ (b) $x^2 + 7x + 5 = 0$ (c) $x^2 + x - 1 = 0$
   (d) $x^2 - 3x - 2 = 0$ (e) $x^2 - 5x - 3 = 0$ (f) $x^2 + 5x + 3 = 0$
   (g) $x^2 + 4x - 10 = 0$ (h) $x^2 - 3x + 1 = 0$ (i) $x^2 + 3x - 2 = 0$

3. Use the quadratic formula to solve these equations.
   (a) $2x^2 + x - 3 = 0$ (b) $2x^2 - 3x + 1 = 0$ (c) $5x^2 + 2x - 3 = 0$

4. Solve these equations, giving the answers correct to 2 decimal places.
   (a) $3x^2 - 5x + 1 = 0$ (b) $2x^2 - 7x + 4 = 0$ (c) $3x^2 + 23x + 8 = 0$
   (d) $3x^2 - 4x - 5 = 0$ (e) $2z^2 + 6z + 3 = 0$ (f) $3x^2 - 2x - 20 = 0$

Quadratic equations and the discriminant

If $ax^2 + bx + c = 0$ and $a \neq 0$ then $b^2 - 4ac$ is called the **discriminant**.

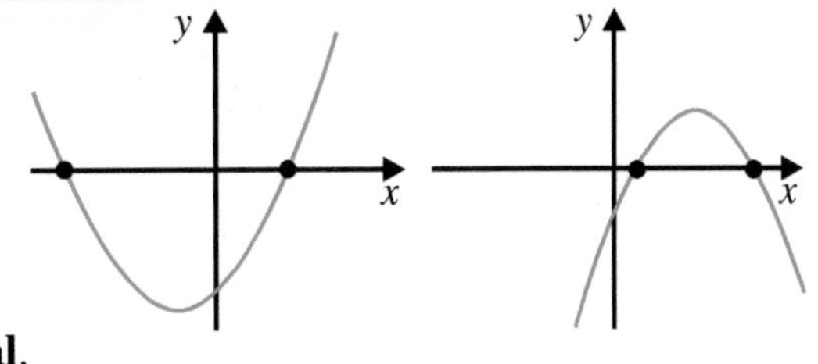

Positive discriminant

$b^2 - 4ac > 0$

Quadratic equation has **two real solutions**.

If the discriminant is a perfect square, the roots are **rational**.
If the discriminant is not a perfect square, the roots are **irrational**.

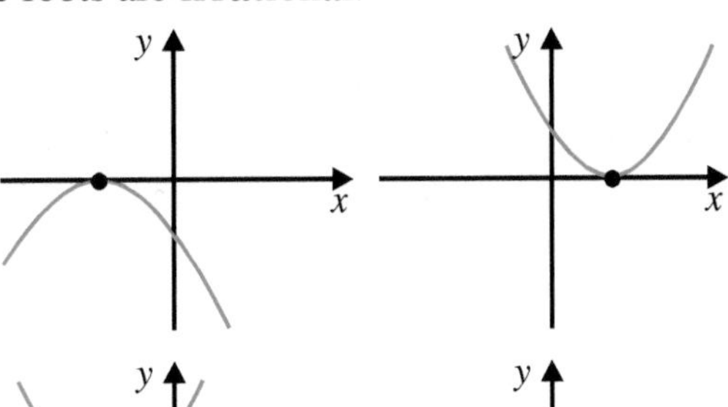

Discriminant = 0

$b^2 - 4ac = 0$

Quadratic equation has **one real solution**.

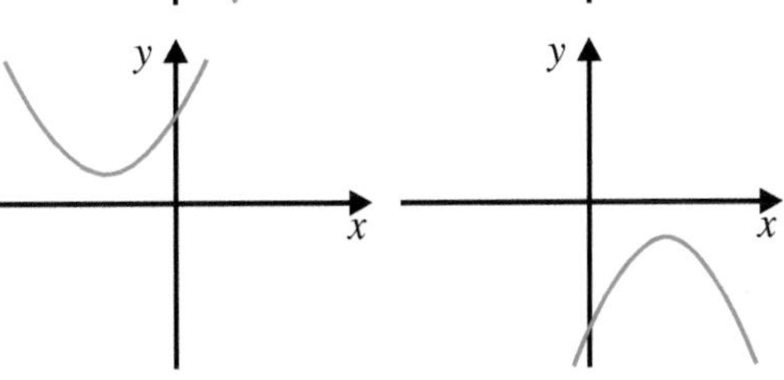

Negative discriminant

$b^2 - 4ac < 0$

Quadratic equation has **no real solutions**.

Example 6

Find p, given that $2x^2 + 4x + p = 0$ has no real roots.

The equation has no real roots, so the discriminant is **negative**.

$b^2 - 4ac < 0$

Substitute $a = 2$, $b = 4$ and $c = p$.

$(4)^2 - 4(2)(p) < 0$

$16 - 8p < 0$

Add $8p$ to both sides.

$16 < 8p$

Divide both sides by 8.

$2 < p$

So, $p > 2$

Example 7

Find q, given that $x^2 + 2qx + 25 = 0$ has one real root.

The equation has one real root, so the discriminant is **zero**.

$b^2 - 4ac = 0$

Substitute $a = 1$, $b = 2q$ and $c = 25$.

$(2q)^2 - 4(1)(25) = 0$

$4q^2 - 100 = 0$

Take out a common factor of 4.

$4(q^2 - 25) = 0$

Factorise.

$4(q - 5)(q + 5) = 0$

Either $q - 5 = 0$ or $q + 5 = 0$

$q = 5$ or $q = -5$

Practice Exercise 14.4

1. Find the discriminant for each of these equations.
   (a) $y = x^2 + 2x - 8$ (b) $y = x^2 - 6x + 9$ (c) $y = 2x^2 + 3x + 2$

2. Find the discriminant for each of the following equations and state the nature of the roots.
   (a) $x^2 + 5x + 3 = 0$ (b) $x^2 + 6x + 9 = 0$ (c) $x^2 - x - 6 = 0$
   (d) $x^2 + 4x + 5 = 0$ (e) $x^2 - 6x + 5 = 0$ (f) $5x^2 + 12x + 15 = 0$
   (g) $3x^2 - 16x + 7 = 0$ (h) $2x^2 - 5x - 3 = 0$ (i) $6x^2 + 10 = 19x$
   (j) $24x^2 = 9x$ (k) $10x^2 = 4x + 1$ (l) $2x - 1 = x^2$

3. The equation $x^2 - 8x + p = 0$ has two real roots. Find the value of p.

4. The equation $x^2 + qx + 25$ has one real root. Find the two values of q.

5. The equation $x^2 - 2x + p = 0$ has no real roots. Find the value of p.

6. The quadratic equation $ax^2 + ax + 1 = 0$ has one real root. Find the value of a.

7. Find the value of p if $3x^2 + 2px + 4 = 0$ has two real roots.

8. Find the value of p if $px^2 - 4x + 1 = 0$ has two real roots.

9. The roots of the equation $2x^2 + ax + 2a = 0$ are real and equal. Find the value of a.

10. (a) Draw axes marked from -4 to 3 for x and from -5 to 10 for y.
    Draw the graph of $y = x^2 + 2x - 3$.
    Use your graph to solve the equation $x^2 + 2x - 3 = 0$.
    (b) Factorise the quadratic expression $x^2 + 2x - 3$.
    Use your answer to solve the equation $x^2 + 2x - 3 = 0$.
    (c) Use the quadratic formula to solve the equation $x^2 + 2x - 3 = 0$.

11. Rearrange the following to form quadratic equations.
    Then solve the quadratic equations, giving answers correct to 2 decimal places.
    (a) $x^2 = 6x - 3$
    (b) $2x^2 - 6 = 3x$
    (c) $3x = 2x^2 - 4$
    (d) $2 = x + 5x^2$
    (e) $3(x + 1) = x^2 - x$
    (f) $x(2x + 5) = 10$
    (g) $3x(x - 4) = 2(x - 1)$
    (h) $(3x - 2)^2 = (x + 2)^2 + 12$
    (i) $5 = 2x(x - 1)$

Problems solved using quadratic equations

Some mathematical problems involve the forming of equations which are quadratic.
The equation is solved and the solutions are analysed to answer the original problem.

Example 8

A rectangular lawn, 10 m by 8 m, is surrounded by a path which is x m wide.
The total area of the path and lawn is 143 m².
Calculate the width of the path.

Drawing a **sketch diagram**, to which labels can be added, may help you to understand and solve problems.

(10 + 2x) m
x
x
x
10 m
8 m
(8 + 2x) m
x

Total length $= (10 + 2x)$ m
Total width $= (8 + 2x)$ m
Total area $= 143$ m²

Total length × total width = total area $\quad (10 + 2x)(8 + 2x) = 143$

Expand the brackets. $\quad 80 + 20x + 16x + 4x^2 = 143$

Rearrange to the form $ax^2 + bx + c = 0$. $\quad 4x^2 + 36x - 63 = 0$

$$(2x - 3)(2x + 21) = 0$$

Either $2x - 3 = 0$ or $2x + 21 = 0$

$2x = 3$ or $2x = -21$

$x = 1.5$ or $x = -10.5$

The quadratic equation has two solutions.
However, $x \neq -10.5$, as it is impossible to have a path of negative width in the context of the problem.
So, $x = 1.5$
The width of the path is 1.5 m.
Check: Does $(10 + 2 \times 1.5) \times (8 + 2 \times 1.5) = 143$?

Practice Exercise 14.5

1. This rectangle has an area of $21\,\text{cm}^2$.
   (a) Form an equation in x.
   (b) By solving your equation, find the value of x.

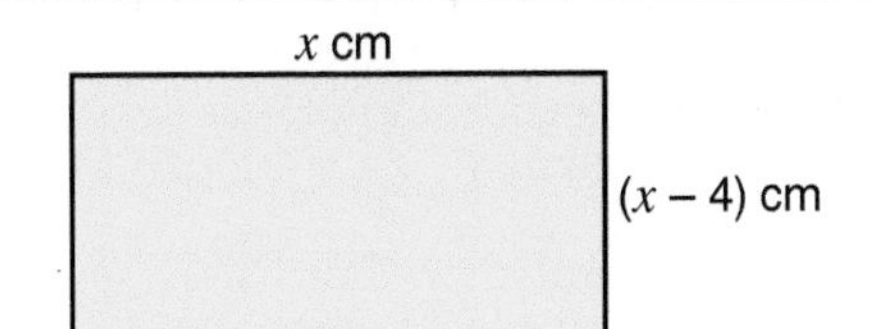

2. A right-angled triangle is cut from the corner of a rectangle as shown.
   The shaded area is $7.5\,\text{cm}^2$.

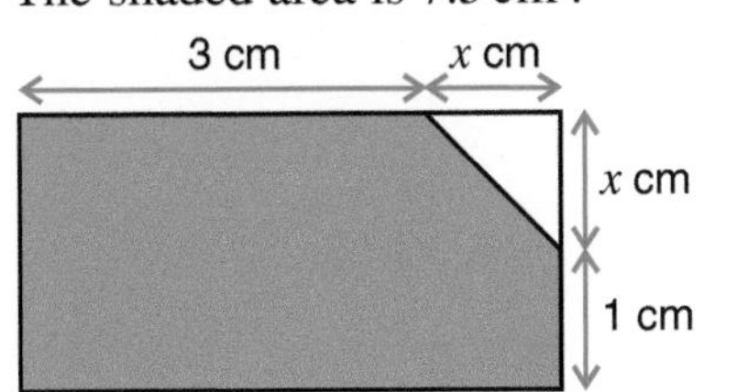

   (a) Form an equation in x and show that it simplifies to $x^2 + 8x - 9 = 0$.
   (b) By solving the equation $x^2 + 8x - 9 = 0$, find the value of x.

3. An open box is made from a square of card of side x cm, by removing squares of side 2 cm from each corner.
   (a) Show that the volume, $V\text{cm}^3$, of the box is given by $V = 2x^2 - 16x + 32$.
   (b) Calculate the value of x when $V = 60\,\text{cm}^3$.
   Give your answer correct to one decimal place.

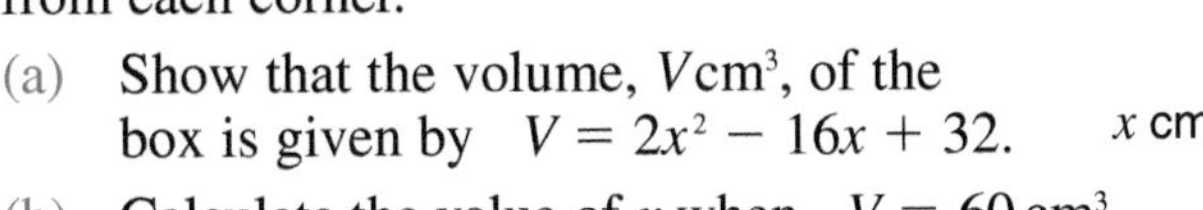

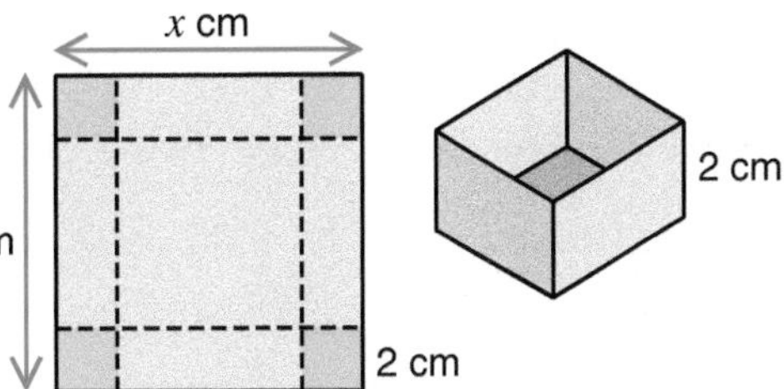

4. Liquid from a storage tank is being pumped out at such a rate that the volume, V litres, of liquid remaining in the tank t minutes after the start is given by the formula $V = 2t^2 - 24t + 72$.
   (a) How much liquid is in the tank at the start when $t = 0$?
   (b) How much liquid is in the tank after 2 minutes?
   (c) How long does it take to empty the tank?

5. An object is slid down a slope so that its distance s metres from the top of the slope after t seconds is given by the equation $s = 5 + 3t + t^2$.
   (a) How far is the object from the top of the slope at the starting time?
   (b) How far is the object from the top after 1 second?
   (c) The slope is 25 m long.
   To the nearest 0.1 second, how long will it take the object to reach the bottom of the slope?

Key Points

- **Quadratic equations** can be solved:
  by factorising, graphically or by using the quadratic formula.
- The general form for a **quadratic equation** is $ax^2 + bx + c = 0$, where a cannot be zero.
- The solutions to a quadratic equation can be found using the **quadratic formula**.

  If $ax^2 + bx + c = 0$ and $a \neq 0$ then $x = \dfrac{-b \pm \sqrt{b^2 - 4ac}}{2a}$

- The **discriminant** of the quadratic equation $ax^2 + bx + c = 0$ is given by $b^2 - 4ac$.

  If $b^2 - 4ac > 0$, the quadratic equation has **two** real and distinct roots.
  If $b^2 - 4ac = 0$, the quadratic equation has **one** repeated real root.
  If $b^2 - 4ac < 0$, the quadratic equation has **no** real roots.

Review Exercise 14

1. (a) Draw the graph of $y = x^2 + 2x - 2$ for values of x from -5 to 3.
   (b) Use your graph to solve the equation $x^2 + 2x - 2 = 0$.

2. Solve the equation $(x - 2)(x + 1) = 0$.

3. (a) Factorise $x^2 - 2x - 15$.
   (b) Hence, or otherwise, solve the equation $x^2 - 2x - 15 = 0$.

4. (a) (i) Factorise $t^2 - 4t$. (ii) Hence, solve $t^2 - 4t = 0$.
   (b) (i) Factorise $y^2 + 3y + 2$. (ii) Hence, solve $y^2 + 3y + 2 = 0$.

5. Find the solutions of the equation $y^2 - 6y - 3 = 0$.
   Give your answers correct to two decimal places.

6. Solve these equations.
   (a) $x^2 - 9 = 0$, (b) $x^2 - 9x = 0$, (c) $x^2 - 8x - 9 = 0$,
   (d) $x^2 - 9x + 8 = 0$, (e) $x^2 - 9x - 8 = 0$, giving your answers to 1 decimal place.

7. Solve the equation $3x^2 + 2x - 2 = 0$.
   Give your answer correct to two decimal places.

8. These two rectangles have the same area.

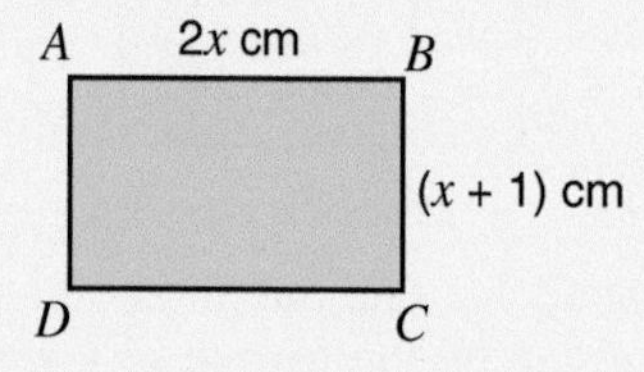

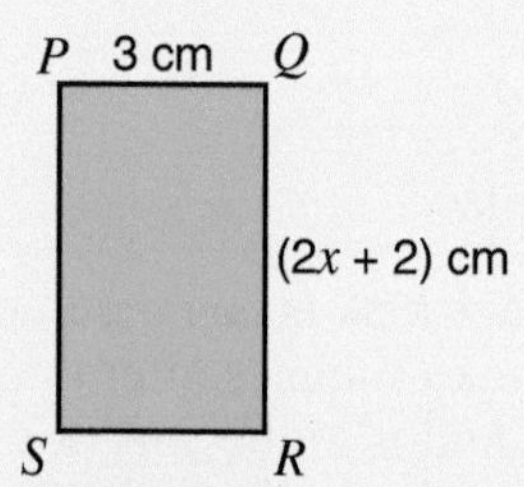

   (a) Form an equation in x and show that it can be simplified to $2x^2 - 4x - 6 = 0$.
   (b) Solve the equation $2x^2 - 4x - 6 = 0$ and find the length of BC.

9. The solutions of the equation $x^2 - 6x + 1 = 0$ can be written in the form $x = a \pm b\sqrt{2}$, where a and b are whole numbers.
   Find the values of a and b.

10. A rectangular lawn, 20 m by 16 m, is surrounded by a concrete path.
    The width of the path is x m.
    The area of the lawn is twice the total area of the path.
    Form an equation in x and solve it to find the width of the path.

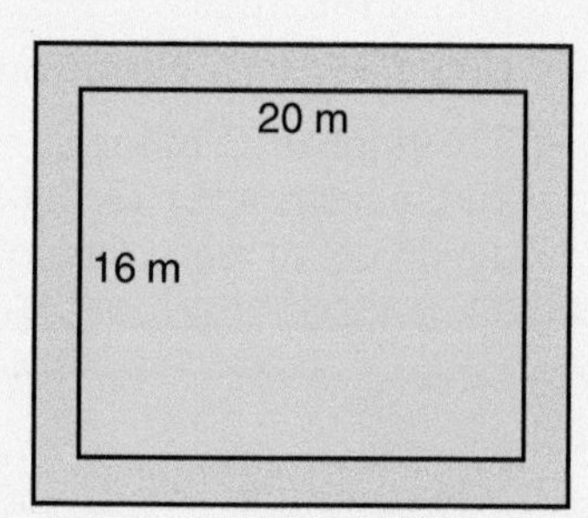

11. If $x = -4$ is the solution of equation $2x^2 + 5x + k = 0$, find the numerical value of k.

12. (a) Find the value of p if $x^2 - 4x - p = 0$ has two real roots.
    (b) Find the value of q if $x^2 + 2qx + 36 = 0$ has one real root.

13. A rectangle has length $(x + 2)$ cm and width $(x + 1)$ cm.
    The rectangle has an area of 6 cm^2.
    (a) Form an equation and show that it can be simplified to $x^2 + 3x - 4 = 0$.
    (b) Find the discriminant for $x^2 + 3x - 4 = 0$ and state the nature of the roots.
    (c) Solve the equation $x^2 + 3x - 4 = 0$ and find the possible dimensions of the rectangle.

15 Working with Arcs *and* Sectors

Length of arc

Circumference of a circle is given by $C = \pi d$ or $C = 2\pi r$.
An **arc** is part of the circumference.
The length of an arc is proportional to the angle at the centre of the circle.

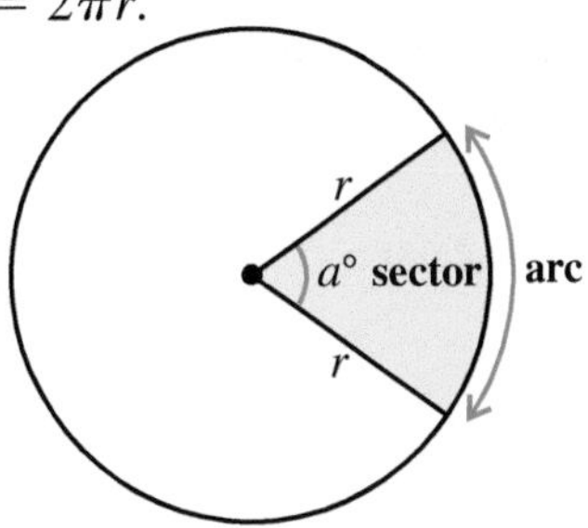

For a sector with angle $a°$:
Length of arc $= \frac{a}{360} \times \pi d$

Example 1

This shape is a sector of a circle with radius 9 cm and angle 80°.
Calculate the length of arc AB,

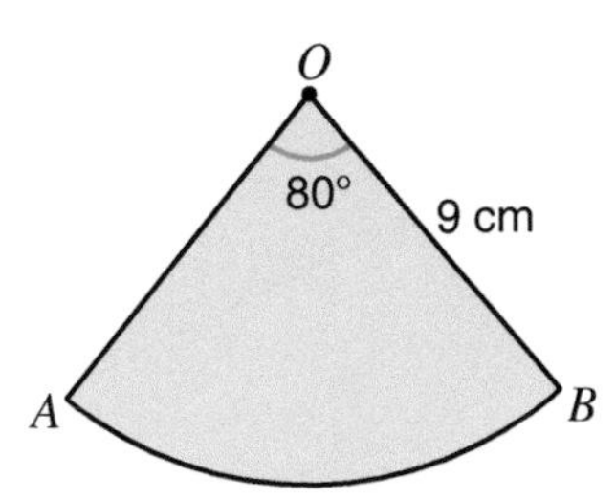

Length $AB = \frac{a}{360} \times \pi d$

$= \frac{80}{360} \times \pi \times 18$

$= 12.56...$

$= 12.6$ cm, to 3 s.f.

Practice Exercise 15.1

Take π to be 3.14 or use the π key on your calculator.
Give answers correct to three significant figures.

1. Calculate the length of the arc of each sector.

(a)

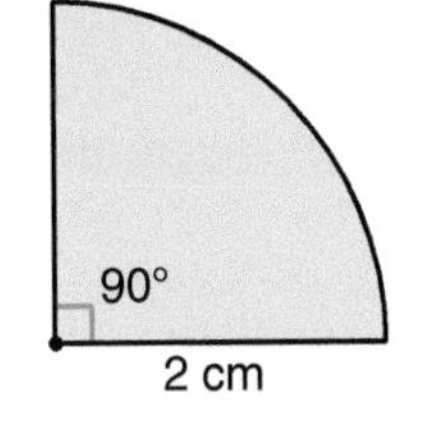

(b)

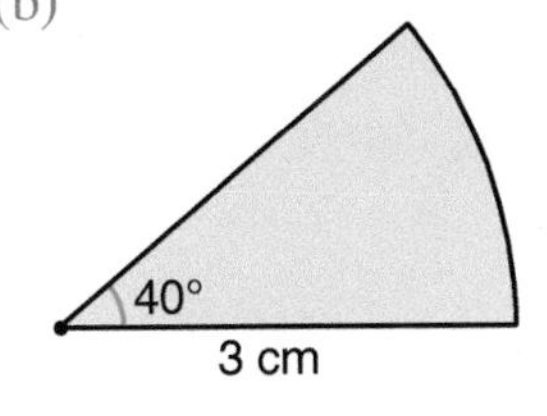

(c)

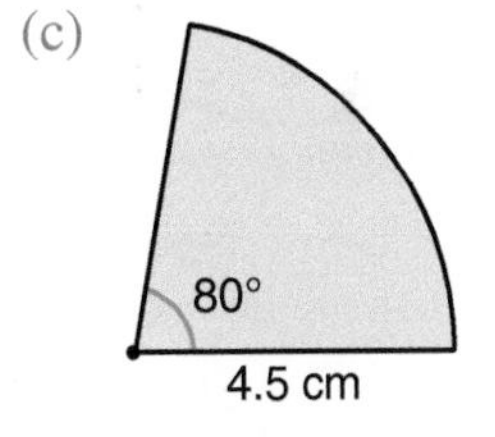

(d)

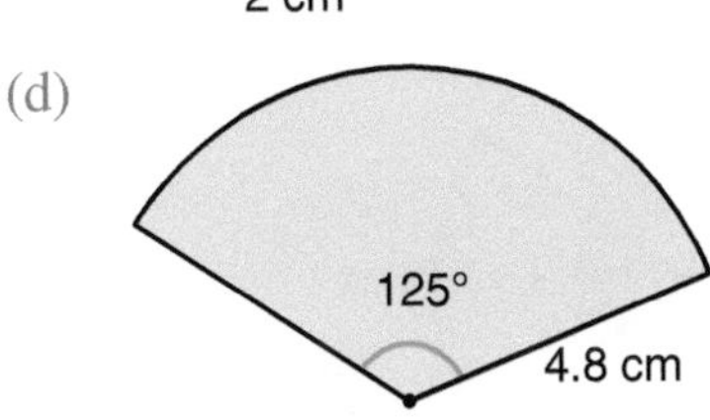

(e)

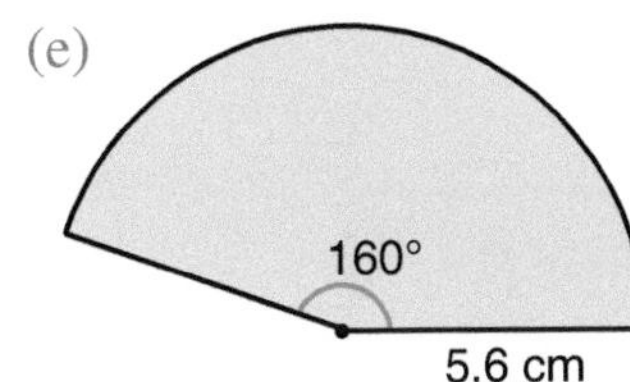

(f)

2. A and B are two points on the circumference of a circle with centre O.
Calculate length of arc AB if:

(a) $\angle AOB = 20°$ and $OA = 5.4$ cm,
(b) $\angle AOB = 140°$ and $OA = 9.3$ cm,
(c) $\angle AOB = 45°$ and $OA = 3.7$ cm,
(d) $\angle AOB = 120°$ and $OA = 12.7$ cm,
(e) $\angle AOB = 270°$ and $OA = 2.5$ cm.

Give your answers correct to 3 significant figures.

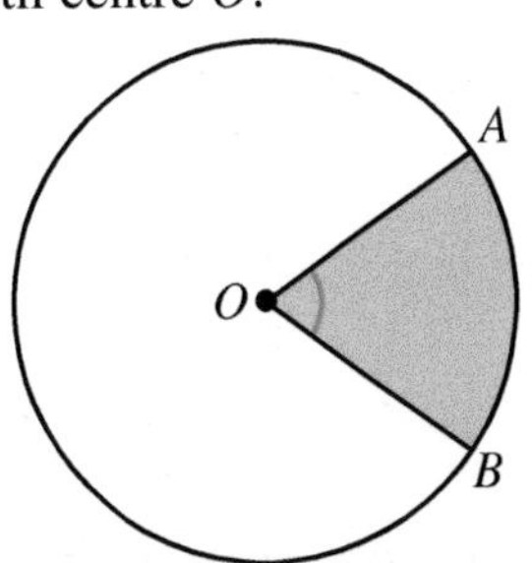

3. In the diagram, O is the centre of the circle.
What fraction of the circumference of the circle is the length of the arc AB, if:

(a) $\angle AOB = 30°$,

(b) $\angle AOB = 45°$,

(c) $\angle AOB = 144°$?

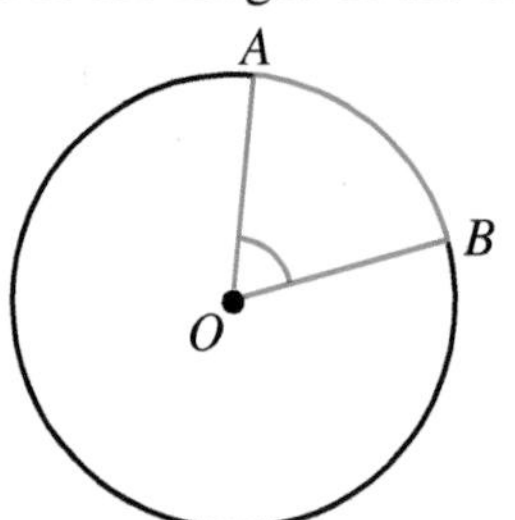

What is the size of $\angle AOB$ if:

(d) arc $AB = \frac{3}{8}$ of the circumference,

(e) arc $AB = \frac{2}{3}$ of the circumference?

4. 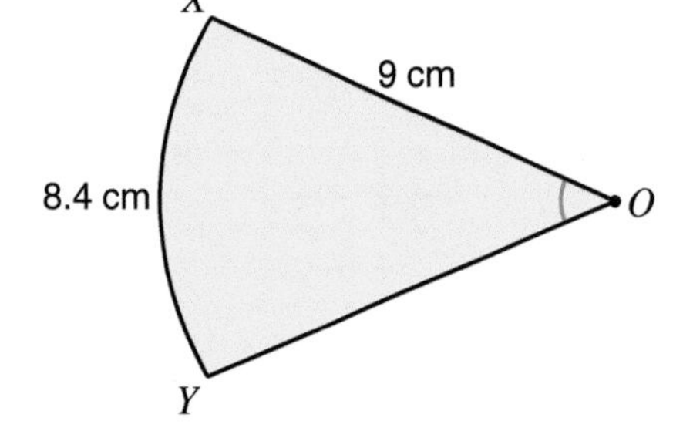

OXY is a sector of a circle, centre O, with radius 9 cm.
The arc length of the sector is 8.4 cm.

Calculate the size of angle XOY.

5. OPQ is a sector of a circle, centre O.
The arc length of the sector is 15.6 cm.
Angle $POQ = 135°$.

Calculate the radius of the circle.

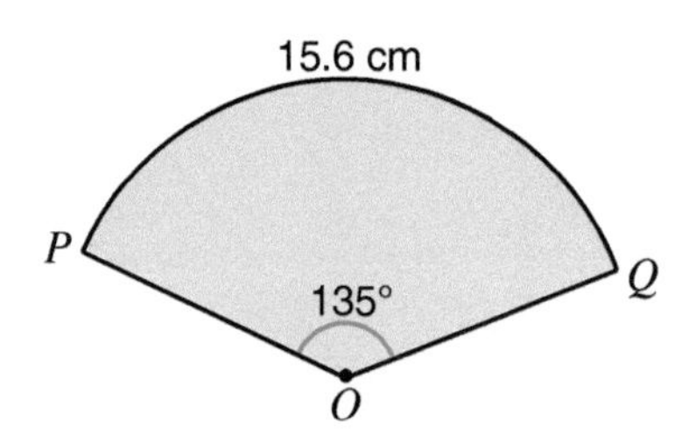

6. 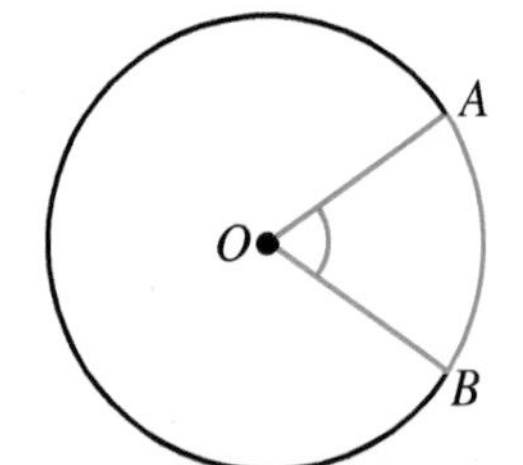

O is the centre of the circle.

(a) If $\angle AOB = 108°$ and the length of the arc $AB = 4.2$ cm, find the radius of the circle.

(b) If the length of the arc $AB = 8.0$ cm and $OA = 6.2$ cm, find the size of $\angle AOB$.

Area of a sector

Area of a circle is given by $A = \pi r^2$.
A **sector** of a circle is formed by two radii.
The area of a sector is proportional to the angle at the centre of the circle.

For a sector with angle $a°$:
Area of sector $= \frac{a}{360} \times \pi r^2$

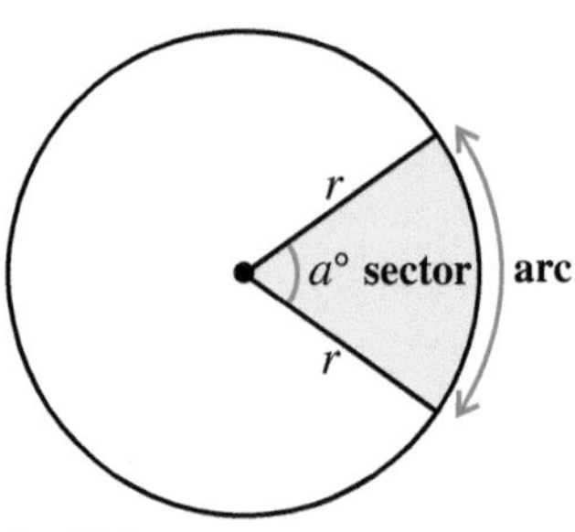

Example 2

This shape is a sector of a circle with radius 9 cm and angle 80°.
Calculate the area of sector OAB.

Area $OAB = \frac{a}{360} \times \pi r^2$

$= \frac{80}{360} \times \pi \times 9^2$

$= 56.54\ldots$

$= 56.5\text{ cm}^2$, to 3 s.f.

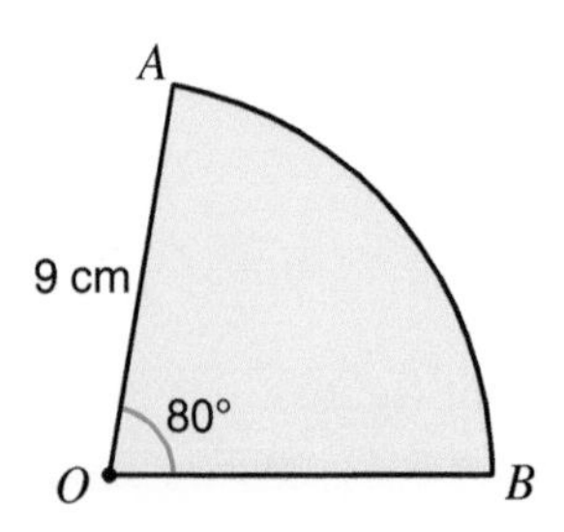

Practice Exercise 15.2

Take π to be 3.14 or use the π key on your calculator.
Give answers correct to three significant figures.

1. Calculate the area of each sector.

(a)

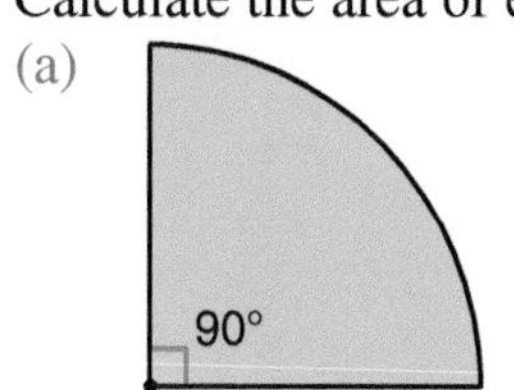

(b) 50°, 4 cm

(c)

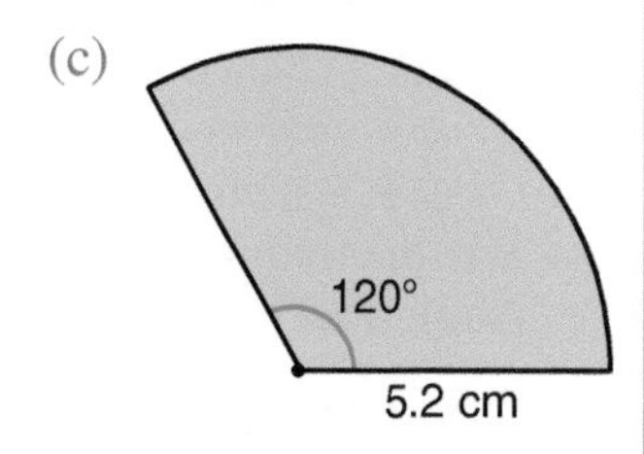

(d) 195°, 6.3 cm

(e)

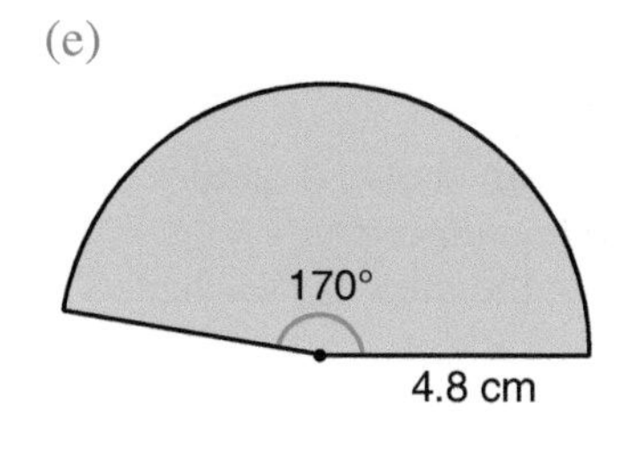

(f)

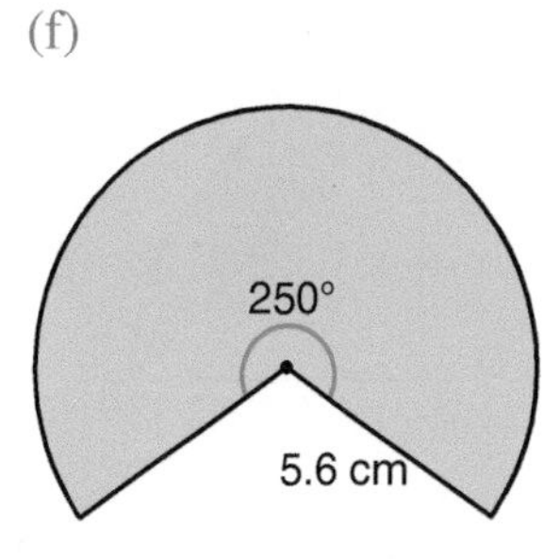

2. In the diagram, O is the centre of the circle.
What fraction of the area of the circle is the area of the shaded sector, if:

(a) $\angle AOB = 135°$,
(b) $\angle AOB = 72°$,
(c) $\angle AOB = 210°$?

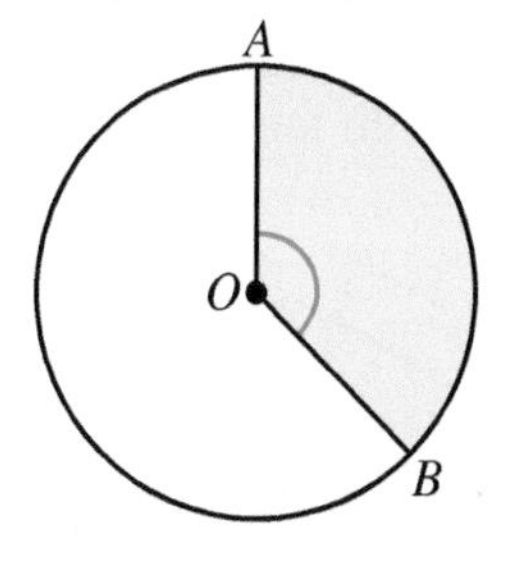

What is the size of $\angle AOB$ if:

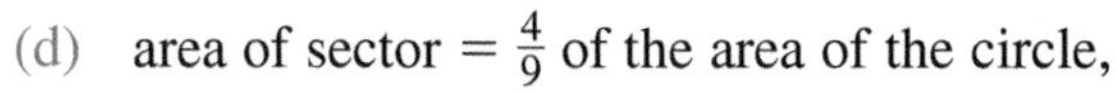

(d) area of sector $= \frac{4}{9}$ of the area of the circle,
(e) area of sector $= \frac{5}{6}$ of the area of the circle,
(f) area of sector $= \frac{3}{10}$ of the area of the circle?

3.

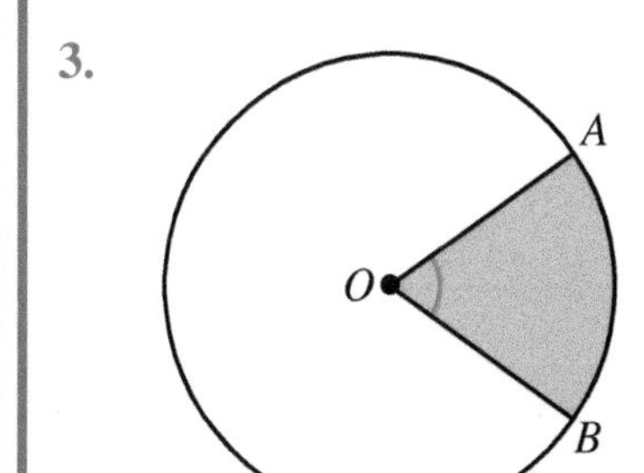

A and B are two points on the circumference of a circle with centre O.
Calculate the area of sector AOB if:

(a) $\angle AOB = 36°$ and $OA = 7.5$ cm,
(b) $\angle AOB = 57°$ and $OA = 4.8$ cm,
(c) $\angle AOB = 74°$ and $OA = 9.3$ cm,
(d) $\angle AOB = 135°$ and $OA = 6.7$ cm,
(e) $\angle AOB = 240°$ and $OA = 15$ cm.

Give your answers correct to 3 significant figures.

4. OAB is a sector of a circle, centre O, with radius 9 cm.
The sector has an area of 13.5 cm².

Calculate the size of angle a.

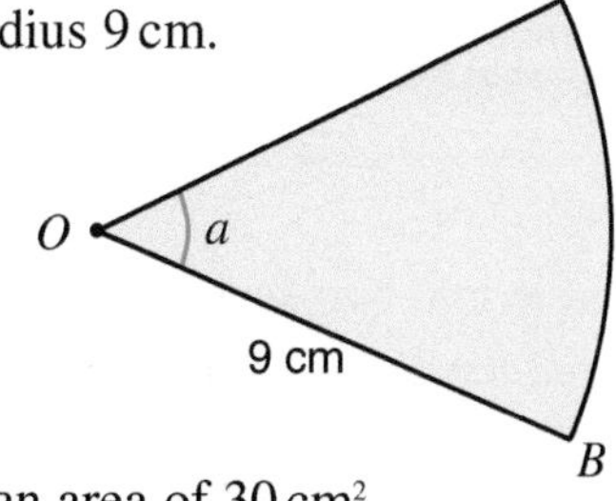

5.

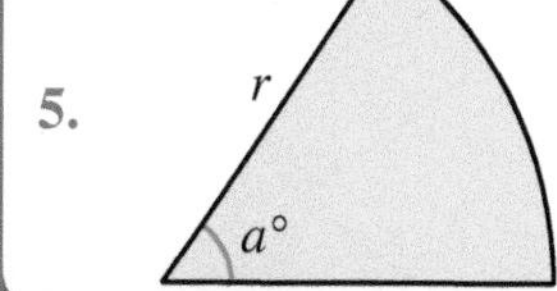

A sector of a circle has an area of 30 cm².
Angle $a = 72°$.

Calculate the radius of the circle.

Compound shapes

Shapes formed by joining different shapes together are called **compound shapes**.
To find the area of a compound shape we must first divide the shape up into rectangles, triangles, circles, etc, and then find the area of each part.
Shapes can be divided in different ways, but they should all give the same answer.

Practice Exercise 15.3

1. OPQ is a sector of a circle with radius 20 cm.
   X is the midpoint of OP and Y is the midpoint of OQ.
   Angle $POQ = 65°$.

   Calculate (a) the area of $XPQY$,
   (b) the perimeter of $XPQY$.

2. 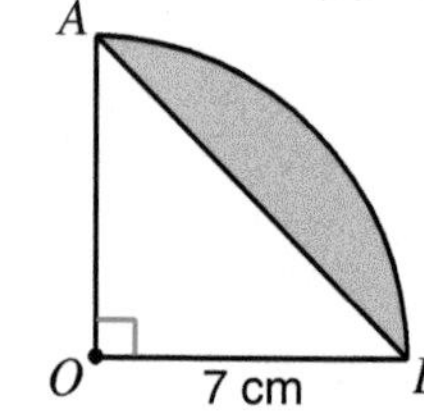

   The diagram shows a quadrant of a circle, centre O.
   Calculate the area of the shaded segment.

3. A lampshade is made from a piece of material $ABCD$, as shown.
   Angle $AOB = 35°$, $OA = OB = 50$ cm,
   $OC = OD = 75$ cm.
   (a) Calculate the area of the sector OAB.
   (b) Calculate the area of $ABCD$.
   (c) Calculate the arc length AB.

4. 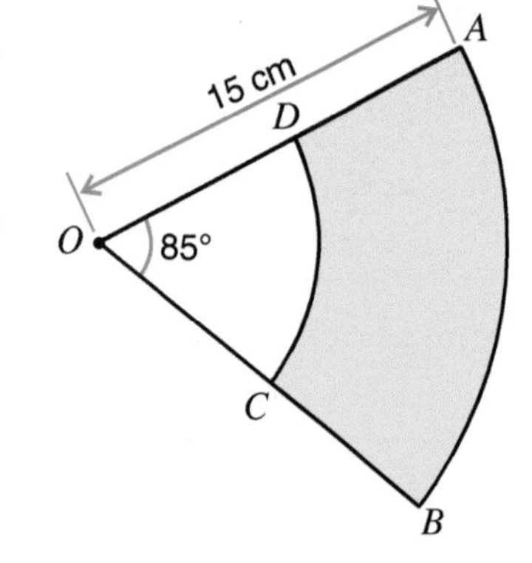

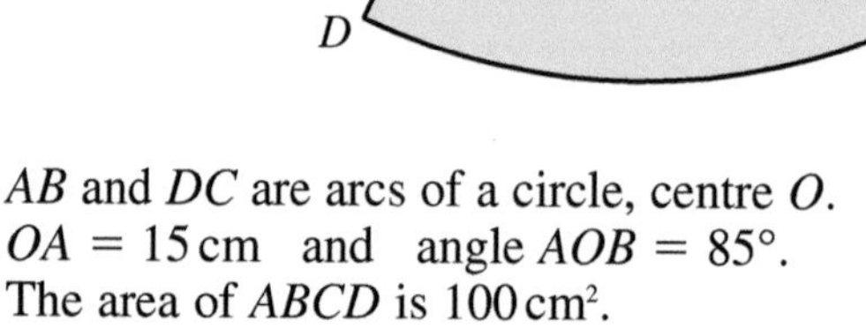

   AB and DC are arcs of a circle, centre O.
   $OA = 15$ cm and angle $AOB = 85°$.
   The area of $ABCD$ is 100 cm².

   Calculate (a) the length of OD,
   (b) the length of the arc DC.

5. A piece of jewellery is made from a metal square $ABCD$ of side 5 cm.
   The two curves are parts of circles with centres at A and C.
   The piece of jewellery is formed from the shaded area on the diagram.
   Find the area of the piece of jewellery.

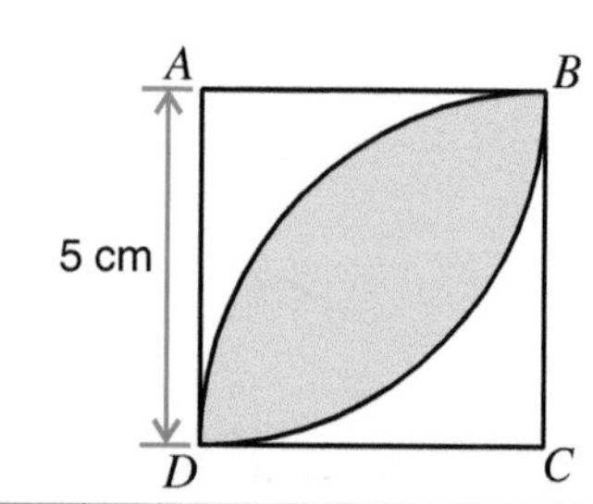

Key Points

- Two **radii** divide a circle into two **sectors**.
- The **lengths of arcs** and the **areas of sectors** are proportional to the angle at the centre of the circle.

For a sector with angle $a°$:
Length of arc $= \frac{a}{360} \times \pi d$ Area of sector $= \frac{a}{360} \times \pi r^2$

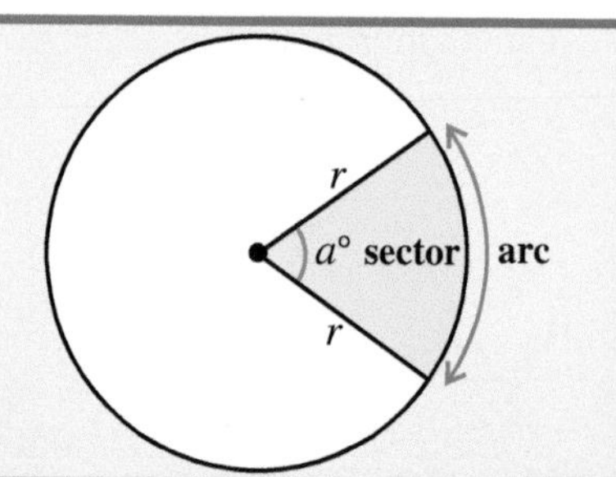

Review Exercise 15

1. O is the centre of the circle.

   (a) Find the length of the arc AB if the radius is 4.5 cm and $\angle AOB = 40°$.

   (b) Find the area of the sector AOB if the radius is 3 cm and $\angle AOB = 150°$.

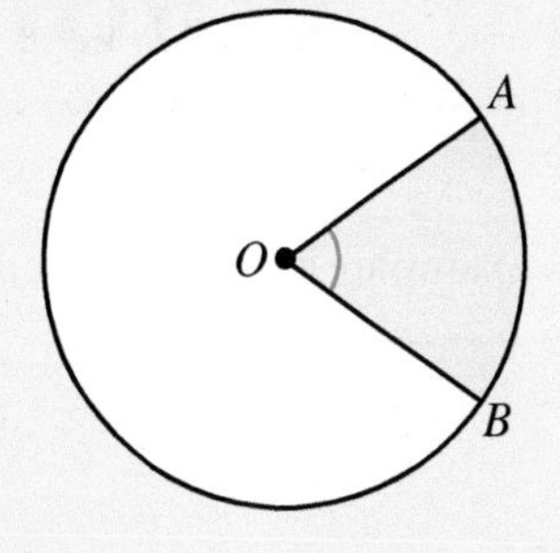

2. 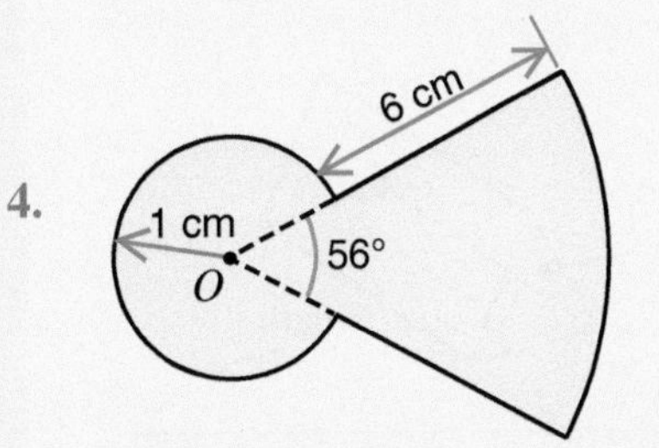

   The diagram shows the cross-section of a building. The roof is an arc of a circle, radius 50 m, centre O. Angle $AOB = 90°$.

   (a) Calculate the arc length AB of the roof.

   (b) Calculate the area of the cross-section.

3. A wiper blade on a windscreen cleans the clear area, as shown.

   Calculate the area of the windscreen cleaned by the wiper.

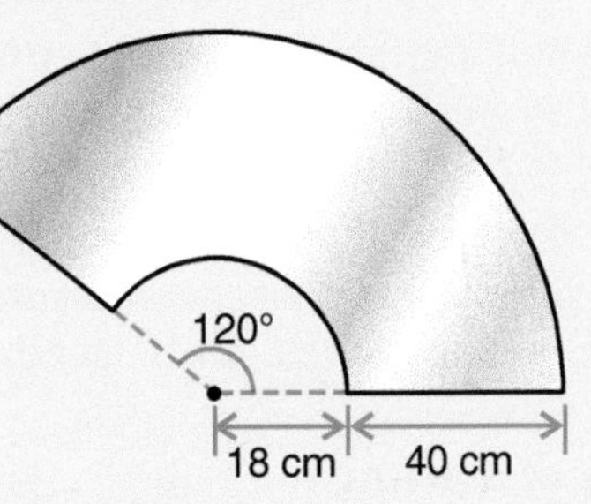

4. 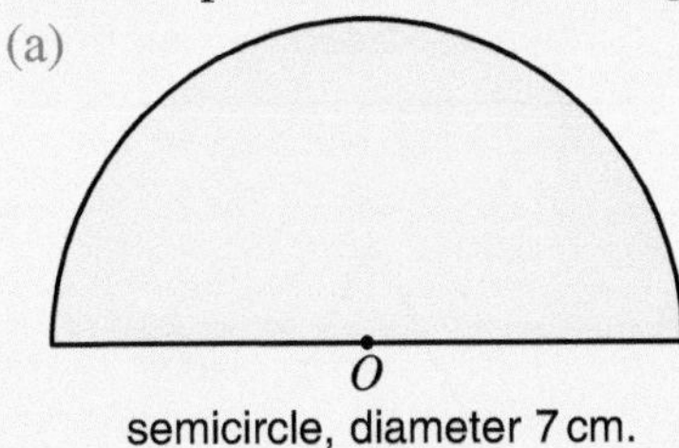

   (a) Calculate the total area of this shape.

   (b) Calculate the perimeter of this shape.

5. Find the perimeters of these figures, giving the answers to the nearest mm.

   (a)

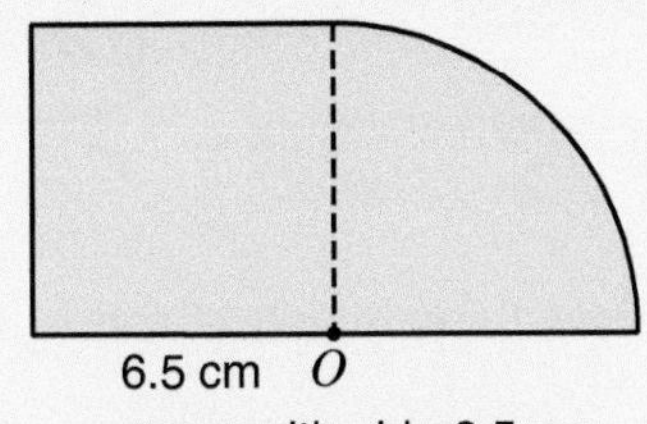

   semicircle, diameter 7 cm.

   (b)

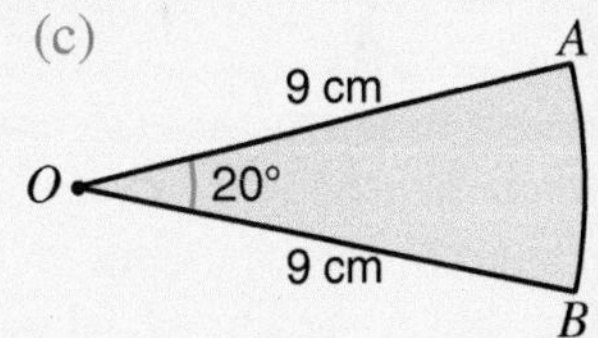

   (c)

   square with side 6.5 cm, quadrant, centre O.

   arc AB has centre O.

6. OAB is a sector of radius 9.5 cm. The length of the arc, AB, is 7.2 cm.

   Calculate angle AOB.

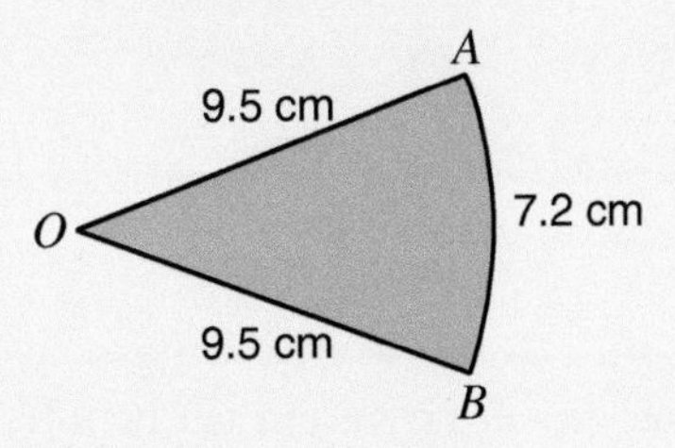

7. 

   O is the centre of the circle.

   (a) The circle has circumference 400 metres. The length of arc $AB = 60$ metres. Calculate angle AOB.

   (b) Find the radius of the circle if $\angle AOB = 150°$ and the length of arc AB is 12.6 cm.

16 Volumes of Solids

These are all examples of 3-dimensional objects or solids.

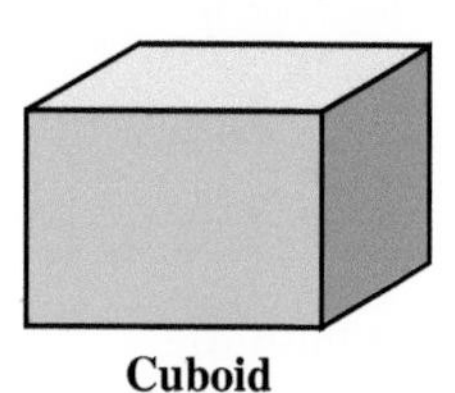

Cuboid

Cylinder

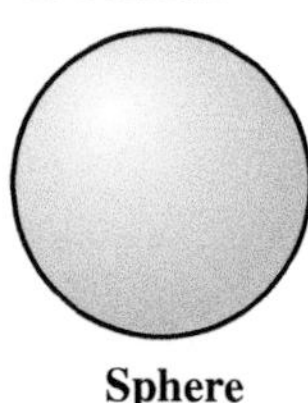

Sphere

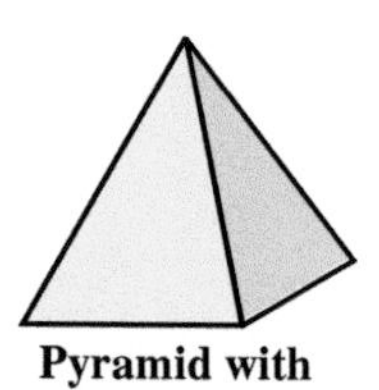

Pyramid with square base

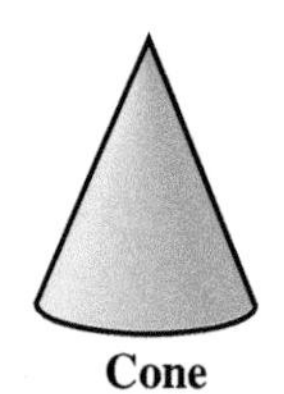

Cone

What other 3-dimensional objects do you know?

Volume

Volume is the amount of space occupied by a three-dimensional object.

The formula for the volume of a **cuboid** is:
Volume = length × breadth × height.
$V = l \times b \times h$

height
breadth
length

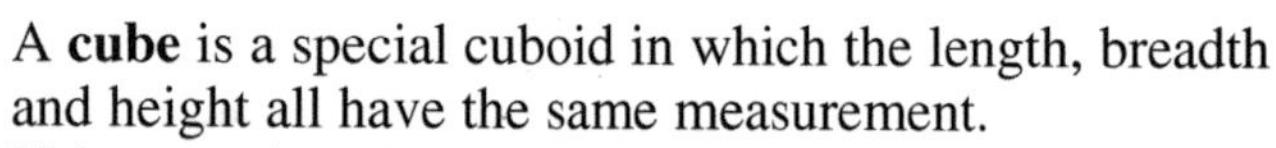

A **cube** is a special cuboid in which the length, breadth and height all have the same measurement.
Volume = length × length × length.
$V = l^3$

Volume of a prism

If you make a cut at right angles to the length of a prism you will always get the same cross-section.
Volume of a prism = area of cross-section × length.

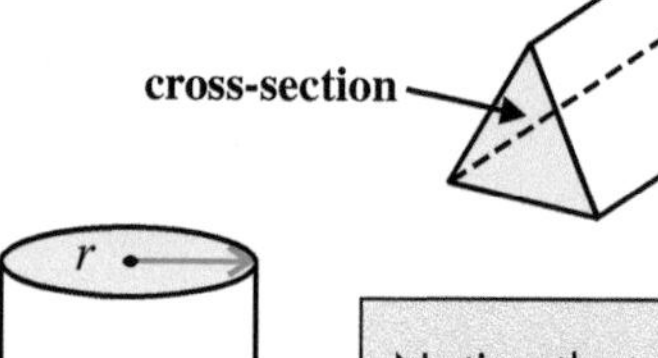

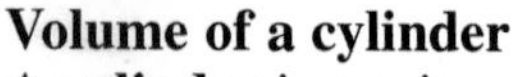

Volume of a cylinder

A **cylinder** is a prism.
The **volume of a cylinder** can be written as:
Volume = area of cross-section × height
$V = \pi r^2 h$

r
h

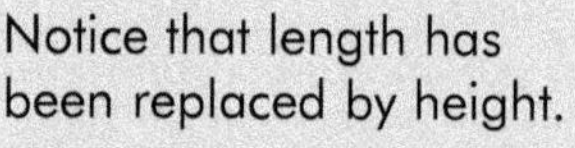

Notice that length has been replaced by height.

Cones, pyramids and spheres

The diagram shows a cone with:
circular base, radius r, slant height l, perpendicular height h.
Using Pythagoras' Theorem, $l^2 = r^2 + h^2$.

> Volume of a cone $= \frac{1}{3} \times$ base area $\times$ perpendicular height
> The area of the circular base $= \pi r^2$.
> $V = \frac{1}{3} \pi r^2 h$

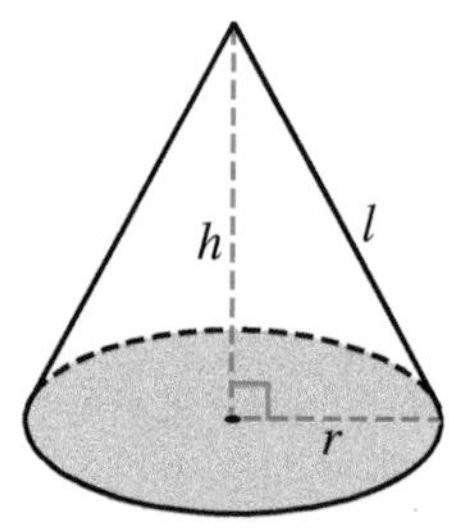

> The volume of a pyramid is given by:
> $V = \frac{1}{3} \times$ base area $\times$ perpendicular height

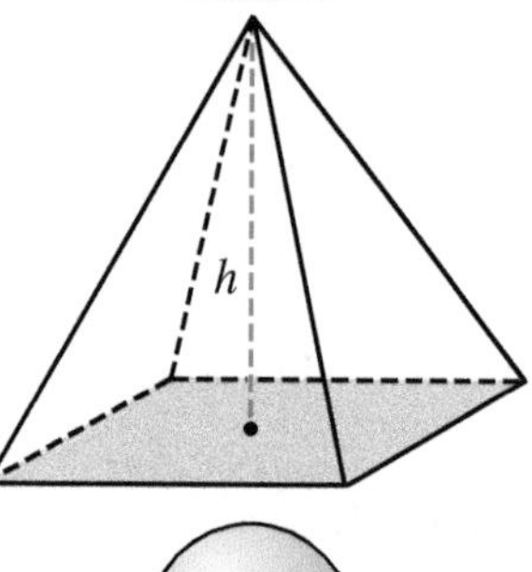

> The volume of a sphere is given by:

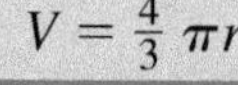

> $V = \frac{4}{3} \pi r^3$

Example 1

This model was formed by joining together a pyramid and a cuboid.
The total height of the model is 5.5 cm.
Find the total volume of the model.

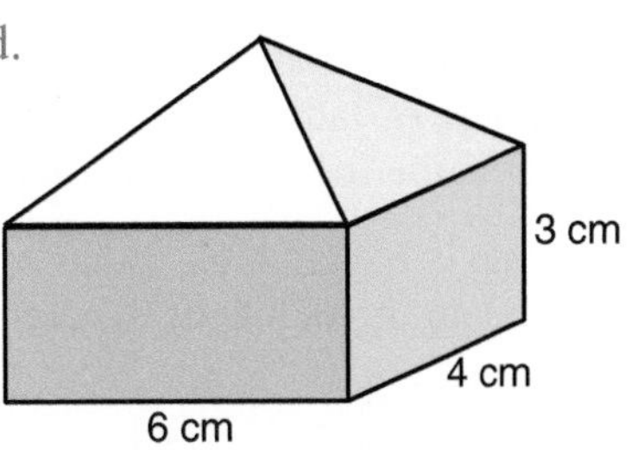

Total volume = volume of pyramid + volume of cuboid.

Volume of pyramid = $\frac{1}{3}$ × base area × perpendicular height.

The base of the pyramid is a rectangle measuring 6 cm by 4 cm.
The height of the pyramid = 5.5 cm − 3 cm = 2.5 cm.

$$\text{Volume of pyramid} = \tfrac{1}{3} \times 6 \times 4 \times 2.5$$
$$= 20\,\text{cm}^3$$
$$\text{Volume of cuboid} = lbh$$
$$= 6 \times 4 \times 3$$
$$= 72\,\text{cm}^3$$
$$\text{Total volume} = 20 + 72$$
$$= 92\,\text{cm}^3$$

Practice Exercise 16.1

Use the π key on your calculator and give answers correct to 3 significant figures where appropriate.

1. Find the volumes of these solids.

(a)

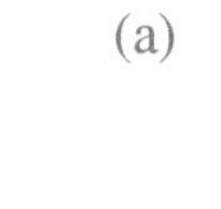

(b)

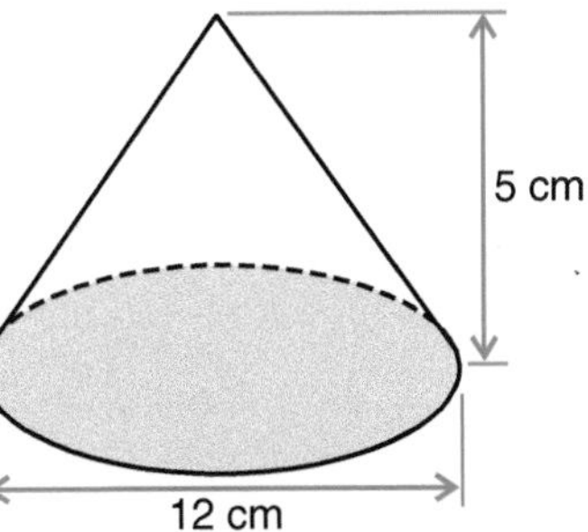

(c)

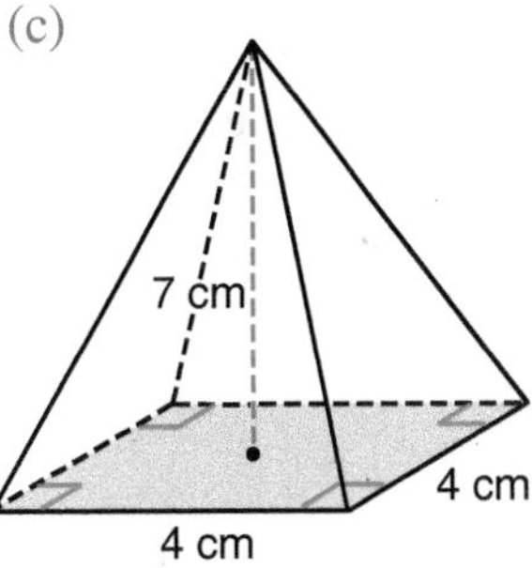

(d)

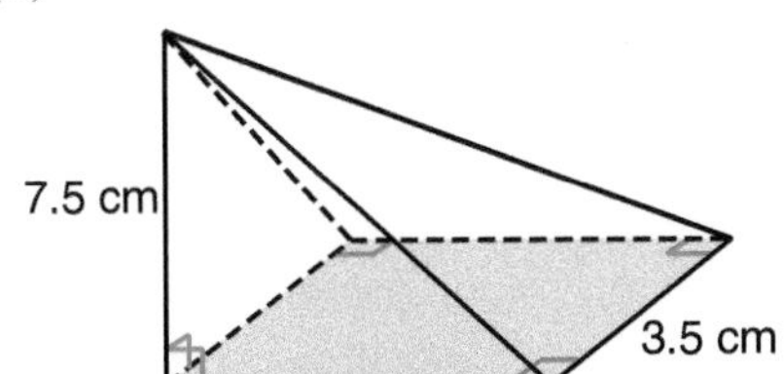

(e)

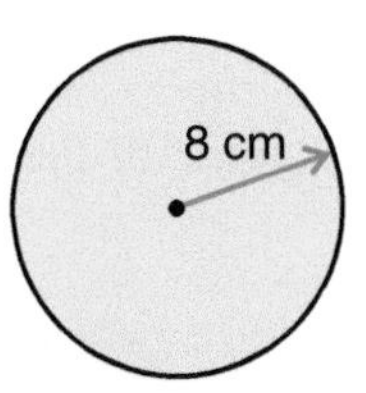

Sphere, radius 8 cm

(f)

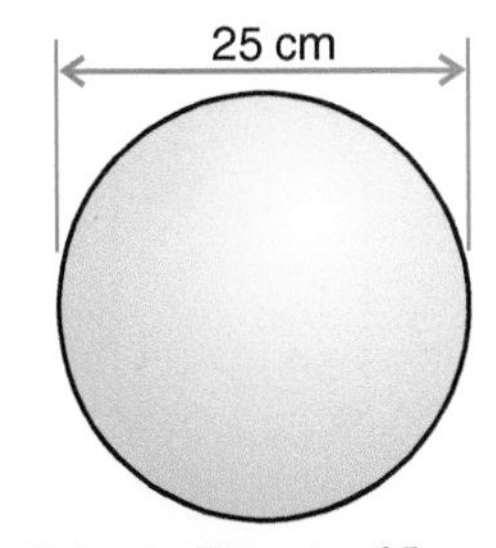

Sphere, diameter 25 cm

2. Which of these containers has the greater volume?
Show all your working.

A hemispherical bowl of radius 6 cm

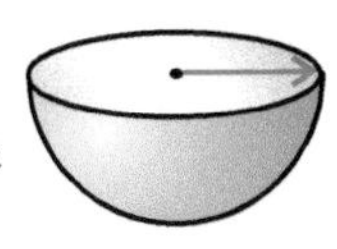

A cone 15 cm high with radius 5 cm

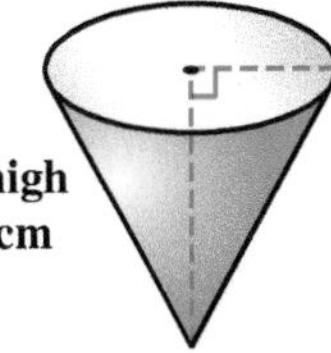

3. A pyramid with base area 20 cm² has volume 250 cm³.
What is the height of the pyramid?

4. A metal cylinder is melted down and made into balls for a game.
The cylinder is 15 cm high and has radius 6 cm. The balls each have radius 1 cm.
How many balls can be made?

5. One hundred ball bearings with radius 5 mm are dropped into a cylindrical can, which is half full of oil. The height of the cylinder is 20 cm and the radius is 8 cm.
By how much does the level of the oil rise?

6. A cylinder of radius 4 cm contains water to a height of 4 cm.
A sphere of radius 3.6 cm is placed in the cylinder.
What is the increase in the depth of the water?

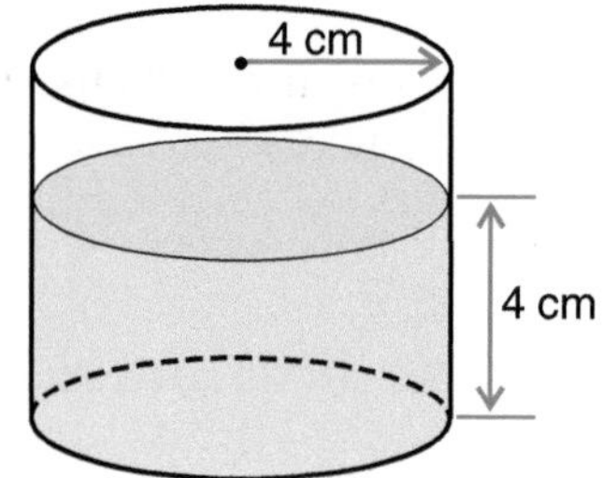

7. A cone is 6.4 cm high. It has a volume of 150 cm^3.
Calculate the radius of the cone.

8. A sphere has a volume of 58 cm^3.
Calculate the radius of the sphere.

9. A quadrant of a circle is cut out of paper.
A cone is made by joining the edges OA and OB, with no overlaps.
Calculate (a) the length of the arc AB,
(b) the base radius of the cone,
(c) the height of the cone,
(d) the volume of the cone.

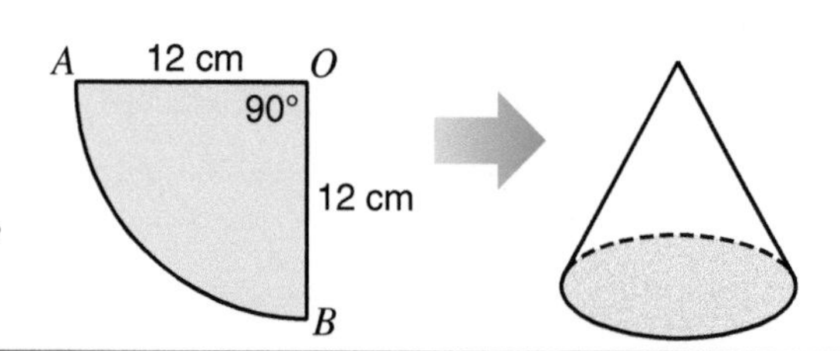

Practice Exercise 16.2

Use the π key on your calculator and give answers correct to 3 significant figures where appropriate.

1. Find the total volumes of these solids.

(a)

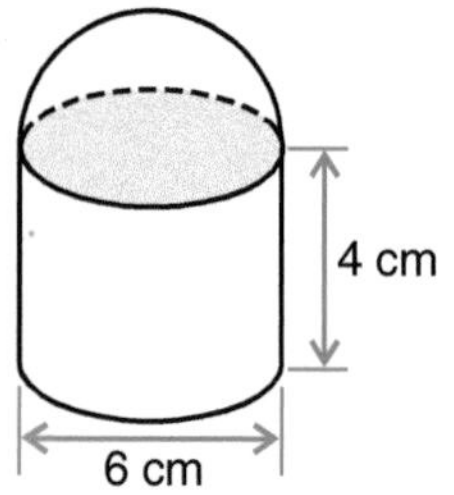

(b)

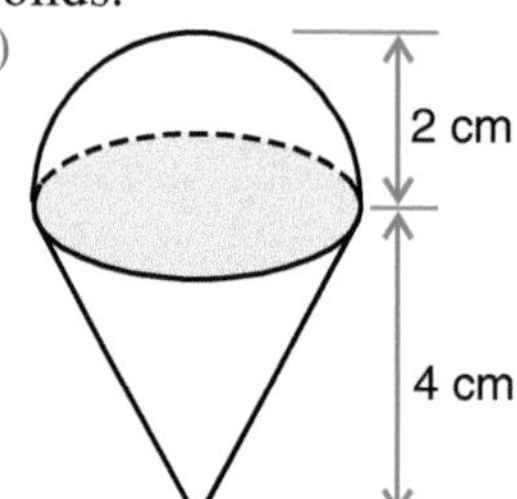

(c)

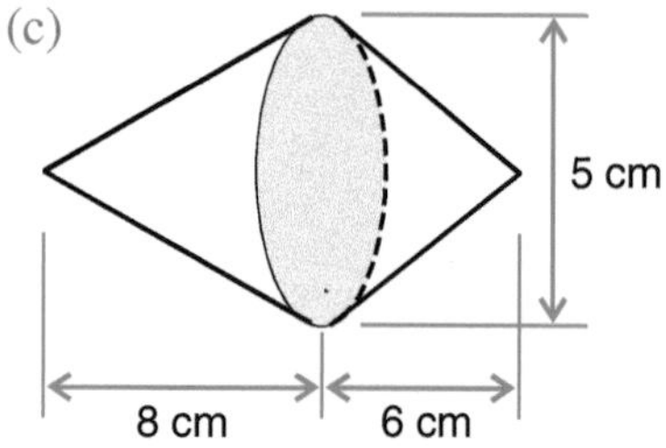

2.

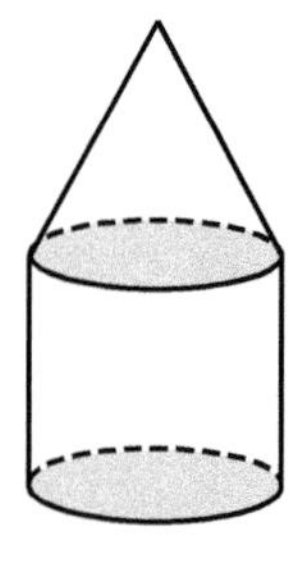

A container is made by joining a cylinder and a cone.
The cylinder has a radius 15 cm and height 25 cm.
The cone has height 25 cm.
Calculate the volume of the container.

3. Rubber bungs are made by removing the tops of cones.
Starting with a cone of radius 10 cm and height 16 cm, a rubber bung is made by cutting a cone of radius 5 cm and height 8 cm from the top.
Find the volume of the rubber bung.

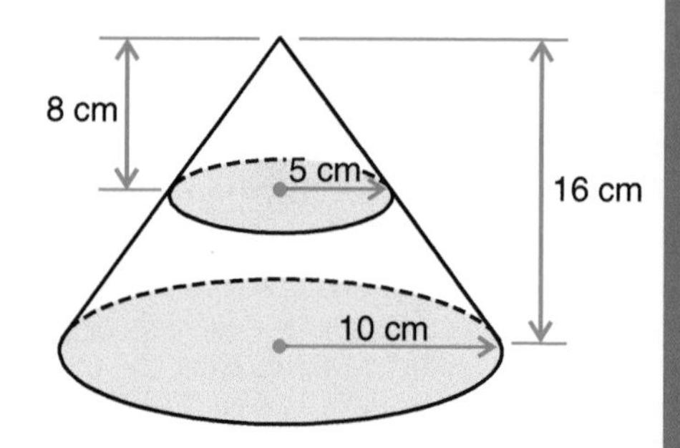

4.

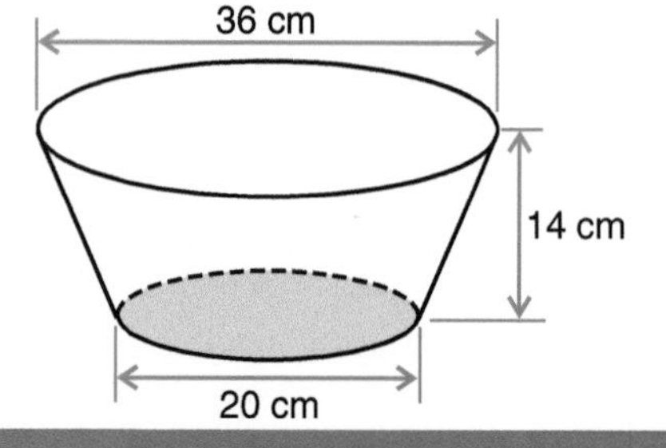

The dimensions of a steel bowl are shown.
Calculate the volume of the bowl.

5. A child's toy is made from a cone with base radius 3 cm and height 4 cm mounted on a hemisphere with radius 3 cm. Calculate the volume of the toy.

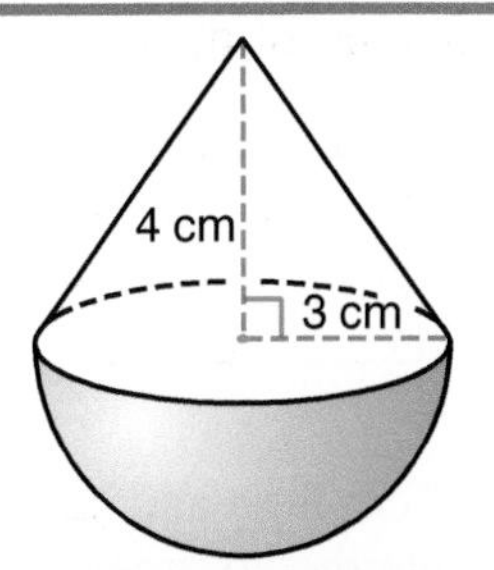

Key Points

- **Volume** is the amount of space occupied by a 3-D object.
- The formula for the volume of a **cuboid** is:

  Volume = length × breadth × height

  $V = l \times b \times h$

height
breadth
length

- Volume of a **cube** is: Volume = $(\text{length})^3$

  $V = l^3$

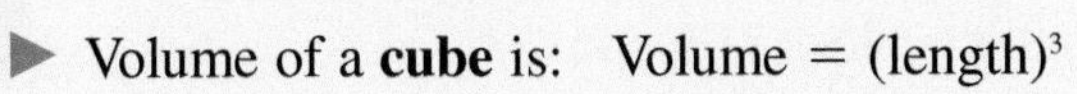

- A prism is a 3-D object with the same cross-section throughout its length.

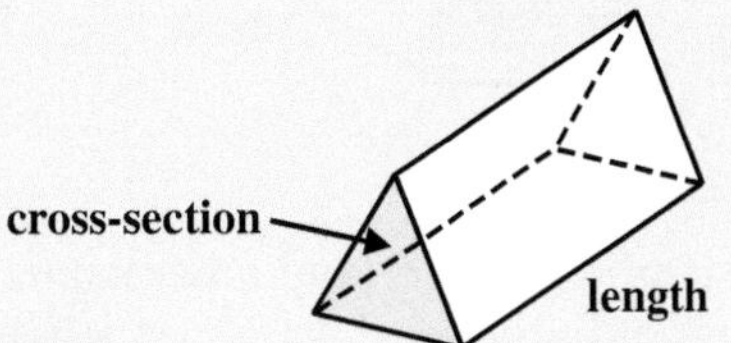

- Volume of a prism = area of cross-section × length
- A **cylinder** is a prism.

  Volume of a cylinder is: $V = \pi \times r^2 \times h$

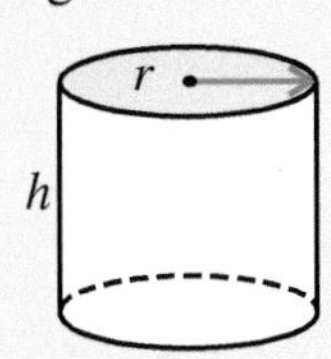

- These formulae are used in calculations involving **cones, pyramids,** and **spheres.**

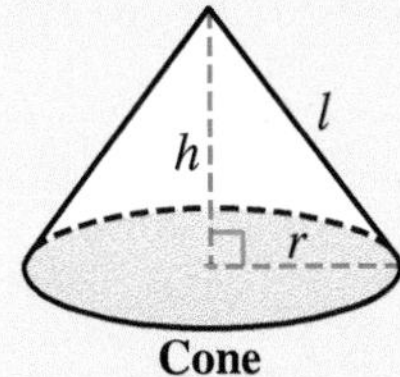

Cone

$V = \frac{1}{3} \times \text{base area} \times \text{height}$

$V = \frac{1}{3}\pi r^2 h$

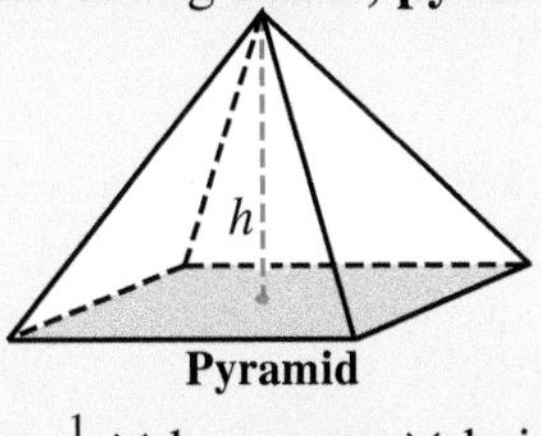

Pyramid

$V = \frac{1}{3} \times \text{base area} \times \text{height}$

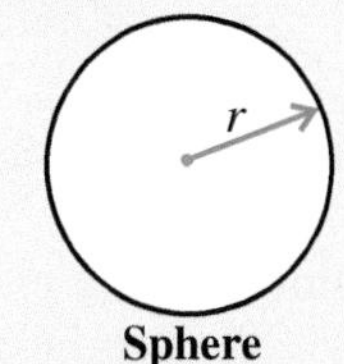

Sphere

Volume = $\frac{4}{3}\pi r^3$

Review Exercise 16

Use the π key on your calculator and give answers correct to 3 significant figures where appropriate.

1. Find the volumes of these solids.

(a)

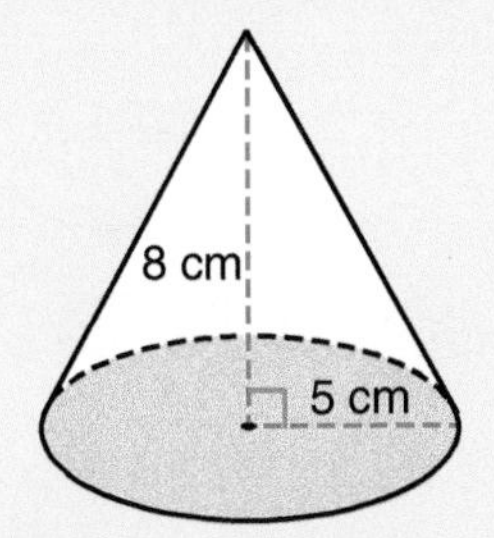

(b)

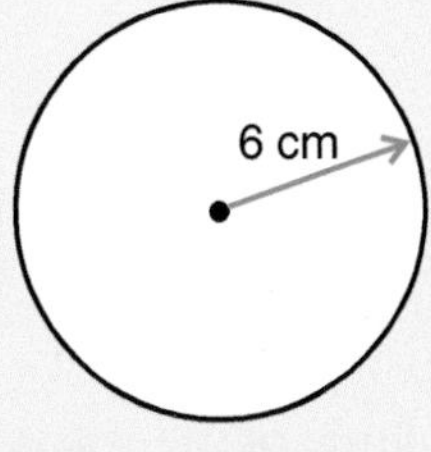

Sphere, radius 6 cm

(c)

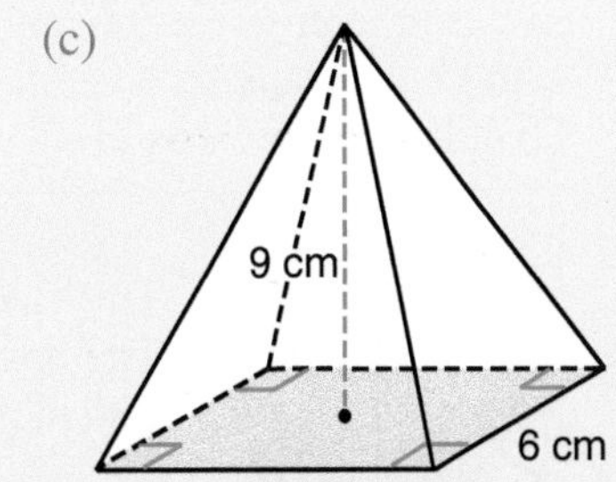

2. A projectile consists of a conical nose cap, length 1.2 m, diameter 48 cm; a cylindrical body, length 3 m, diameter 48 cm; and a hemispherical tail, diameter 48 cm. Find the volumes of the nose, the body and the tail, and, hence, find the total volume of the projectile.

3. A cone is 10 cm high and has a base radius of 6 cm.
   (a) Calculate the volume of the cone.

   The top of the cone is cut off to leave a frustum 8 cm high.
   (b) Calculate the volume of the frustum.

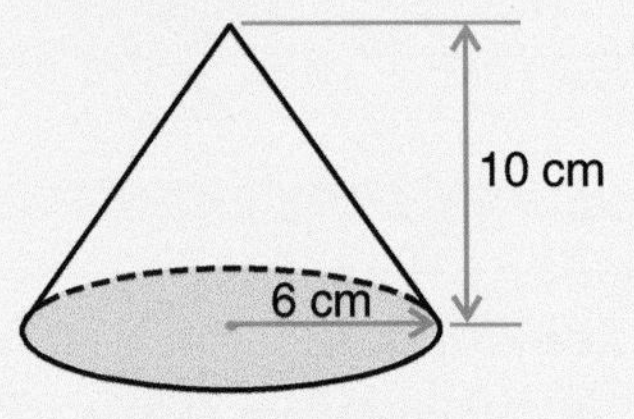

4. The diagram shows a block of wood. The block is a cuboid measuring 8 cm by 13 cm by 16 cm. A cylindrical hole of radius 5 cm is drilled through the block of wood. Find the volume of wood remaining.

5. The diagram shows a pepper pot. The pot consists of a cylinder and a hemisphere. The cylinder has a diameter of 4 cm and a height of 6 cm. The pepper takes up half the **total** volume of the pot. Find the depth of pepper in the pot, marked x in the diagram.

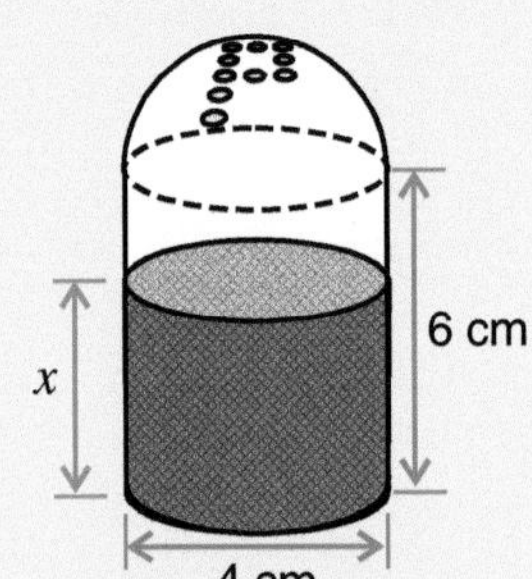

6. The diagram shows a cuboid which is just big enough to hold six tennis balls. Each tennis ball has a diameter of 6.8 cm. Calculate the volume of the cuboid.

7. The diagram shows a triangular pyramid. Angle $ABC = 90°$, $AB = 5$ cm and $BC = 3$ cm. The volume of the pyramid is 28 cm³. Calculate the height of the pyramid.

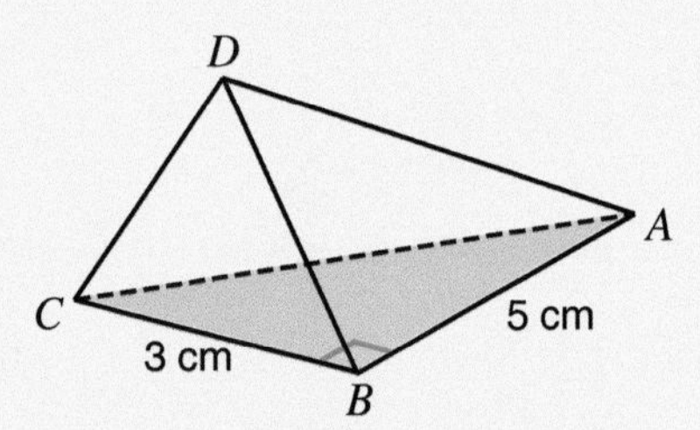

8. 

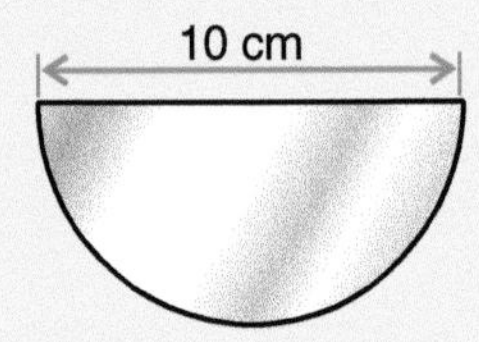

A cone is formed from a semicircular sheet of foil. The foil has a diameter of 10 cm.
Find (a) the radius of the base of the cone,
(b) the volume of the cone.

9. A sector of a circle of radius 12 cm is cut out of card and used to create a cone by joining OX to OY, with no overlaps, as shown. Calculate the volume of the cone.

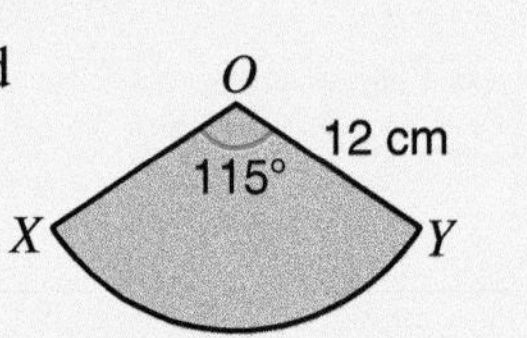

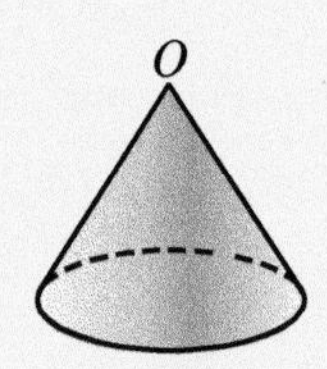

10. 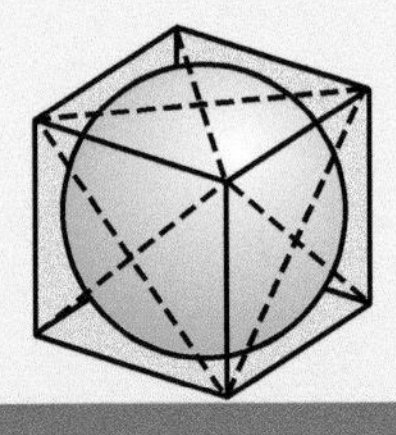

This diagram shows a sphere of radius 4.5 cm fitting tightly inside a box. The box is a cube. Calculate the volume of the space around the sphere, inside the cube.

17 Pythagoras' Theorem

The longest side in a right-angled triangle is called the **hypotenuse**.
In any right-angled triangle it can be proved that:

"The square on the hypotenuse is equal to the sum of the squares on the other two sides."

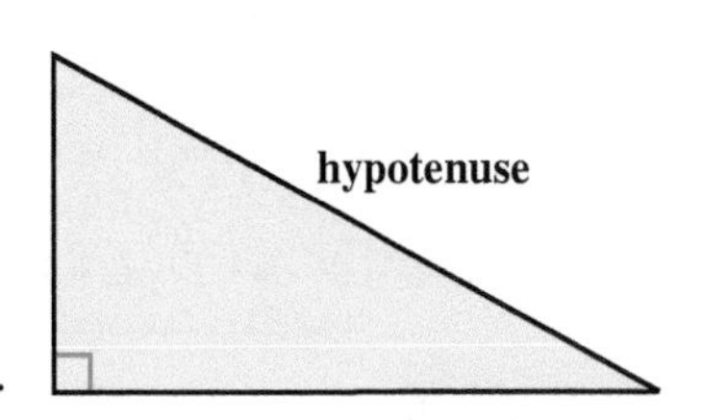

This is known as the **Theorem of Pythagoras**, or **Pythagoras' Theorem**.

When we know the lengths of two sides of a right-angled triangle, we can use the Theorem of Pythagoras to find the length of the third side.

$a^2 = b^2 + c^2$
Rearranging gives: $b^2 = a^2 - c^2$
$c^2 = a^2 - b^2$

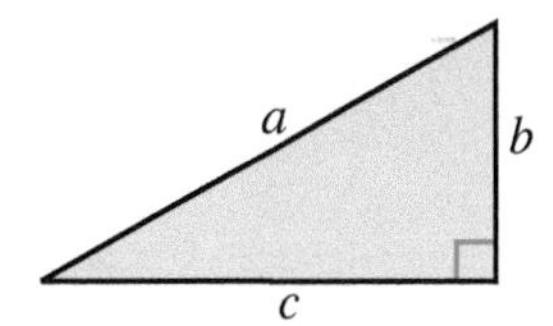

Finding the hypotenuse

Example 1

The roof of a house is 12 m above the ground.
What length of ladder is needed to reach the roof, if the foot of the ladder has to be placed 5 m away from the wall of the house?

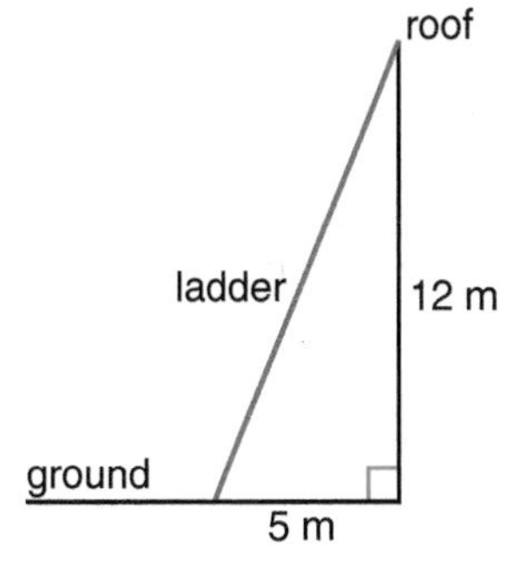

Using Pythagoras' Theorem.
$l^2 = 5^2 + 12^2$
$l^2 = 25 + 144$
$l^2 = 169$

Take the square root of both sides.
$l = \sqrt{169}$
$l = 13$ m
The ladder needs to be 13 m long.

Finding one of the shorter sides

To find one of the shorter sides we can rearrange the Theorem of Pythagoras.

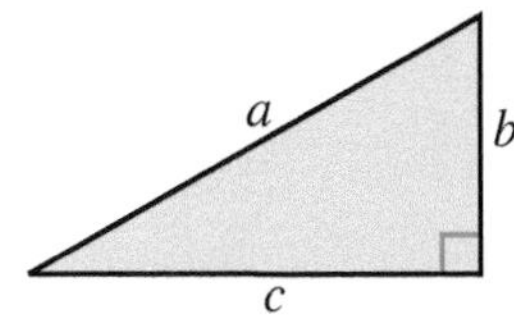

To find b we use: $b^2 = a^2 - c^2$
To find c we use: $c^2 = a^2 - b^2$

To find the length of a shorter side of a right-angled triangle:
Subtract the square of the known short side from the square on the hypotenuse.
Take the square root of the result.

Example 2

A wire used to keep a radio aerial steady is 9 metres long.
The wire is fixed to the ground 4.6 metres from the base of the aerial.
Find the height of the aerial, giving your answer correct to one decimal place.

Using Pythagoras' Theorem. $9^2 = h^2 + 4.6^2$

Rearranging this we get: $h^2 = 9^2 - 4.6^2$
$h^2 = 81 - 21.16$
$h^2 = 59.84$

Take the square root of both sides. $h = \sqrt{59.84}$
$h = 7.735...$
$h = 7.7$ m, correct to 1 d.p.

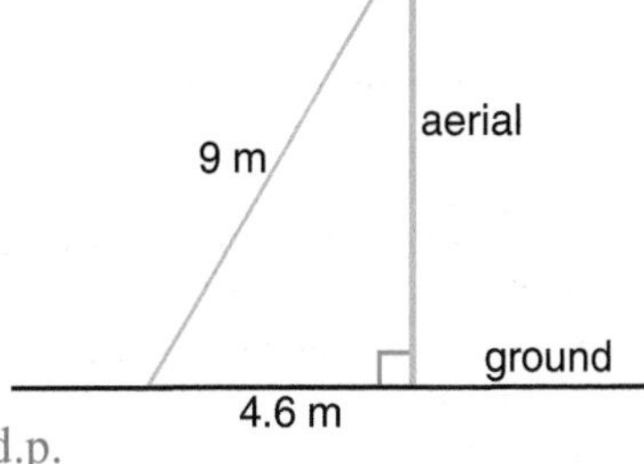

The height of the aerial is 7.7 m, correct to 1 d.p.

The converse of Pythagoras' Theorem

Pythagoras' Theorem is about a right-angled triangle.

The opposite way round for this statement is:
If, in a triangle with sides of length p, q, and r, $r^2 = p^2 + q^2$, then the angle opposite to side r is a right angle.
This result is called the **converse** of Pythagoras' Theorem.

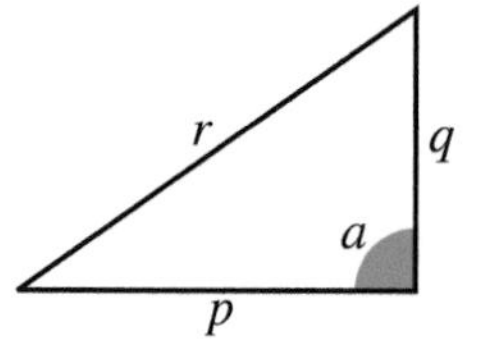

If $r^2 = p^2 + q^2$, then $a = 90°$.
If $r^2 < p^2 + q^2$, then $a < 90°$.
If $r^2 > p^2 + q^2$, then $a > 90°$.

Example 3

In triangle ABC, $AB = 12\,\text{cm}$, $BC = 35\,\text{cm}$ and $AC = 37\,\text{cm}$.
Is angle B a right angle?

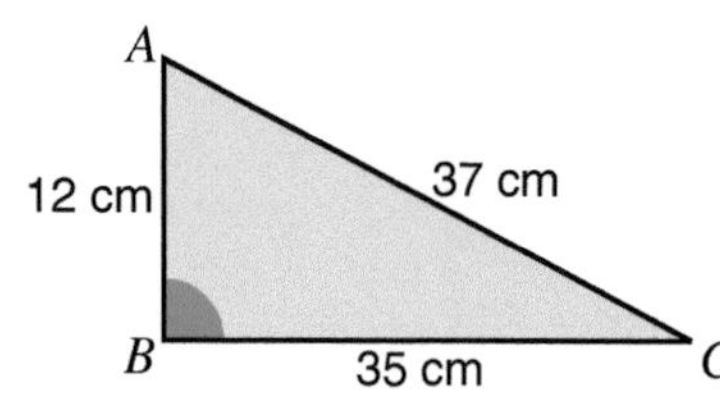

The longest side of the triangle is AC.
$AC = 37 \quad AC^2 = 37^2 = 1369$
The other two sides are AB and BC.

$$AB^2 + BC^2 = 12^2 + 35^2 = 144 + 1225 = 1369$$

So, $AC^2 = AB^2 + BC^2$
And, $\angle B = 90°$ (using the converse of Pythagoras' Theorem).

Practice Exercise 17.1

1. These triangles are right-angled. Calculate the length of the side marked with a letter. Give your answers correct to one decimal place, where appropriate.

(a)

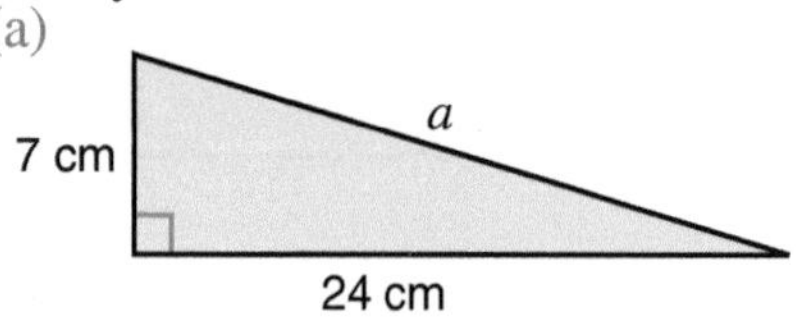

(b)

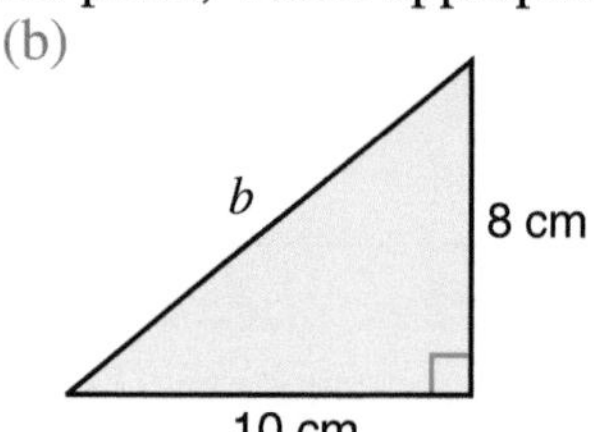

(c)

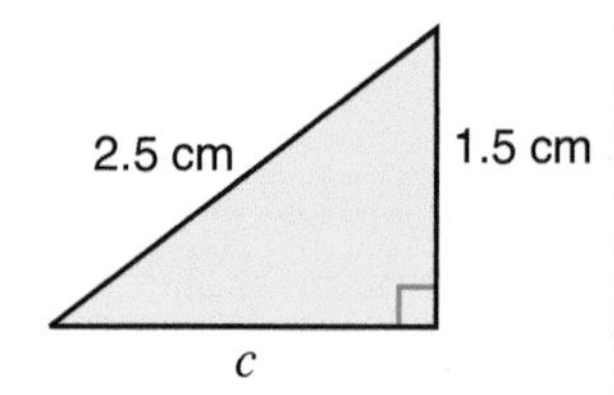

2. AB and CD are line segments, drawn on a centimetre-squared grid.
Calculate the exact length of
(a) AB,
(b) CD.

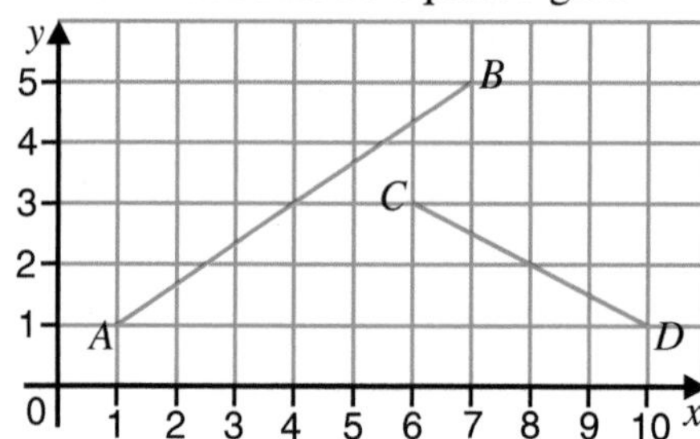

3. Calculate the distance between the following points.
(a) $A(2, 0)$ and $B(6, 3)$.
(b) $C(6, 3)$ and $D(0, 10)$.
(c) $E(2, 2)$ and $F(-3, -10)$.
(d) $G(-2, -2)$ and $H(-6, 5)$.
(e) $I(3, -1)$ and $J(-3, -5)$.

4. The coordinates of the vertices of a parallelogram are $P(1, 1)$, $Q(3, 5)$, $R(x, y)$ and $S(7, 3)$.
(a) Find the coordinates of R.
(b) X is the midpoint of PQ. Find the coordinates of X.
(c) Y is the midpoint of PS. Find the coordinates of Y.
(d) Calculate the distance XY.

5. 

Two boats A and B are 360 m apart.
Boat A is 120 m due east of a buoy.
Boat B is due north of the buoy.
How far is boat B from the buoy?

6. The diagram shows a right-angled triangle, ABC, and a square, $ACDE$.

$AB = 2.5\,\text{cm}$ and $BC = 6.5\,\text{cm}$.
Calculate the area of the square $ACDE$.

7. 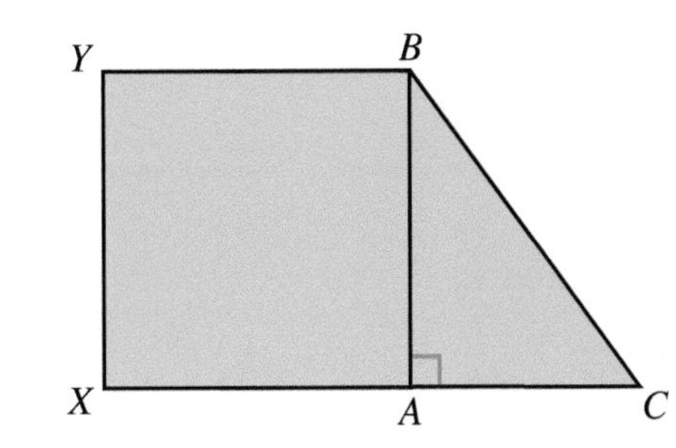

The diagram shows a right-angled triangle, ABC, and a square, $XYBA$.
$BC = 6\,\text{cm}$.
The square $XYBA$ has an area of $23.04\,\text{cm}^2$.
Calculate the length of AC.

8. Use the converse of Pythagoras' Theorem to find whether these triangles are right-angled.

(a) 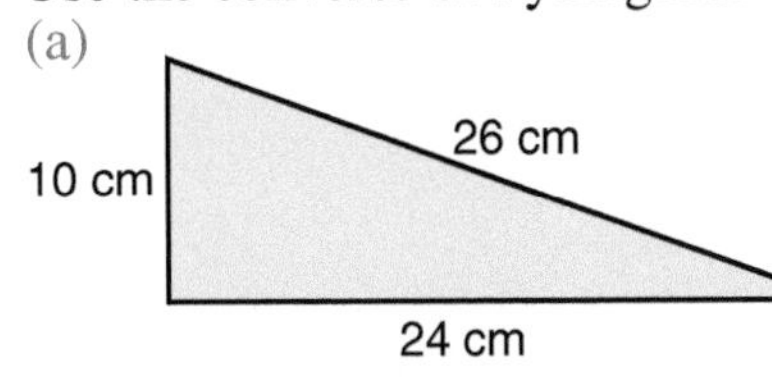

(b) 4 cm, 5 cm, 6 cm

(c) 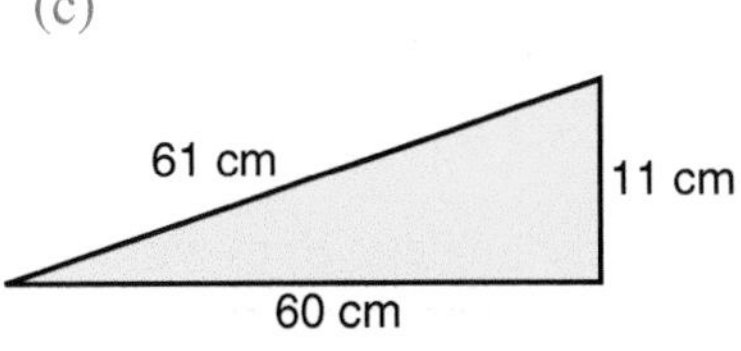

9. The points $A(3, 3)$, $B(5, 5)$ and $C(9, 1)$ form a triangle ABC.
Show that triangle ABC is right-angled and name the right angle.

10. In the diagram, BD is perpendicular to AC.
$AD = 4\,\text{cm}$, $BD = 2\,\text{cm}$ and $CD = 1\,\text{cm}$.

Prove that triangle ABC is a right-angled triangle.

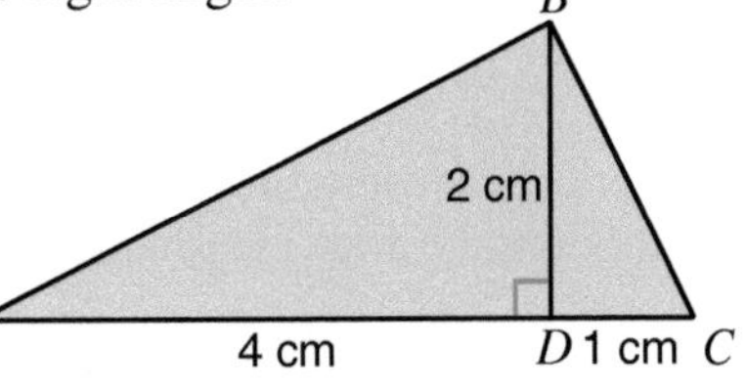

11. $P(-3, 5)$, $Q(2, 2)$ and $R(-2, -2)$.
Is triangle PQR right-angled?
You must show your working.

Problems involving the use of Pythagoras' Theorem

Questions leading to the use of Pythagoras' Theorem often involve:

Understanding the problem.
What information is given? What are you required to find?

Drawing diagrams.
In some questions a diagram is not given.
Drawing a diagram may help you to understand the problem.

Selecting a suitable right-angled triangle.
In more complex problems you will have to select a right-angled triangle which can be used to answer the question. It is a good idea to draw this triangle on its own, especially if it has been taken from a three-dimensional drawing.

Example 4

The diagram shows the side view of a swimming pool.
It slopes steadily from a depth of 1 m to 3.6 m.
The pool is 20 m long.
Find the length of the sloping bottom of the pool, giving the answer correct to three significant figures.

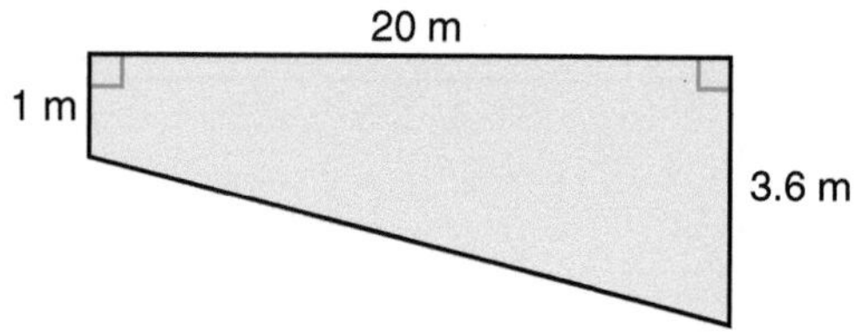

20 m
1 m
3.6 m

ΔCDE is a suitable right-angled triangle.

$CD = 3.6 - 1 = 2.6$ m

Using Pythagoras' Theorem in ΔCDE.

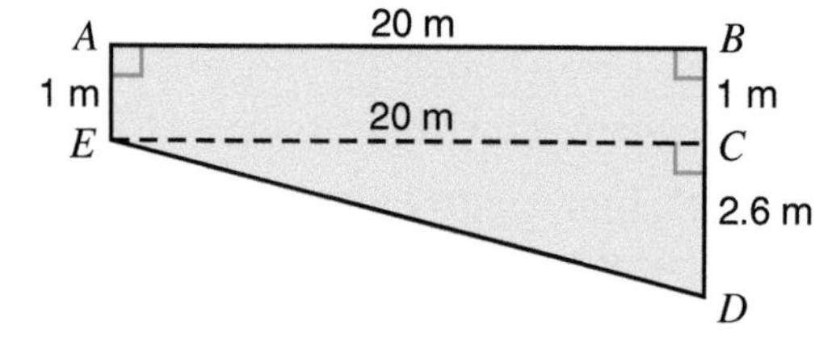

$DE^2 = CD^2 + CE^2$

$DE^2 = 2.6^2 + 20^2$

$DE^2 = 6.76 + 400$

$DE^2 = 406.76$

$DE = \sqrt{406.76}$ m

$DE = 20.1682...$ m

The length of the sloping bottom of the pool is 20.2 m, correct to 3 sig. figs.

Practice Exercise 17.2

1. In each of the following, work out the length of the side marked x.

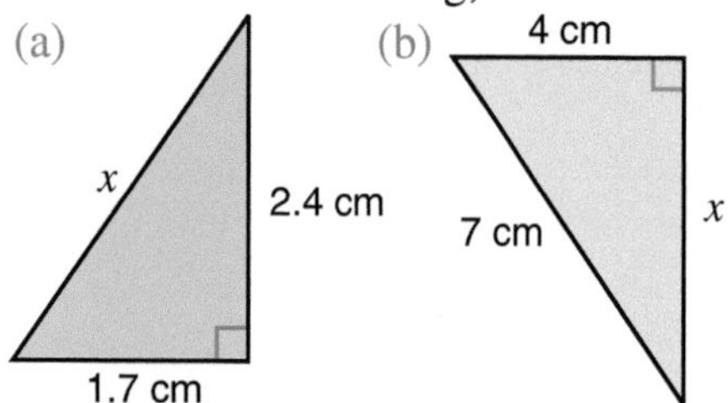

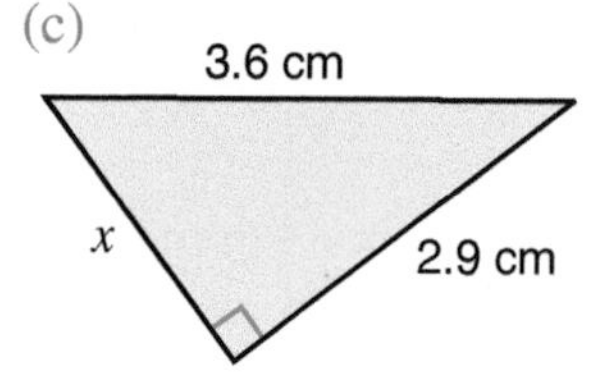

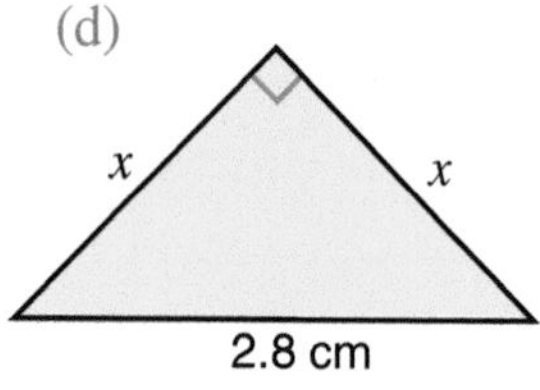

2. A rectangle is 8 cm wide and 15 cm long.
   Work out the length of its diagonals.

3. The length of a rectangle is 24 cm.
   The diagonals of the rectangle are 26 cm.
   Work out the width of the rectangle.

4. A square has sides of length 6 cm.
   Work out the length of its diagonals.

5. The diagonals of a square are 15 cm.
   Work out the length of its sides.

6. The height of an isosceles triangle is 12 cm.
   The base of the triangle is 18 cm.
   Work out the length of the equal sides.

7. An equilateral triangle has sides of length 8 cm.
   Work out the height of the triangle.

8. The diagram shows the side view of a car ramp.
   The ramp is 110 cm long and 25 cm high.
   The top part of the ramp is 40 cm long.
   Calculate the length of the sloping part of the ramp.

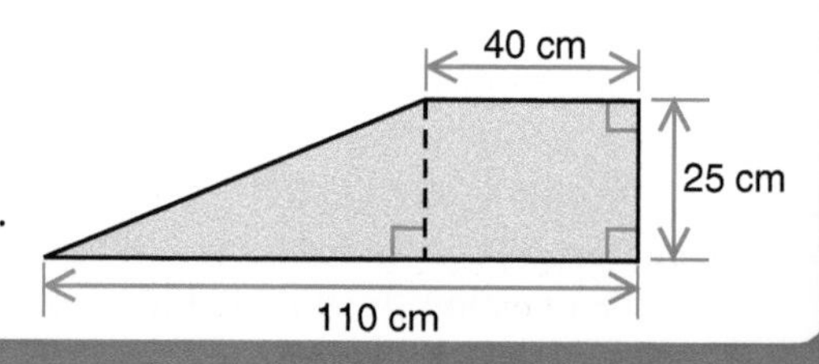

Solving problems in three dimensions

When we solve problems in three dimensions we often need to use more than one triangle to solve the problem.

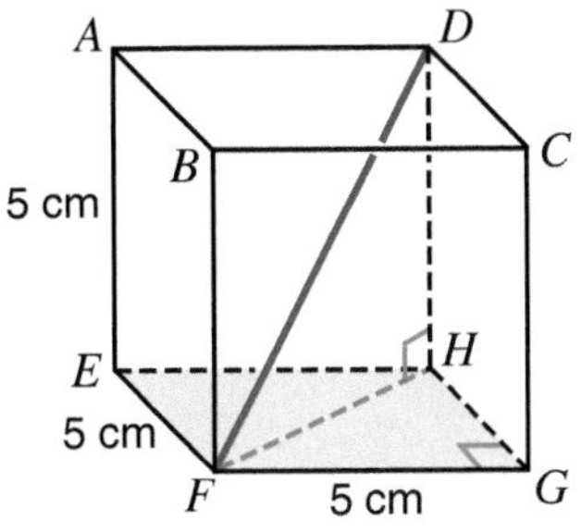

Example 5

This cube has sides of length 5 cm.
What is the distance from D to F?
Give the answer correct to 1 decimal place.

ΔDFH is a suitable right-angled triangle, but we only know the length of the side DH.
The length of FH can be found by using ΔFGH.

Using Pythagoras' Theorem in ΔFGH.

$FH^2 = FG^2 + GH^2$
$FH^2 = 5^2 + 5^2$
$FH^2 = 25 + 25$
$FH^2 = 50$
$FH = \sqrt{50}$ cm

Using Pythagoras' Theorem in ΔDFH.

$DF^2 = DH^2 + FH^2$
$DF^2 = 5^2 + (\sqrt{50})^2$
$DF^2 = 25 + 50$
$DF^2 = 75$
$DF = \sqrt{75}$ cm
$DF = 8.66\ldots$ cm
$DF = 8.7$ cm, correct to 1 d.p.

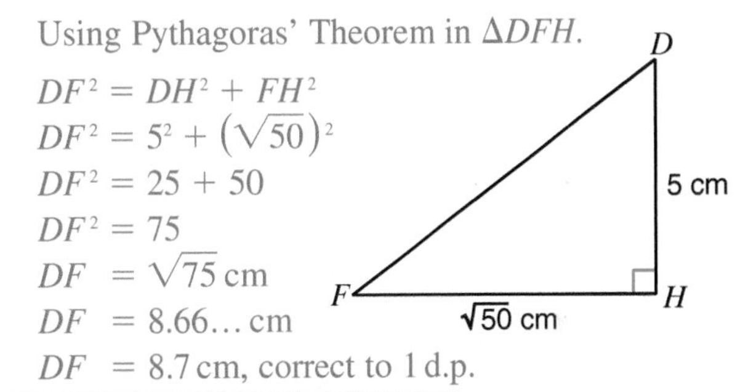

For accuracy, the exact value of FH, $(\sqrt{50})$, should be used.

Practice Exercise 17.3

1. The top of a lampshade has a diameter of 10 cm.
The bottom of the lampshade has a diameter of 20 cm.
The height of the lampshade is 12 cm.
Calculate the length, l, of the sloping sides.

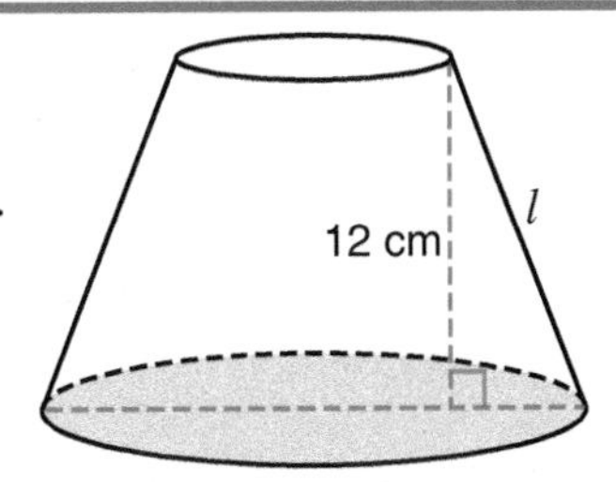

2. The top of a bucket has a diameter of 30 cm.
The bottom of the bucket has a diameter of 16 cm.
The sloping sides are 25 cm long.
How deep is the bucket?

3. The cube has sides of length 6 cm.
Calculate the length of the diagonal PQ.

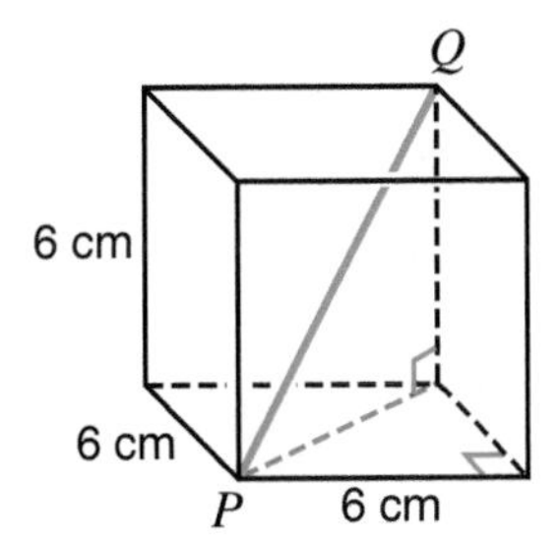

4. The diagram shows a wedge $ABCDEF$.
$AD = 13$ cm, $DE = 15$ cm and $EC = 6$ cm.
Calculate the length of the line AC.

5. $ABCDEFGH$ is a cuboid.
$AE = 5\,\text{cm}$, $EH = 9\,\text{cm}$ and $HG = 6\,\text{cm}$.
Calculate the length of the line AG.

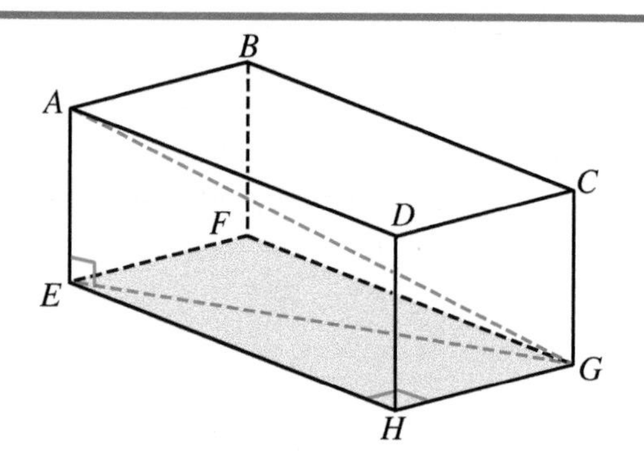

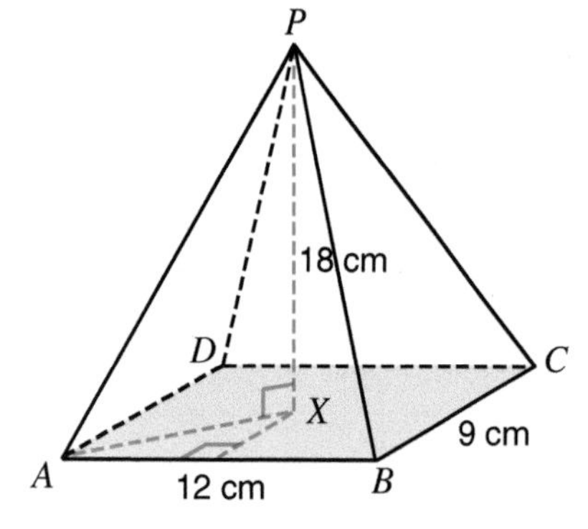

6. The diagram shows a pyramid $PABCD$.
X is at the centre of the base.
The base $ABCD$ is a rectangle with
$AB = CD = 12\,\text{cm}$ and $BC = DA = 9\,\text{cm}$.
The height of the pyramid, PX, is $18\,\text{cm}$.
Calculate the length of the edge PA.

7. The diagram of a cuboid is shown.

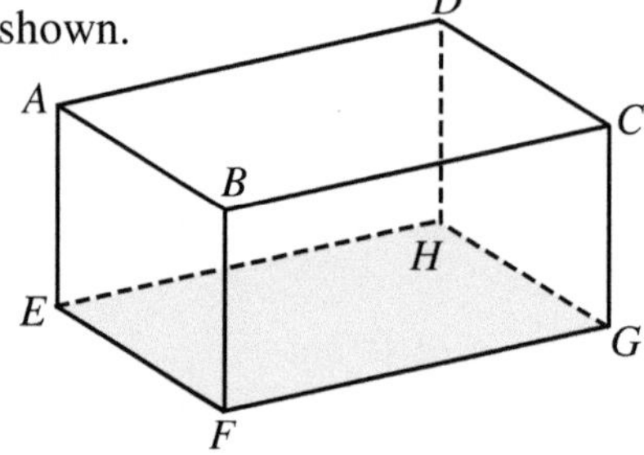

(a) Calculate the length of EC, when $EF = 6\,\text{cm}$, $FG = 8\,\text{cm}$ and $CG = 5\,\text{cm}$.
(b) Calculate the length of AG, when $AD = 10\,\text{cm}$, $DC = 8\,\text{cm}$ and $CG = 6\,\text{cm}$.
(c) Calculate the length of FD, when $FG = 7\,\text{cm}$, $GH = 5\,\text{cm}$ and $HD = 4\,\text{cm}$.

8. The diagram shows a square-based pyramid.
All edges are $7\,\text{cm}$ in length.
Calculate the height of the pyramid, PX.

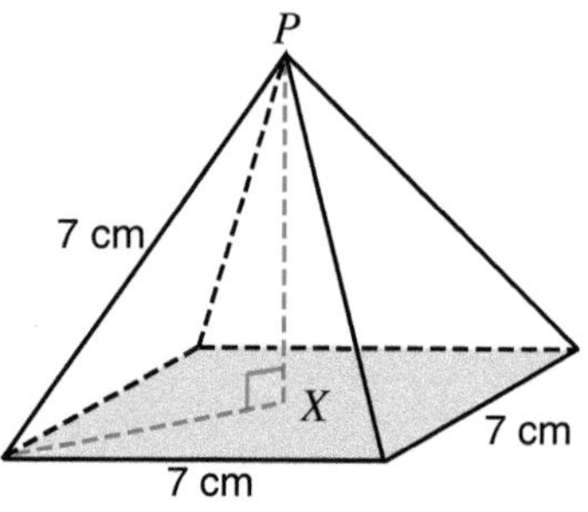

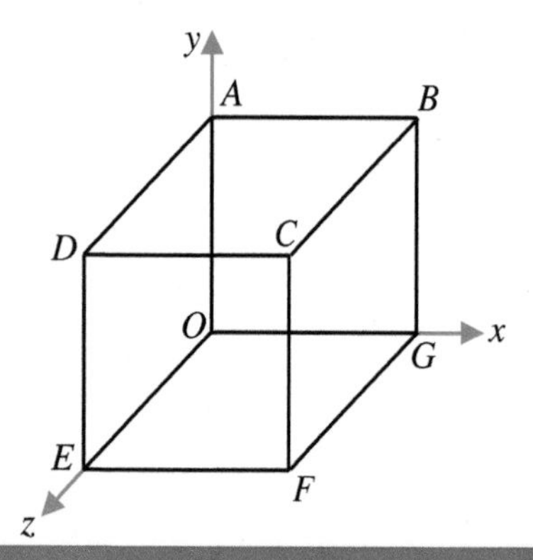

9. The diagram shows a cuboid drawn in 3 dimensions.
Using the x, y and z axes shown:
Point A has coordinates $(0, 3, 0)$.
Point B has coordinates $(2, 3, 0)$.
Point C has coordinates $(2, 3, 4)$.

Find the lengths of the following.

(a) AB (b) BC (c) AC
(d) DF (e) AE (f) OF

Give your answers correct to one decimal place.

Key Points

- The longest side in a right-angled triangle is called the **hypotenuse**.
- The **Theorem of Pythagoras** states:

  "In any right-angled triangle the square on the hypotenuse is equal to the sum of the squares on the other two sides."

  $a^2 = b^2 + c^2$

  Rearranging gives:

  $b^2 = a^2 - c^2$

  $c^2 = a^2 - b^2$

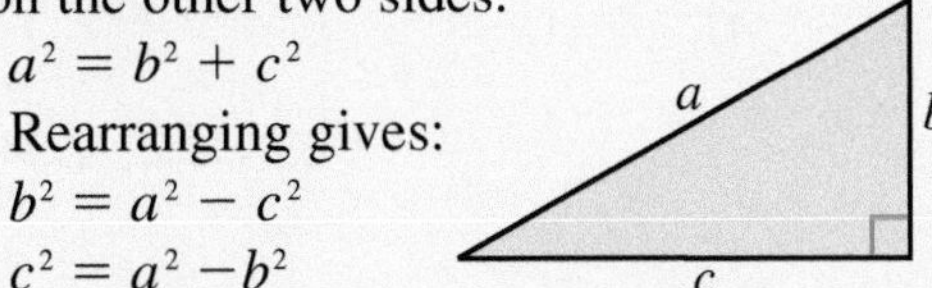

- When we know the lengths of two sides of a right-angled triangle, we can use the Theorem of Pythagoras to find the length of the third side.
- You should be able to solve problems in three dimensions.
- We can use the **converse** of Pythagoras' Theorem to see if a triangle is right-angled. If the sum of the squares on the two shorter sides of a triangle is equal to the square on the longest side of the triangle, then the triangle is right-angled.

  If $r^2 = p^2 + q^2$, then $a = 90°$.

  If $r^2 < p^2 + q^2$, then $a < 90°$.

  If $r^2 > p^2 + q^2$, then $a > 90°$.

Review Exercise 17

1. The diagram shows a right-angled triangle, PQR.
   $PR = 12$ cm and $QR = 9$ cm
   Calculate the length of PQ.

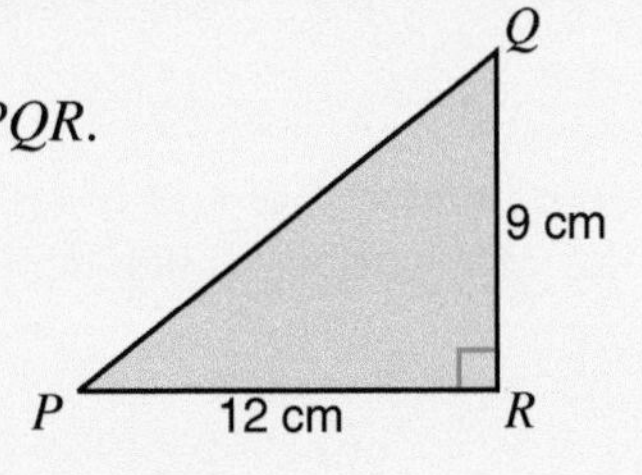

2. The coordinates of the points A and B are (4, 8) and (−1, 1).
   Work out the length of AB.

3. The diagram shows a sketch of triangle PQR.
   Show that PQR is a right-angled triangle.

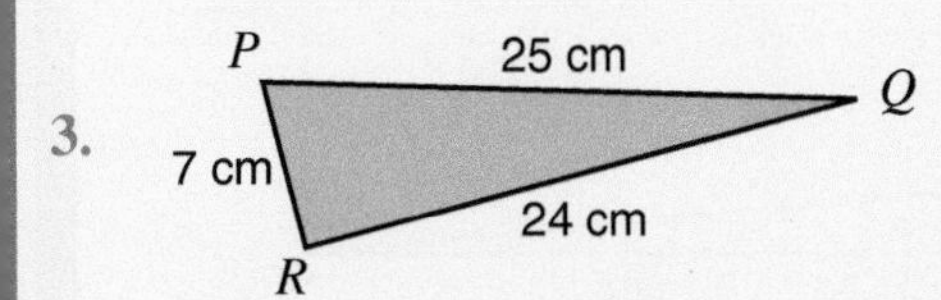

4. Mike is standing 200 m due west of a power station and 300 m due north of a pylon.
   Calculate the distance of the power station from the pylon.

5. The diagram shows a triangle PQR.
   $PQ = 35$ cm, $QR = 15$ cm and $PR = 40$ cm.
   Show that angle $PQR > 90°$.

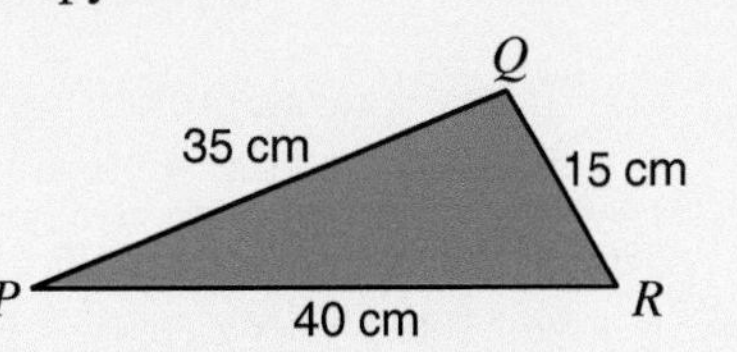

6. The cube has sides of length 10 cm.
   Calculate the length of the diagonal PQ.

7. The diagram shows a cuboid drawn in 3 dimensions.
Using the axes x, y and z shown:

Point A is given as $(-2, 4, 1)$.
Point B is given as $(4, 4, 1)$.
Point C is given as $(4, 4, 9)$.

The length of AB is 6 units.
Find the exact lengths of the lines AC, AE and AF.

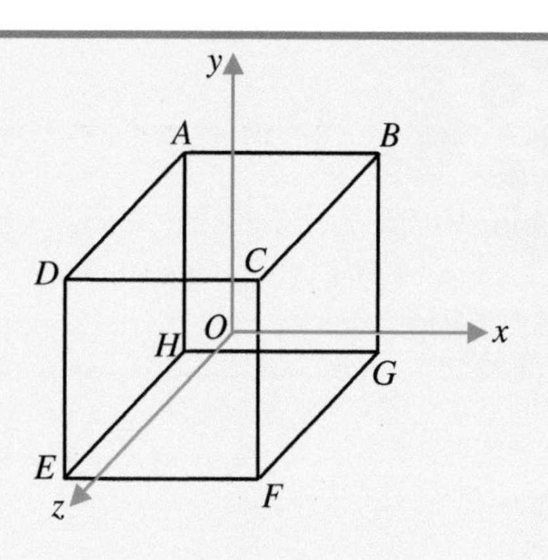

8.

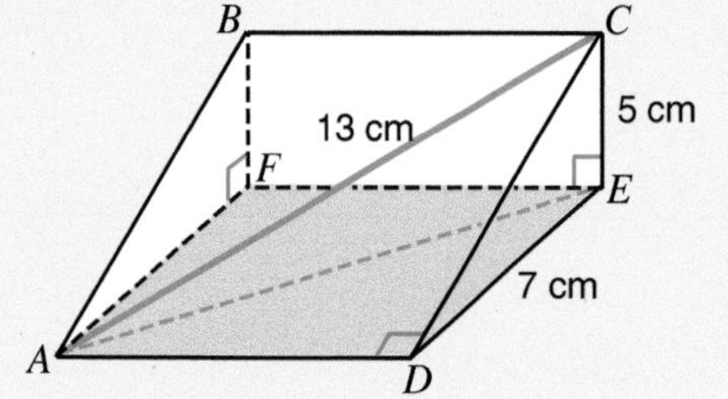

The diagram shows a wedge $ABCDEF$.
$AC = 13\,\text{cm}$, $DE = 7\,\text{cm}$ and $EC = 5\,\text{cm}$.
Calculate the length of the line AD.

9. The diagram shows a square-based pyramid.
All edges are 8 cm in length.
Calculate the height of the pyramid, PX.

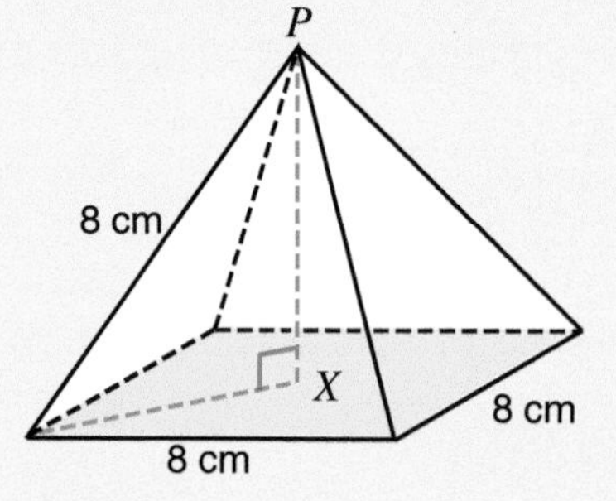

10. The 3-dimensional coordinates of M and N are $(2, 0, -1)$ and $(-1, 2, 3)$.
Calculate the length of MN, correct to one decimal place.

11. $ABCD$ is a rectangle measuring 11 cm by 6 cm.
F is a point on AB such that $AF = 5\,\text{cm}$.
E is a point on AD such that $DE = 1\,\text{cm}$.
Prove that angle CFE is a right angle.

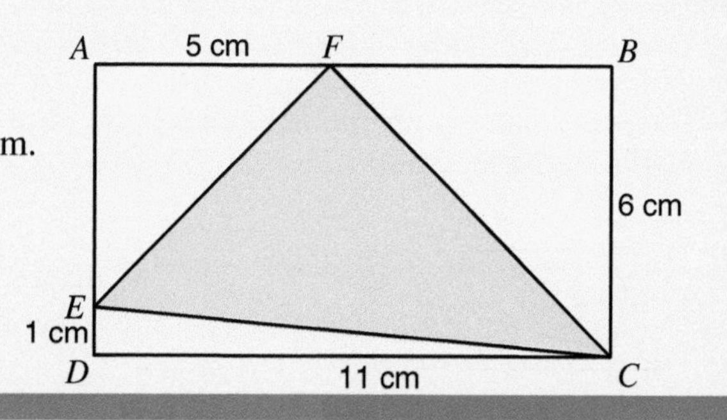

(a) Investigate the relationship between the areas of the semicircles A, B and C.

(b) What happens if the semicircles are replaced by regular polygons?

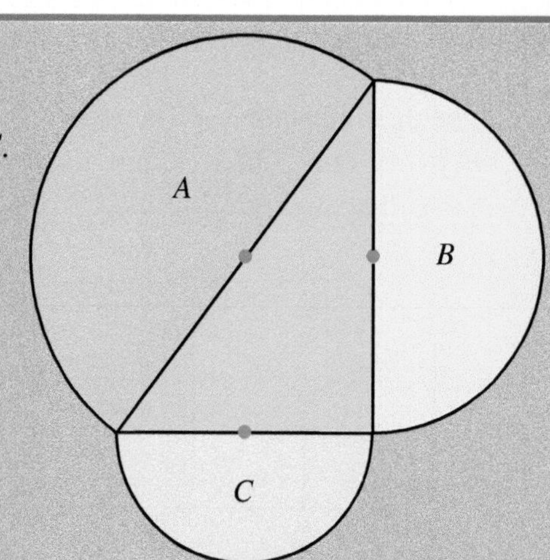

18 Properties of Shapes

Triangles

A **triangle** is a shape made by three straight lines.
The sum of the three angles in a triangle is 180°.

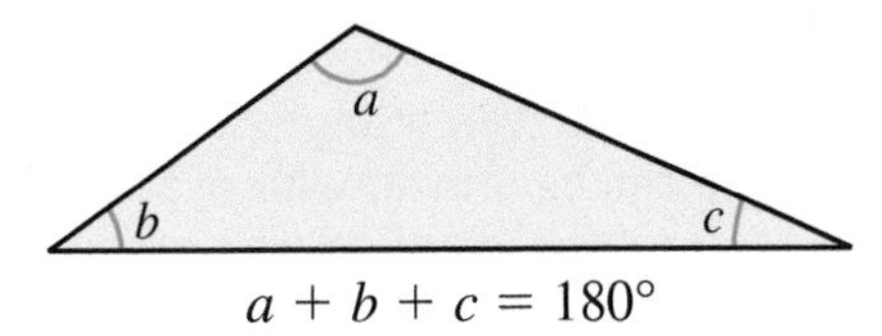

$a + b + c = 180°$

Types of triangles

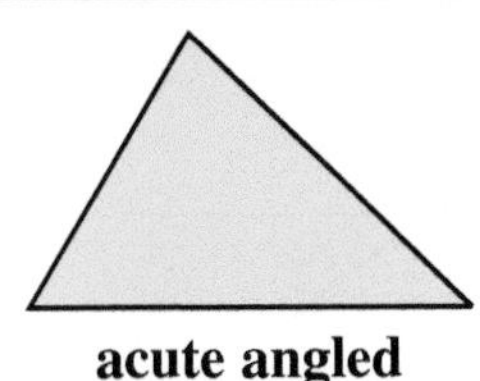

acute angled
Triangles with three acute angles.

obtuse angled
Triangles with an obtuse angle.

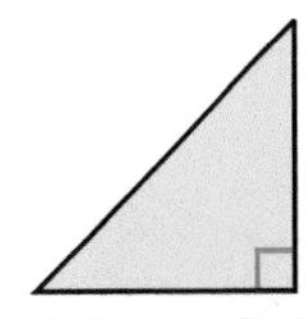

right angled
Triangles with a right angle.

Special triangles

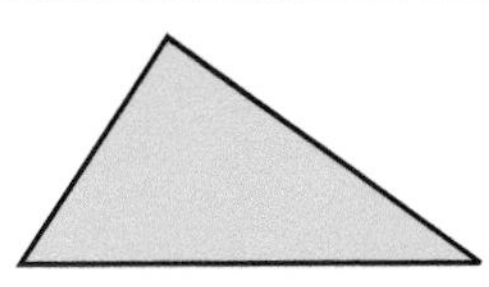

scalene triangle
Sides have different lengths.
Angles are all different.

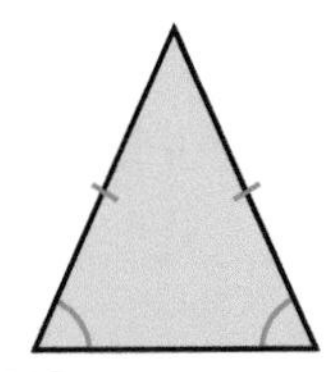

isosceles triangle
Two equal sides.
Two angles equal.

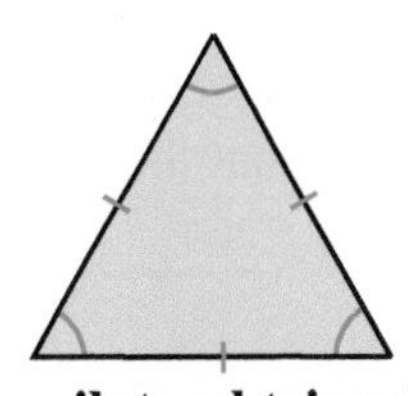

equilateral triangle
Three equal sides.
All angles are 60°.

Exterior angle of a triangle

When one side of a triangle is extended, as shown, the angle formed is called an **exterior angle**.

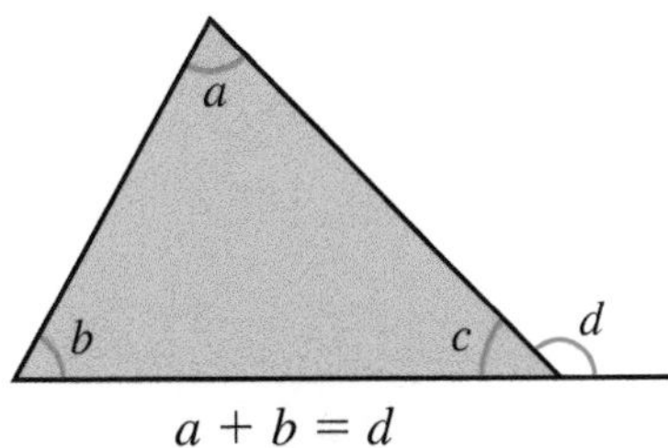

$a + b = d$

This result can easily be proved.
$a + b + c = 180°$
(sum of angles in a triangle)
$c + d = 180°$
(supplementary angles)
$a + b + c = c + d$
$a + b = d$

In any triangle the exterior angle is always equal to the sum of the two opposite interior angles.
Check this by measuring the angles a, b and d in the diagram.

Example 1

Find the sizes of the angles marked a and b.

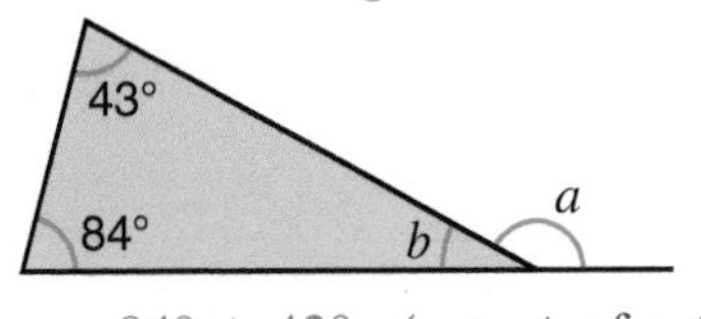

$a = 84° + 43°$ (ext. ∠ of a Δ)
$a = 127°$
$b + 127° = 180°$ (supp. ∠'s)
$b = 180° - 127°$
$b = 53°$

Short but accurate:
In Geometry we often abbreviate words and use symbols to provide the reader with full details using the minimum amount of writing.
Δ is short for triangle.
ext. ∠ of a Δ means exterior angle of a triangle.
supp. ∠'s means supplementary angles.

Practice Exercise 18.1

The diagrams in this exercise have not been drawn accurately.

1. Is it possible to draw triangles with the following types of angles?
   Give a reason for each of your answers.
   (a) three acute angles,
   (b) one obtuse angle and two acute angles,
   (c) two obtuse angles and one acute angle,
   (d) three obtuse angles,
   (e) one right angle and two acute angles,
   (f) two right angles and one acute angle.

2. Is it possible to draw a triangle with these angles?
   If a triangle can be drawn, what type of triangle is it?
   Give a reason for each of your answers.
   (a) 95°, 78°, 7°
   (b) 48°, 62°, 90°
   (c) 48°, 62°, 70°
   (d) 90°, 38°, 52°
   (e) 130°, 35°, 15°

3. Work out the size of the marked angles.

(a) (b)

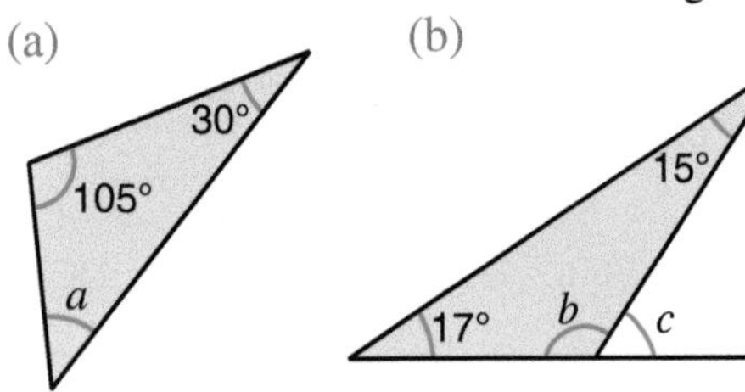

(c) (d)

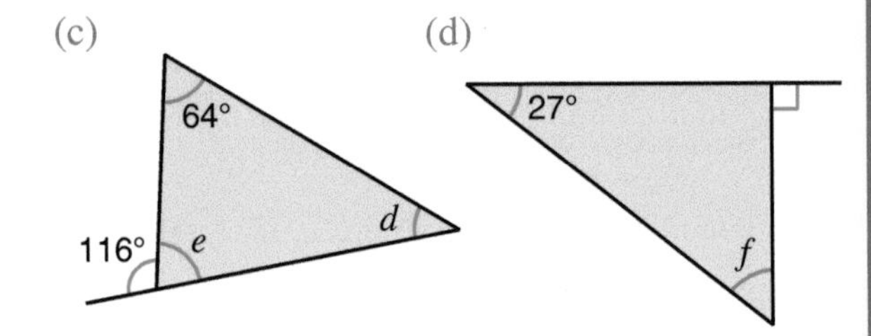

4. Work out the size of the angles marked with letters.

(a)

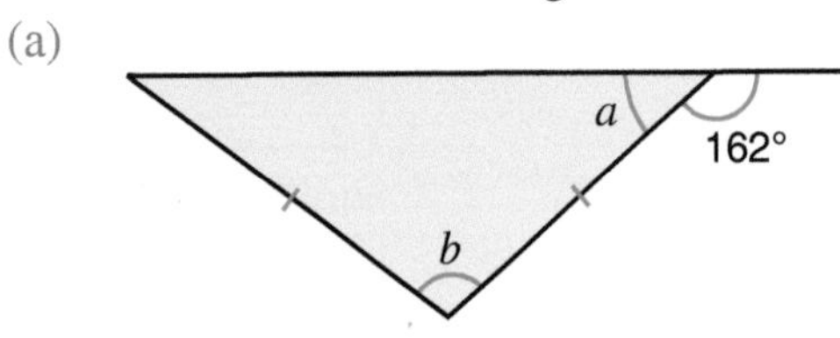

(b)

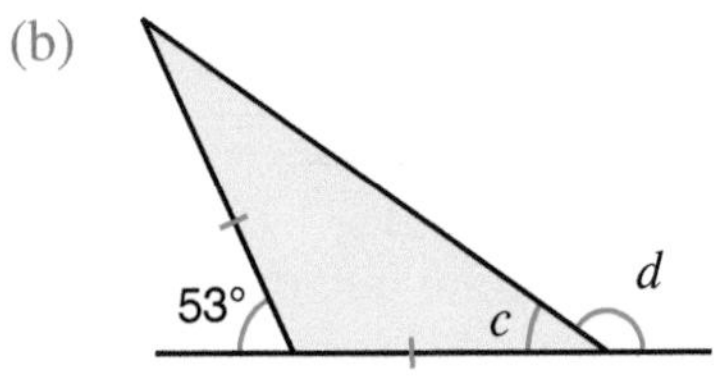

5. In the diagram:
   $AB = BD = DA$ and $BC = CD$.
   CD is extended to E.
   (a) What type of triangle is BCD?
   (b) What is the size of angle BDC?
   (c) Work out the size of angle ADE.

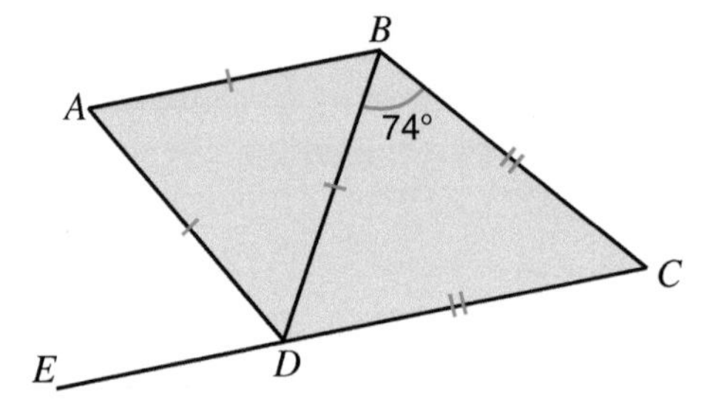

6. Work out the size of the required angles.

(a)

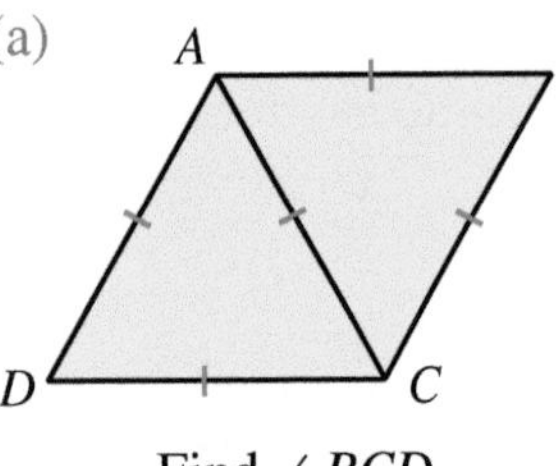

Find $\angle BCD$.

(b)

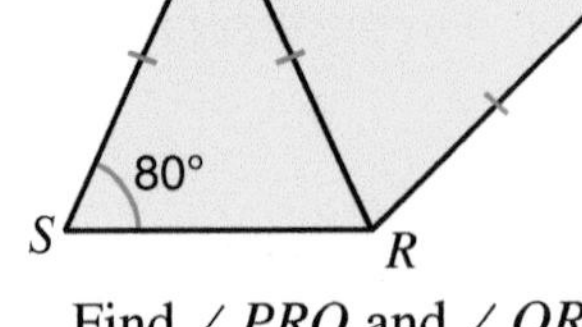

Find $\angle PRQ$ and $\angle QRS$.

(c)

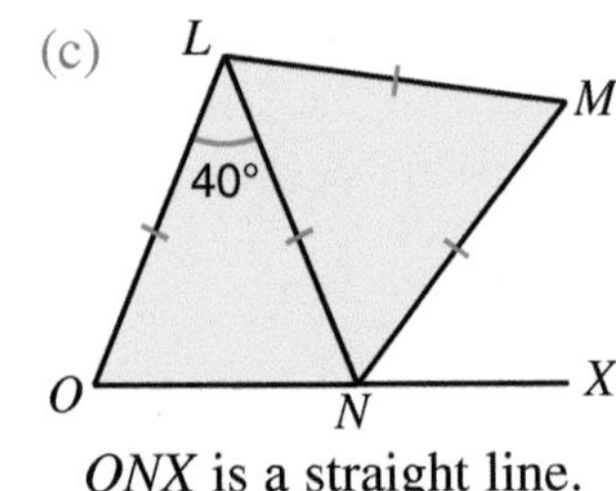

ONX is a straight line.
Find $\angle MNX$.

Quadrilaterals

A **quadrilateral** is a shape made by four straight lines.
The sum of the four angles of a quadrilateral is 360°.

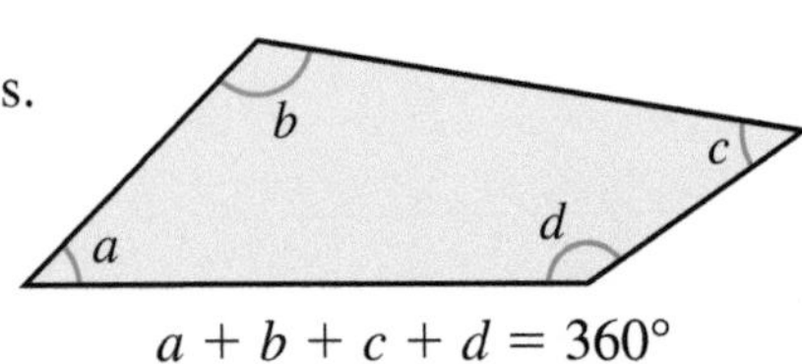

$a + b + c + d = 360°$

Special quadrilaterals

Square

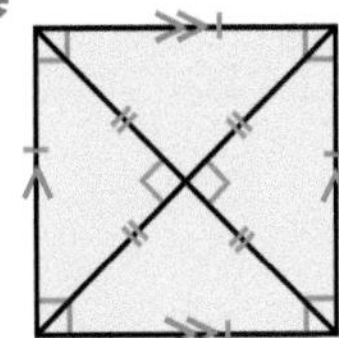

Four equal sides.
Opposite sides parallel.
Angles of 90°.
Diagonals bisect each other at 90°.

Parallelogram

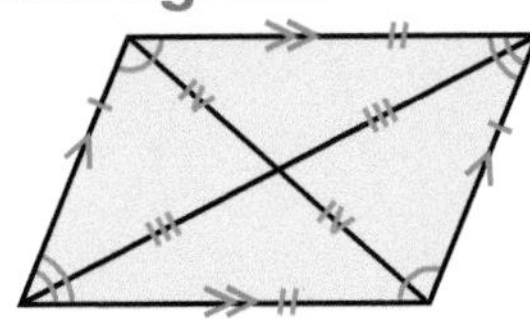

Opposite sides equal and parallel.
Opposite angles equal.
Diagonals bisect each other.

Trapezium

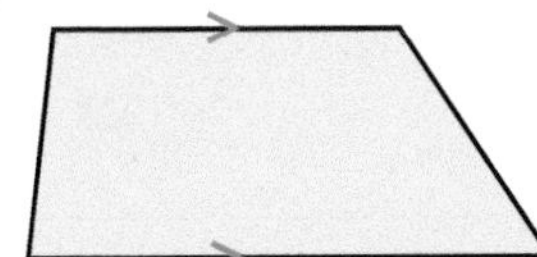

One pair of parallel sides.

Rectangle

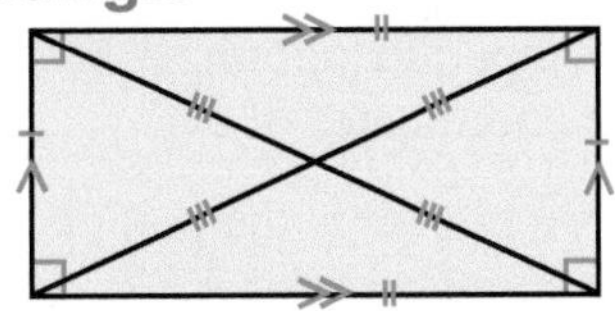

Opposite sides equal and parallel.
Angles of 90°.
Diagonals bisect each other.

Rhombus

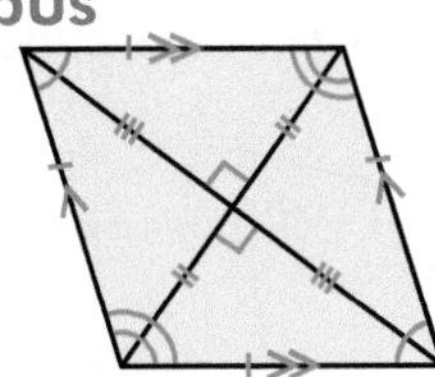

Four equal sides.
Opposite sides parallel.
Opposite angles equal.
Diagonals bisect each other at 90°.

Kite

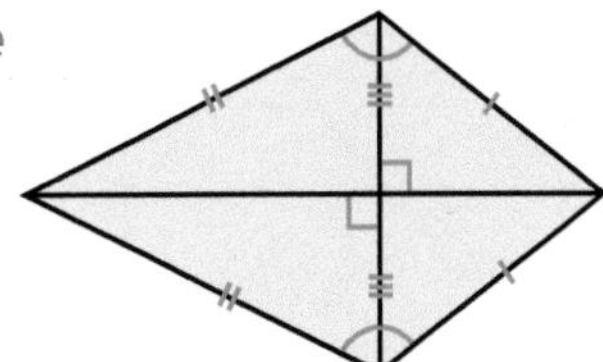

Two pairs of adjacent sides equal.
One pair of opposite angles equal.
One diagonal bisects the other at 90°.

Parallel lines and angles

Here is a reminder of angle properties linked with parallel lines.
Each diagram shows two parallel lines crossed by another straight line called a **transversal**.

Corresponding angles

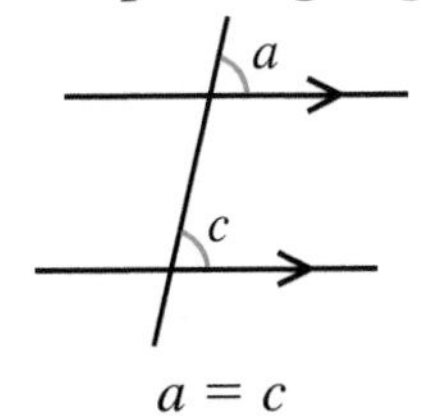

$a = c$

Corresponding angles are always on the same side of the transversal.

Alternate angles

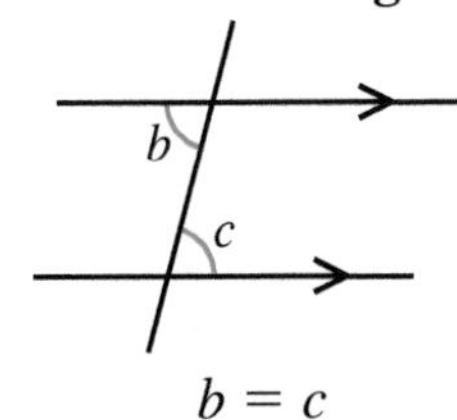

$b = c$

Alternate angles are always on opposite sides of the transversal.

Allied angles

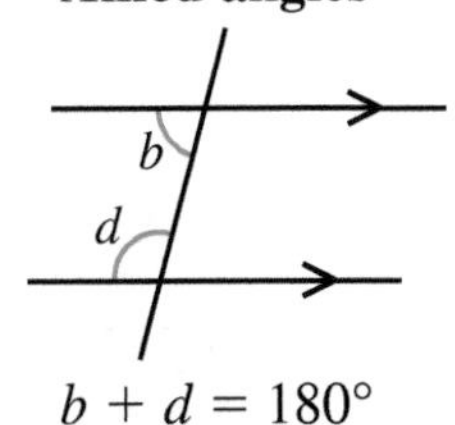

$b + d = 180°$

Allied angles are always between parallels on the same side of the transversal.

Example 2

PQRS is a parallelogram. Work out the size of the angle marked *x*.

The opposite angles of a parallelogram are equal.

$55° + 55° + x + x = 360°$

$110° + 2x = 360°$

$2x = 360° - 110°$

$2x = 250°$

$x = 125°$

P Q S R 55° x

Show how allied angles can be used to find the size of angle x.

Practice Exercise 18.2

Use squared paper to answer questions 1 to 4.

1. *PQRS* is a rectangle. *P* is the point (1, 3), *Q* (4, 6), *R* (6, 4).
   Find the coordinates of *S*.

2. *WXYZ* is a parallelogram. *W* is the point (1, 0), *X* (4, 1), *Z* (3, 3).
   Find the coordinates of *Y*.

3. *OABC* is a kite. *O* is the point (0, 0), *B* (5, 5), *C* (3, 1).
   Find the coordinates of *A*.

4. *STUV* is a square with *S* at (1, 3) and *U* at (5, 3).
   Find the coordinates of *T* and *V*.

5. The following diagrams have not been drawn accurately.
   Work out the size of the angles marked with letters.

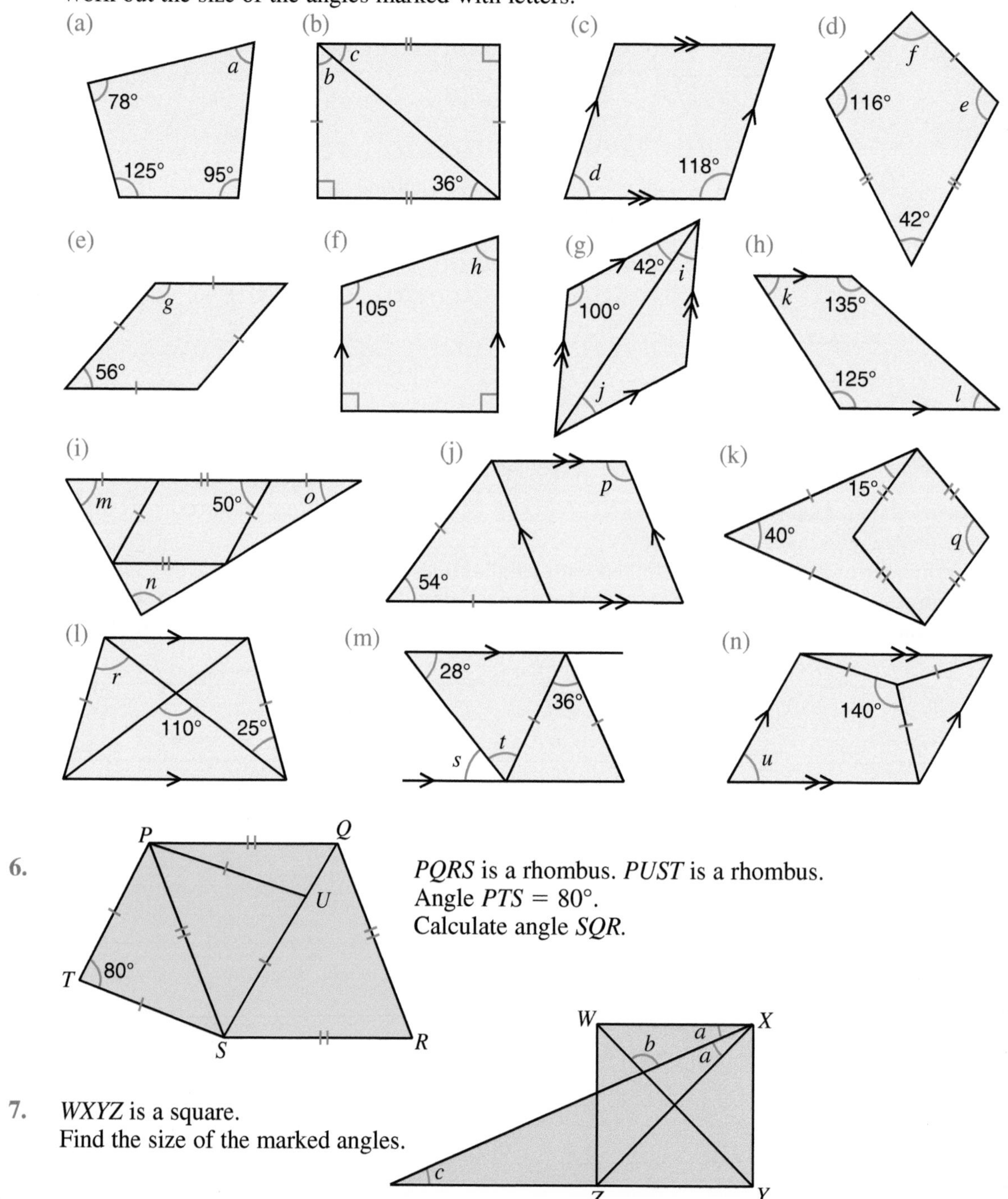

6. *PQRS* is a rhombus. *PUST* is a rhombus.
   Angle $PTS = 80°$.
   Calculate angle *SQR*.

7. *WXYZ* is a square.
   Find the size of the marked angles.

Polygons

A **polygon** is a shape made by straight lines.
A three-sided polygon is a **triangle**.
A four-sided polygon is called a **quadrilateral**.

A polygon is a many-sided shape.
Look at these polygons.

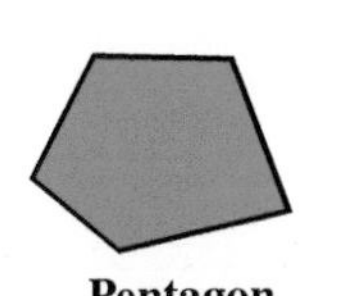

Pentagon
5 sides

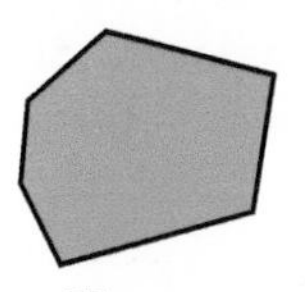

Hexagon
6 sides

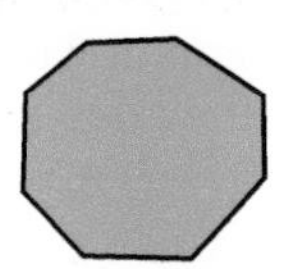

Octagon
8 sides

Interior and exterior angles of a polygon

Angles formed by sides inside a polygon are called **interior angles**.

When a side of a polygon is extended, as shown, the angle formed is called an **exterior angle**.

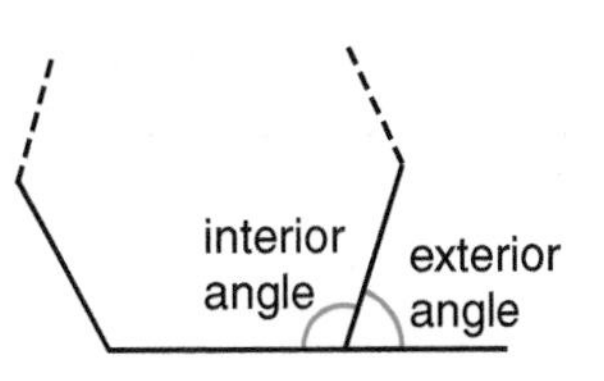

At each vertex of the polygon:
interior angle + exterior angle = 180°

Sum of the interior angles of a polygon

The diagram shows a polygon with the diagonals from one vertex drawn.

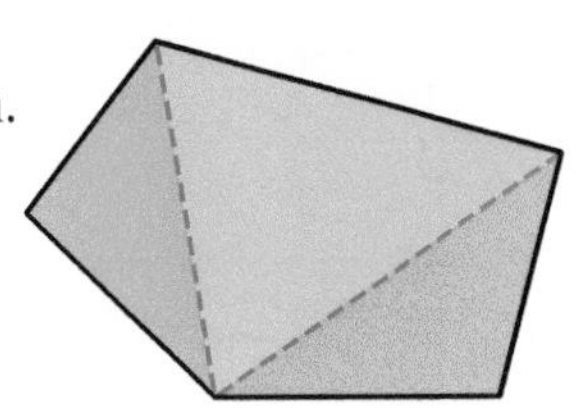

The polygon has 5 sides and is divided into 3 triangles.
So, the sum of the interior angles is $3 \times 180° = 540°$.

In general, for any n-sided polygon, the sum of the interior angles is $(n - 2) \times 180°$.

Sum of the exterior angles of a polygon

The sum of the exterior angles of **any** polygon is 360°.

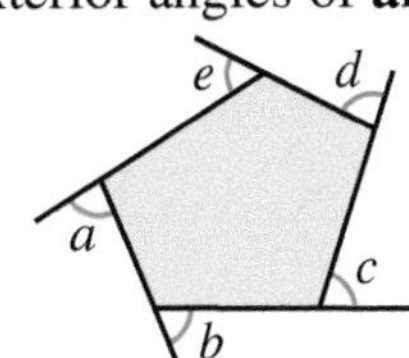

$$a + b + c + d + e = 360°$$

Example 3

Find the size of angle x.

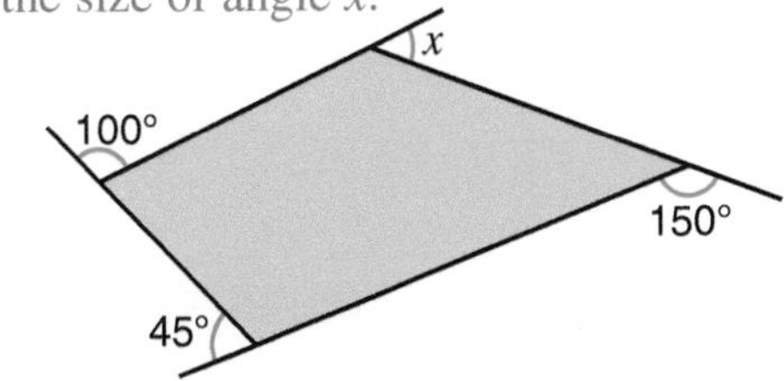

The sum of the exterior angles is 360°.

$x + 100° + 45° + 150° = 360°$
$x + 295° = 360°$
$x = 360° - 295°$
$x = 65°$

Example 4

Find the sum of the interior angles of a pentagon.

To find the sum of the interior angles of a pentagon substitute $n = 5$ into $(n - 2) \times 180°$.

$(5 - 2) \times 180°$
$= 3 \times 180°$
$= 540°$

Example 5

Find the size of the angles marked a and b.

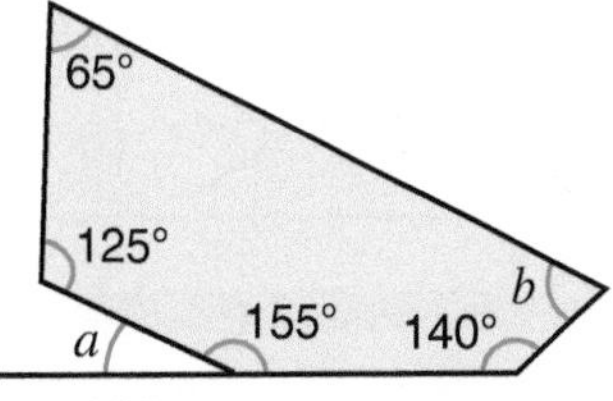

$155° + a = 180°$
(int. angle + ext. angle = 180°)
$a = 180° - 155°$
$a = 25°$

The sum of the interior angles of a pentagon is 540°.

$b + 140° + 155° + 125° + 65° = 540°$
$b + 485° = 540°$
$b = 540° - 485°$
$b = 55°$

Practice Exercise 18.3

The diagrams in this exercise have not been drawn accurately.

1. In the diagram, ABC is a straight line.
   (a) Explain why angle $x = 50°$.
   (b) Show that angle $y = 60°$.

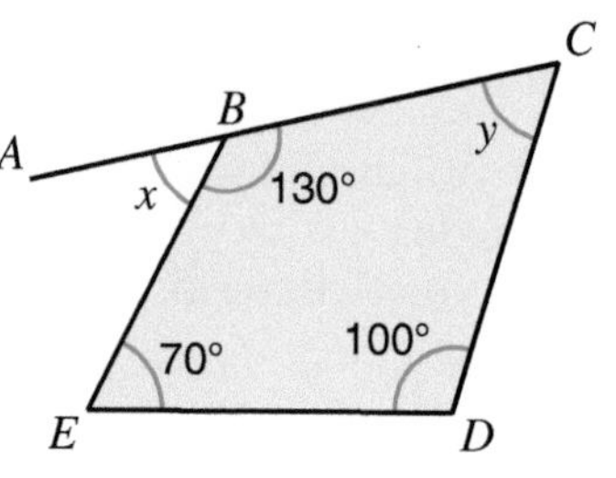

2. Work out the size of the angles marked with letters.

(a)

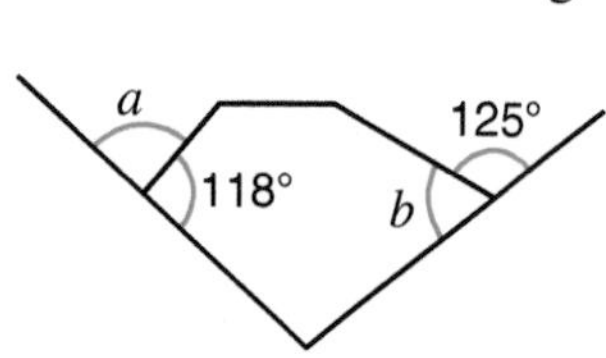

(b)

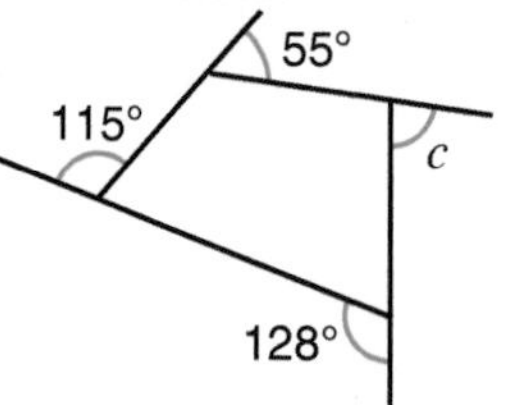

(c)

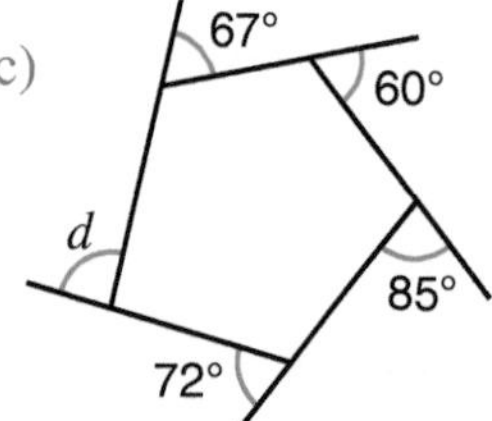

3. Work out the sum of the interior angles of these polygons.

(a)

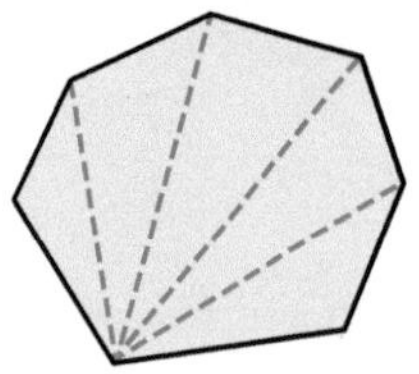

(b)

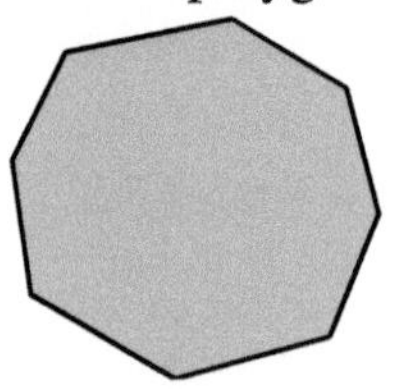

(c)

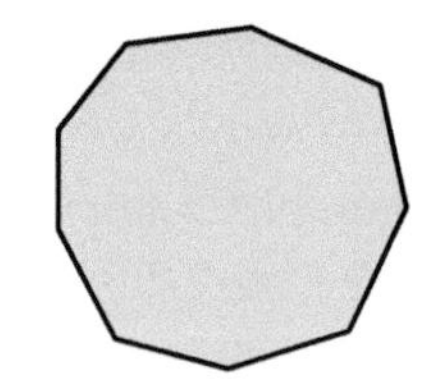

4. Work out the size of the angles marked with letters.

(a)

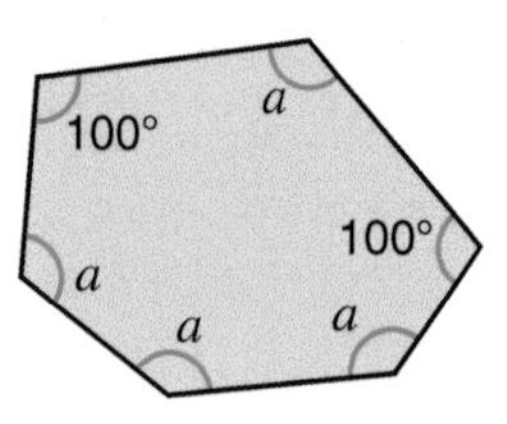

(b)

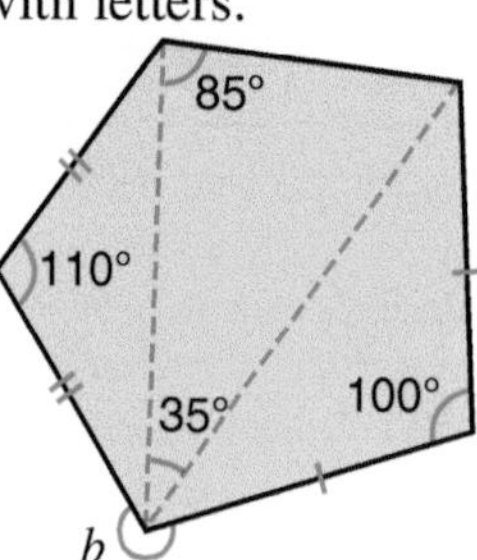

(c)

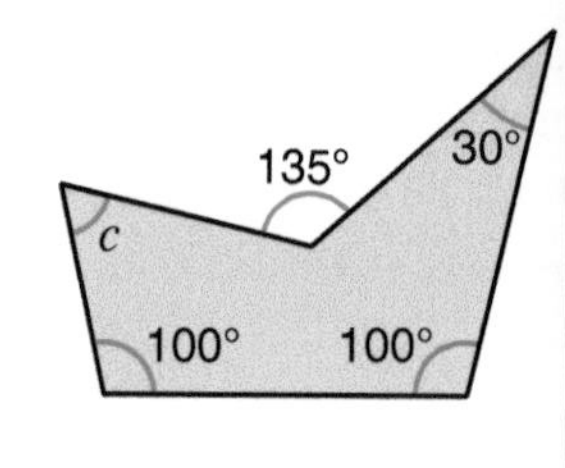

5.

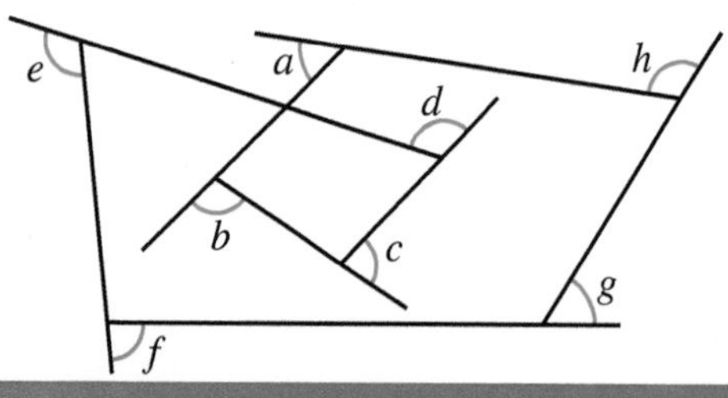

The figure is made up of straight lines.
Find the sum of the angles a, b, c, d, e, f, g, h.

Regular polygons

A polygon with all sides equal and all angles equal is called a **regular polygon**.

A regular triangle is usually called an **equilateral triangle**.
A regular quadrilateral is usually called a **square**.

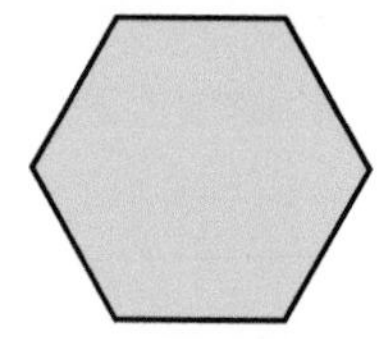

Regular hexagon

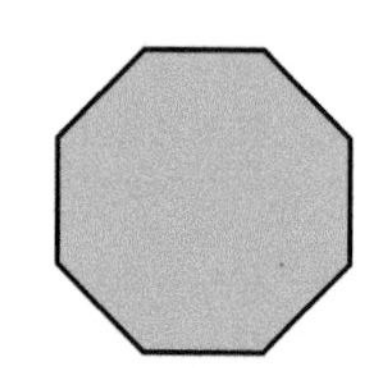

Regular octagon

Exterior angles of regular polygons

In general, for any regular n-sided polygon: exterior angle $= \dfrac{360°}{n}$

By rearranging the formula we can find the number of sides, n, of a regular polygon when we know the exterior angle.

$$n = \frac{360°}{\text{exterior angle}}$$

Example 6

A regular polygon has an exterior angle of 30°.

(a) How many sides has the polygon?

(b) What is the size of an interior angle of the polygon?

> **Remember:**
> It is a good idea to write down the formula you are using.

(a) $n = \frac{360°}{\text{exterior angle}}$

$n = \frac{360°}{30°} = 12$

The polygon has 12 sides.

(b) interior angle + exterior angle = 180°

int. ∠ + 30° = 180°

int. ∠ = 180° − 30°

interior angle = 150°

Tessellations

Covering a surface with identical shapes produces a pattern called a **tessellation**.

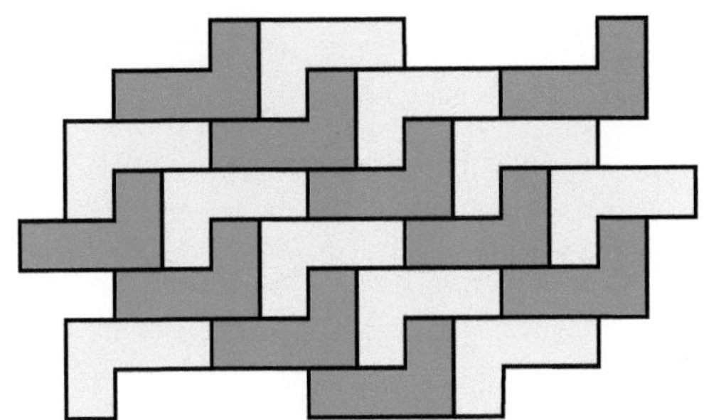

To tessellate the shape must not overlap and there must be no gaps.

Regular tessellations

This pattern shows a tessellation of regular hexagons.

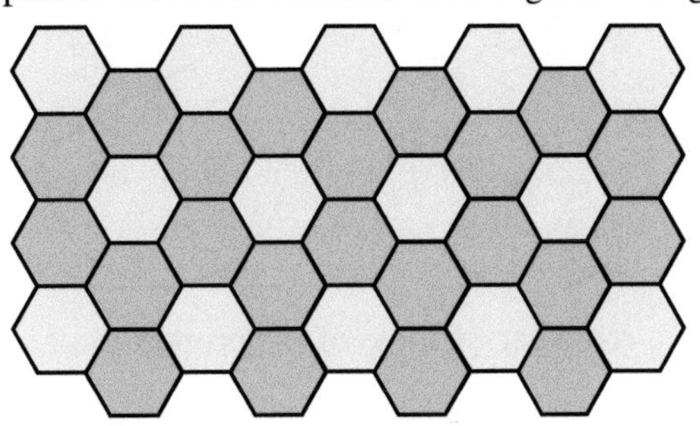

This pattern is called a **regular tessellation** because it is made by using a single regular polygon.

Practice Exercise 18.4

1. Calculate (a) the exterior angle and (b) the interior angle of these regular polygons.

(i)

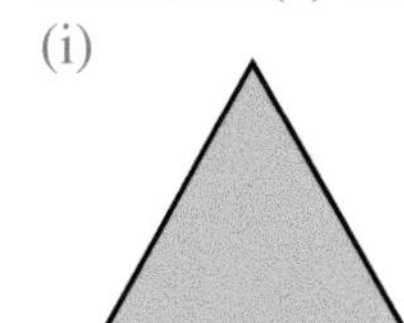

(ii)

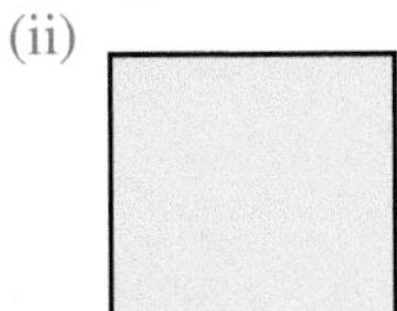

(iii)

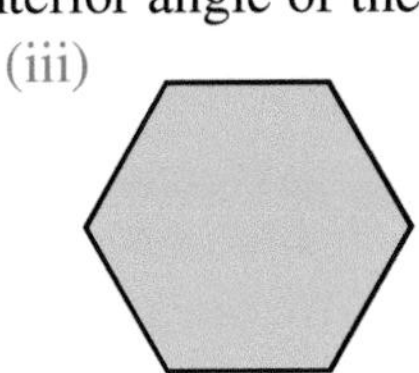

(iv)

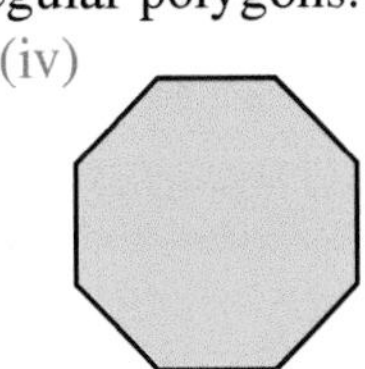

2. A regular polygon has an exterior angle of 18°. How many sides has the polygon?

3. A regular polygon has an interior angle of 135°. How many sides has the polygon?

4. (a) Calculate the size of an exterior angle of a regular pentagon.
   (b) What is the size of an interior angle of a regular pentagon?
   (c) What is the sum of the interior angles of a pentagon?

5. The following diagrams are drawn using regular polygons. Work out the values of the marked angles.

(a)

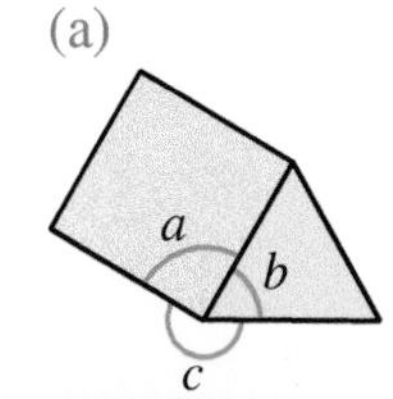

(b)

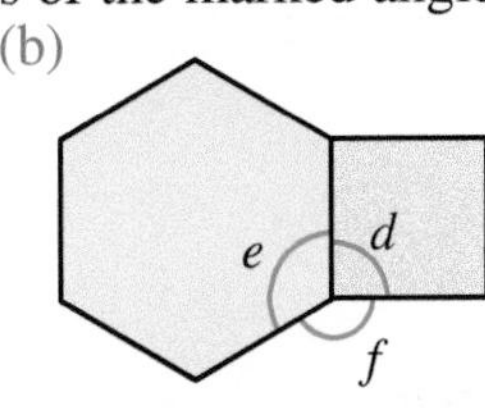

(c)

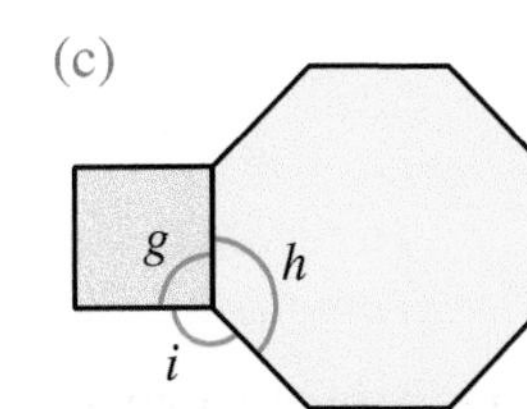

(d)

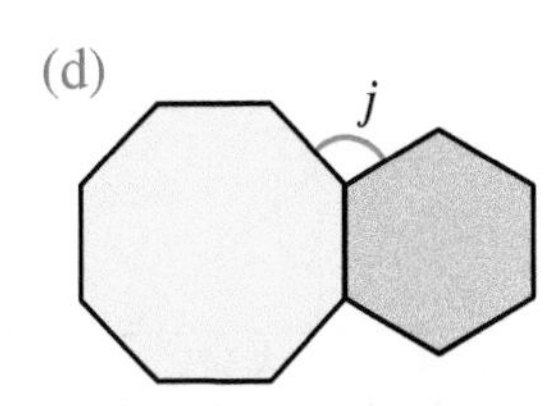

6. AB and BC are two sides of a regular octagon.
   PAB is an isosceles triangle with $AP = PB$.
   Angle $APB = 36°$.
   Calculate angle PBC.

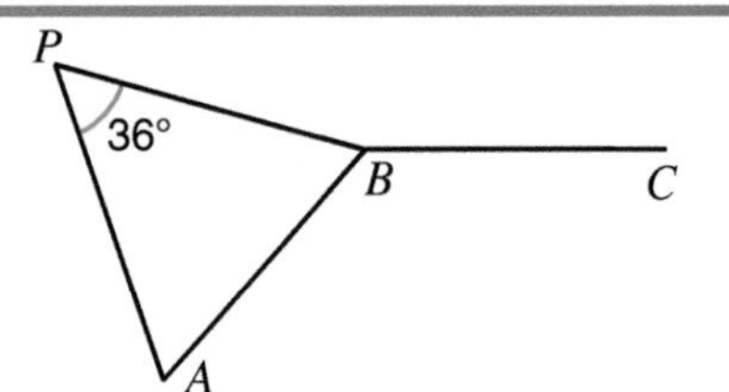

7. A regular polygon has 12 sides.
   Calculate the size of an interior angle.

8. Explain why regular pentagons will not tessellate.

9. Any triangle can be used to make a tessellation.
   Draw a triangle of your own, make copies, and show that it will tessellate.

Circle terms and properties

A **circle** is the shape drawn by keeping a pencil the same distance from a fixed point on a piece of paper.
Here is a reminder of some of the terms that are used about circles.

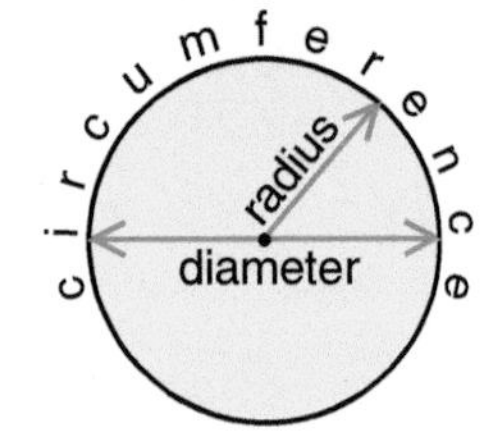

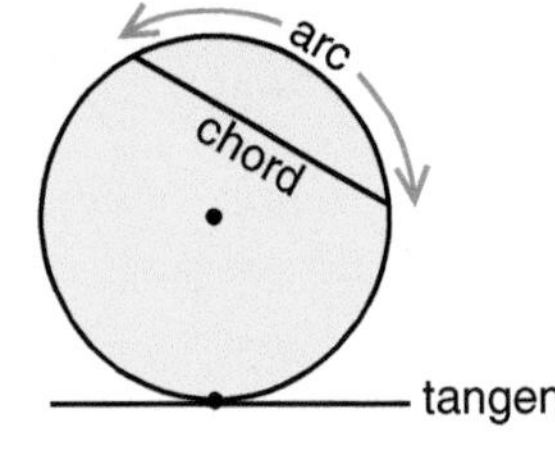

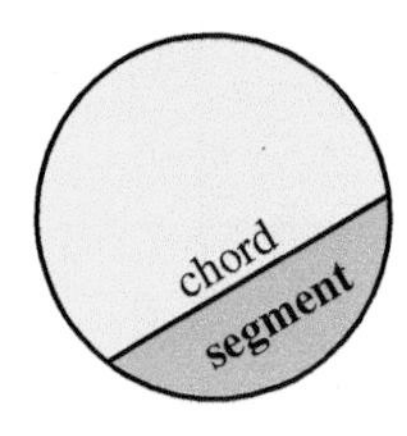

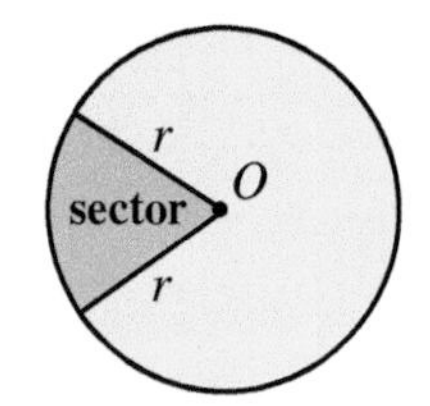

Here is a reminder of some of the properties of circles.

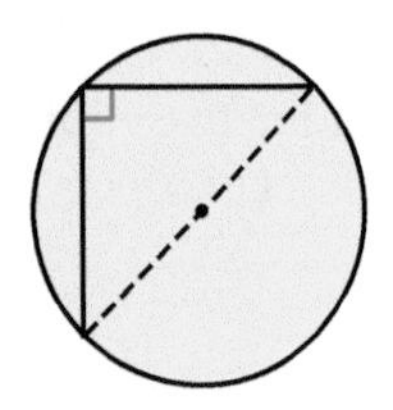

The angle in a semicircle is a right angle.

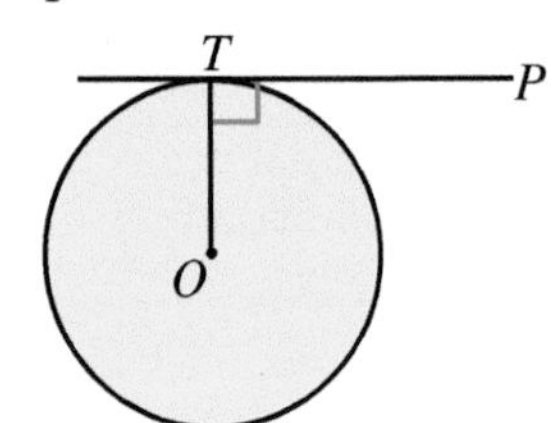

A tangent is perpendicular to the radius at the point of contact.

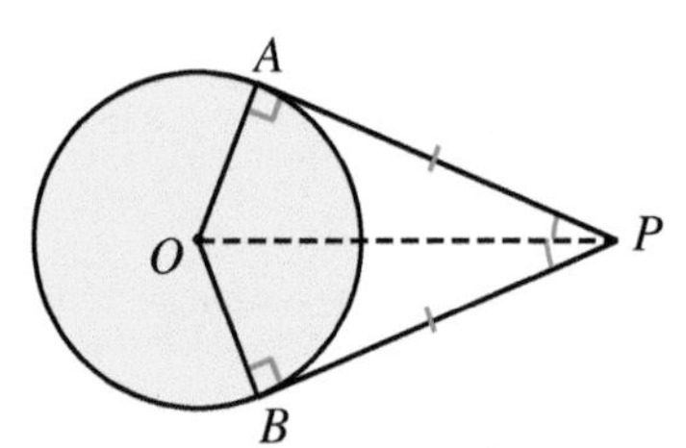

Tangents drawn to a circle from the same point are equal.

Perpendicular bisector of a chord

The perpendicular bisector of a chord, AB, always passes through the centre, O, of a circle.
The triangle formed by the chord, AB, and the two radii, AO and BO, is **isosceles**.

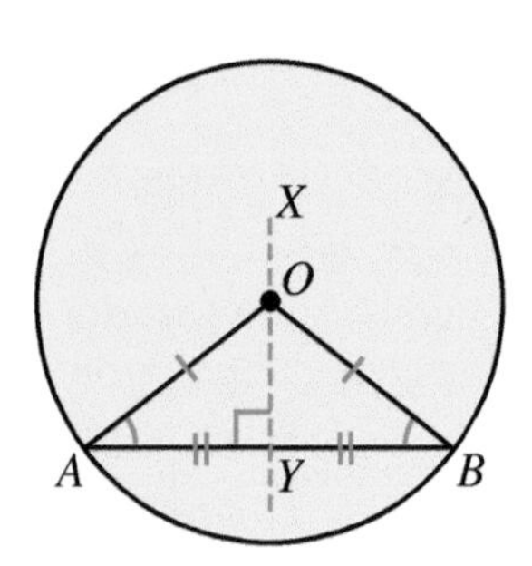

Example 7

AB is a chord 16 cm long, in a circle, centre O.
The radius of the circle is 10 cm.
Find the length of OC.

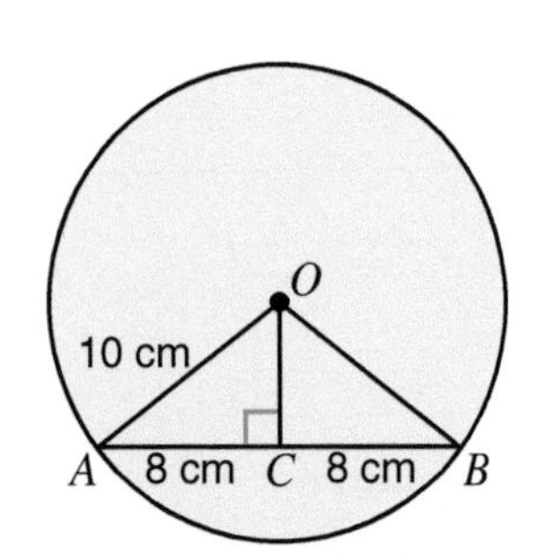

OC is the perpendicular bisector of AB.
So, $AC = AB = 8$ cm.

Using Pythagoras' Theorem in ΔOAC.

$OC^2 = OA^2 - AC^2$
$OC^2 = 10^2 - 8^2$
$OC^2 = 100 - 64$
$OC^2 = 36$
$OC = \sqrt{36}$
$OC = 6$ cm

Practice Exercise 18.5

1. The following diagrams show triangles drawn in semicircles.
   Work out the size of the marked angles.

   (a)
   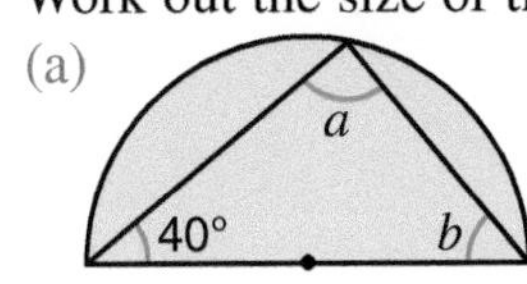

   (b) c

   (c)
   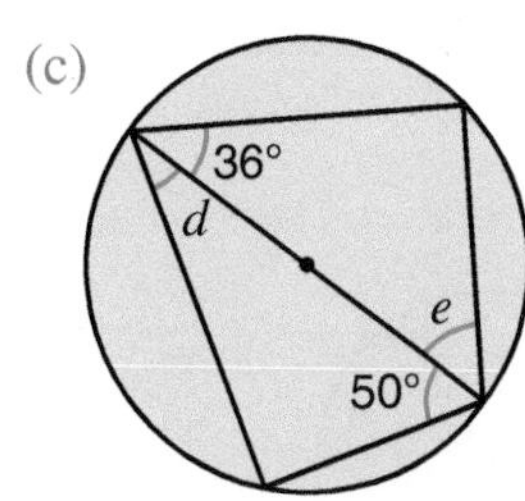

   (d) 70° f

2. O is the centre of the circle. Work out the size of the marked angles.

   (a)
   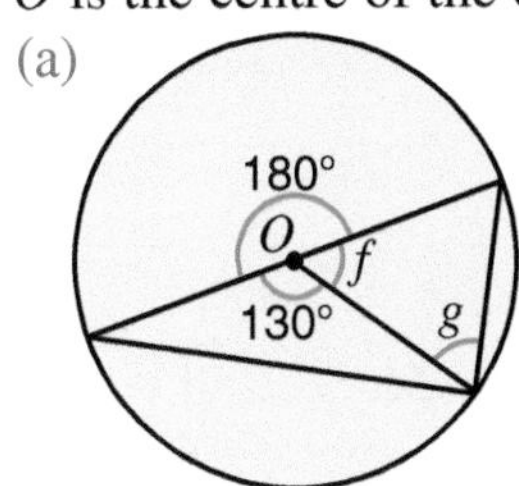

   (b) 84° O i h 25°

   (c)
   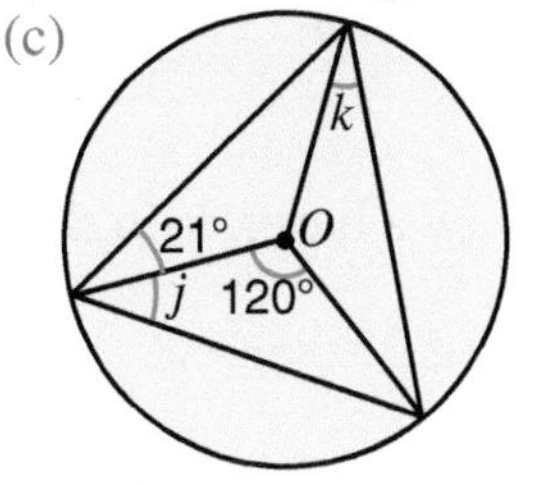

3. O is the centre of the circle. Work out the size of the marked angles.

   (a)
   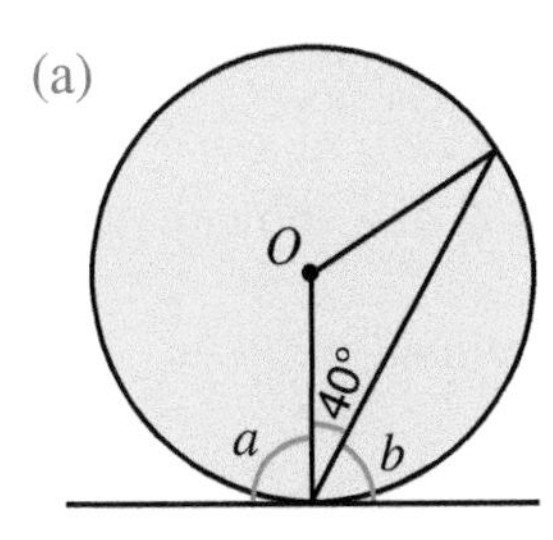

   (b)
   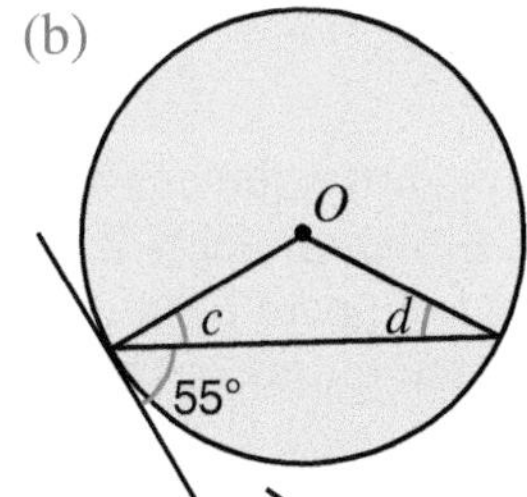

   (c)
   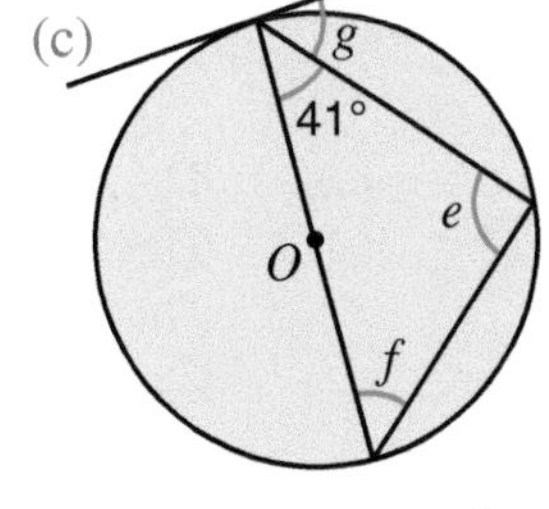

   (d)
   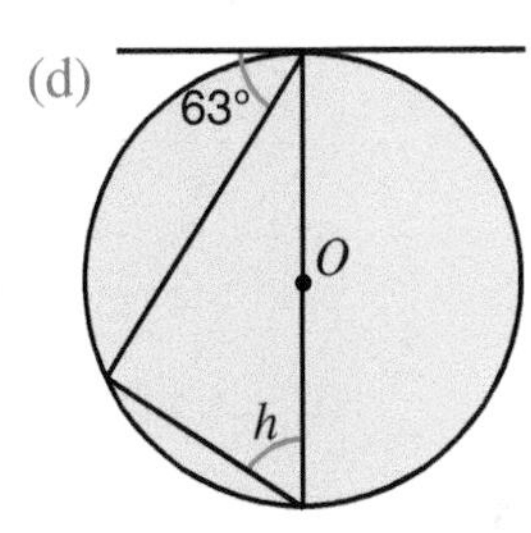

   (e)
   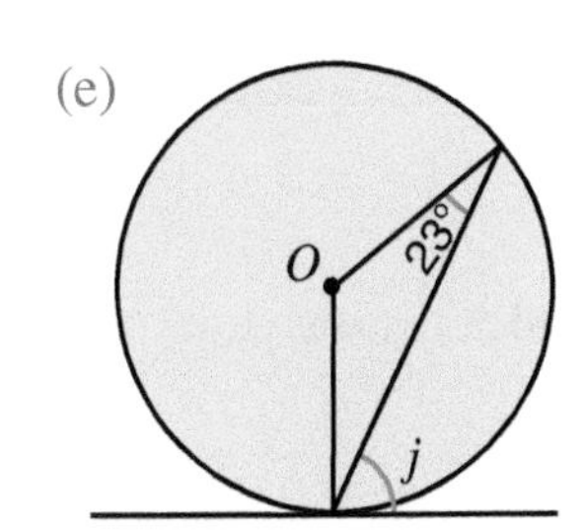

   (f)
   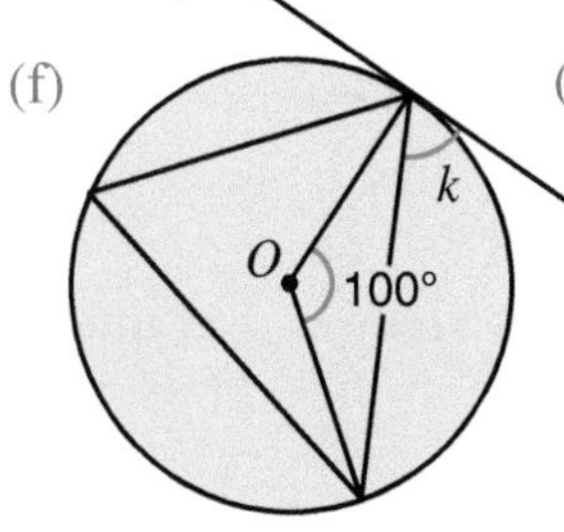

   (g)
   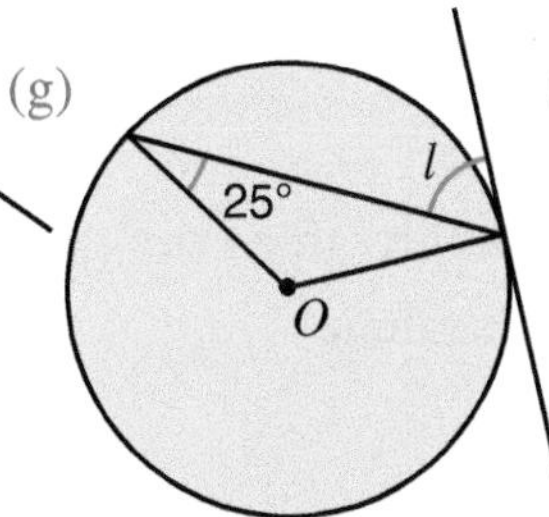

   (h)
   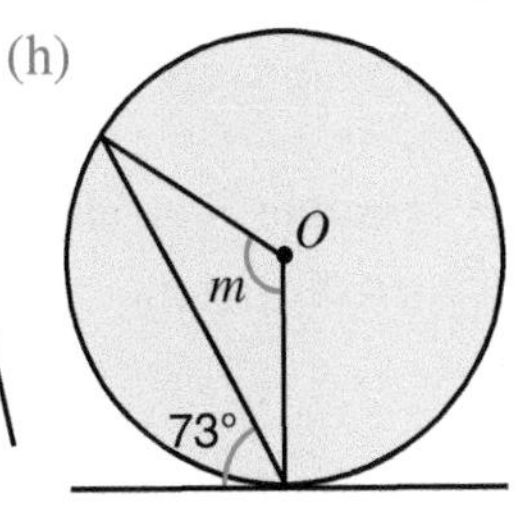

4. AB is a chord 30 cm long, in a circle, centre O.
   The radius of the circle is 17 cm.
   Find the length of OC.

   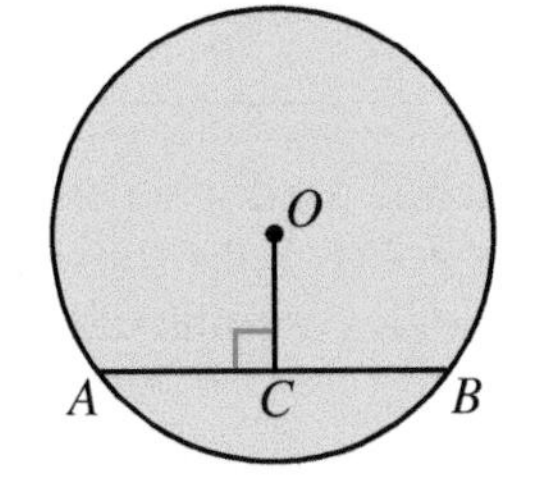

5. 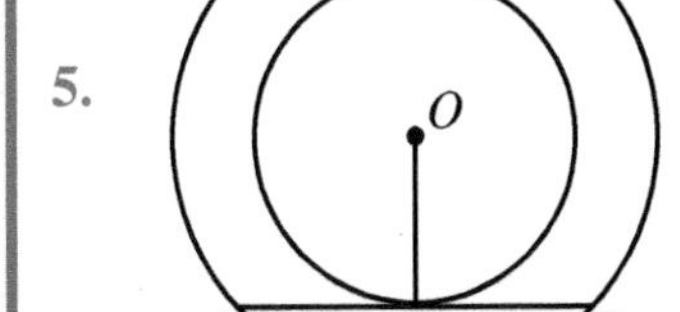

   The two circles each have centre O.
   AB is a tangent to the smaller circle, touching at X.
   (a) What is the size of $\angle AXO$?
   (b) Explain why $AX = XB$.
   (c) If the circles have radii 7.2 cm and 12 cm, find the length of AB.

6. A circular cake has diameter 30 cm.
A segment of cake is taken by cutting along line AB, 5 cm from the edge of the cake.
Calculate the length of AB.

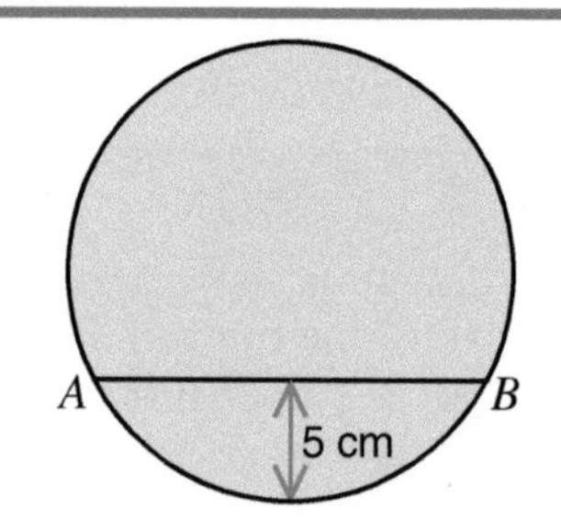

7. 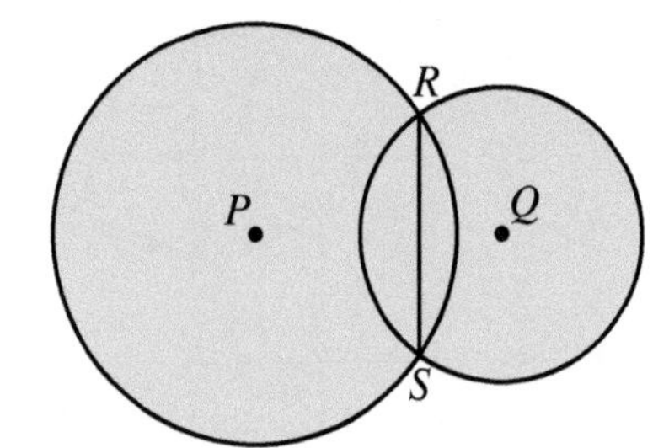

Circles with centres at P and Q intersect at points R and S.
The radius of the circle with centre P is 15 cm.
The radius of the circle with centre Q is 13 cm.
$RS = 24$ cm.
Calculate the length of PQ.

8. A dough scraper, used by a baker, is in the shape of a circle with a segment removed.
The radius of the circle is 25 cm.
$AB = 48$ cm.
Find the height, h, of the dough scraper.

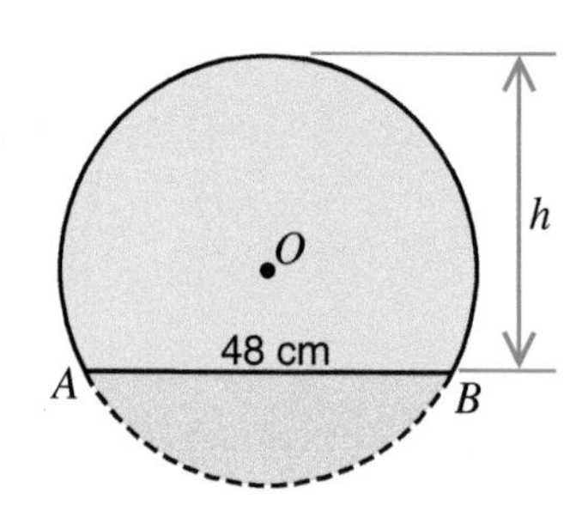

9. 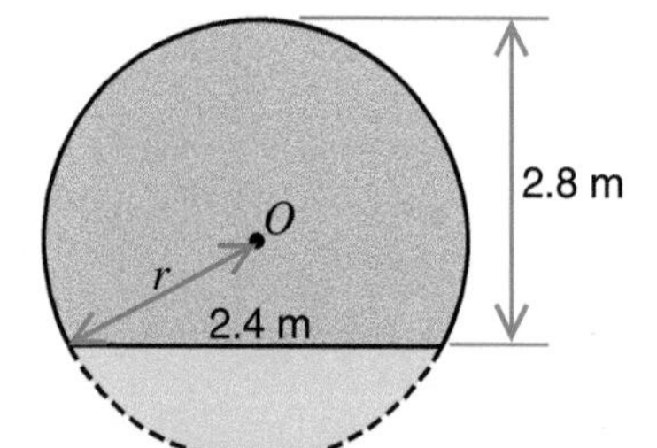

The diagram shows the cross-section of a subway tunnel.
The height of the tunnel is 2.8 m.
The width of the path through the tunnel is 2.4 m.
Calculate r, the radius of the tunnel.

Key Points

- A **triangle** is a shape made by three straight sides.
- Types of triangle: **Scalene** **Isosceles** **Equilateral**

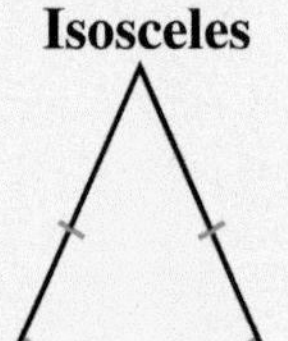

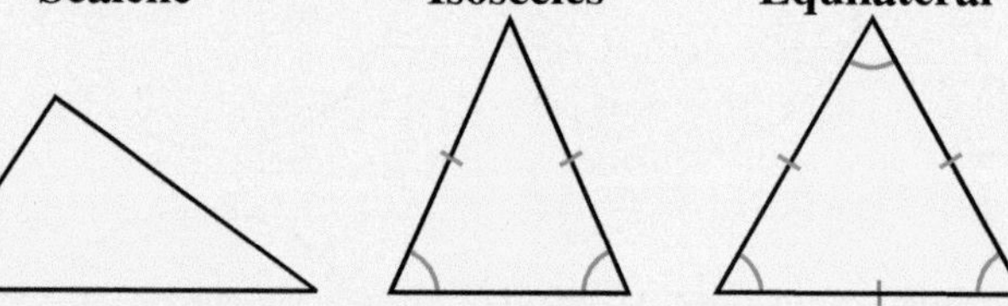

A **sketch** is used when an accurate drawing is not required.
Dashes across lines show sides that are equal in length.
Equal angles are marked using arcs.

- The sum of the angles in a triangle is 180°.

  $a + b + c = 180°$

  $a + b = d$

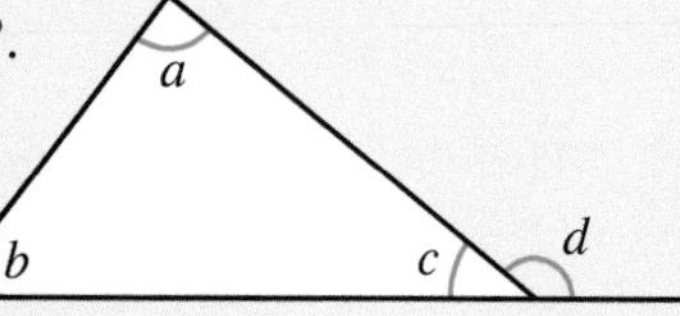

- A **quadrilateral** is a shape made by four straight lines.
- The sum of the angles in a quadrilateral is 360°.

- Facts about these special quadrilaterals:

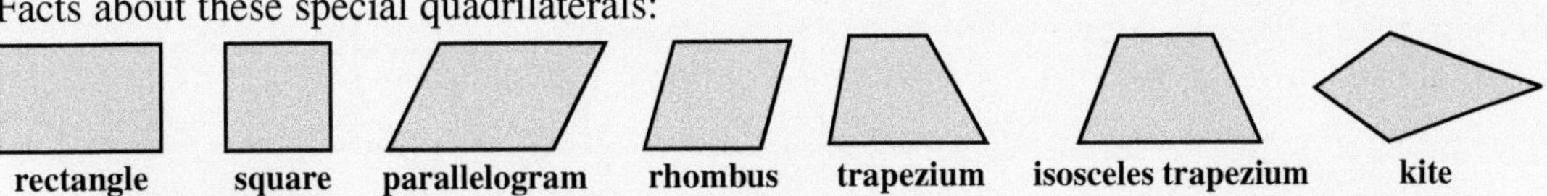

Quadrilateral	Sides	Angles	Diagonals
Rectangle	Opposite sides equal and parallel	All 90°	Bisect each other
Square	4 equal sides, opposite sides parallel	All 90°	Bisect each other at 90°
Parallelogram	Opposite sides equal and parallel	Opposite angles equal	Bisect each other
Rhombus	4 equal sides, opposite sides parallel	Opposite angles equal	Bisect each other at 90°
Trapezium	1 pair of parallel sides		
Isosceles trapezium	1 pair of parallel sides, non-parallel sides equal	2 pairs of equal angles	Equal in length
Kite	2 pairs of adjacent sides equal	1 pair of opposite angles equal	One bisects the other at 90°

- When two parallel lines are crossed by a **transversal** the following pairs of angles are formed.

Corresponding angles

a

c

$a = c$

Alternate angles

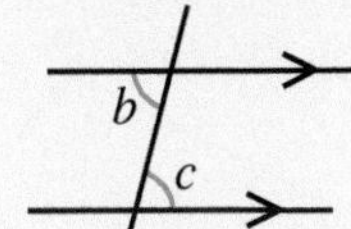

$b = c$

Allied angles

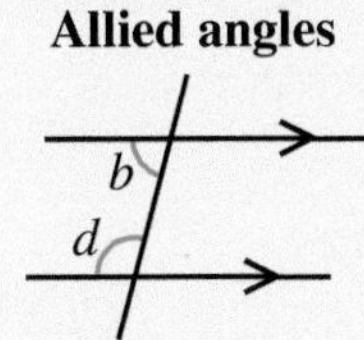

$b + d = 180°$

Arrowheads are used to show that lines are **parallel**.

- A **polygon** is a many-sided shape made by straight lines.
- A polygon with all sides equal and all angles equal is called a **regular polygon**.
- The sum of the exterior angles of any polygon is 360°.
- At each vertex of a polygon: interior angle + exterior angle = 180°.
- The sum of the interior angles of an n-sided polygon is given by: $(n - 2) \times 180°$.
- For a regular n-sided polygon: exterior angle $= \frac{360°}{n}$.
- A shape will **tessellate** if it covers a surface without overlapping and leaves no gaps.
- A **circle** is the shape drawn by keeping a pencil the same distance from a fixed point on a piece of paper.
- You should know the meaning of the words shown on the diagrams below.

interior angle

exterior angle

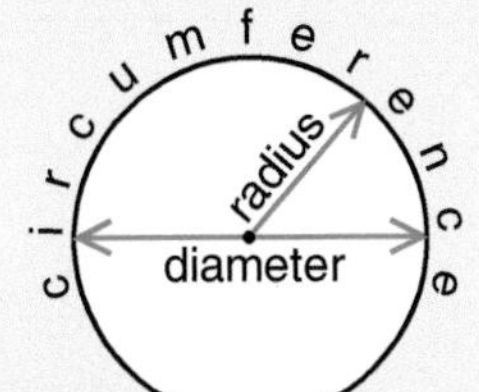

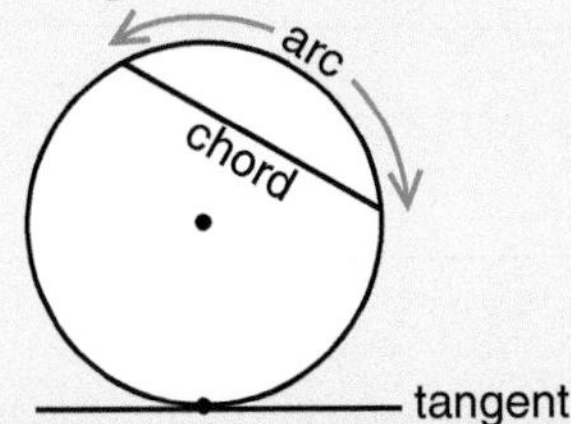

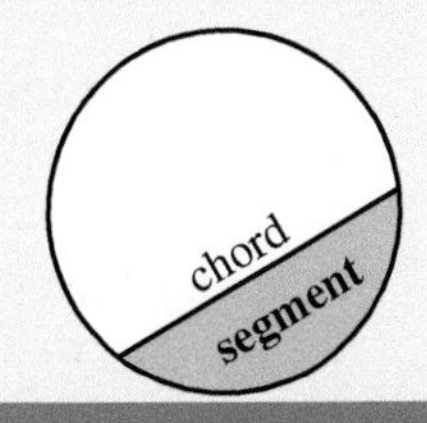

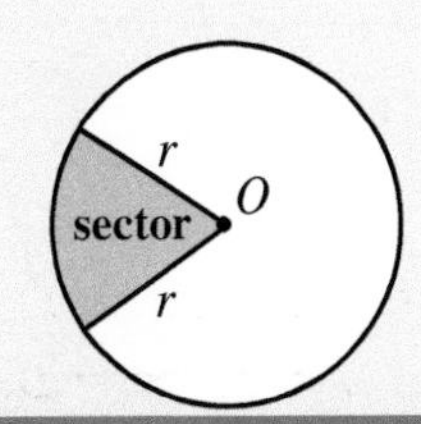

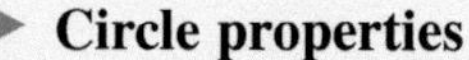

▶ **Circle properties**

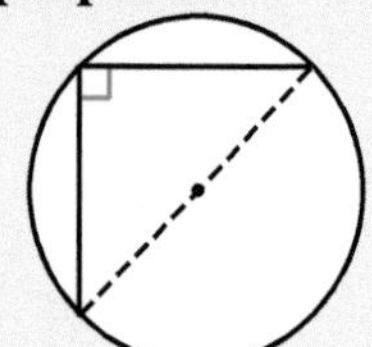

The angle in a semicircle is a right angle.

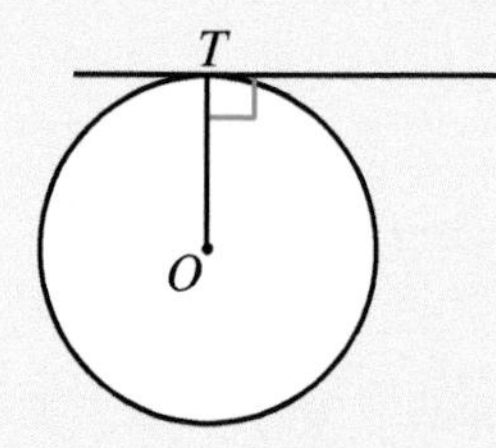

A tangent is perpendicular to the radius at the point of contact.

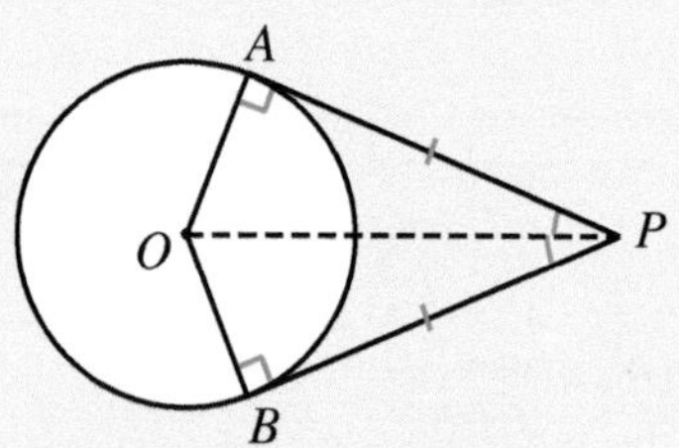

Tangents drawn to a circle from the same point are equal.

▶ The perpendicular bisector of a chord, AB, always passes through the centre, O, of a circle. The triangle formed by the chord, AB, and the two radii, AO and BO, is **isosceles**.

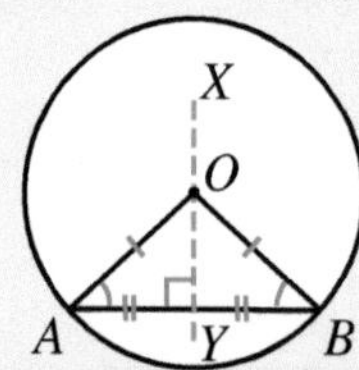

Review Exercise 18

1. PQR is an equilateral triangle.
PSR is an isosceles triangle. $\angle SPQ = 40°$.
Work out the size of angle x, giving a reason for your answer.

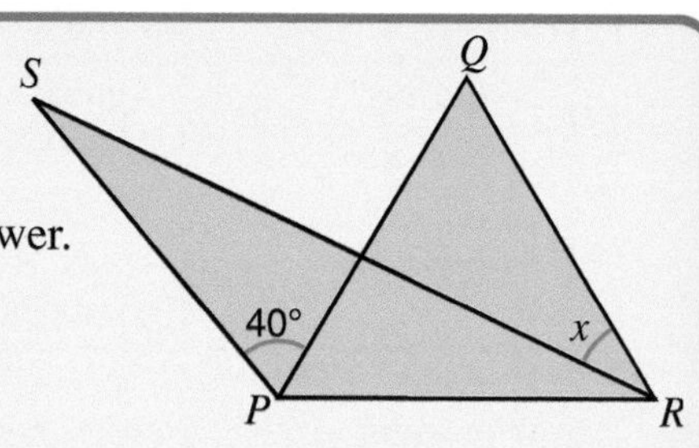

2. In the figure, the bisector of the angle ABC meets AC at D.
$BD = DC$ and angle $BAD = 78°$.
Calculate angle BDC.

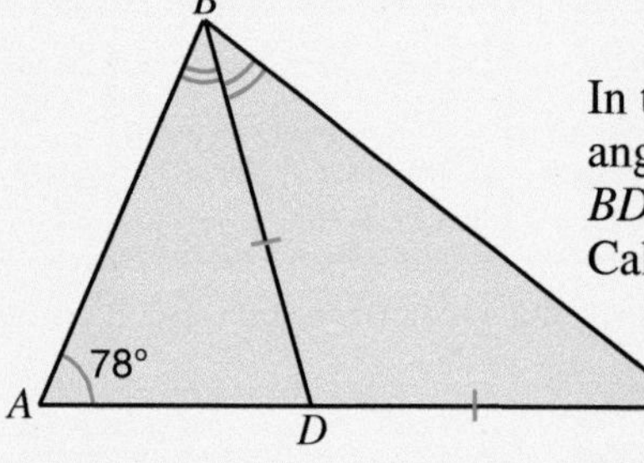

3. The diagram shows a kite $PQRS$.
(a) How many lines of symmetry has $PQRS$?

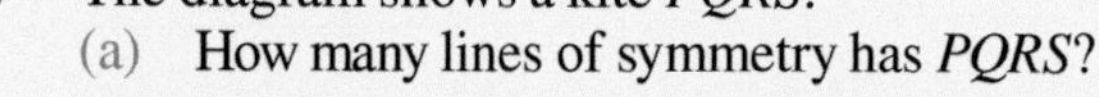

Angle $QRS = 97°$ and angle $PQR = 112°$.
(b) Work out the size of angle PSR.

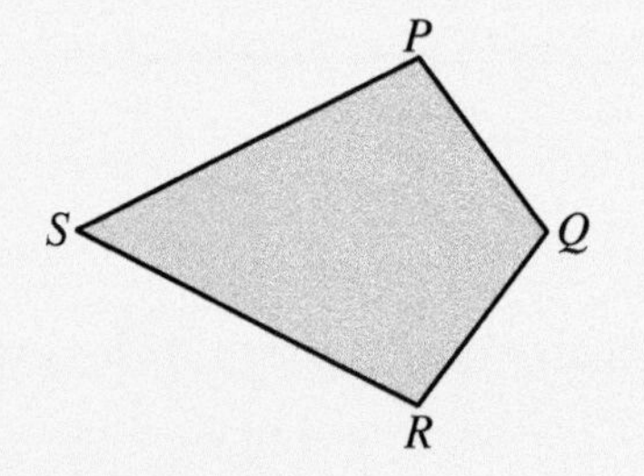

4. The diagram shows a trapezium.
Find the size of angle a and angle b.

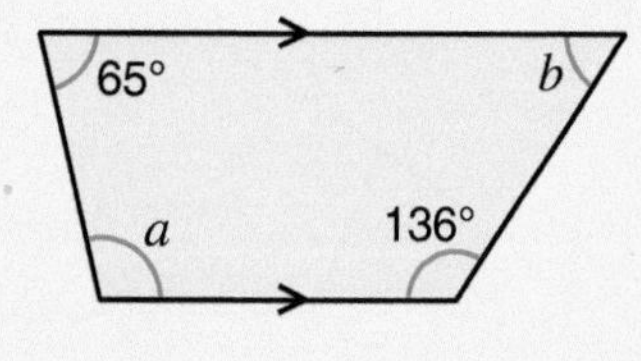

5. Part of a regular polygon is shown.
(a) What is the size of angle q?
(b) How many sides has the polygon?

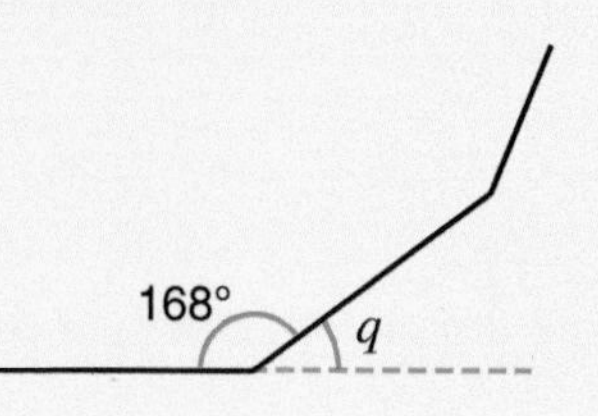

6. $ABCDE$ is a regular pentagon.
Work out
(a) angle x,
(b) angle y.

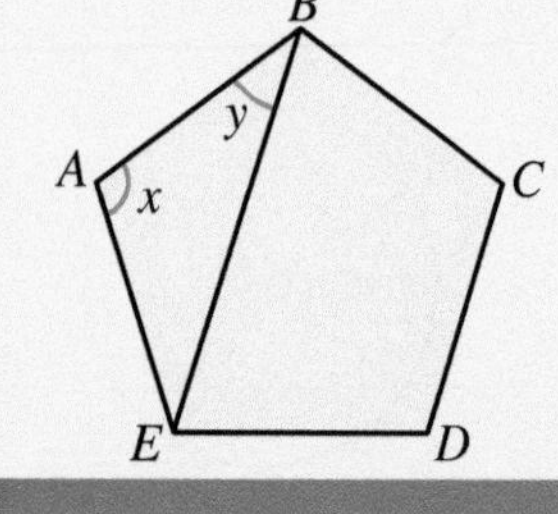

7. PQ is a chord in the circle with centre O.
The circle has radius 13 cm.
$\angle OMP = 90°$ and $OM = 5$ cm.
Find the length of PQ.

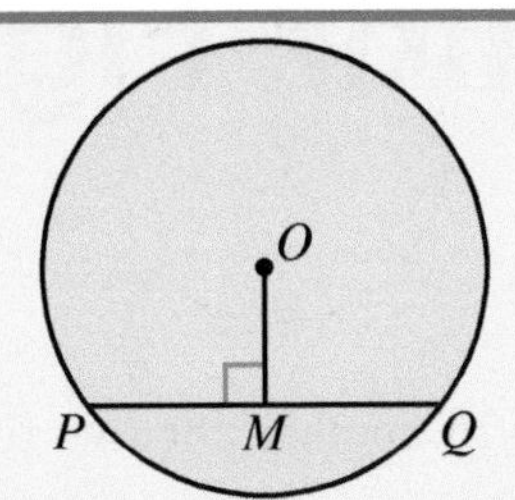

8.

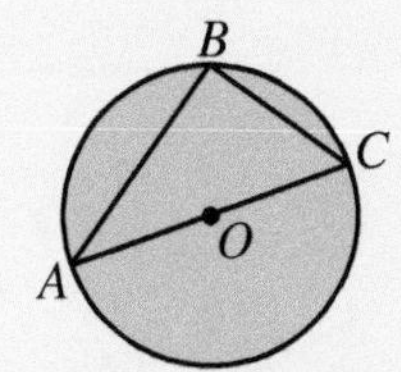

AC is a diameter of the circle, centre O.
$\angle BAC = (2x - 5)°$ and $\angle ACB = (2x + 15)°$.
Find the value of x.

9. O is the centre of the circle.
AT is a diameter.
TD is a tangent touching the circle at T.
$\angle CBT = 58°$.

Find the size of:
(a) $\angle ATD$,
(b) $\angle OAC$,
(c) $\angle AOC$,
(d) $\angle ODT$.

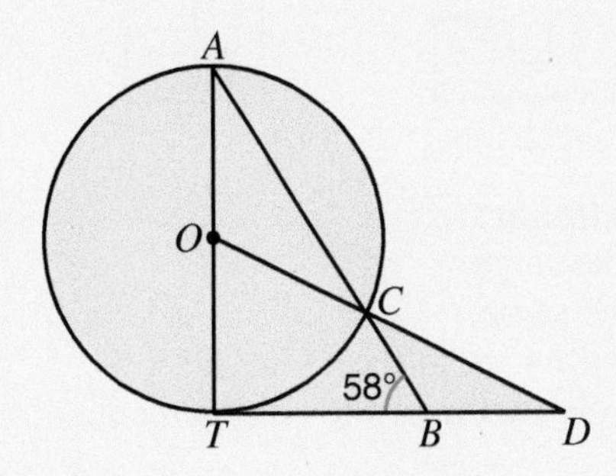

10.

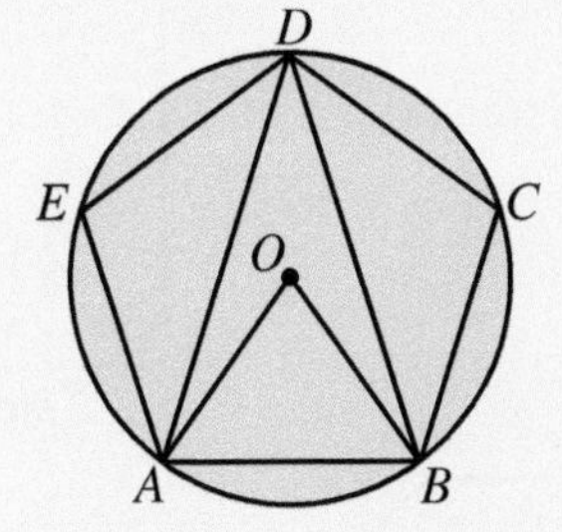

A regular pentagon, $ABCDE$, is inscribed in a circle, centre O.
(a) Find the size of $\angle AOB$.
(b) Find the size of $\angle ADB$.
(c) What fraction of $\angle CDE$ is $\angle ADB$?

11. In the diagram, O is the centre of the circle and TR is the tangent to the circle at T.
Calculate x.

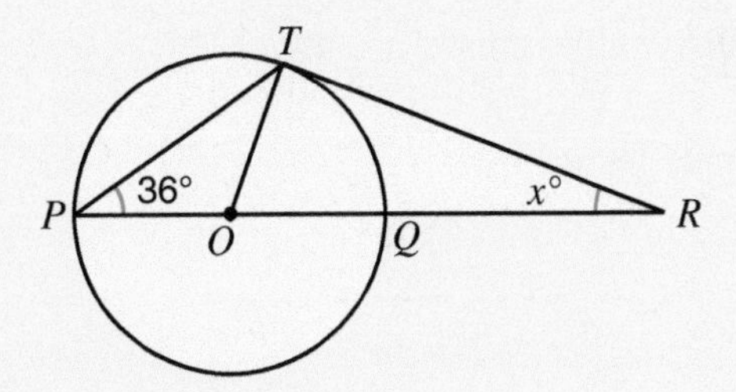

12.

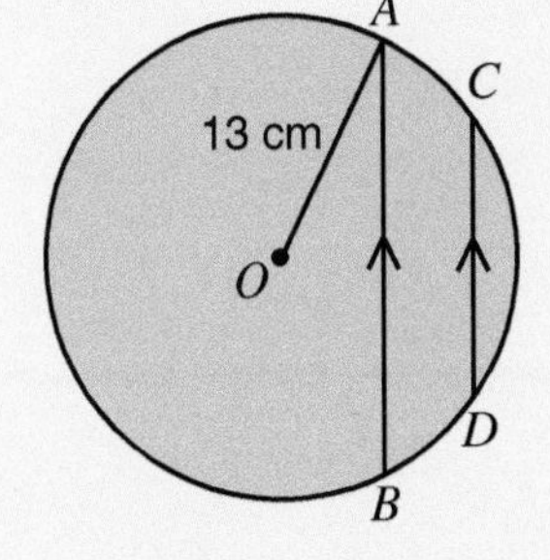

Circle, centre O, has radius 13 cm.
Chords AB and CD are parallel.
$AB = 24$ cm and $CD = 10$ cm.
Calculate the distance between AB and CD.

13. The diagram shows a regular decagon which has been divided into three parts.
(a) Work out the size of angle AED.
(b) What is the size of angle AEI?
(c) Show that angle $EAJ = 90°$.

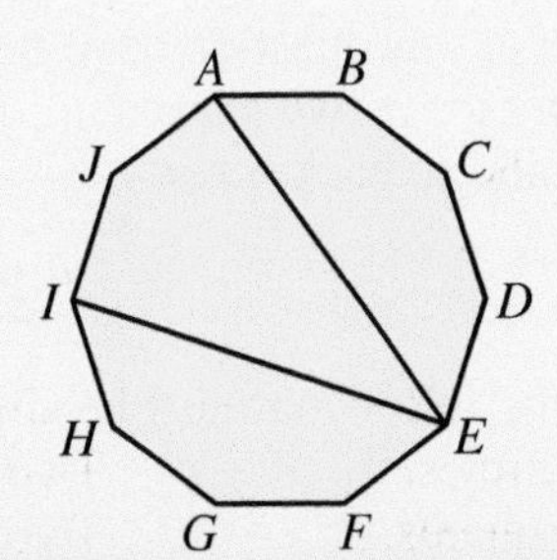

19 Similar Figures

Similar figures

When one figure is an enlargement of another, the two figures are **similar**.

Sometimes one of the figures is rotated or reflected.
For example:

Figures **C** and **E** are enlargements of figure **A**.
Figures **A**, **C** and **E** are similar.

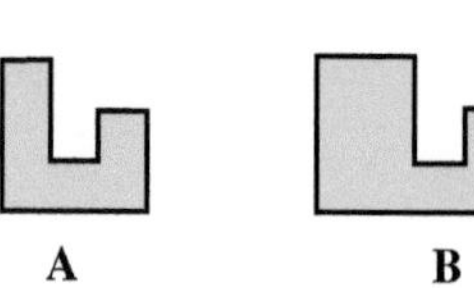

C

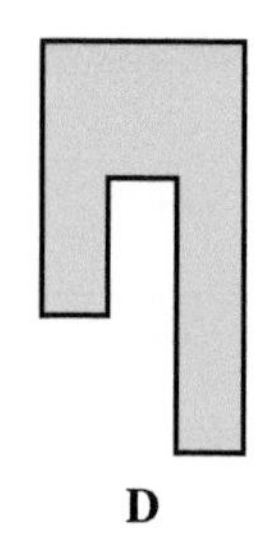

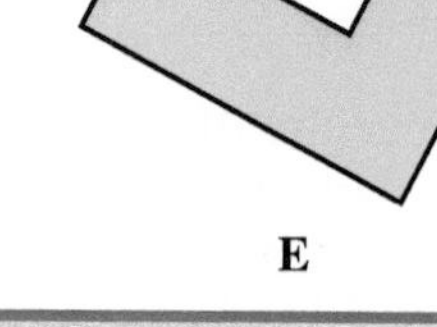

When two figures are **similar**:
their **shapes** are the same,
their **angles** are the same,
corresponding **lengths** are in the same ratio,
this ratio is the **scale factor** of the enlargement.

> $\text{Scale factor} = \dfrac{\text{new length}}{\text{original length}}$
>
> This can be rearranged to give
> new length = original length × scale factor.

Example 1

A photo has width 6 cm and height 9 cm.
An enlargement is made, which has width 8 cm.
Calculate the height of the enlargement.

Scale factor $= \frac{8}{6}$

$h = 9 \times \frac{8}{6}$

$h = 12\,\text{cm}$

Example 2

These two figures are similar.
Calculate the lengths of x and y.
Write down the size of the angle marked a.

1.4 cm, y, 62°, 3 cm; x, 2.7 cm, a, 4.5 cm

The scale factor $= \frac{4.5}{3} = 1.5$

Lengths in the large figure are given by: length in small figure × scale factor

$x = 1.4 \times 1.5$

$x = 2.1\,\text{cm}$

Lengths in the small figure are given by: length in large figure ÷ scale factor

$y = 2.7 \div 1.5$

$y = 1.8\,\text{cm}$

The angles in similar figures are the same, so, $a = 62°$.

Practice Exercise 19.1

1. The shapes in this question have been drawn accurately.
   Explain why these two shapes are not similar to each other.

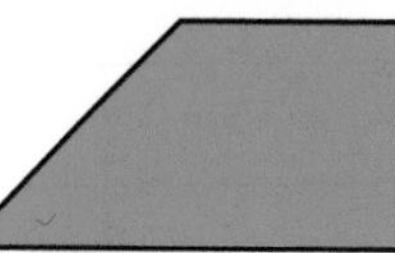

2. Which of the following must be similar to each other?
   (a) Two circles.   (b) Two kites.   (c) Two parallelograms.
   (d) Two squares.   (e) Two rectangles.

3. These rectangles are all similar. The diagrams have not been drawn accurately.
Work out the lengths of the sides marked a and b.

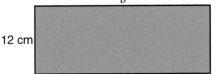

4. These two kites are similar.
(a) What is the scale factor of their lengths?
(b) Find the length of the side marked x.
(c) What is the size of angle a?

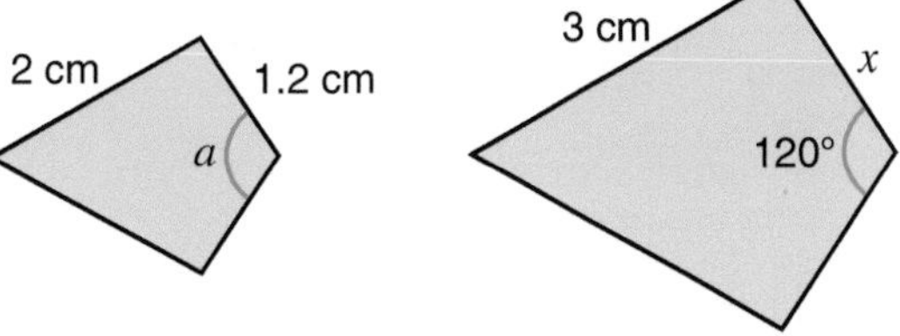

5. A shape has width 8 cm and length 24 cm.
It is enlarged to give a new shape with width 10 cm.
Calculate the length of the new shape.

6. In each part, the two figures are similar. Lengths are in centimetres.
Calculate the lengths and angles marked with letters.

(a)

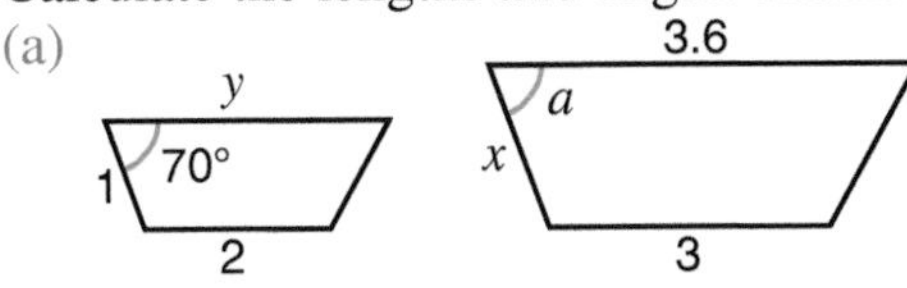

(b)

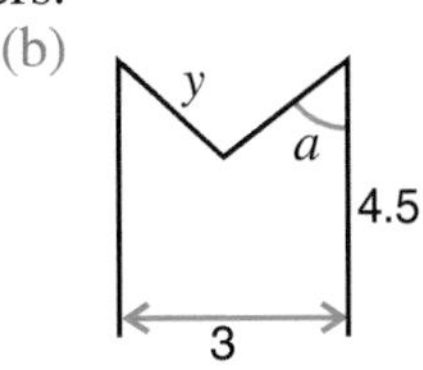

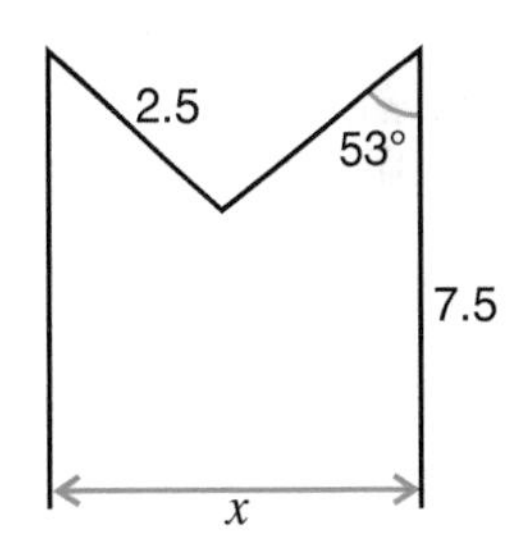

(c)

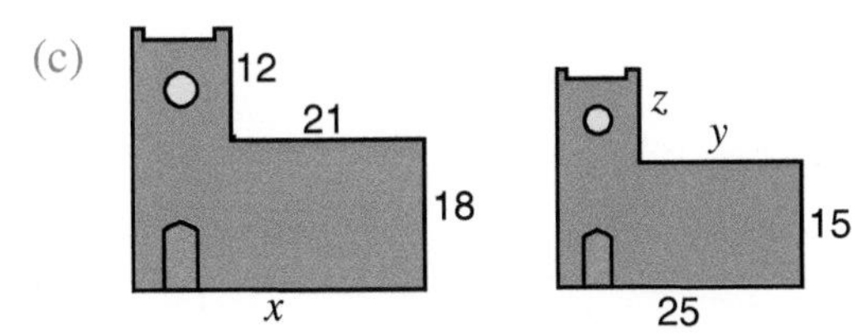

7. These two tubes are similar.
The width of the small size is 2.4 cm and the height of the small size is 10 cm.
The width of the large size is 3.6 cm.
Calculate the height of the large size.

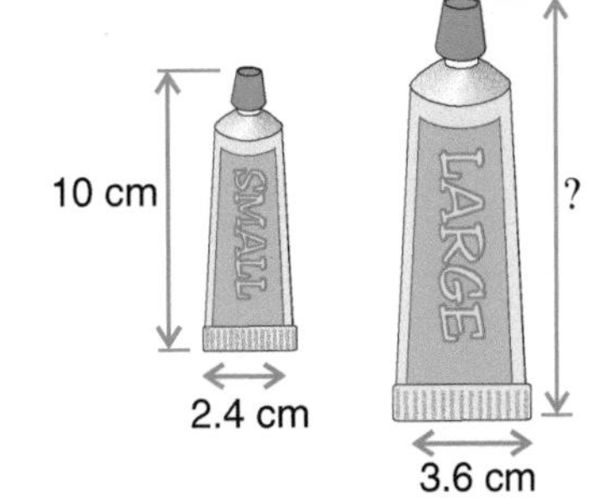

8. A motor car is 4.2 m long and 1.4 m high.
A scale model of the car is 8.4 cm long.
What is the height of the model?

9. The smallest angle in triangle T is 18°.
Triangle T is enlarged by a scale factor of 2.
How big is the smallest angle in the enlarged triangle?

10. A castle has height 30 m.
The height of the castle wall is 6 m.
A scale model of the castle has height 25 cm.
Calculate the height of the castle wall in the scale model.

11. The dimensions of three sizes of paper are given. All the sizes are similar.

Length (cm)	24	30	y
Width (cm)	x	20	32

Calculate the values of x and y.

Similar triangles

For any pair of similar triangles:
corresponding lengths are opposite equal angles, the scale factor is the ratio of corresponding sides.

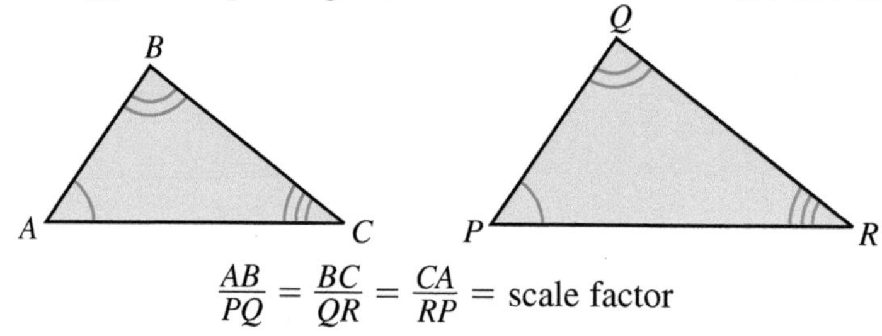

$$\frac{AB}{PQ} = \frac{BC}{QR} = \frac{CA}{RP} = \text{scale factor}$$

Example 3

These two triangles are similar, with the equal angles marked.
Calculate the lengths x and y.

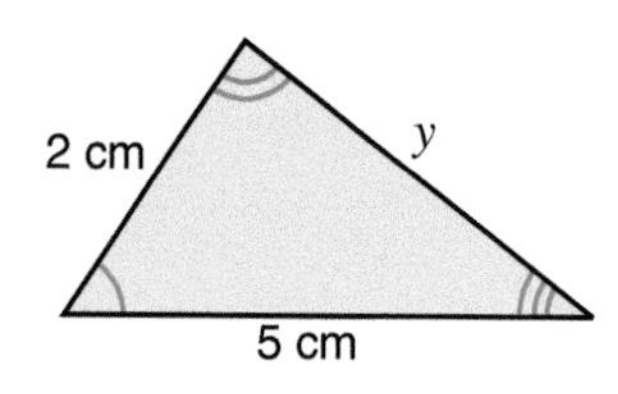

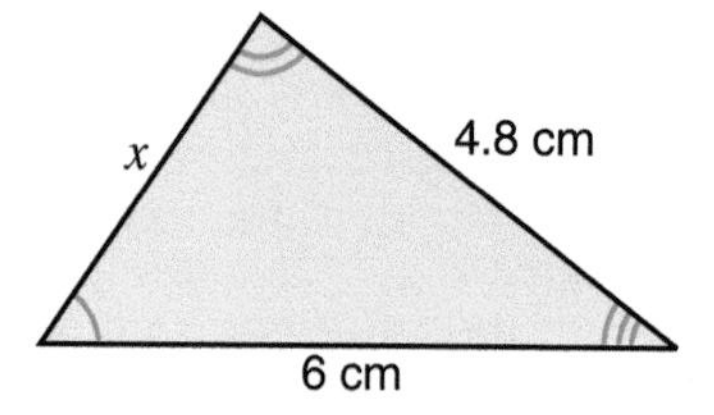

Scale factor $= \frac{6}{5} = 1.2$

$x = 2 \times 1.2$

$x = 2.4\,\text{cm}$

$y = 4.8 \div 1.2$

$y = 4\,\text{cm}$

Example 4

Triangles ABC and PQR are similar.
Calculate the lengths of AC and PQ.

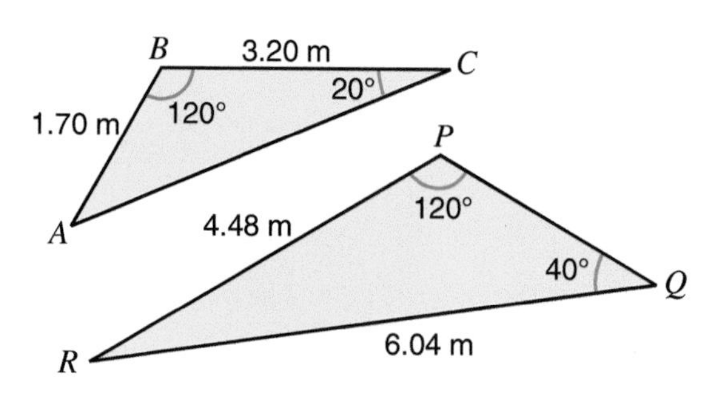

BC and PR are corresponding sides.

Scale factor $= \frac{4.48}{3.20} = 1.4$

$AC = 6.04 \div 1.4$

$AC = 4.31$ m, correct to 2 d.p.

$PQ = 1.70 \times 1.4$

$PQ = 2.38\,\text{m}$

Practice Exercise 19.2

Question 1 should be done without a calculator.

1. In each part, the triangles are similar, with equal angles marked. Lengths are in centimetres. Calculate lengths x and y.

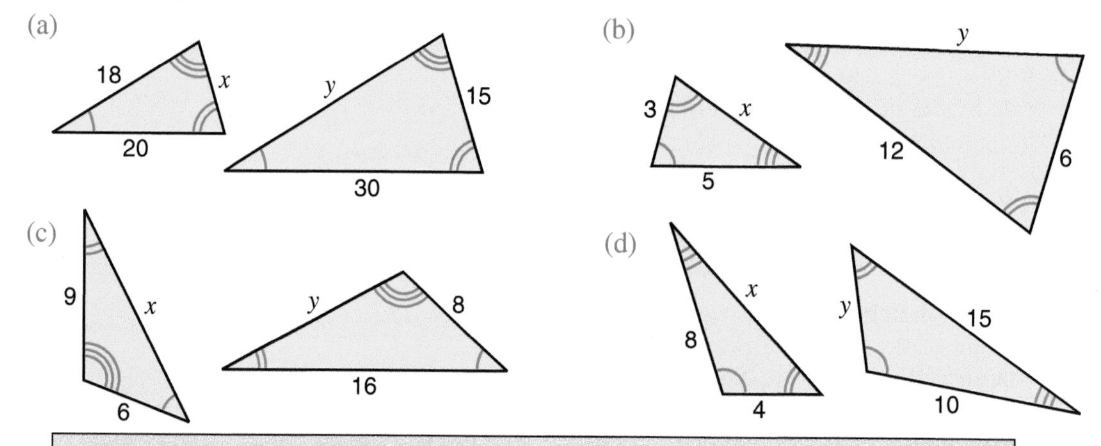

To show that two triangles are similar you have to show that:
either they have equal angles
or corresponding lengths are all in the same ratio.
If you can show that one of these conditions is true then the other one is also true.

2. BC is parallel to PQ.

Show that triangles ABC and APQ are similar and calculate the required lengths.

(a)

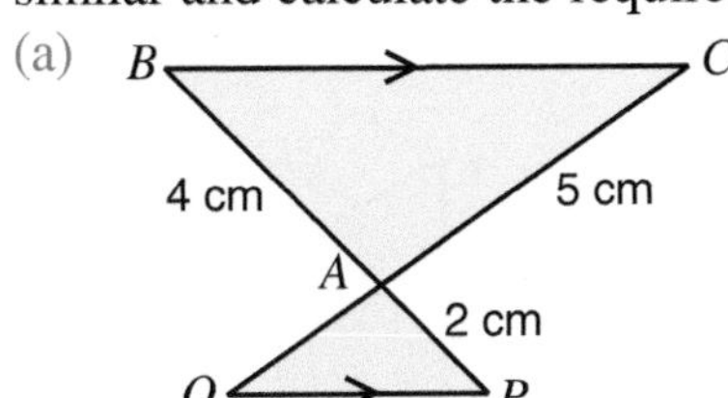

Calculate AQ and BC.

(b)

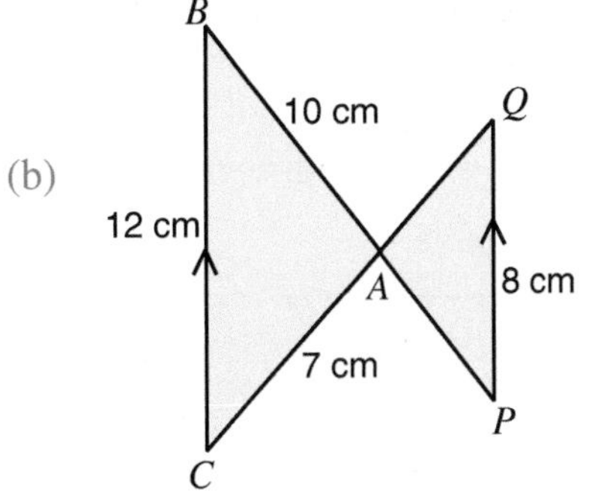

Calculate AQ and BP.

3. Show that these pairs of triangles are similar and find angle x.

(a)

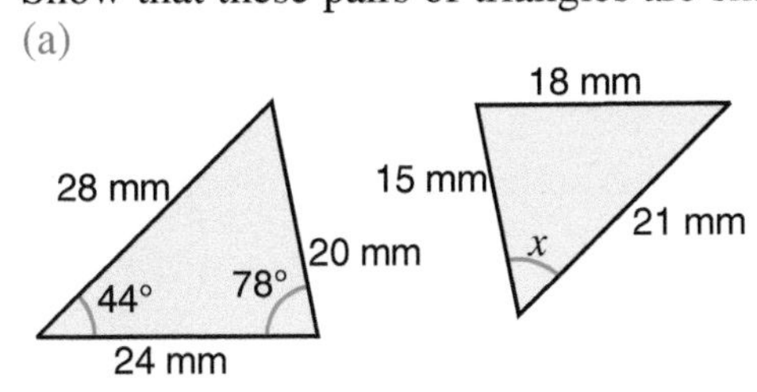

(b)

2 cm, 3.6 cm, 3 cm, 34°

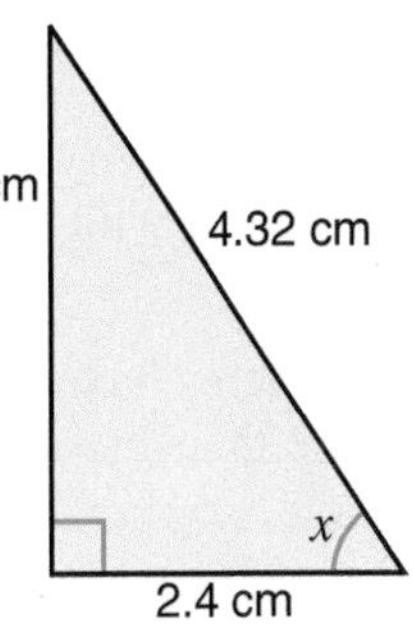

4. All marked lengths are in centimetres.

(a) In each part show that triangles ABC and APQ are similar and find angle x.

(i)

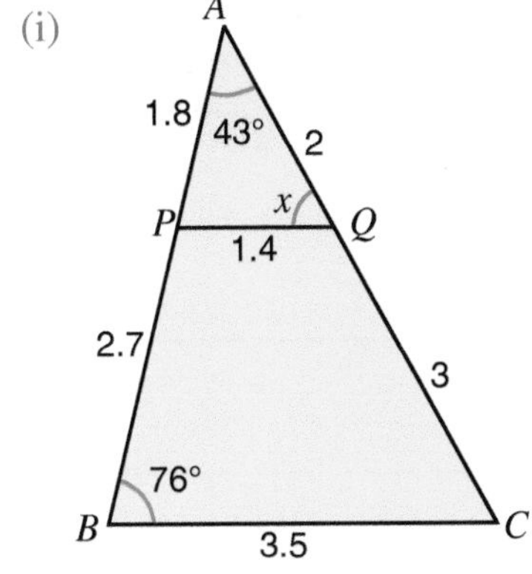

(ii)

A, 9, 104°, Q, 3, x, C, 6, 12, 47°, P, 16, 2, B

(b) In each part find the perimeters of triangles ABC and APQ.
What do you notice about the ratio of the perimeters of the triangles and the ratio of the lengths of corresponding sides?

5. (a) Explain why triangles ABC and PQR are similar.

(b) Calculate the length of AB.

(c) The perimeter of triangle ABC is 7.5 cm.
Find the perimeter of triangle PQR.

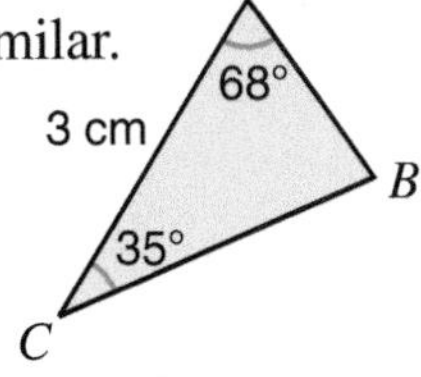

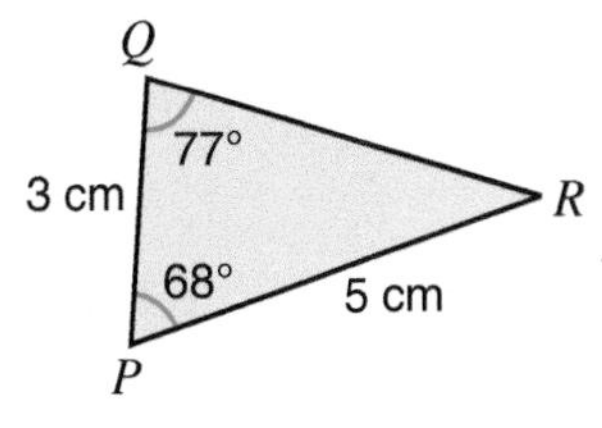

6.

Y, Q, 4.5 cm, X, P, 4 cm, Z

Triangles XYZ and PQZ are similar.

$\angle ZXY = \angle ZPQ = 90°$.

$PZ = 4\,\text{cm}$ and $XY = 4.5\,\text{cm}$.

The ratio $QZ : YZ$ is $2 : 3$.

Calculate the area of triangle XYZ.

Lengths, areas and volumes of similar figures

Activity

Some cubes have side 2 cm.
They are built together to make a larger cube with side 6 cm.
This represents an enlargement with scale factor 3.

Copy and complete the table.

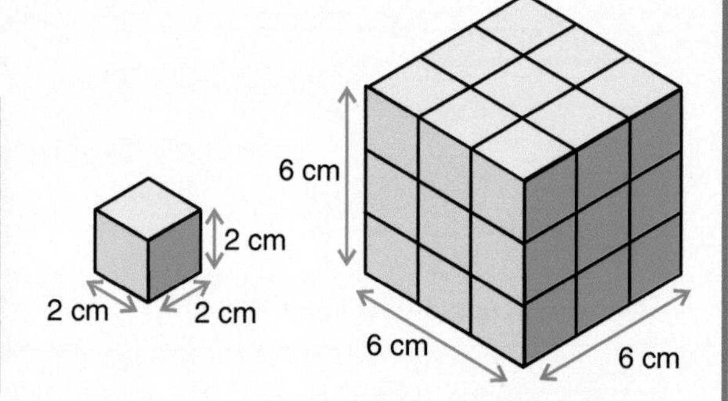

	Length of side (cm)	Area of face (cm²)	Volume of cube (cm³)
Small cube	2		8
Large cube	6	36	
Scale factor	$\frac{6}{2} = 3$	$\frac{36}{} =$	$\frac{}{8} =$

What do you notice about the three scale factors?

Repeat this activity with a scale factor of 4.
Now repeat it with a scale factor of k.

Spheres

Three spheres have radii r cm, $2r$ cm and $3r$ cm.
Compare the surface areas and volumes of the three spheres.

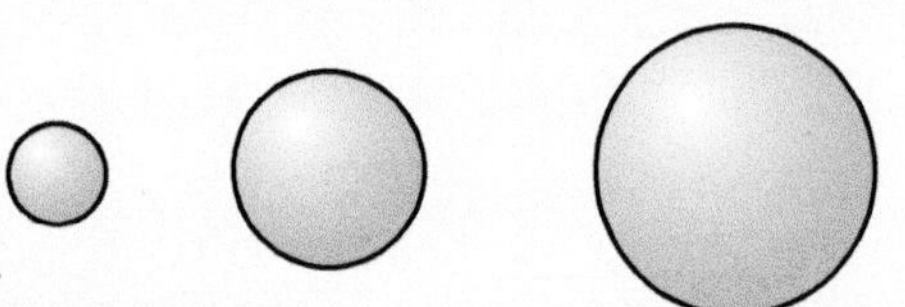

Scale factors for length, area and volume

When the **length** scale factor $= k$
the **area** scale factor $= k^2$
the **volume** scale factor $= k^3$

For solids made of the same material, **mass is proportional to volume**.
So, the mass scale factor is k^3.

Example 5

A prototype for a new plane is made.
The real plane will be an enlargement of the prototype with scale factor 5.

It can be assumed that "scale factor" refers to the length scale factor, unless specified differently in a question.

(a) The area of the windows on the prototype is 0.18 m².
Find the area of the real windows.

(b) The volume of the real fuel tank is 4000 litres.
Find the volume of the fuel tank on the prototype.

(a) Area scale factor $= 5^2 = 25$.
Real area $= 0.18 \times 25 = 4.5\text{ m}^2$

(b) Volume scale factor $= 5^3 = 125$.
Prototype volume $= \frac{4000}{125} = 32$ litres.

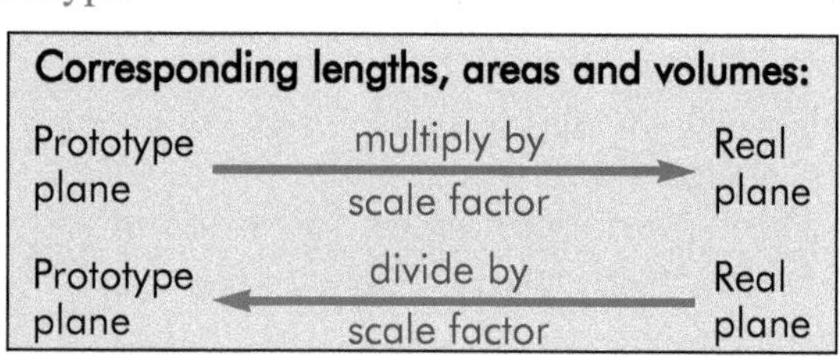

Corresponding lengths, areas and volumes:
Prototype plane
multiply by
scale factor
Real plane
Prototype plane
divide by
scale factor
Real plane

Example 6

A metal ingot has volume 20 000 cm³.
It is melted down and made into identical smaller ingots.
Each small ingot is similar to the original ingot and has volume 25 cm³.
The length of the large ingot is 50 cm.
Calculate the length of the small ingots.

Volume scale factor $= \frac{20\,000}{25} = 800$. Length scale factor $= \sqrt[3]{800} = 9.28\ldots$

Length of small ingot $= \frac{50}{\sqrt[3]{800}} = 5.386\ldots = 5.39$ cm, correct to 3 s.f.

Practice Exercise 19.3

1. A model of a train is 60 cm long. It is made on a scale of 1 to 50.
   What is the length of the actual train in metres?
2. A rectangle has length 8 cm and width 6 cm. A similar rectangle has length 12 cm.
   (a) What is the scale factor of their lengths?
   (b) What is the width of the larger rectangle?
3. A motor car is 4.2 m long and 1.4 m high. A scale model of the car is 8.4 cm long.
   (a) What is the scale of the model?
   (b) What is the height of the model?
4. The lengths of the sides of a kite are doubled. What happens to its area?
5. A rectangular vegetable plot needs 10 kg of fertiliser.
   How much fertiliser is needed for a plot with double the dimensions?
6. Circle P has a radius of 3.5 cm. Circle Q has a radius of 35 cm.
   How many times larger than the area of circle P is the area of circle Q?
7. A company logo is printed on all its stationery.
   On small sheets of paper the logo is 1.2 cm high and covers an area of 3.5 cm^2.
   On large sheets of paper the logo covers an area of 14 cm^2.
   What is the height of the logo on large sheets of paper?
8. The lengths of the sides of a square are halved.
   What happens to its area?
9. A picture is 30 cm high and has an area of 360 cm^2.
   Another print of the same picture is 15 cm high.
   What is its area?

10. A king-size photograph is 18 cm long and 12 cm wide.
    A standard size photograph is 12 cm long.
    (a) What is the width of a standard size photograph?
    (b) What is the area of a standard size photograph?
11. Two rugs are similar. The larger rug is 3.6 m in length and has an area of 9 m^2.
    The smaller rug has an area of 4 m^2. What is the length of the smaller rug?
12. The scale of a map is 1 to 50 000.
    (a) The distance between two junctions on the map is 3 cm.
    What is the actual distance between the junctions, in kilometres?
    (b) A lake covers 20 cm^2 on the map.
    How many square kilometres does the lake actually cover?
13. A map has a scale of 1 : 25 000.
    (a) The length of a road is 3.5 km. Calculate its length, in centimetres, on the map.
    (b) The area of a field on the map is 12 cm^2. Calculate the true area in square metres.
    (c) A park has an area of 120 000 m^2. Calculate the area of the park on the map.
14. The measurements of a rabbit hutch are all doubled.
    How many times bigger is its volume?
15. A teapot has a volume of 500 ml.
    A similar teapot is double the height.
    What is the volume of this teapot?

16. A box of height 4 cm has a surface area of 220 cm^2 and a volume of 200 cm^3.
    (a) What is the surface area and volume of a similar box of height 8 cm?
    (b) What is the surface area and volume of a similar box of height 2 cm?

17. Two garden ponds are similar.
The dimensions of the larger pond are three times as big as the smaller pond.
The smaller pond holds 200 litres of water.
How many litres of water does the larger pond hold?

18. Two solid spheres are made of the same material.
The smaller sphere has a radius of 4 cm and weighs 1.5 kg.
The larger sphere has a radius of 8 cm.
How much does it weigh?

19. The measurements of a box are each halved. What happens to its volume?

20. A cylinder with a height of 6 cm has a volume of 200 cm^3.
What is the volume of a similar cylinder with a height of 3 cm?

21. A box with a height of 4 cm has a volume of 120 cm^3.
What is the volume of a similar box with a height of 6 cm?

22. The volumes of two similar statues are 64 cm^3 and 125 cm^3.
The height of the smaller statue is 8 cm.
(a) What is the height of the larger statue?
(b) What is the ratio of their surface areas?

23. Two fish tanks are similar. The smaller tank is 12 cm high and has a volume of 3.6 litres.
The larger tank has a volume of 97.2 litres. What is the height of the larger tank?

24. Pop and Fizzo come in similar cans.
Cans of Pop are 8 cm tall and cans of Fizzo are 10 cm tall.
A can of Pop holds 200 ml.
How much does a can of Fizzo hold?

25. (a) A scale model of a house is made using a scale of 1 : 200.
The roof area on the model is 82 cm^2. Find the real roof area, in square metres.
(b) The volume of the real roof space is 360 m^3.
Find the volume of the roof space in the model, in cubic centimetres.

26. Jane makes a scale model of her village.
A fence of length 12 m is represented by a length of 4 cm on her model.
(a) Calculate the scale which Jane is using.
(b) Calculate, in cubic centimetres, the volume on the model of a pool which has a volume of 200 m^3.
(c) Calculate the actual area of a playground which has an area of 320 cm^2 on the model.

27. Joe wants to enlarge a picture so that its area is doubled.
What length scale factor should he use?

28. Two balls have radii of 2 cm and 5 cm.
(a) Calculate the volumes of the two balls and show that the ratio of the volumes is 8 : 125.
(b) Show that the surface area of the smaller ball is 16% of the surface area of the larger ball.

29. Coffee filters are paper cones.
The cones are made in these similar sizes: small, medium and large.
The slant height of a small cone is 5 cm and the surface area is 15π cm^2.
A large cone has a surface area of 135π cm^2.
Calculate the length of its slant height.

30. A bronze sculpture is made in two sizes.
The taller sculpture is 15 cm high and the shorter one is 9 cm high.
The taller sculpture weighs 3.75 kg.
What is the weight of the shorter sculpture?

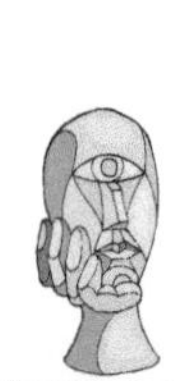
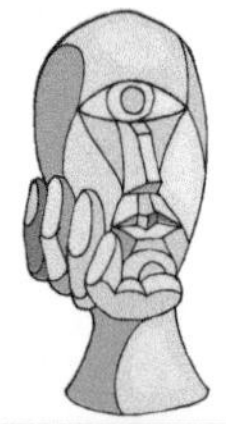

Key Points

- When two figures are **similar**:
  their **shapes** are the same,
  their **angles** are the same,
  corresponding **lengths** are in the same ratio, this ratio is the **scale factor** of the enlargement.

$$\text{Scale factor} = \frac{\text{new length}}{\text{original length}}$$

- All circles are similar to each other.
- All squares are similar to each other.
- For **similar triangles**:
  corresponding lengths are opposite equal angles, the scale factor is the ratio of the corresponding sides.
- You should be able to find corresponding lengths in similar triangles.

$$\frac{AB}{PQ} = \frac{BC}{QR} = \frac{CA}{RP} = \text{scale factor}$$

- You should be able to find corresponding lengths, areas and volumes in **similar figures**.

For **similar areas** and **volumes**:
When the **length** scale factor $= k$
the **area** scale factor $= k^2$
the **volume** scale factor $= k^3$

Review Exercise 19

1. These pairs of figures are similar.
   Calculate the lengths of the unknown sides marked with letters.

   (a)

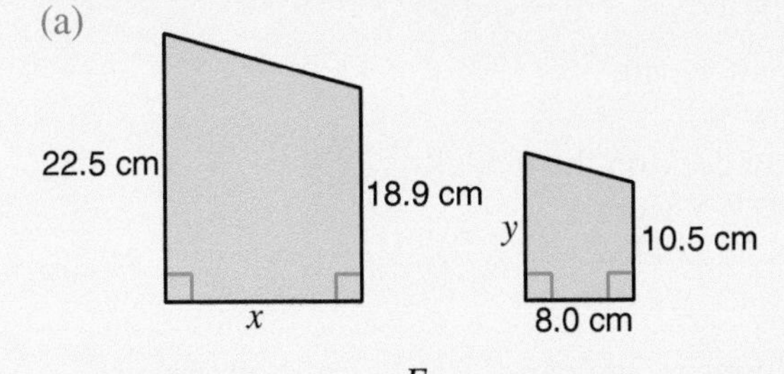

   (b)

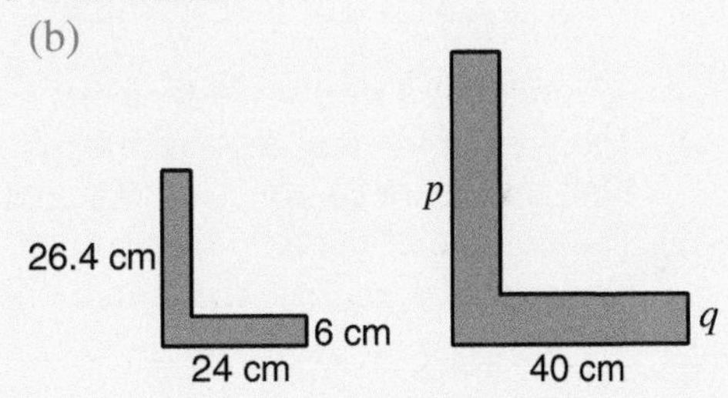

2. ΔDEF is an enlargement of ΔABC.
   (a) What is the scale factor of the enlargement?
   (b) How many times greater is the area of ΔDEF than the area of ΔABC?
   (c) Which angle is equal in size to $\angle ACB$?

3. A stick 2 m long is placed vertically so that its top is in line with the top of a cliff from a point, A, on the ground 3 m from the stick and 120 m from the cliff.
   How high is the cliff?

4. A map is drawn using a scale of 2 cm to represent 1 km.
   (a) Write the scale of the map in ratio form.
   (b) Two villages are 8.4 cm apart on the map.
   What is the actual distance between the two villages?

5. A hall has dimensions 20 metres by 13.5 metres.
   A plan of the hall is drawn using a scale of 1 : 250.
   What are the dimensions of the hall on the plan?

6. These cylinders are similar.
   The radius, r, of the smaller cylinder is 5.4 cm.
   The radius of the larger cylinder is 6.3 cm.
   (a) What is the ratio of corresponding lengths?
   (b) What is the height of the 2nd cylinder?
   (c) What is the ratio of their total surface areas?
   (d) What is the ratio of their volumes?

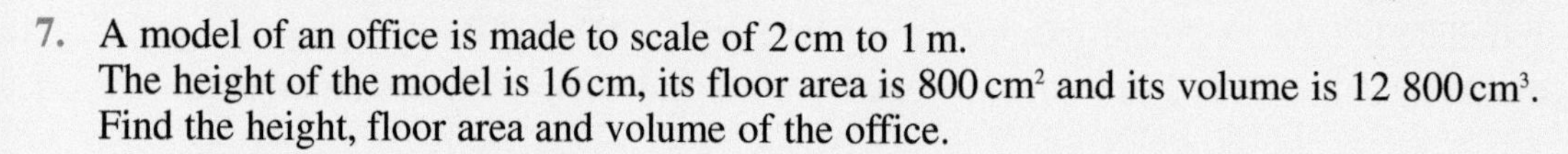

7. A model of an office is made to scale of 2 cm to 1 m.
   The height of the model is 16 cm, its floor area is 800 cm² and its volume is 12 800 cm³.
   Find the height, floor area and volume of the office.

8. Two similar solid statues weigh 560 g and 1890 g.
   The lighter one has a height of 10 cm.
   If they are made of similar material, what is the height of the other one?

9. These candles are similar.
   P has a surface area of 24 cm².
   Q has a surface area of 54 cm².
   P is 4 cm high.
   (a) How high is **Q**?
   (b) What is the ratio of their volumes?

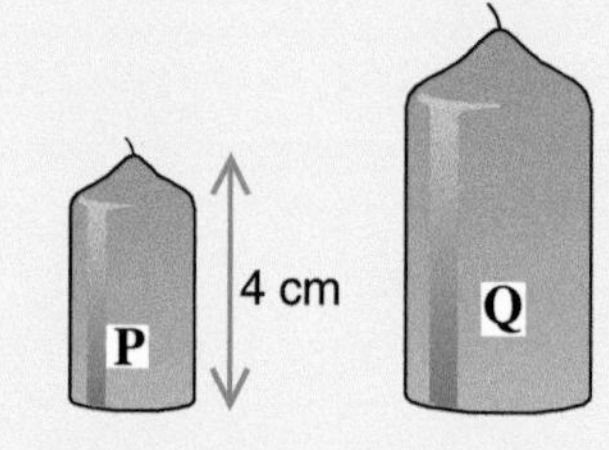

10. A scale model of a ship is made, using a scale of 1 : 40.
    (a) The area of the real deck is 500 m².
    Find, in square centimetres, the area of the deck on the model.
    (b) The volume of the hold on the model is 187 500 cm³.
    Find, in cubic metres, the volume of the real hold.

11. The scale of a map is 5 cm to 1 km.
    (a) What is the actual distance between two points which are 15.5 cm apart on the map?
    (b) What is the actual area of a field which is represented on the map by an area of 5 cm²?

12. Two spheres have radii 15 cm and 20 cm.
    What is the ratio of (a) their diameters, (b) their surface areas, (c) their volumes?

13. Two similar cones have volumes in the ratio 27 : 125.
    (a) What is the ratio of their radii? (b) What is the ratio of the areas of their bases?

14. The length of the shadow of a vertical post 2.4 m high is 3.3 m.
    At the same time and place, the length of the shadow of a vertical flagpole is 17.6 m.
    Find the height of the flagpole.

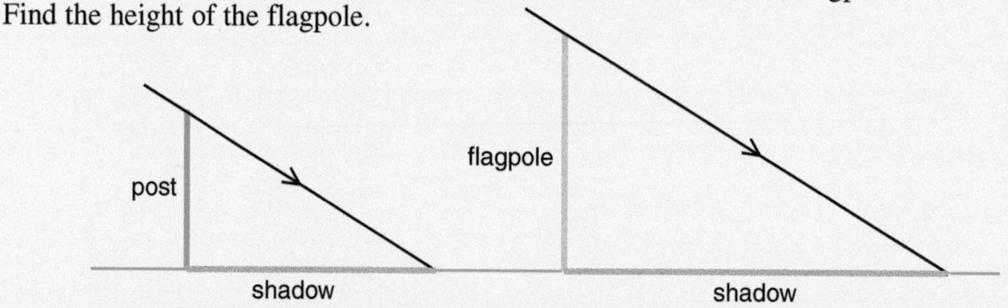

You can assume that because the sun is so far away, its rays form parallel lines.

20 Vectors *and* 3D Coordinates

Vectors and scalars

Quantities, which have both **size** and **direction**, are called **vectors**.
Examples of vector quantities are:
Displacement - A combination of distance and direction.
For example, a train travels 30 km due south.
Velocity - A combination of speed and direction.
For example, the wind is blowing south at 20 km/h, a river is flowing at 10 km/h northwards.

Quantities which have size only are called **scalars**.
Examples of scalar quantities are: distance, area, speed, mass, volume, temperature, etc.

Vector notation

Column vectors

The diagram shows the translation of a triangle by the vector $\begin{pmatrix} 3 \\ 2 \end{pmatrix}$.

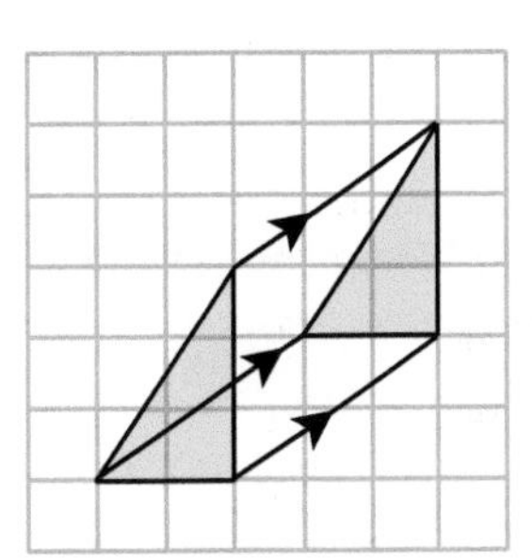

The triangle has been displaced 3 units to the right and 2 units up.

$\begin{pmatrix} 3 \\ 2 \end{pmatrix}$ gives information about the size and direction of the displacement.

The vector $\begin{pmatrix} 3 \\ 2 \end{pmatrix}$ is sometimes called a **column vector**, because it consists of one column of numbers.

> The numbers 3 and 2 are called the components of the vector $\begin{pmatrix} 3 \\ 2 \end{pmatrix}$.
> 3 is the x-component.
> 2 is the y-component.

Directed line segments

Vectors can be represented in diagrams using **line segments**.
The **length** of a line represents the **size of the vector**.
The **direction of the vector** is shown by an **arrow**.

Labelling vectors

The notation $\overrightarrow{AB}, \overrightarrow{CD}, \overrightarrow{OX}, \ldots$ is often used.

$\overrightarrow{AB}$ indicates the displacement from A to B.

$\overrightarrow{AB}$ should be read as "vector AB".

Bold lower case letters such as **a**, **b**, **c**, … are also used.

> In handwritten work **a** should be written as $\underline{a}$.
> **a** and $\underline{a}$ should both be read as "vector a".

Example 1

Draw and label the following vectors.

(a) $\mathbf{a} = \begin{pmatrix} -2 \\ 3 \end{pmatrix}$ (b) $\mathbf{b} = \begin{pmatrix} 3 \\ -2 \end{pmatrix}$ (c) $\overrightarrow{AB} = \begin{pmatrix} -4 \\ -2 \end{pmatrix}$

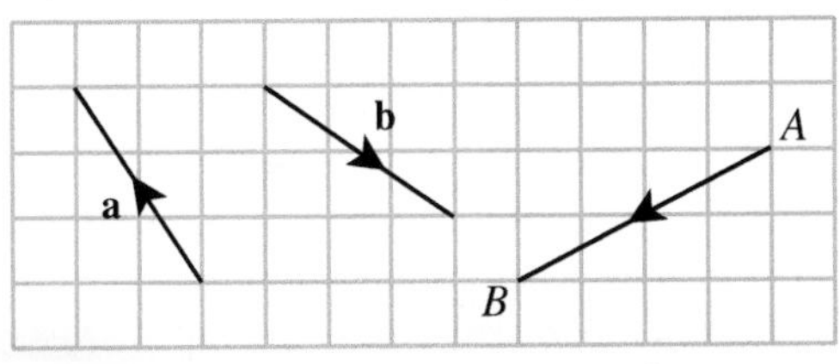

> **Remember:**
> The top number describes the **horizontal** part of the movement:
> + = to the right
> − = to the left
> The bottom number describes the **vertical** part of the movement:
> + = upwards
> − = downwards

Practice Exercise 20.1

1. On squared paper, draw and label the following vectors.

   (a) $\mathbf{a} = \begin{pmatrix} -5 \\ 2 \end{pmatrix}$ (b) $\mathbf{b} = \begin{pmatrix} 5 \\ -2 \end{pmatrix}$ (c) $\mathbf{c} = \begin{pmatrix} -5 \\ -2 \end{pmatrix}$ (d) $\mathbf{d} = \begin{pmatrix} 5 \\ 2 \end{pmatrix}$

2. Mark the point O on squared paper.
   Draw and label the following vectors.

   (a) $\overrightarrow{OA} = \begin{pmatrix} 2 \\ 4 \end{pmatrix}$ (b) $\overrightarrow{OB} = \begin{pmatrix} 4 \\ 2 \end{pmatrix}$ (c) $\overrightarrow{OC} = \begin{pmatrix} 4 \\ -2 \end{pmatrix}$ (d) $\overrightarrow{OD} = \begin{pmatrix} 2 \\ -4 \end{pmatrix}$

   (e) $\overrightarrow{OE} = \begin{pmatrix} -2 \\ -4 \end{pmatrix}$ (f) $\overrightarrow{OF} = \begin{pmatrix} -4 \\ -2 \end{pmatrix}$ (g) $\overrightarrow{OG} = \begin{pmatrix} -4 \\ 2 \end{pmatrix}$ (h) $\overrightarrow{OH} = \begin{pmatrix} -2 \\ 4 \end{pmatrix}$

3. Write each of these vectors as column vectors.

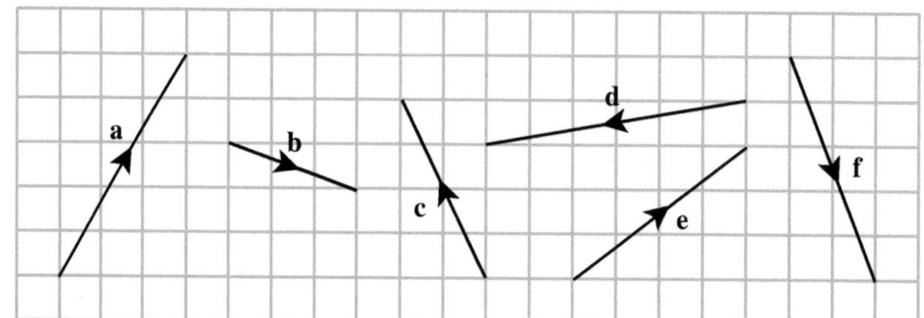

Equal vectors

The diagram shows three vectors, $\overrightarrow{AB}$, $\overrightarrow{PQ}$ and $\overrightarrow{XY}$.

$\overrightarrow{AB} = \begin{pmatrix} 5 \\ 2 \end{pmatrix}$ $\overrightarrow{PQ} = \begin{pmatrix} 5 \\ 2 \end{pmatrix}$ $\overrightarrow{XY} = \begin{pmatrix} 5 \\ 2 \end{pmatrix}$

The column vectors are equal, so, $\overrightarrow{AB} = \overrightarrow{PQ} = \overrightarrow{XY}$.

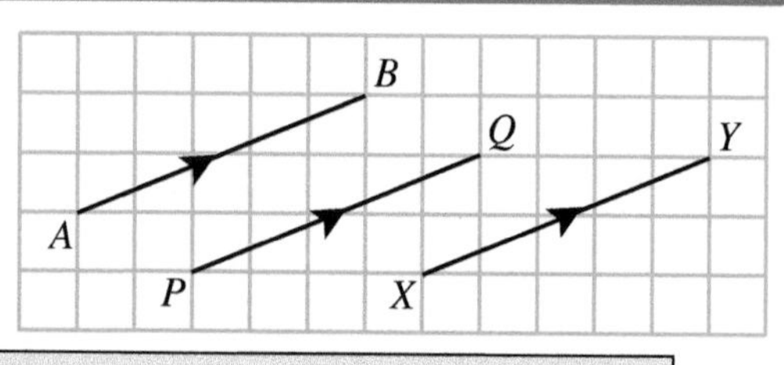

Vectors are **equal** if: they have the same length, **and** they are in the same direction.

Multiplying a vector by a scalar

The diagram shows three vectors, **a**, **b** and **c**.

$\mathbf{a} = \begin{pmatrix} 4 \\ -2 \end{pmatrix}$ $\mathbf{b} = \begin{pmatrix} 8 \\ -4 \end{pmatrix}$ $\mathbf{c} = \begin{pmatrix} 2 \\ -1 \end{pmatrix}$

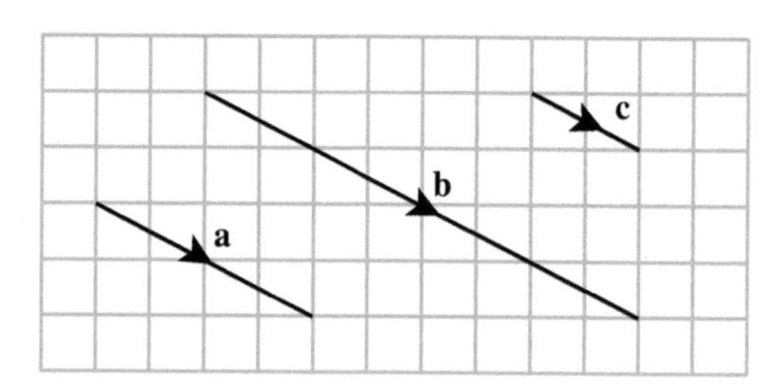

a and **b** are in the same direction.

$\mathbf{b} = \begin{pmatrix} 8 \\ -4 \end{pmatrix} = \begin{pmatrix} 2 \times 4 \\ 2 \times -2 \end{pmatrix}$

This can be written as $\mathbf{b} = 2 \times \begin{pmatrix} 4 \\ -2 \end{pmatrix}$

So, $\mathbf{b} = 2\mathbf{a}$.

This means that **b** is twice the length of **a**.

b and **c** are in the same direction.

$\mathbf{b} = \begin{pmatrix} 8 \\ -4 \end{pmatrix} = 4 \times \begin{pmatrix} 2 \\ -1 \end{pmatrix} = 4\mathbf{c}$

This means that **b** is 4 times as long as **c**.

a and **c** are in the same direction.

$\mathbf{c} = \frac{1}{2}\mathbf{a}$.

This means that **c** is half the length of **a**.

a and n**a** are vectors in the same direction.
The length of vector n**a** $= n \times$ the length of vector **a**.

Vectors in opposite directions

The diagram shows two vectors, **a** and **b**.
The vectors are equal in length, **but** are in opposite directions.

$\mathbf{a} = \begin{pmatrix} 5 \\ 2 \end{pmatrix}$ $\mathbf{b} = \begin{pmatrix} -5 \\ -2 \end{pmatrix}$ So, $\mathbf{b} = -\mathbf{a}$

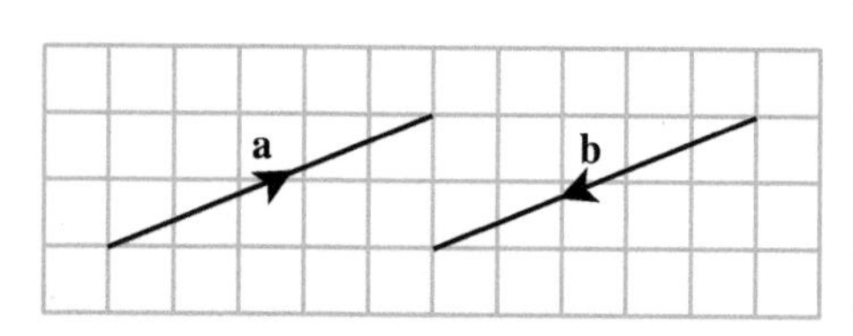

Vectors **a** and −**a** have the same length, **but** are in opposite directions.

Example 2

The diagram shows three vectors: **a**, **b** and **c**.

(a) Write each of **a**, **b** and **c** as column vectors.
(b) Express **b** and **c** in terms of **a**.

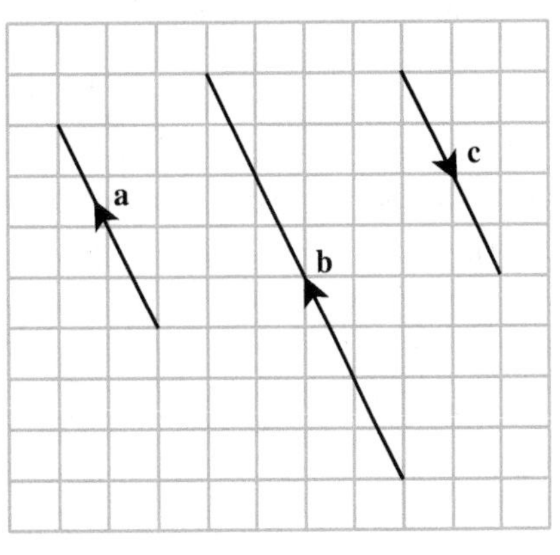

(a) $\mathbf{a} = \begin{pmatrix} -2 \\ 4 \end{pmatrix}$ $\quad \mathbf{b} = \begin{pmatrix} -4 \\ 8 \end{pmatrix}$ $\quad \mathbf{c} = \begin{pmatrix} 2 \\ -4 \end{pmatrix}$

(b) **b** is twice the length of **a** and in the same direction.

$\mathbf{b} = 2\mathbf{a}$

c is the same length as **a** but in the opposite direction.

$\mathbf{c} = -\mathbf{a}$

Magnitude of a vector

The length of a vector **a** is sometimes called its **magnitude**.
The magnitude of vector **a** can be written using the notation $|\mathbf{a}|$.
If **a** is written as a column vector, or drawn on a square grid, we can use Pythagoras' Theorem to calculate the magnitude of **a**.

In general, if $\mathbf{a} = \begin{pmatrix} x \\ y \end{pmatrix}$, then, $|\mathbf{a}| = \sqrt{x^2 + y^2}$.

Example 3

The diagram shows $\mathbf{a} = \begin{pmatrix} 3 \\ 4 \end{pmatrix}$ and $\mathbf{b} = \begin{pmatrix} -5 \\ 8 \end{pmatrix}$.

Find $|\mathbf{a}|$ and $|\mathbf{b}|$.

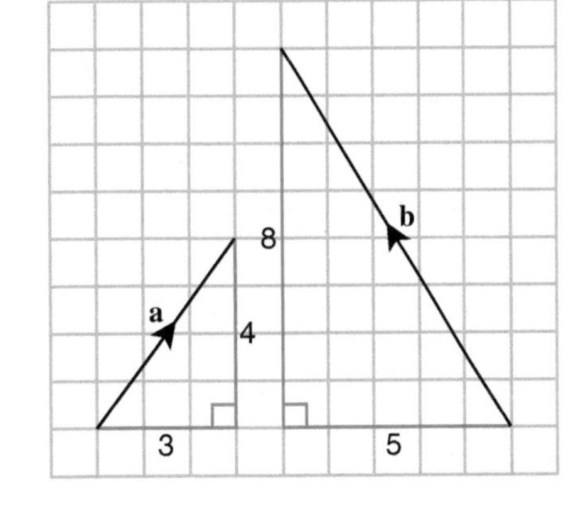

$|\mathbf{a}| = \sqrt{3^2 + 4^2}$

$= \sqrt{9 + 16}$

$= \sqrt{25}$

$|\mathbf{a}| = 5$

$|\mathbf{b}| = \sqrt{(-5)^2 + 8^2}$

$= \sqrt{25 + 64}$

$|\mathbf{b}| = \sqrt{89}$

$\sqrt{89}$ cannot be simplified.
To preserve accuracy, it is common practice to leave answers in surd form.

Practice Exercise 20.2

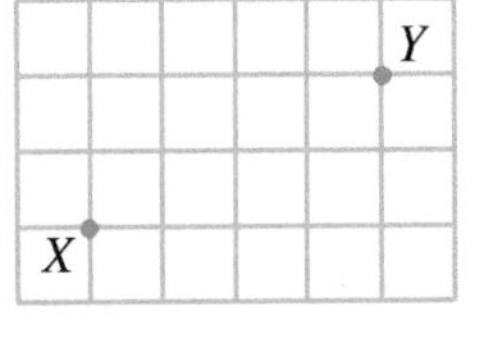

1. The points X and Y are marked on the grid.
   On squared paper, draw each of the vectors **a** to **d**, where
   $\mathbf{a} = 2\overrightarrow{XY}$, $\quad \mathbf{b} = -\overrightarrow{XY}$, $\quad \mathbf{c} = \frac{1}{2}\overrightarrow{XY}$, $\quad \mathbf{d} = \overrightarrow{YX}$.

2. 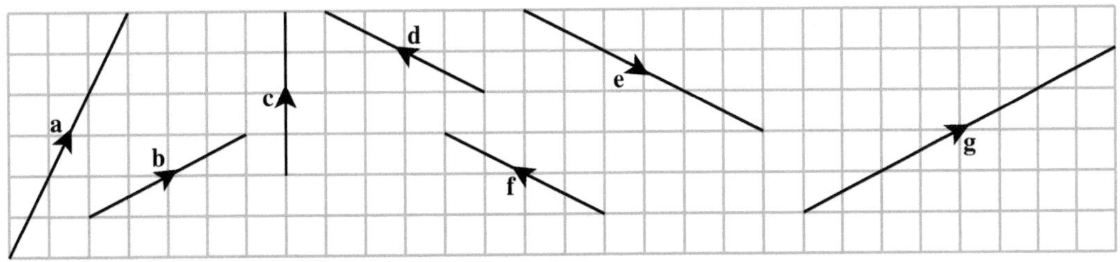

   (a) Write down vectors **a**, **b**, **c**, **d**, **e**, **f**, and **g** as column vectors.
   (b) Which two vectors are equal?
   (c) Which vector is equal to **a** in size but not in direction?
   (d) Which vector is equal to 2**b**?

3. This diagram shows three vectors, labelled **p**, **q** and **r**.

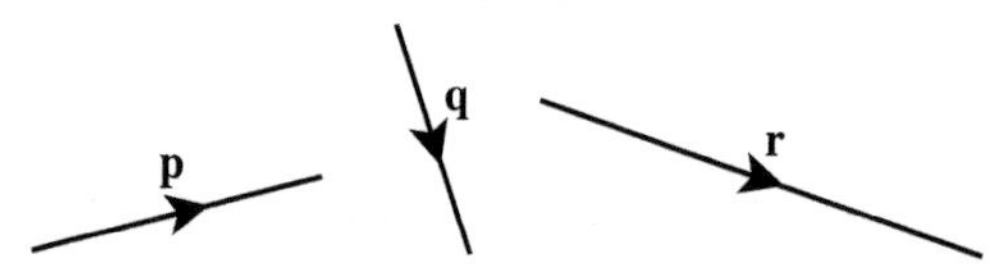

In the diagram below, which vectors can be expressed in the form:

(a) $n\mathbf{p}$, (b) $n\mathbf{q}$, (c) $n\mathbf{r}$?

For each vector give the value of n.

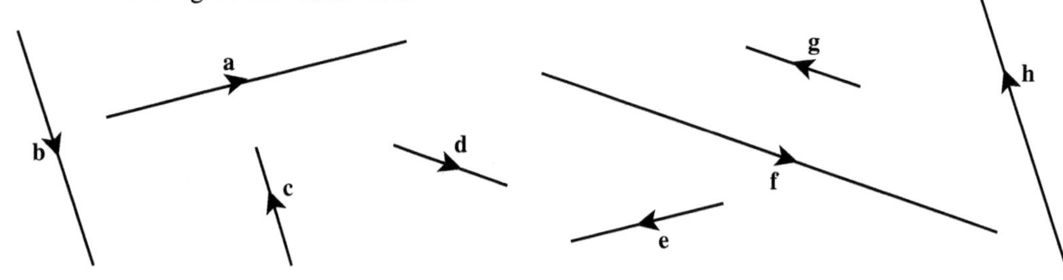

4. $\mathbf{a} = \begin{pmatrix} 2 \\ -1 \end{pmatrix}$ $\mathbf{b} = \begin{pmatrix} -1 \\ 4 \end{pmatrix}$ $\mathbf{c} = \begin{pmatrix} -3 \\ -1 \end{pmatrix}$ $\mathbf{d} = \begin{pmatrix} -1 \\ 2 \end{pmatrix}$

Write column vectors to represent the following.

(a) $2\mathbf{a}$ (b) $2\mathbf{b}$ (c) $3\mathbf{a}$ (d) $-\mathbf{a}$
(e) $2\mathbf{c}$ (f) $-2\mathbf{d}$ (g) $-4\mathbf{b}$ (h) $5\mathbf{d}$

5. $\mathbf{a} = \begin{pmatrix} 5 \\ 12 \end{pmatrix}$, $\mathbf{b} = \begin{pmatrix} -3 \\ 4 \end{pmatrix}$ and $\mathbf{c} = \begin{pmatrix} 15 \\ -8 \end{pmatrix}$.

Find the magnitude of vectors **a**, **b** and **c**.

6. Which of the following vectors are equal in magnitude?

$\mathbf{d} = \begin{pmatrix} 7 \\ -9 \end{pmatrix}$, $\mathbf{e} = \begin{pmatrix} 10 \\ 6 \end{pmatrix}$, $\mathbf{f} = \begin{pmatrix} -6 \\ 10 \end{pmatrix}$, $\mathbf{g} = \begin{pmatrix} -8 \\ -5 \end{pmatrix}$, $\mathbf{h} = \begin{pmatrix} -10 \\ -6 \end{pmatrix}$.

7. If $\mathbf{a} = \begin{pmatrix} 4 \\ 7 \end{pmatrix}$ and $\mathbf{b} = \begin{pmatrix} -3 \\ 5 \end{pmatrix}$, find

(a) $|\mathbf{a}|$ (b) $|\mathbf{b}|$

Leave your answers in surd form.

Vector addition

The diagram shows the points A, B and C.
Start at A and move to B.
Then move from B to C.

The combination of these two displacements is equivalent to a total displacement from A to C.

This can be written using vectors as:

$$\overrightarrow{AB} + \overrightarrow{BC} = \overrightarrow{AC}$$

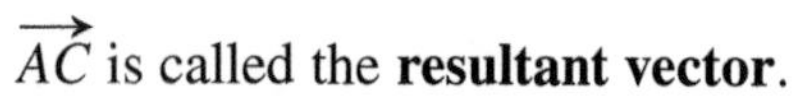

$\overrightarrow{AC}$ is called the **resultant vector**.

Using column vectors: $\overrightarrow{AB} = \begin{pmatrix} 5 \\ 3 \end{pmatrix}$, $\overrightarrow{BC} = \begin{pmatrix} 3 \\ -4 \end{pmatrix}$ and $\overrightarrow{AC} = \begin{pmatrix} 8 \\ -1 \end{pmatrix}$.

The resultant vector, $\overrightarrow{AC}$, is a combination of: $\begin{pmatrix} \text{5 units right followed by 3 units right} \\ \text{3 units up followed by 4 units down} \end{pmatrix} = \begin{pmatrix} \text{8 units right} \\ \text{1 unit down} \end{pmatrix}$

$\overrightarrow{AB} + \overrightarrow{BC} = \overrightarrow{AC}$ $\begin{pmatrix} 5 \\ 3 \end{pmatrix} + \begin{pmatrix} 3 \\ -4 \end{pmatrix} = \begin{pmatrix} 5 + 3 \\ 3 + -4 \end{pmatrix} = \begin{pmatrix} 8 \\ -1 \end{pmatrix}$

Use vector addition to show that: $\overrightarrow{AB} + \overrightarrow{BC} = \overrightarrow{BC} + \overrightarrow{AB} = \overrightarrow{AC}$.

Vector diagrams

Diagrams can be drawn to show the combination of any number of vectors.

To combine vectors when the vectors are not joined, draw equal vectors so that the second vector joins the end of the first vector.

Note:
With the arrow = +
Against the arrow = −

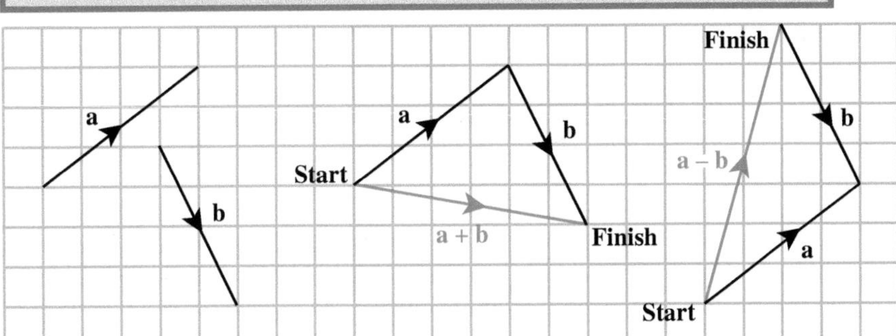

With the arrow, +**a**.
With the arrow, +**b**.
Resultant vector, **a** + **b**.

With the arrow, +**a**.
Against the arrow, −**b**.
Resultant vector, **a** − **b**.

Example 4

a, **b** and **c** are three vectors, as shown in the diagram.

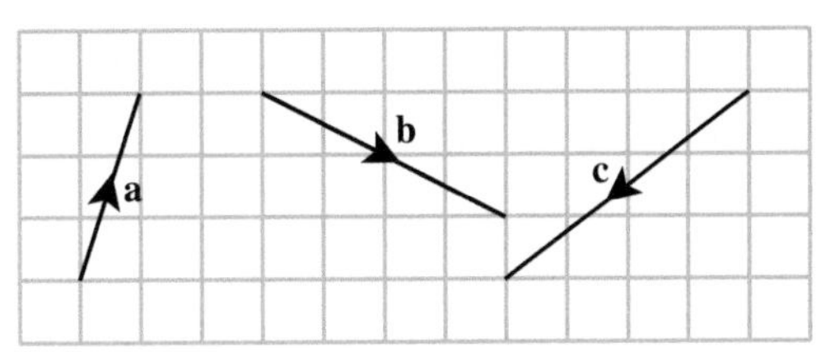

Draw diagrams to show:

(a) **a** + 2**b**

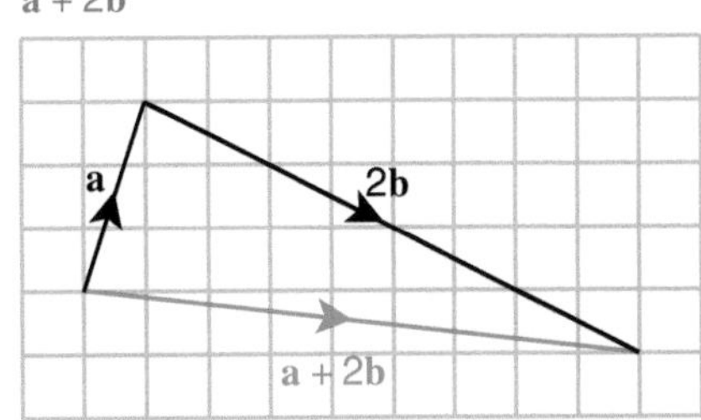

(b) **a** − **c**

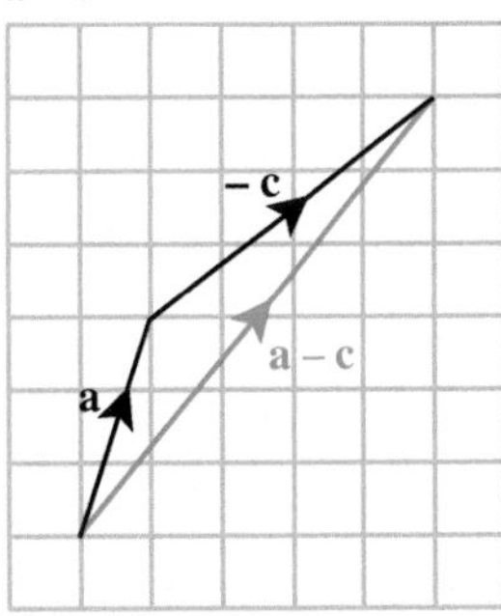

Draw a vector diagram to show that: **a** + **b** = **b** + **a**.

Practice Exercise 20.3

1. (a) Write $\overrightarrow{AB}$ and $\overrightarrow{BC}$ in column form.
   (b) The result of $\overrightarrow{AB} + \overrightarrow{BC} = \overrightarrow{AC}$.
   Write $\overrightarrow{AC}$ as a column vector.

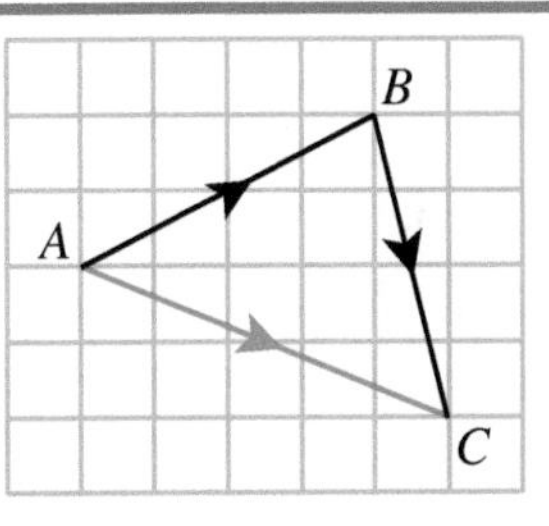

2. Copy the diagrams on squared paper.

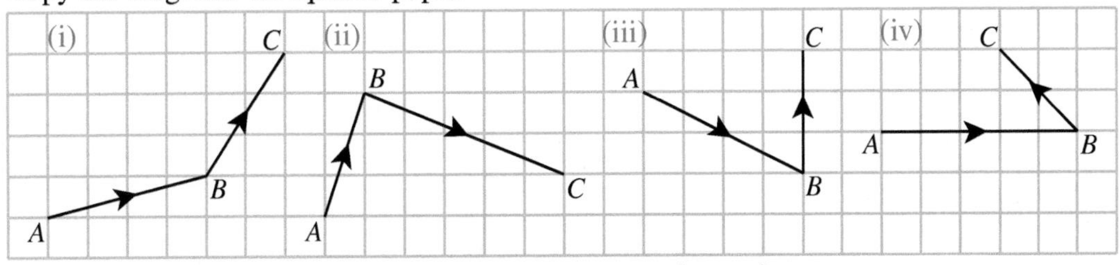

   (a) For each diagram draw a vector representing $\overrightarrow{AB} + \overrightarrow{BC}$.
   (b) Use vectors in column form to find $\overrightarrow{AB} + \overrightarrow{BC}$ for each diagram.
   (c) Use your diagram for part (a) to check your answers to part (b).

3. Copy the diagram on squared paper.

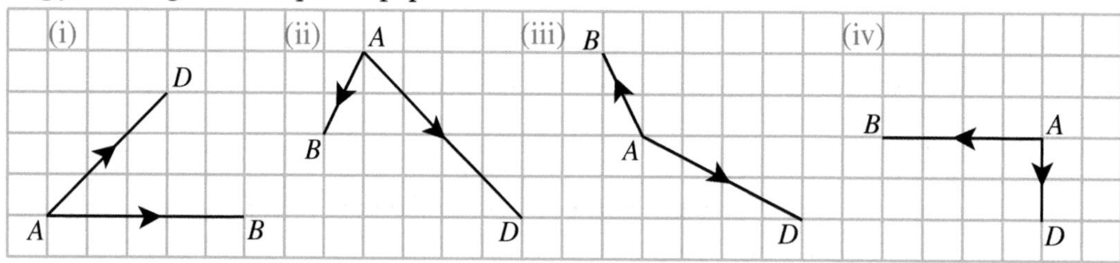

(a) For each diagram draw a vector representing $\overrightarrow{AB} + \overrightarrow{AD}$.

(b) Write the vectors $\overrightarrow{AB}$ and $\overrightarrow{AD}$ in column form and find $\overrightarrow{AB} + \overrightarrow{AD}$.

4. Copy the vectors of question 2 again.

(a) For each diagram show a vector representing $\overrightarrow{AB} - \overrightarrow{BC}$.

(b) Use vectors in column form to find $\overrightarrow{AB} - \overrightarrow{BC}$.

5. Copy the vectors of question 3 again.

(a) For each diagram show a vector representing $\overrightarrow{AB} - \overrightarrow{AD}$.

(b) Use vectors in column form to find $\overrightarrow{AB} - \overrightarrow{AD}$.

6. $\overrightarrow{OA} = \begin{pmatrix} -1 \\ 4 \end{pmatrix}$, $\overrightarrow{OB} = \begin{pmatrix} 4 \\ 3 \end{pmatrix}$, $\overrightarrow{OC} = \begin{pmatrix} 5 \\ 1 \end{pmatrix}$.

Write the following vectors in column form.

(a) $\overrightarrow{OD}$, where $\overrightarrow{OD} = \overrightarrow{OA} + \overrightarrow{OB}$.

(b) $\overrightarrow{OE}$, where $\overrightarrow{OE} = \overrightarrow{OC} - \overrightarrow{OB}$.

(c) $\overrightarrow{OF}$, where $\overrightarrow{OF} = \overrightarrow{OA} + \overrightarrow{OB} + \overrightarrow{OC}$.

(d) $\overrightarrow{OG}$, where $\overrightarrow{OG} = \overrightarrow{OB} - \overrightarrow{OA}$.

(e) $\overrightarrow{OH}$, where $\overrightarrow{OH} = \overrightarrow{OA} + \overrightarrow{OA}$.

$\overrightarrow{OA} + \overrightarrow{OA}$ can be written as $2\overrightarrow{OA}$.

7. The diagram below shows vectors **a**, **b** and **c**.

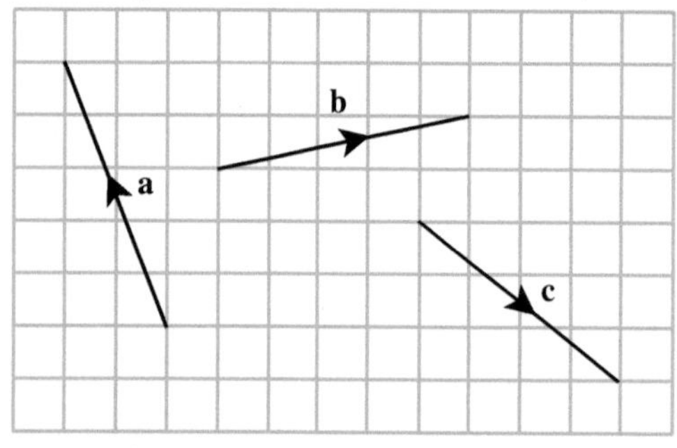

Match the following vectors to those in the diagram opposite.

(a) **a** + **b**

(b) **a** − **b**

(c) **b** + **c**

(d) **b** − **c**

(e) **a** + **b** − **c**

(f) **a** + 2**b**

(g) −2**a** − 3**c**

(h) 2**a** + **c**

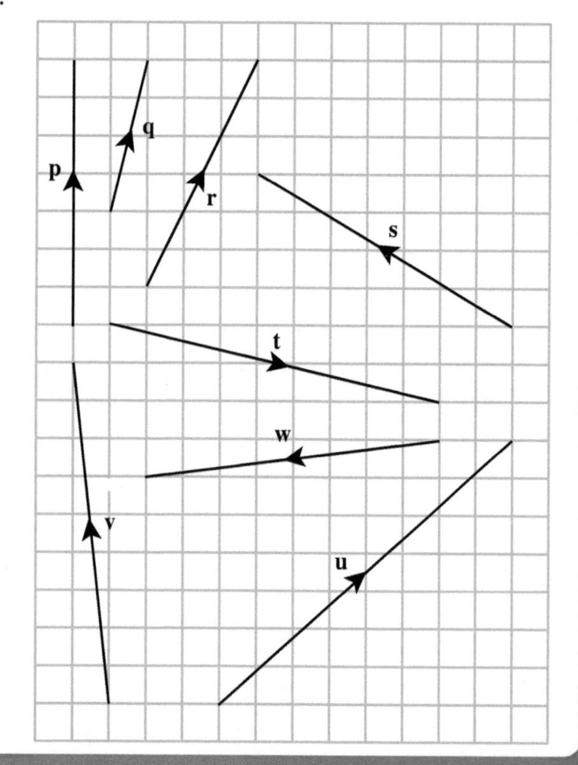

8. If $\mathbf{a} = \begin{pmatrix} 3 \\ 4 \end{pmatrix}$ and $\mathbf{b} = \begin{pmatrix} 2 \\ -1 \end{pmatrix}$, find by calculation the vectors representing

(a) $\mathbf{a} + \mathbf{b}$, (b) $\mathbf{a} - \mathbf{b}$, (c) $3\mathbf{a}$, (d) $\mathbf{a} + 4\mathbf{b}$, (e) $2\mathbf{a} - 3\mathbf{b}$.

9. On squared paper, plot the points $A(3, 2)$, $B(5, 6)$, $C(0, 4)$ and $D(-2, 0)$.

(a) Find $\overrightarrow{AD}$ and $\overrightarrow{BC}$ as a column vector.

(b) What kind of quadrilateral is $ABCD$?

10. In the diagram, $\overrightarrow{OA}$ represents vector **a** and
$\overrightarrow{OB}$ represents vector **b**.

M is the midpoint of AB.

Express, in terms of **a** and **b**, in its simplest form:

(a) $\overrightarrow{AO}$, (b) $\overrightarrow{AB}$,

(c) $\overrightarrow{MB}$, (d) $\overrightarrow{MO}$.

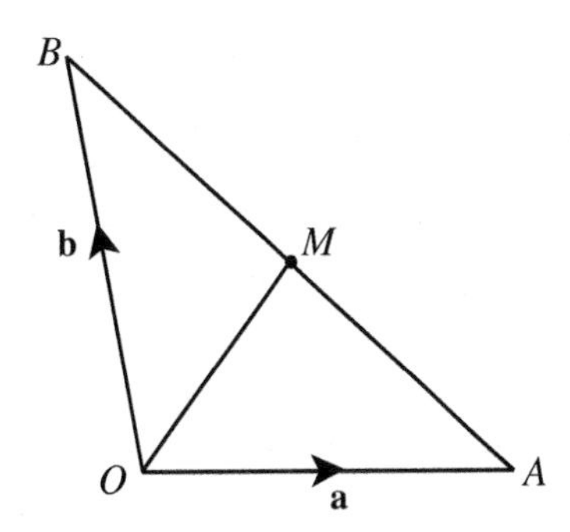

11.

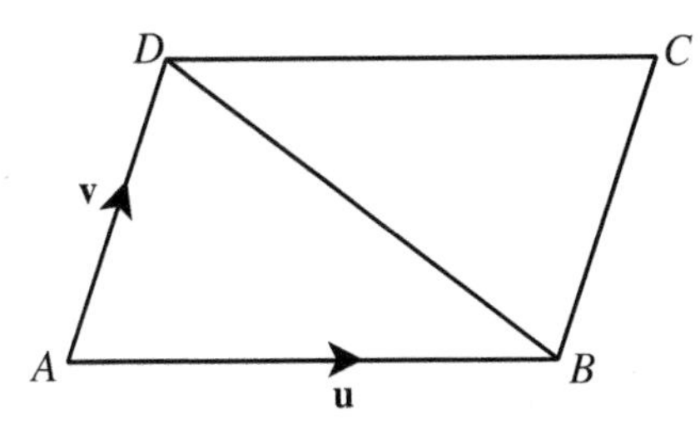

$ABCD$ is a parallelogram.

$\overrightarrow{AB}$ represents vector **u**.

$\overrightarrow{AD}$ represents vector **v**.

(a) Express $\overrightarrow{BD}$ in terms of **u** and **v**.

(b) Express $\overrightarrow{CA}$ in terms of **u** and **v**.

12. $\overrightarrow{OA} = -2\mathbf{a} + 4\mathbf{b}$.

$\overrightarrow{OB} = -\mathbf{a} + 6\mathbf{b}$.

$\overrightarrow{OD} = \mathbf{a} + 2\mathbf{b}$.

$\overrightarrow{OC} = 5\mathbf{a} + 2\mathbf{b}$.

(a) Find, in terms of **a** and **b**,
$\overrightarrow{AD}$ and $\overrightarrow{BC}$.

(b) What do $\overrightarrow{AD}$ and $\overrightarrow{BC}$ and tell you about the quadrilateral $ABCD$?

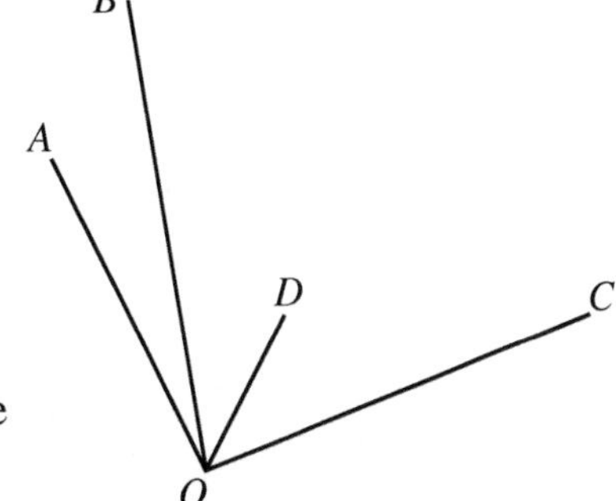

3-dimensional coordinates

One coordinate identifies a point on a **line**.

Two coordinates identify a point on a **plane**.

Three coordinates identify a point in **space**.

The diagram shows a cuboid drawn in 3-dimensions.

Using the axes x, y and z shown:

Point A is given as (2, 0, 0).

Point B is given as (2, 2, 0).

Point F is given as (2, 2, 3).

Give the 3-dimensional coordinates of points C, D, E and G.

Practice Exercise 20.4

1. *ABCDEFGH* is a cuboid.
   A is the point (2, 1, 0).
   B is the point (8, 1, 0).
   D is the point (2, 4, 0).
   E is the point (2, 1, 6).
   Find the coordinates of the points *C*, *F*, *G* and *H*.

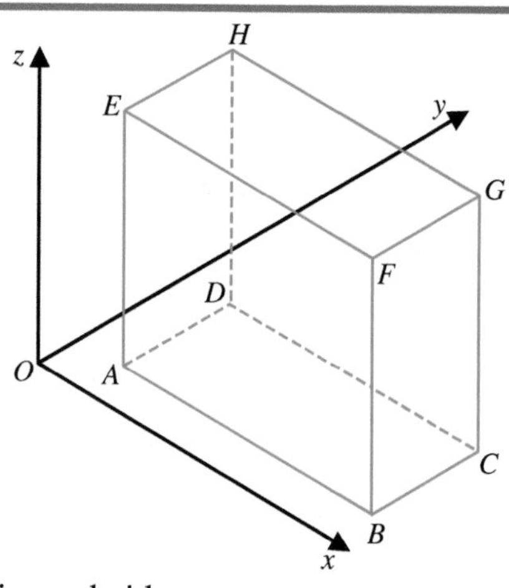

2. 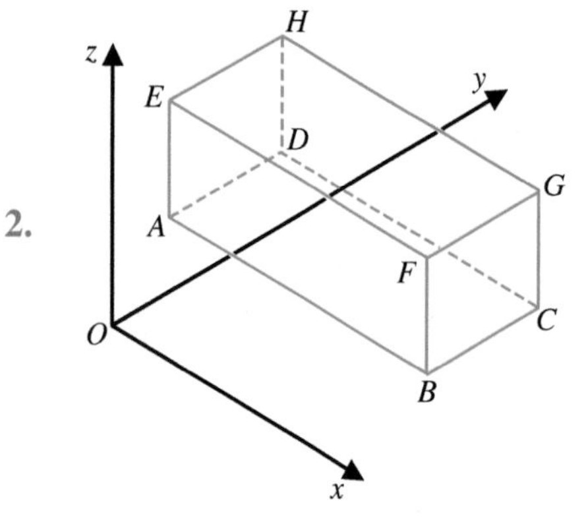

   ABCDEFGH is a cuboid.
   A is the point (1, 4, 5).
   G is the point (12, 9, 9).
   If *AB*, *AD* and *AE* are parallel to the *x*, *y* and *z* axes respectively, find the coordinates of *B*, *C*, *D*, *E*, *F* and *H*.

3. *ABCDEFGH* is a cuboid.
   A is the point (2, 3, 4).
   AB has length 7 units and is parallel to the *x* axis.
   AD has length 1 unit and is parallel to the *y* axis.
   AE has length 3 units and is parallel to the *z* axis.

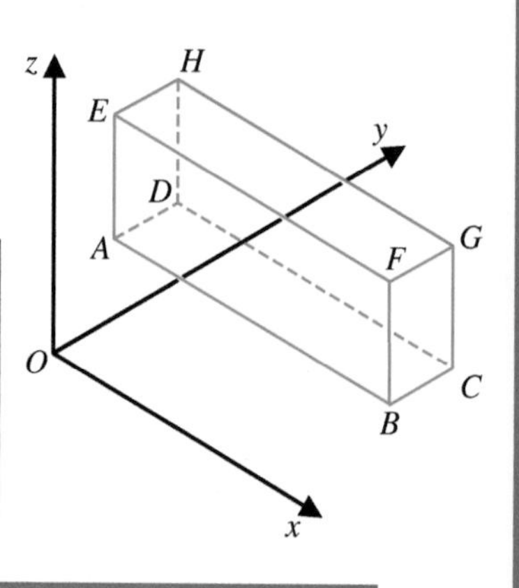

> *A* has coordinates (2, 3, 4).
> We can use a column vector to represent the position of *A* as:
> 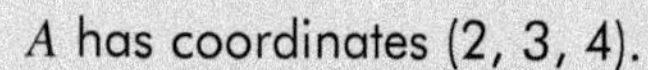
> $\overrightarrow{OA} = \begin{pmatrix} 2 \\ 3 \\ 4 \end{pmatrix}$, where O is the origin.
> The **position vector** for B is $\overrightarrow{OB} = \begin{pmatrix} 9 \\ 3 \\ 4 \end{pmatrix}$.

   (a) Write position vectors for points *C*, *D*, *E*, *F*, *G* and *H*.

> We can add and subtract position vectors to find other vectors.
> To find $\overrightarrow{AB}$, go from A to O, and then from O to B.
> 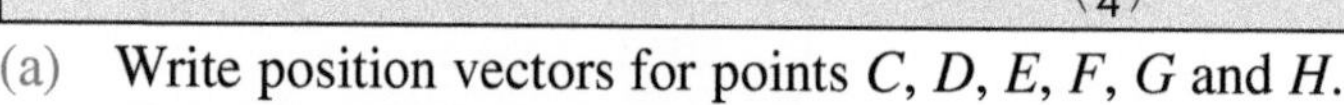
> In vector form, this is written as $\overrightarrow{AB} = -\overrightarrow{OA} + \overrightarrow{OB}$.
> $\overrightarrow{AB} = -\overrightarrow{OA} + \overrightarrow{OB}$ is the same as $\overrightarrow{AB} = \overrightarrow{OB} - \overrightarrow{OA}$.
> $$\overrightarrow{AB} = \begin{pmatrix} 9 \\ 3 \\ 4 \end{pmatrix} - \begin{pmatrix} 2 \\ 3 \\ 4 \end{pmatrix} = \begin{pmatrix} 9-2 \\ 3-3 \\ 4-4 \end{pmatrix} = \begin{pmatrix} 7 \\ 0 \\ 0 \end{pmatrix}$$
>
> **Remember:**
> **With** the arrow = +
> **Against** the arrow = −

   (b) Find $\overrightarrow{BF}$, where $\overrightarrow{BF} = \overrightarrow{OF} - \overrightarrow{OB}$.
   (c) Find $\overrightarrow{AG}$, where $\overrightarrow{AG} = \overrightarrow{OG} - \overrightarrow{OA}$.

4. If $\mathbf{a} = \begin{pmatrix} 2 \\ 1 \\ 3 \end{pmatrix}$ and $\mathbf{b} = \begin{pmatrix} -2 \\ 4 \\ -1 \end{pmatrix}$, find
   (a) $2\mathbf{a}$ (b) $3\mathbf{b}$ (c) $-2\mathbf{a}$ (d) $-\mathbf{b}$ (e) $\mathbf{a} + \mathbf{b}$

5. $\mathbf{u} = \begin{pmatrix} -1 \\ 2 \\ 3 \end{pmatrix}$ $\mathbf{v} = \begin{pmatrix} 0 \\ 3 \\ -2 \end{pmatrix}$ $\mathbf{w} = \begin{pmatrix} 3 \\ -2 \\ 4 \end{pmatrix}$.
   Express as column vectors.
   (a) $\mathbf{u} + \mathbf{v}$ (b) $\mathbf{v} + 2\mathbf{w}$ (c) $\mathbf{u} - \mathbf{w}$ (d) $2\mathbf{u} + 3\mathbf{v}$
   (e) $2\mathbf{w} - \mathbf{u}$ (f) $4\mathbf{v} - 3\mathbf{w}$ (g) $2\mathbf{u} - 3\mathbf{w}$ (h) $2\mathbf{u} + \mathbf{w} - 3\mathbf{v}$

6. Find $\overrightarrow{AC}$ if $\overrightarrow{AB} = \begin{pmatrix} 2 \\ 3 \\ -1 \end{pmatrix}$ and $\overrightarrow{BC} = \begin{pmatrix} -3 \\ 0 \\ 4 \end{pmatrix}$.

7. *ABCDEFGH* is a cuboid.
G is the point (7, 9, 4).
HG has length 15 units and is parallel to the x axis.
FG has length 13 units and is parallel to the y axis.
CG has length 7 units and is parallel to the z axis.

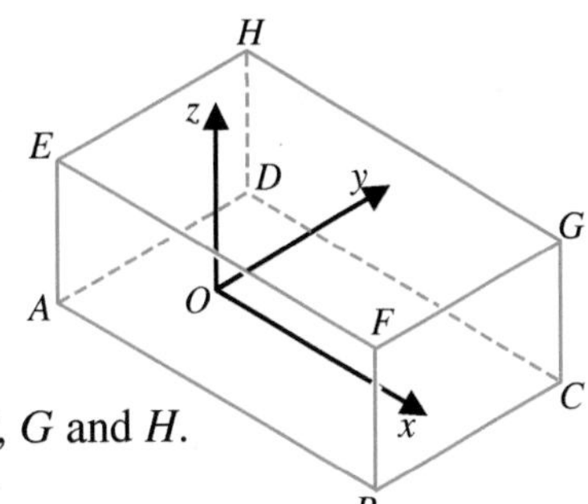

(a) Write down position vectors for points A, B, C, D, E, F, G and H.

(b) Find $\overrightarrow{AG}$ as a column vector.

> In general, the magnitude (length) of a vector can be found by: adding the squares of its components and taking the square root.
>
> For vectors in 2 dimensions, $\begin{pmatrix} x \\ y \end{pmatrix}$ length $= \sqrt{x^2 + y^2}$.
>
> For vectors in 3 dimensions, $\begin{pmatrix} x \\ y \\ z \end{pmatrix}$ length $= \sqrt{x^2 + y^2 + z^2}$.

(c) Calculate the length of vector $\overrightarrow{AG}$. Leave your answer in surd form.

(d) Find $\overrightarrow{BH}$ and then calculate the length of vector $\overrightarrow{BH}$.

(e) Comment on your answers to (c) and (d).

Key Points

- **Vector quantities**
  Quantities which have both **size** and **direction** are called **vectors**.

- **Vector notation**
  Vectors can be represented by **column vectors** or by **directed line segments**.
  Vectors can be labelled using:
  capital letters to indicate the start and finish of a vector, bold lower case letters.

  In a **column vector**:
  The top number describes the **horizontal** part of the movement:
  $+$ = to the right
  $-$ = to the left
  The bottom number describes the **vertical** part of the movement:
  $+$ = upwards
  $-$ = downwards

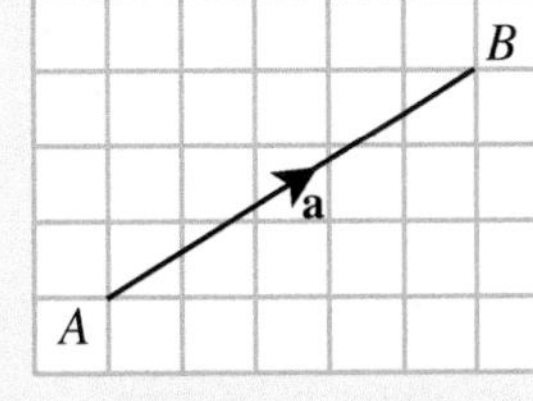

$\overrightarrow{AB} = \mathbf{a} = \begin{pmatrix} 5 \\ 3 \end{pmatrix}$

- Vectors are **equal** if they have the same length **and** they are in the same direction.
  Vectors **a** and $-$**a** have the same length **but** are in **opposite directions**.
  The vector $n\mathbf{a}$ is parallel to the vector **a**.
  The length of vector $n\mathbf{a} = n \times$ the length of vector **a**.

- **Vector addition**
  The combination of the displacement from A to B followed by the displacement from B to C is equivalent to a total displacement from A to C.
  This can be written using vectors as $\overrightarrow{AB} + \overrightarrow{BC} = \overrightarrow{AC}$.
  $\overrightarrow{AC}$ is called the **resultant vector**.

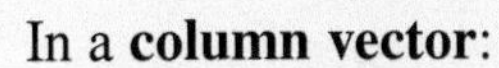

- Combinations of vectors can be shown on **vector diagrams**.

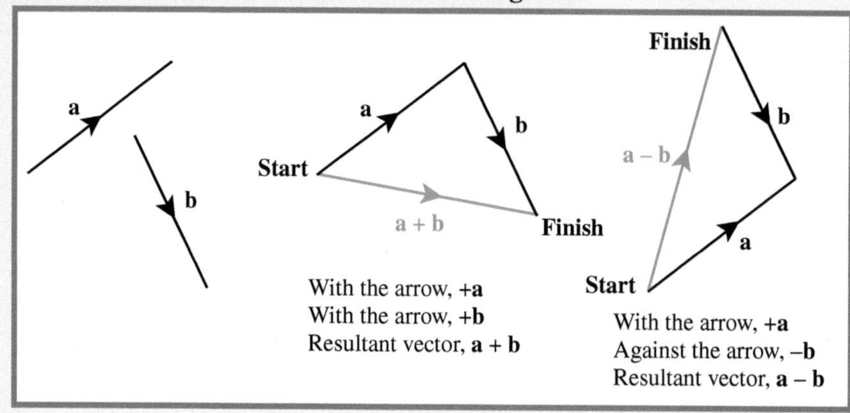

- 3-dimensional coordinates identify points in space.
  A point $P(x, y, z)$ can be written as a **position vector**, $\overrightarrow{OP} = \begin{pmatrix} x \\ y \\ z \end{pmatrix}$.
- The same rules applied to vectors in two dimensions can be applied to vectors in three dimensions.
- The **magnitude (length)** of a vector is found by adding the squares of its components and taking the square root.

For vectors in 2 dimensions, e.g., $\begin{pmatrix} x \\ y \end{pmatrix}$ length $= \sqrt{x^2 + y^2}$.

For vectors in 3 dimensions, e.g., $\begin{pmatrix} x \\ y \\ z \end{pmatrix}$ length $= \sqrt{x^2 + y^2 + z^2}$.

Review Exercise 20

1. On squared paper, draw vectors to represent the following.

   $\mathbf{a} = \begin{pmatrix} 2 \\ -1 \end{pmatrix}$ $\mathbf{b} = \begin{pmatrix} -1 \\ 4 \end{pmatrix}$ $\mathbf{c} = \begin{pmatrix} -3 \\ -1 \end{pmatrix}$ $\mathbf{d} = \begin{pmatrix} -1 \\ 2 \end{pmatrix}$

2. The diagram below shows vectors **a**, **b**, **c** and **d**.

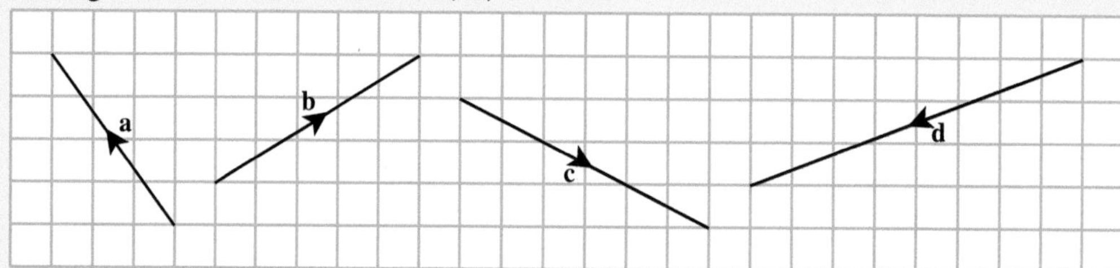

   Express the following as column vectors.

   (a) **a** + **b** (b) **a** − **b** (c) **b** + **d** (d) **d** − **a**
   (e) 2**a** (f) **b** + 2**c** (g) **a** + **b** + **d** (h) 2**c** − 3**b**
   (i) 2**b** + 3**c** − **d** (j) **a** + **b** + **c** + **d**
   (k) Find the magnitudes (lengths) of vectors **a**, **b**, **c** and **d**.

3. $\overrightarrow{OA} = \begin{pmatrix} 2 \\ 5 \end{pmatrix}$, $\overrightarrow{OB} = \begin{pmatrix} -3 \\ 4 \end{pmatrix}$ and $\overrightarrow{OC} = \begin{pmatrix} 2 \\ 7 \end{pmatrix}$.

   Write the following vectors in column form.

   (a) $\overrightarrow{OD}$, where $\overrightarrow{OD} = \overrightarrow{OA} + \overrightarrow{OC}$.
   (b) $\overrightarrow{OE}$, where $\overrightarrow{OE} = \overrightarrow{OB} - \overrightarrow{OA}$.
   (c) $\overrightarrow{OF}$, where $\overrightarrow{OF} = \overrightarrow{OA} + \overrightarrow{OB} + \overrightarrow{OC}$.

4. In the diagram:
$\overrightarrow{AB}$ is represented by vector **a**,
$\overrightarrow{BC}$ is represented by vector **b**,
$\overrightarrow{CD}$ is represented by vector **c**.

Express, in terms of **a**, **b** and **c**:

(a) $\overrightarrow{AC}$, (b) $\overrightarrow{AD}$, (c) $\overrightarrow{DB}$.

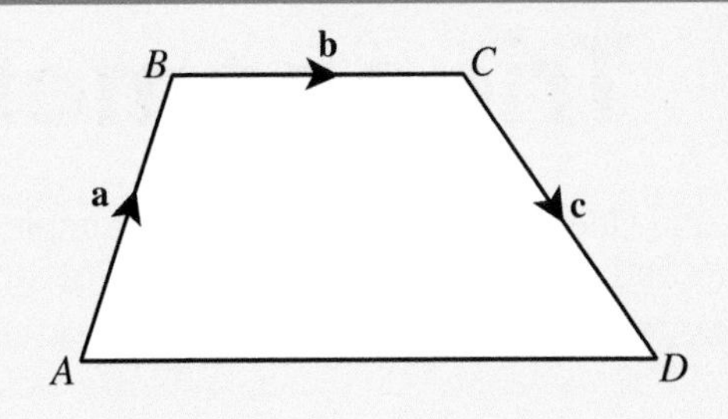

5. The diagram shows triangle ABC.
Points P and Q are the midpoints of lines AB and AC.

If vector **u** represents $\overrightarrow{AB}$ and vector **v** represents $\overrightarrow{AC}$, find, in terms of **u** and **v**:

(a) $\overrightarrow{BC}$,

(b) $\overrightarrow{PQ}$.

(c) Make one comment about the direction of $\overrightarrow{PQ}$.

(d) Make one comment about the length of $\overrightarrow{PQ}$.

6. $ABCDEFGH$ is a cuboid.
A is the point $(-2, -3, -1)$.
AB has length 10 units and is parallel to the x axis.
BC has length 9 units and is parallel to the y axis.
AE has length 6 units and is parallel to the z axis.

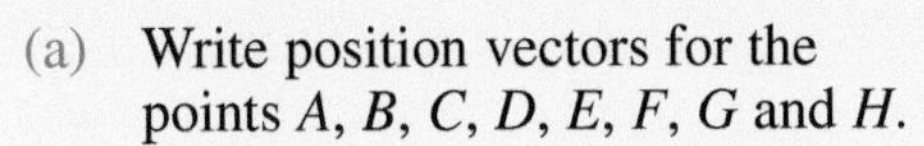

(a) Write position vectors for the points A, B, C, D, E, F, G and H.

(b) Find $\overrightarrow{AF}$ as a column vector.

(c) Find the length of vector $\overrightarrow{AF}$.

(d) Find the length of vector $\overrightarrow{BH}$.

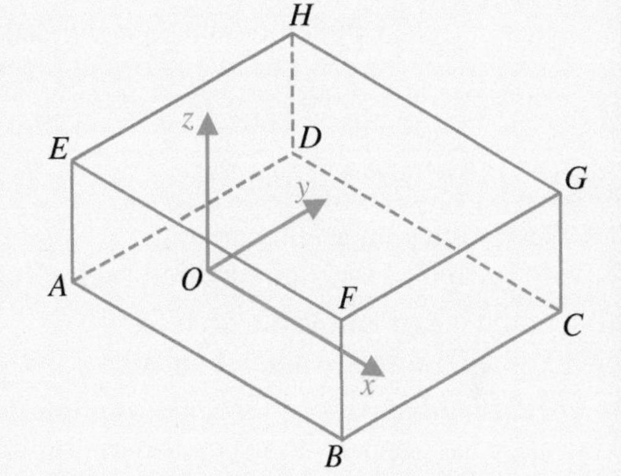

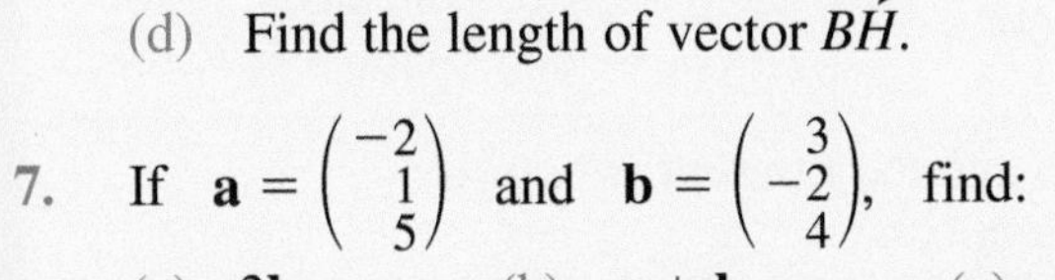

7. If $\mathbf{a} = \begin{pmatrix} -2 \\ 1 \\ 5 \end{pmatrix}$ and $\mathbf{b} = \begin{pmatrix} 3 \\ -2 \\ 4 \end{pmatrix}$, find:

(a) $3\mathbf{b}$, (b) $\mathbf{a} + \mathbf{b}$, (c) $\mathbf{a} - \mathbf{b}$, (d) $2\mathbf{a} + 3\mathbf{b}$, (e) $3\mathbf{b} - \mathbf{a}$.

8. Find $\overrightarrow{AC}$ if $\overrightarrow{AB} = \begin{pmatrix} 1 \\ 3 \\ 2 \end{pmatrix}$ and $\overrightarrow{BC} = \begin{pmatrix} -3 \\ 4 \\ -2 \end{pmatrix}$.

Reuleaux curves

To draw a Reuleaux curve, start by drawing an equilateral triangle, ABC. Then, set your compasses to length AB and draw arcs BC, AC and AB from centres A, B and C respectively.

(a) If $AB = x$ cm, find the length of the arc AB in terms of π and x.

(b) Prove that the Reuleaux curve has the same perimeter as a circle with diameter AB.

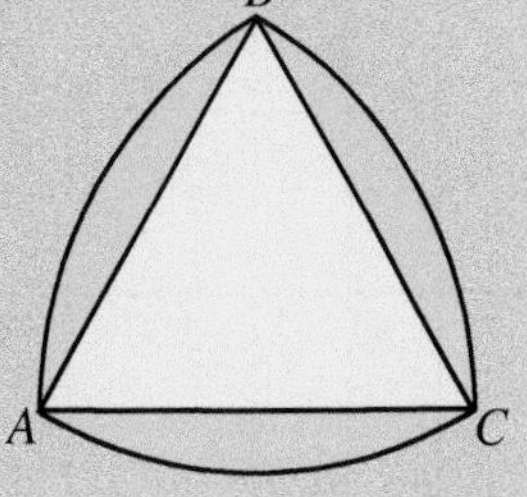

20p and 50p coins are Reuleaux curves.
Constant diameters, but lousy wheels.

Investigate.

Graphs of Trigonometric Functions

So far we have found sines, cosines and tangents of angles between 0° and 90°, but it is possible to find the sine, cosine and tangent of any angle.

Definitions of the trigonometric functions

Point $P(x, y)$ lies on the circumference of a circle, centre O, radius r.
Starting from the x axis, P moves in an anticlockwise direction.

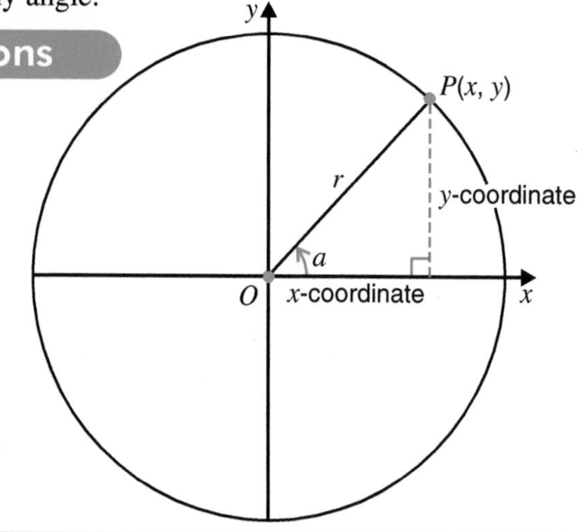

$$\sin a = \frac{y\text{-coordinate}}{r}$$
$$\cos a = \frac{x\text{-coordinate}}{r}$$
$$\tan a = \frac{y\text{-coordinate}}{x\text{-coordinate}}$$

Graphs of trigonometric functions

Activity

The sine function

Copy and complete this table. Use your calculator to find the value of $\sin x°$, correct to two decimal places, where necessary.

$x°$	0	30	60	90	120	150	180	210	240	270	300	330	360
$\sin x°$	0	0.5	0.87										

Plot these values on graph paper.
Draw the graph of the sine function by joining the points with a smooth curve.
What is the maximum value of $\sin x°$?
What is the minimum value of $\sin x°$?
Describe what happens to the sine function when $x > 360°$.
Describe what happens to the sine function when $x < 0°$.

You might also try this activity using a graphical calculator, or a computer and suitable software.

In a similar way, draw the graphs of the **cosine function** and the **tangent function**.
What do you notice about the graphs of each of these functions?

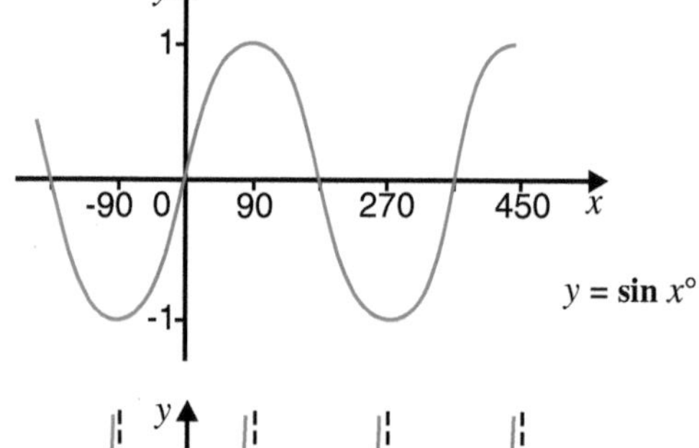

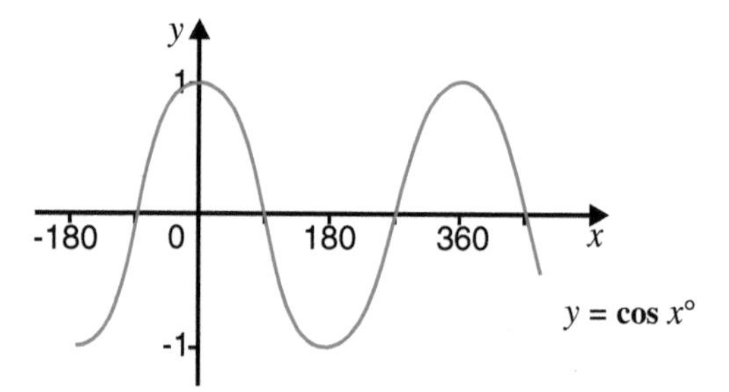

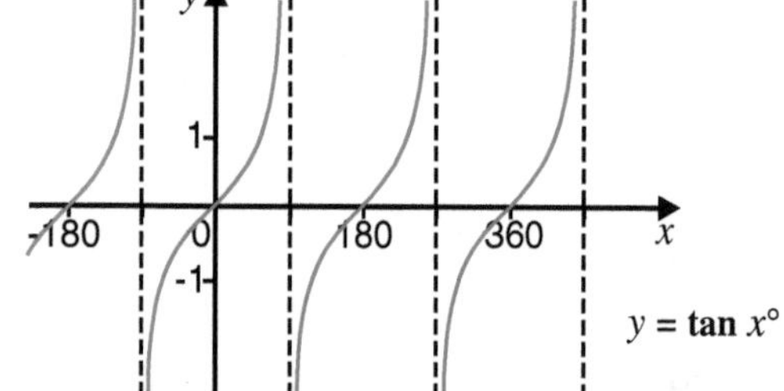

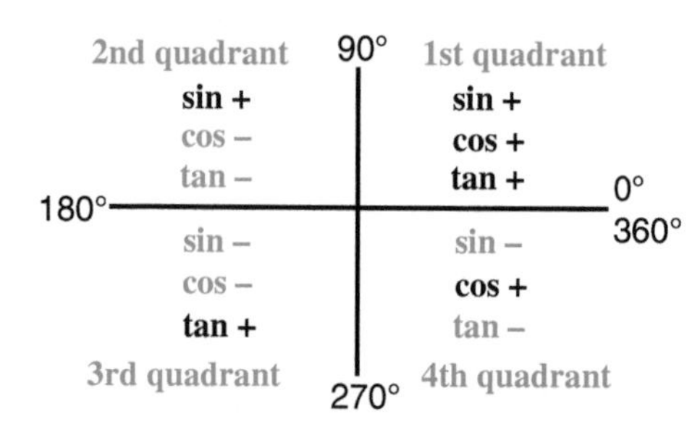

The graph of a **periodic function** is repetitive.
The **period** of a function is the number of degrees between repeats.
The sine and cosine functions both have periods of 360°.
The tangent function has a period of 180°.

The **range** of values of the sine and cosine functions lies between -1 and 1.
The graphs go **one unit below** and **one unit above** their midlines (in this case, the x axis).
The value "1" is called the **amplitude**.

Transforming graphs

Transformations, such as **translations** and **stretches**, can be used to change the position and size of a graph of a trigonometric function.
We first looked at transforming graphs in Chapter 13.

In general, if $y = \mathrm{f}(x)$ where $\mathrm{f}(x) = \sin x, \cos x, \tan x$, etc., then:

$y = \mathrm{f}(x) + c$	means	If c is **positive**, the graph of $\mathrm{f}(x)$ moves c units **up**. If c is **negative**, the graph of $\mathrm{f}(x)$ moves c units **down**.
$y = \mathrm{f}(x + p)$	means	If p is **positive**, the graph of $\mathrm{f}(x)$ moves p units **left**. If p is **negative**, the graph of $\mathrm{f}(x)$ moves p units **right**. The term p is called the **phase angle**.
$y = a\mathrm{f}(x)$	means	The y-coordinates on the graph of $\mathrm{f}(x)$ are **multiplied** by a. The **amplitude** of the graph is a.
$y = \mathrm{f}(bx)$	means	The x-coordinates on the graph of $\mathrm{f}(x)$ are **divided** by b. The **period** of the graph is $\frac{360°}{b}$.

Example 1

Compare the graph of $y = \cos x$ with:

(a) $\cos x + 2$ and $\cos x - 1$
(b) $2 \cos x$
(c) $\cos 2x$
(d) $\cos \frac{1}{2}x$
(e) $\cos (x - 45)$
(f) $\cos (x + 30)$

(a)

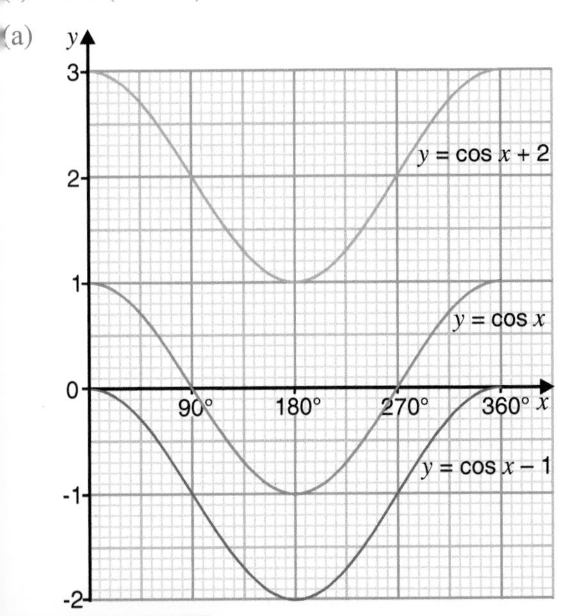

(b)

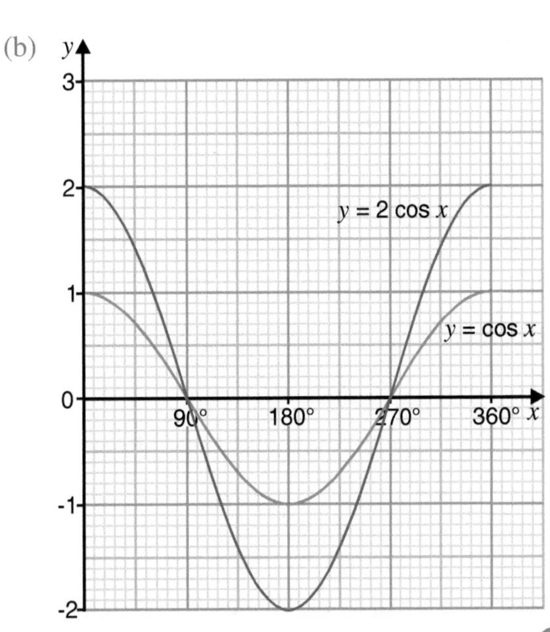

(c)

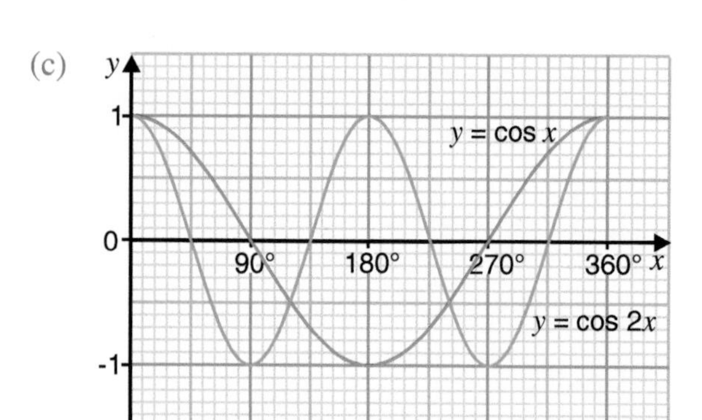

(d)

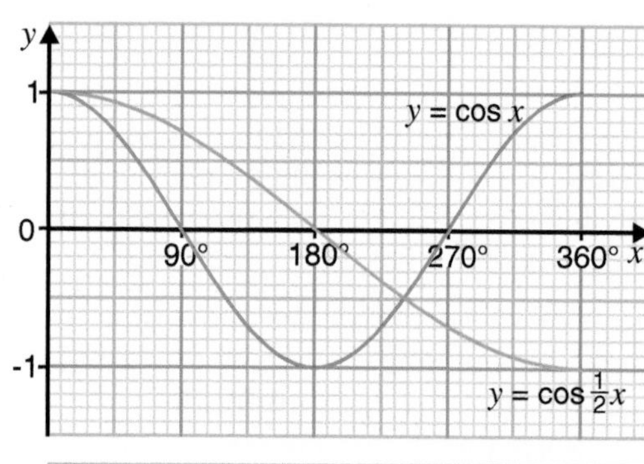

(e)

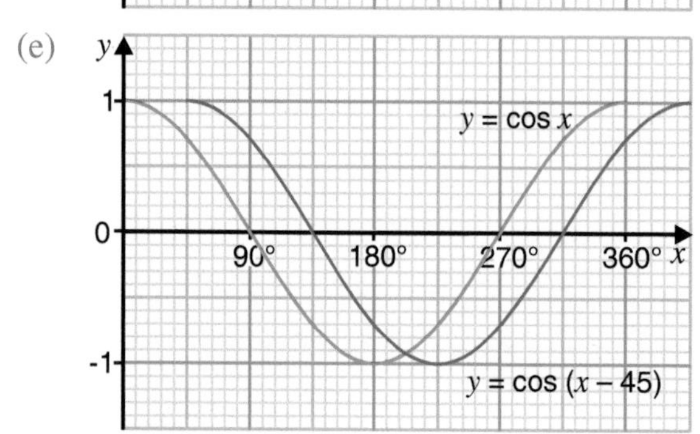

(f)

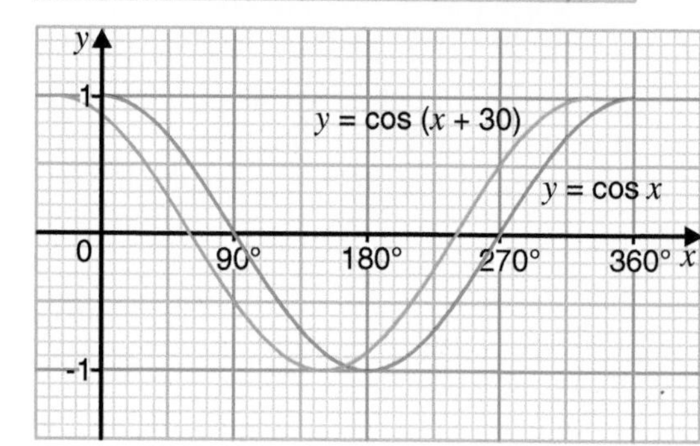

Example 2

State the amplitude and period for the graph of $y = 3 \sin 4x$.

The amplitude of $y = af(x)$ is given by a.

The amplitude of $y = 3 \sin 4x$ is 3.

The period of the graph of $y = f(bx)$ is given by $\frac{360°}{b}$.

The period of the graph of $y = 3 \sin 4x$ is $\frac{360°}{4} = 90°$.

Note that the amplitude does not affect the period of a graph.

Example 3

The diagram shows the graphs of $y = \sin x$ and two related functions **a** and **b**.
Find equations for the functions **a** and **b**.

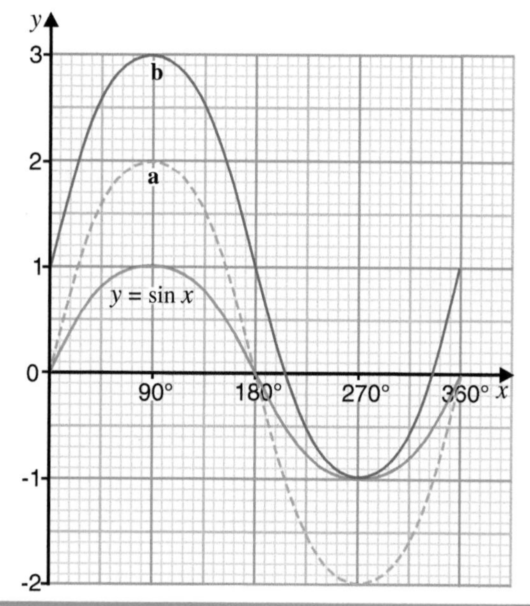

The graph of $y = \sin x$ is transformed to the graph of function **a** by:
a **stretch** from the x axis, parallel to the y axis, scale factor 2.
So, the equation of function **a** is $y = 2 \sin x$.

The graph of function **a** is transformed to the graph of function **b** by:
a **translation** 1 unit up.
So, the equation of function **b** is $y = 2 \sin x + 1$.

Practice Exercise 21.1

1. The graph shows $y = \cos x$.

On separate diagrams, sketch the graph of:

(a) $y = \cos x + 1$,

(b) $y = \cos 3x$,

(c) $y = -\cos x$.

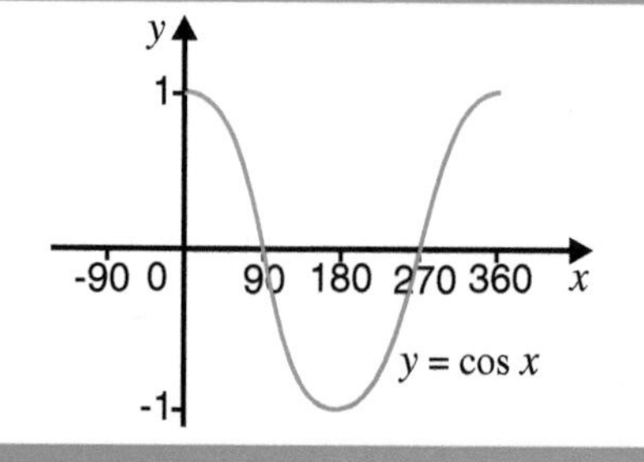

2. State the amplitude and period for each of the following graphs.
(a) $y = 2 \cos 3x$ (b) $y = 5 \sin 6x$ (c) $y = 4 \cos \frac{1}{2} x$ (d) $y = 3 \cos \frac{1}{4} x$

3. Give the amplitude and period for the function $y = 2 \tan 3x$.

4. The diagram shows a sketch of $y = \sin x$ and three related functions.

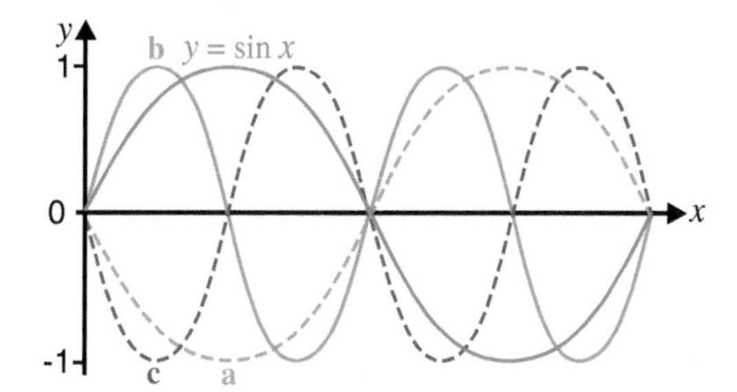

(a) Describe how the graph of $y = \sin x$ is transformed to:
(i) the graph of function **a**,
(ii) the graph of function **b**.
(b) Describe how the graph of function **b** is transformed to the graph of function **c**.
(c) Describe how the graph of $y = \sin x$ is transformed to the graph of function **c**, by using a combination of **two** transformations.
(d) What are the equations of the graphs **a**, **b** and **c**?

5. Using a sketch of the graph $y = \sin x$ to help you, on the same axes:
(a) sketch the graph of $y = -\sin x$ for values of x between 0° and 360°,
(b) sketch the graph of $y = 2 - \sin x$ for values of x between 0° and 360°.

6. The diagram shows the graph of $y = \sin (x - p)$.

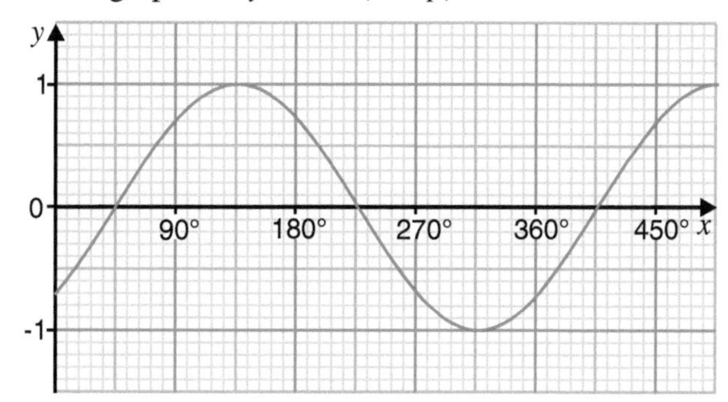

(a) What is the value of the phase angle p?
(b) Sketch the graph of $y = \sin (x + p)$ for values of x between $-360°$ and 360°.

7. Sketch the graph of $y = \cos (x - 90°)$ for values of x between $-360°$ and 360°.

8. The diagram shows the graph of $y = \cos x$.

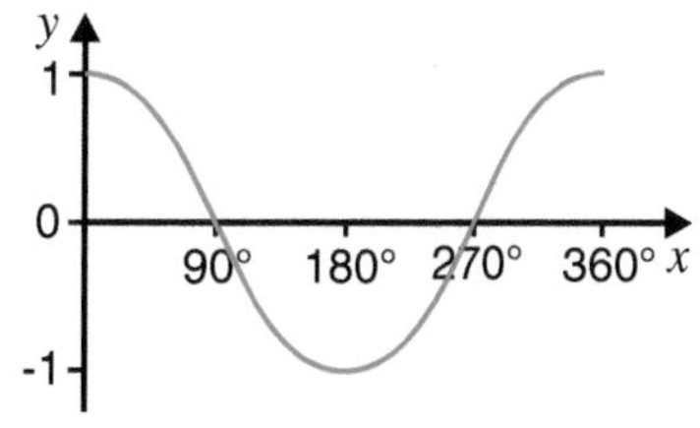

The diagrams below show transformations of the graph of $y = \cos x$.

Write down the equation of each graph.

(a)

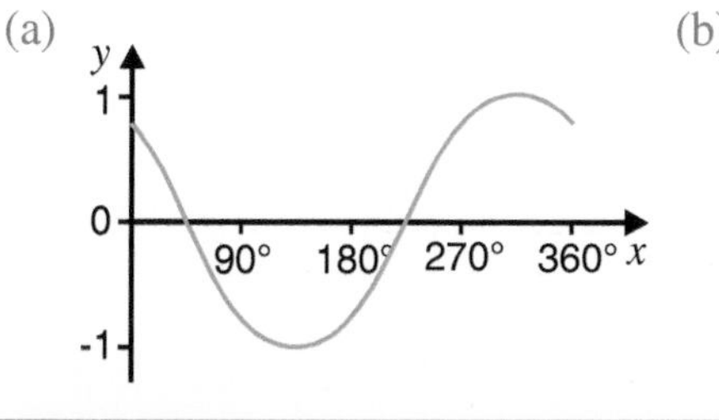

(b)

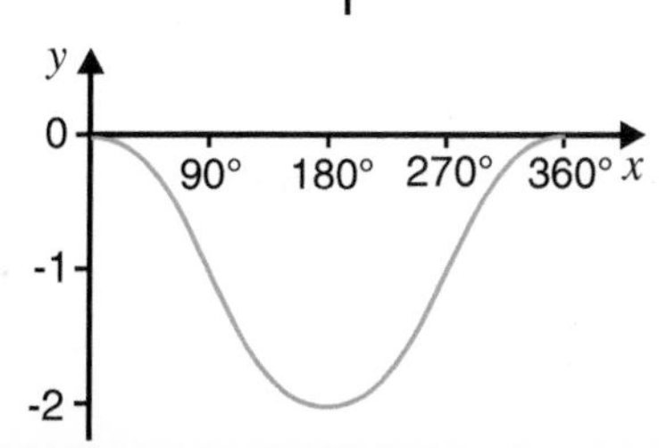

Sines, cosines and tangents of angles between 0° and 360°

Consider a point P. Line OP makes an angle $a°$ with line OX.
If $OP = r$ units, then in triangle OMP:

$\sin a° = \frac{\text{opp}}{\text{hyp}}$	$\cos a° = \frac{\text{adj}}{\text{hyp}}$	$\tan a° = \frac{\text{opp}}{\text{adj}}$
$\sin a° = \frac{PM}{OP}$	$\cos a° = \frac{OM}{OP}$	$\tan a° = \frac{PM}{OM}$
$\sin a° = \frac{y}{r}$	$\cos a° = \frac{x}{r}$	$\tan a° = \frac{y}{x}$

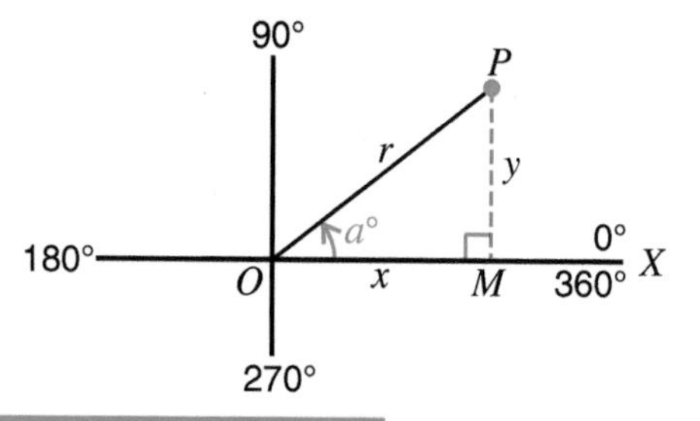

Positive angles are measured **anticlockwise** from the axis OX.

2nd quadrant

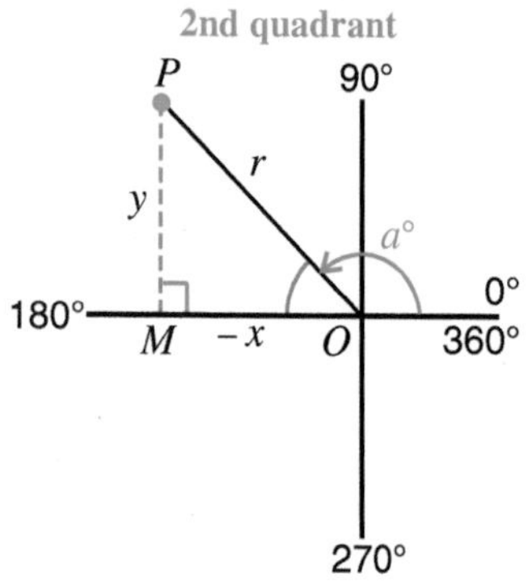

$\angle POM = (180 - a)°$

$\sin (180 - a)° = \frac{y}{r} = \sin a°$

$\cos (180 - a)° = \frac{-x}{r} = -\cos a°$

$\tan (180 - a)° = \frac{y}{-x} = -\tan a°$

Examples

$\sin 150° = \sin (180 - 150)°$
$= \sin 30°$
$= 0.5$

$\cos 110° = -\cos (180 - 110)°$
$= -\cos 70°$
$= -0.342$

$\tan 145° = -\tan (180 - 145)°$
$= -\tan 35°$
$= -0.700$

3rd quadrant

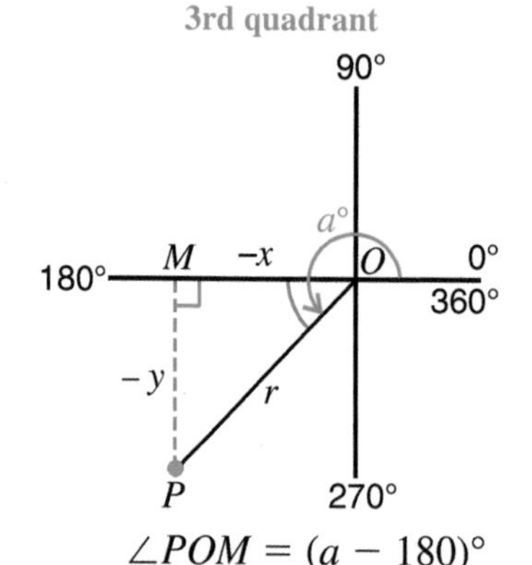

$\angle POM = (a - 180)°$

$\sin (a - 180)° = \frac{-y}{r} = -\sin x$

$\cos (a - 180)° = \frac{-x}{r} = -\cos x$

$\tan (a - 180)° = \frac{-y}{-x} = \tan x$

Examples

$\sin 245° = -\sin (245 - 180)°$
$= -\sin 65°$
$= -0.906$

$\cos 210° = -\cos (210 - 180)°$
$= -\cos 30°$
$= -0.866$

$\tan 200° = \tan (200 - 180)°$
$= \tan 20°$
$= 0.364$

4th quadrant

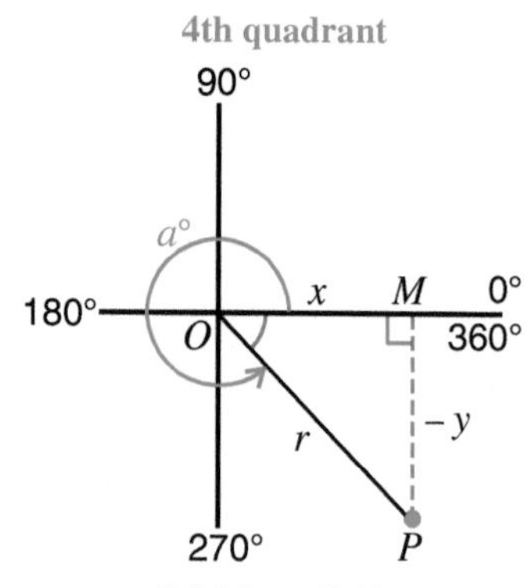

$\angle POM = (360 - a)°$

$\sin (360 - a)° = \frac{-y}{r} = -\sin a°$

$\cos (360 - a)° = \frac{x}{r} = \cos a°$

$\tan (360 - a)° = \frac{-y}{x} = -\tan a°$

Examples

$\sin 350° = -\sin (360 - 350)°$
$= -\sin 10°$
$= -0.174$

$\cos 300° = \cos (360 - 300)°$
$= \cos 60°$
$= 0.5$

$\tan 345° = -\tan (360 - 345)°$
$= -\tan 15°$
$= -0.268$

Example 4

Express the following in terms of the sin, cos or tan of an acute angle.

(a) tan 100° (b) sin 200° (c) cos 320° (d) sin (−10)°
(e) cos (−20)° (f) tan 400° (g) sin 500°

(a) $\tan 100° = -\tan (180 - 100)° = -\tan 80°$
(b) $\sin 200° = -\sin (200 - 180)° = -\sin 20°$
(c) $\cos 320° = \cos (360 - 320)° = \cos 40°$
(d) $\sin (-10)° = -\sin 10°$
(e) $\cos (-20)° = \cos 20°$
(f) $\tan 400° = \tan (400 - 360)° = \tan 40°$
(g) $\sin 500° = \sin (500 - 360)° = \sin 140°$
$= \sin (180 - 140)° = \sin 40°$

Negative angles are measured **clockwise** from the axis OX.

Angles greater than 360°
Subtract 360°, or multiples of 360°, to get the equivalent angle between 0° and 360°.

Practice Exercise 21.2

Do not use a calculator for this exercise.

1. State whether the following are positive or negative.

(a) sin 50°	(b) cos 50°	(c) tan 50°	(d) sin 100°
(e) cos 100°	(f) tan 100°	(g) cos 150°	(h) sin 200°
(i) tan 250°	(j) cos 300°	(k) sin 350°	(l) tan 87°
(m) cos 143°	(n) sin 117°	(o) tan 162°	(p) cos 296°
(q) tan 321°	(r) cos 196°	(s) sin 218°	(t) cos 400°
(u) tan 500°	(v) sin 500°	(w) cos (−30)°	(x) sin (−100)°
(y) cos (−100)°	(z) tan (−100)°		

2. Express each of the following in terms of the sin, cos or tan of an acute angle.

(a) tan 100°	(b) cos 150°	(c) sin 200°	(d) tan 250°
(e) cos 300°	(f) sin 120°	(g) tan 170°	(h) cos 210°
(i) sin 290°	(j) tan 330°	(k) cos 370°	(l) sin 260°
(m) tan 370°	(n) sin 480°	(o) cos 600°	(p) sin (−50)°
(q) cos (−100)°	(r) tan (−150)°		

Finding angles

For which angles, between 0° and 360°, is $\cos p = 0.412$?

$\cos p = 0.412$

$\cos p$ is positive, so, possible angles are in the 1st and 4th quadrants.

This can be shown on a diagram.

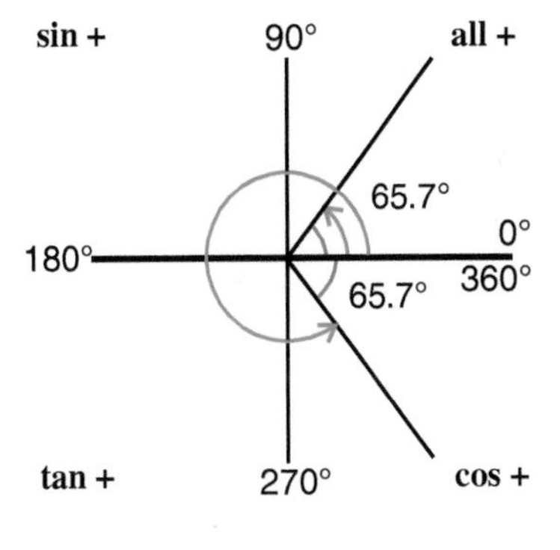

$p = \cos^{-1}(0.412)$

Using a calculator.

$p = 65.669...$

$p = 65.7°$ to 1 d.p.

Because of the symmetries of the graphs of trigonometric functions, angles formed between the lines and the horizontal axis are always **equal**.

Positive angles are measured anticlockwise.

1st quadrant.

$p = 65.7°$

4th quadrant.

$p = 360° - 65.7° = 294.3°$

So, $p = 65.7°$ and $294.3°$, correct to 1 d.p.

Example 5

For which angles, between 0° and 360°, is $\tan p = -1.5$?

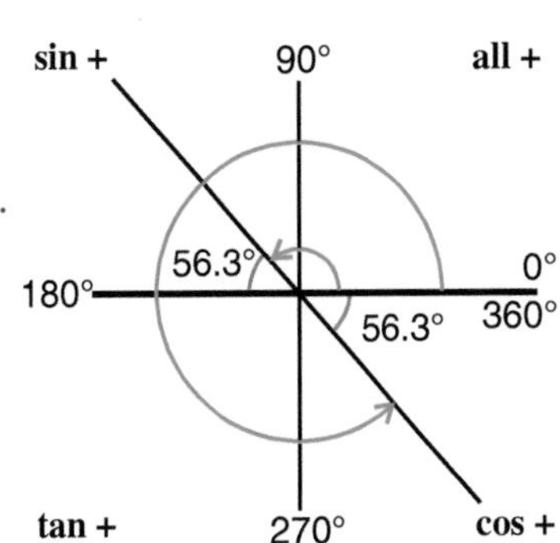

$\tan p = -1.5$

$\tan p$ is negative, so, possible angles are in the 2nd and 4th quadrants.

$p = \tan^{-1}(-1.5)$

$p = -56.309...$

$p = -56.3°$, to 1 d.p.

2nd quadrant.

$p = 180° - 56.3° = 123.7°$

4th quadrant.

$p = 360° - 56.3° = 303.7°$

So, $p = 123.7°$ and $303.7°$, correct to 1 d.p.

Practice Exercise 21.3

You may use a calculator for this exercise.

1. For each of the following, find all the values of p between 0° and 360°.

(a) $\cos p = 0.5$	(b) $\sin p = -0.5$	(c) $\tan p = 1$
(d) $\sin p = 0.5$	(e) $\tan p = -1$	(f) $\cos p = -0.5$
(g) $\sin p = -0.766$	(h) $\cos p = 0.766$	(i) $\sin p = 0.866$
(j) $\tan p = -2.050$	(k) $\tan p = 0.193$	(l) $\cos p = 0.565$
(m) $\sin p = 0.342$	(n) $\cos p = -0.866$	(o) $\tan p = 0.700$

2. Find the values of x, between 0° and 720°, where $\sin x = -0.75$.
Give your answers to one decimal place.

3. One solution of the equation $\cos x = -0.42$ is $x = 115°$, to the nearest degree.
Find **all** the other solutions to the equation $\cos x = -0.42$
for values of the x between $-360°$ and 360°.

4. Solve the equation $\sin x = 0.8$ for $-360° \leqslant x \leqslant 360°$.
Give your answers to the nearest degree.

Key Points

▶ You should know the graphs of the trigonometric functions for $\sin x$, $\cos x$ and $\tan x$.

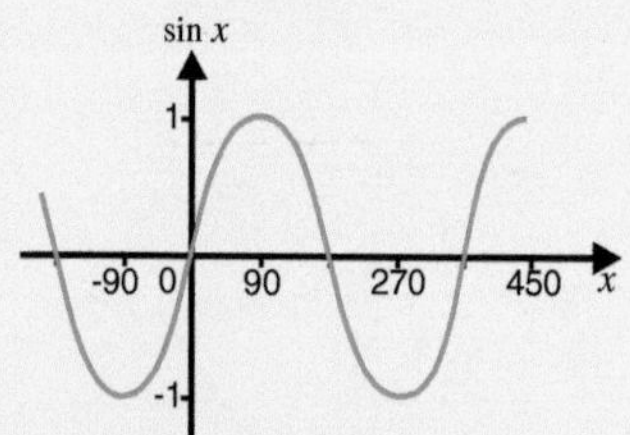

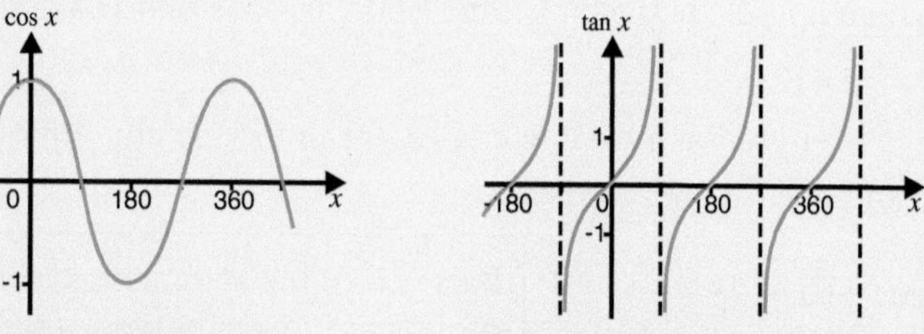

The **sine function** is a periodic function with period 360°.
$-1 \leqslant \sin x \leqslant 1$
Amplitude = 1

The **cosine function** is a periodic function with period 360°.
$-1 \leqslant \cos x \leqslant 1$
Amplitude = 1

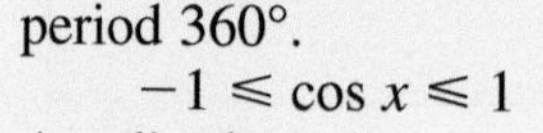

The **tangent function** is a periodic function with period 180°.
Tan x is undefined at 90°, 270°, ...

▶ The graph of a **periodic function** is repetitive.
The **period** is the number of degrees between repeats.
The **amplitude** of a function is the stretch factor of the graph measured from the x axis, in an upward direction parallel to the y axis.

▶ You should be able to identify, and sketch, transformations of periodic functions.
In general, if $y = f(x)$ where $f(x) = \sin x, \cos x, \tan x$, etc., then:

$y = f(x) + c$	means	If c is **positive**, the graph of $f(x)$ moves c units **up**. If c is **negative**, the graph of $f(x)$ moves c units **down**.
$y = f(x + p)$	means	If p is **positive**, the graph of $f(x)$ moves p units **left**. If p is **negative**, the graph of $f(x)$ moves p units **right**. The term p is called the **phase angle**.
$y = af(x)$	means	The y-coordinates on the graph of $f(x)$ are **multiplied** by a. The **amplitude** of the graph is a.
$y = f(bx)$	means	The x-coordinates on the graph of $f(x)$ are **divided** by b. The **period** of the graph is $\frac{360°}{b}$.

▶ You should be able to sketch and compare the graphs of trigonometric functions.

▶ You should be able to find the sine, cosine and tangent of angles between 0° and 360°.
For every angle $x°$, the signs of $\sin x°$, $\cos x°$ and $\tan x°$ can be shown on a diagram.

Positive angles are measured **anticlockwise**.
Negative angles are measured **clockwise**.

For angles greater than 360°: subtract 360°, or multiples of 360°, to get the equivalent angle between 0° and 360°.

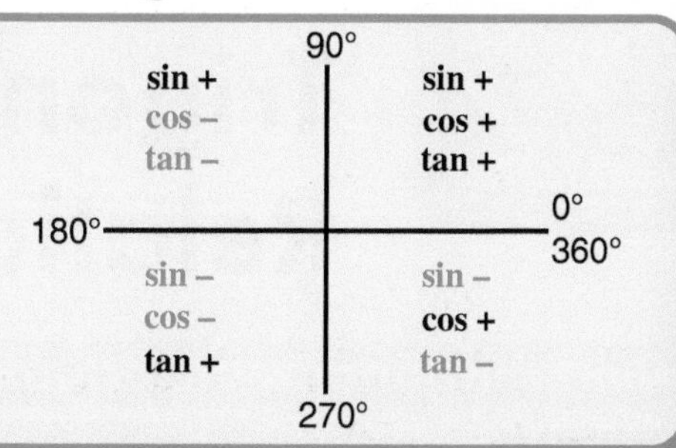

Review Exercise 21

1. The graph of $y = \sin x$ is shown.

   Sketch the graphs of:
   (a) $y = 2 \sin x$, (b) $y = \sin x + 2$,
   (c) $y = \sin 2x$, (d) $y = -\sin x$.

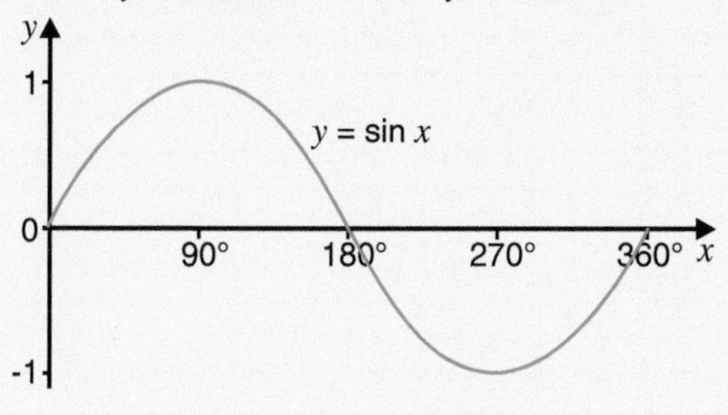

2. The diagram shows a sketch of the graph $y = \cos x$ for $0° \leqslant x \leqslant 360°$.

   Copy the diagram and, on the same axes, sketch the graph of $y = 3 \cos x$ for $0° \leqslant x \leqslant 360°$.

3. State the amplitude and period for each of the following graphs.
   (a) $y = 3 \cos 2x$ (b) $y = 4 \sin \frac{1}{2} x$
   (c) $y = 3 \tan 2x$ (d) $y = 5 \sin \frac{1}{4} x$

4. Angles x and p lie between 0° and 360°.
   (a) Find the **two** values of x which satisfy the equation $\sin x = 0.5$.
   (b) Solve the equation $\cos p = \sin 330°$.

5. (a) Sketch the graph of $y = \sin x$ for $-360° \leqslant x \leqslant 360°$.
   (b) One solution of the equation $\sin x = 0.6$ is $x = 37°$ to the nearest degree. Find all the other solutions to the equation $\sin x = 0.6$ for $0° \leqslant x \leqslant 360°$.

6. Find **two** values of x, between 0° and 360°, when $\cos x = -0.75$.

7. Given that $\tan 45° = 1$, find all the values of x, between 0° and 360°, when
   (a) $\tan x = 1$ (b) $\tan x = 0$ (c) $\tan x = -1$

8. Sketch the graphs of the following.
   (a) $y = \sin x + 1$ (b) $y = \sin x - 2$ for $0° \leqslant x \leqslant 360°$

9. $\sin 30° = 0.5$
   (a) Write down the value of $\sin 150°$.
   (b) If $\sin x = -0.5$, write down the possible values of x between 0° and 360°.

10. $\cos 30° = \frac{\sqrt{3}}{2}$.
    (a) Write down the value of $\cos 150°$.
    (b) $\cos x = \cos 30°$, $0° \leqslant x \leqslant 360°$ and $x \neq 30°$. Write down the value of x.
    (c) $\cos y = -\frac{\sqrt{3}}{2}$. Write down the possible values of y between 0° and 360°.

22 Working with Trigonometric Relationships

Solving trigonometric equations

Example 1

Solve the equation $6 \cos x - 3 = 0$ for values of x between 0° and 360°.

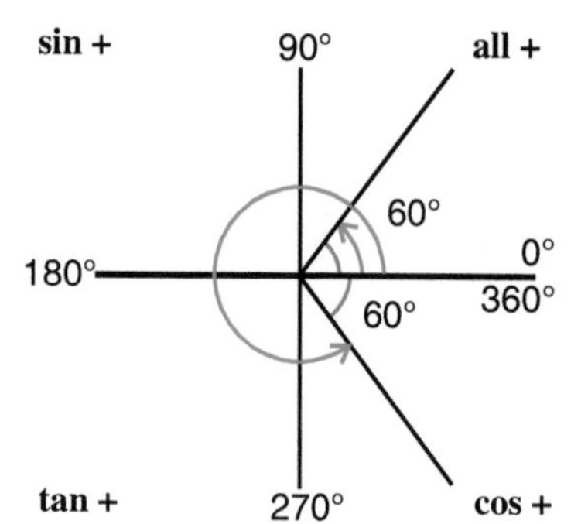

$6 \cos x - 3 = 0$

Add 3 to both sides.

$6 \cos x = 3$

Divide both sides by 6.

$\cos x = \frac{1}{2}$

$x = \cos^{-1}\left(\frac{1}{2}\right)$

Using a calculator, $x = 60°$.

So, $x = 60°$ and $x = 360° - 60°$.

$x = 60°$ and $x = 300°$.

Example 2

Solve the equation $4 \sin x + 3 = 2$ for values of x between 0° and 360°.

$4 \sin x + 3 = 2$

Subtract 3 from both sides.

$4 \sin x = -1$

Divide both sides by 4.

$\sin x = -\frac{1}{4}$

$x = \sin^{-1}\left(-\frac{1}{4}\right)$

Using a calculator, $x = -14.5°$.

So, $x = 180° + 14.5°$ and $x = 360° - 14.5°$.

$x = 194.5°$ and $x = 345.5°$, correct to nearest 1 d.p.

Practice Exercise 22.1

You may use a calculator for this exercise.

1. Solve the equation $\tan x + 1 = 0$ for $0° \leqslant x \leqslant 360°$.
2. Solve algebraically the equation $\cos x - 0.6 = 0$ for $0° \leqslant x \leqslant 360°$.
3. Solve the equation $2 \cos x - 1 = 0$ for $0° \leqslant x \leqslant 360°$.
4. Solve algebraically the equation $2 \tan x - 2 = 0$ for $0° \leqslant x \leqslant 360°$.
5. Solve the equation $5 \cos x + 4 = 0$ for $0° \leqslant x \leqslant 360°$.
6. Solve algebraically the equation $3 \cos x + 2 = 0$ for $0° \leqslant x \leqslant 360°$.
7. Solve the equation $\sqrt{3} \tan x - 4 = 0$ for $0° \leqslant x \leqslant 360°$.
8. Solve the equation $5 \sin x + 3 = \sin x + 1$ for $0° \leqslant x \leqslant 360°$.

9. The diagram shows part of the graph of $y = \tan x + 1$. The lines $y = 2$ and $y = -1$ are also shown.
   (a) Find the coordinates of A and B.
   (b) Explain how you can use your answer to part (a) to find the x-coordinate of C.
   (c) Find the coordinates of P and Q.

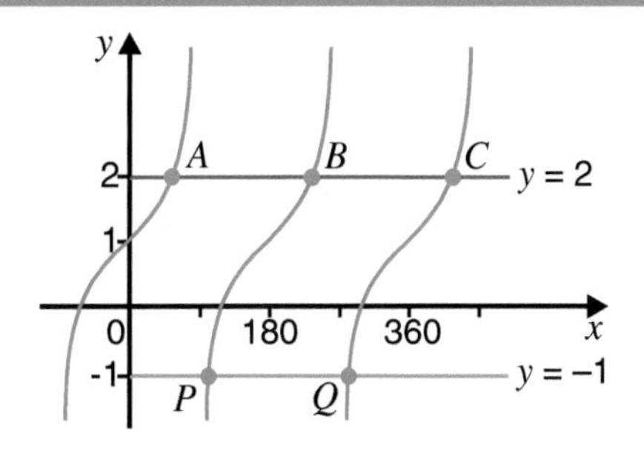

10. A particle is moving along a straight line.
    After t seconds, its distance from a point, O, is y metres, where $y = 6 \cos t - 1$.
    (a) Sketch the graph of $y = 6 \cos t - 1$ to show the movement of the particle during the first 6 minutes.
    (b) For what values of t does $6 \cos t - 1 = 0$?
    Explain what your answer means.
    (c) What is the furthest distance the particle is from O, and at what time is it at this point?

Trigonometric identities

Consider this right-angled triangle.

$\sin a° = \frac{y}{r}, \quad \cos a° = \frac{x}{r}.$

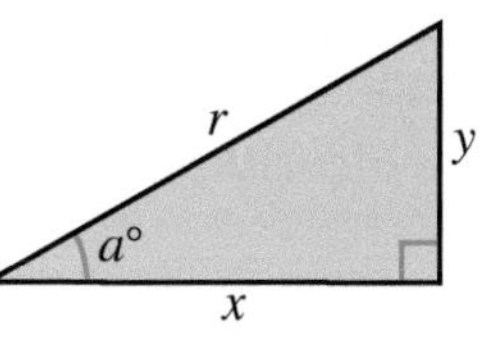

$$\frac{\sin a°}{\cos a°} = \frac{\frac{y}{r}}{\frac{x}{r}} = \frac{y}{x} = \tan a°.$$

$$\sin^2 a° + \cos^2 a° = \left(\frac{y}{r}\right)^2 + \left(\frac{x}{r}\right)^2 = \frac{y^2}{r^2} + \frac{x^2}{r^2} = \frac{y^2 + x^2}{r^2} = \frac{r^2}{r^2} = 1$$

$\tan x = \frac{\sin x}{\cos x}$ and $\sin^2 x + \cos^2 x = 1$ are called **trigonometric identities**.

Example 3

Show that $\cos x \tan x = \sin x$.

LHS $= \cos x \tan x$

Substitute $\tan x = \frac{\sin x}{\cos x}$.

$= \cancel{\cos x} \times \frac{\sin x}{\cancel{\cos x}}$

$= \sin x$

$=$ RHS

LHS = RHS, so, $\cos x \tan x = \sin x$.

ABBREVIATIONS:
LHS = left-hand side.
RHS = right-hand side.
To show that an identity is true, either:
start with the LHS and show that it is equal to the RHS, or
start with the RHS and show that it is equal to the LHS.

Example 4

Show that $\sin x \cos^2 x + \sin^3 x = \sin x$.

LHS $= \sin x \cos^2 x + \sin^3 x$

Substitute $\cos^2 x = (1 - \sin^2 x)$.

$= \sin x (1 - \sin^2 x) + \sin^3 x$

Multiply out the brackets.

$= \sin x - \sin^3 x + \sin^3 x$

$= \sin x$

$=$ RHS

LHS = RHS, so, $\sin x \cos^2 x + \sin^3 x = \sin x$.

$\sin^2 x + \cos^2 x = 1$
can be written as:
$\sin^2 x = 1 - \cos^2 x$
$\cos^2 x = 1 - \sin^2 x$

Practice Exercise 22.2

1. Show that $\cos x \tan x - \sin x = 0$.

2. Show that $\frac{1 - \cos^2 x}{\sin^2 x} = 1$.

3. Show that $\frac{1 - \sin^2 x}{1 - \sin x} = 1 + \sin x$.

4. Show that $\frac{\sin^2 x}{\cos x} - \tan x = \tan x (\sin x - 1)$.

5. Show that $(\sin x - \cos x)^2 + 2 \sin x \cos x = 1$.

6. Show that $\frac{\cos^2 x}{\sin^2 x} = \frac{1}{\tan^2 x}$.

7. Show that $\sin^4 x - \cos^4 x$ can be written as $\sin^2 x - \cos^2 x$.

8. Show that $(\sin x + \cos x)^2 + (\sin x - \cos x)^2 = 2$.

9. Show that $\frac{1 - \sin x}{\cos x} = \frac{\cos x}{1 + \sin x}$.

Key Points

- You should be able to solve trigonometric equations.
- You should know the following **trigonometric identities** involving, sin, cos and tan.

$$\sin^2 x + \cos^2 x = 1 \quad \text{and} \quad \tan x = \frac{\sin x}{\cos x}$$

- You should be able to solve problems involving trigonometric functions.

Review Exercise 22

1. Solve algebraically these equations for $0° \leqslant x \leqslant 360°$.
   (a) $2 \cos x - 1 = 0$ (b) $3 \cos x + 1 = 0$

2. Solve algebraically these equations for $0° \leqslant x \leqslant 360°$.
   (a) $\frac{1}{4} \tan x + 3 = 5$ (b) $1 = 3 - 5 \sin x$ (c) $1 - 3 \cos x = 0.5$

3. Solve algebraically $5 \cos x° + 3 = 0$ for $0 \leqslant x \leqslant 360$.

4. Solve algebraically $4 \sin x° - 1 = 2$ for $0 \leqslant x \leqslant 360$.

5. Show that $\cos x \tan x = \sin x$.

6. Given that $a \sin^2 x + a \cos^2 x = 4$, find the value of a.

7. Show that $\frac{\cos^2 x}{1 - \sin x} = 1 + \sin x$.

8. The depth of the water, H metres, in a tidal marina is given by the formula $H = 8 + 7 \cos 24t$ where t is the number of hours after midnight.
   (a) What is the maximum depth of water recorded in the marina?
   (b) What is the minimum depth of water recorded in the marina?
   (c) What is the first time, after midnight, that the depth of water in the marina is recorded at 10 m?
   (d) What is the next time, in the same day, that the depth of water is recorded at 10 m?

23 Sine Rule *and* Cosine Rule

Area of a triangle

Look at these triangles.
Both are labelled ABC.

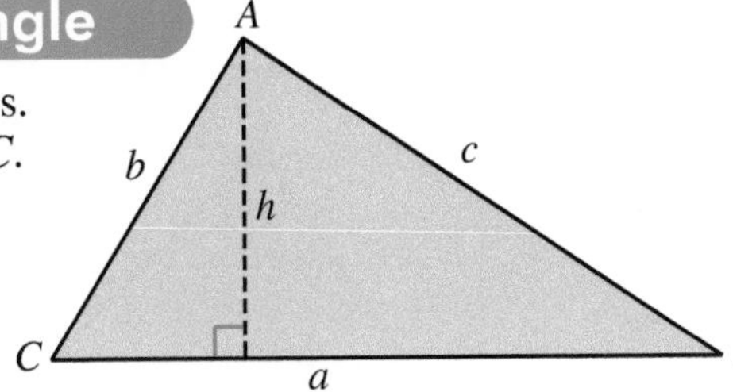

The height of each triangle is labelled h.
In both triangles $h = b \sin C$.

The area of triangle $ABC = \frac{1}{2} \times \text{base} \times \text{height}$.

$= \frac{1}{2} \times a \times b \sin C$.

$= \frac{1}{2} ab \sin C$.

For any triangle ABC: Area $= \frac{1}{2} ab \sin C$

Can you find a rule which gives the area of a parallelogram?

To use this formula to find the area of a triangle we need:

two sides of known length, **and** the size of the angle between the known sides.

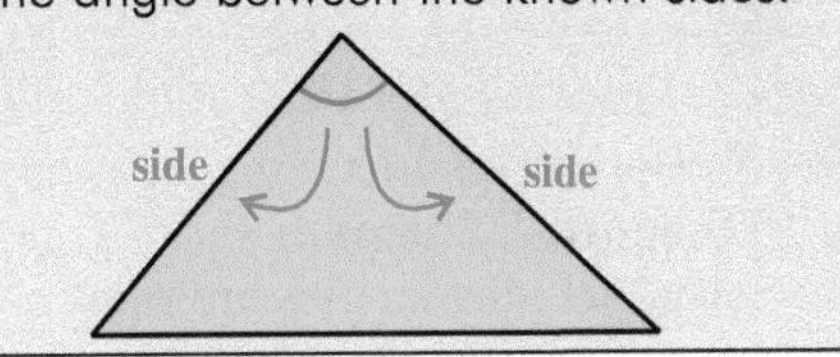

Example 1

Calculate the area of triangle PQR.

Area $PQR = \frac{1}{2} \times p \times q \times \sin R$
$= 0.5 \times 2.4 \times 4.5 \times \sin 67°$
$= 4.9707...$
$= 4.97\ \text{cm}^2$, correct to 3 s.f.

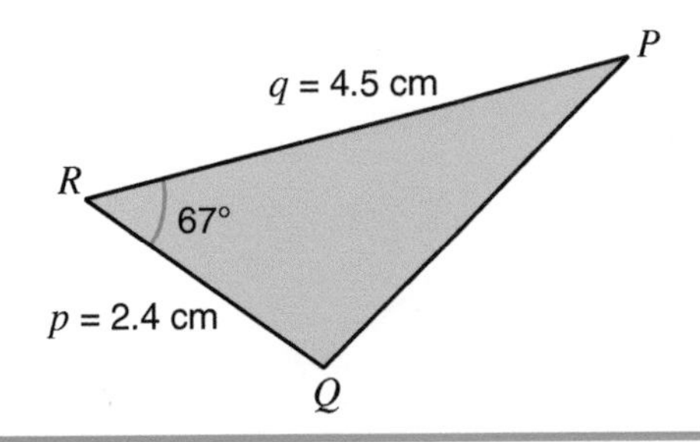

Practice Exercise 23.1

1. Calculate the areas of these triangles.
   Give your answers correct to three significant figures.

(a)

(b)

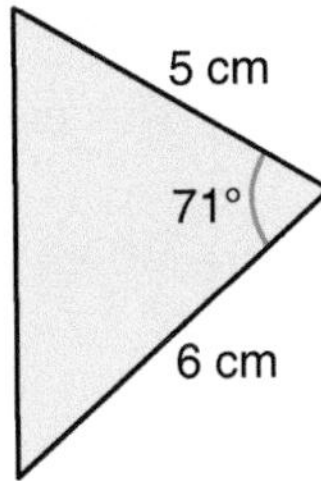

(c) (d)

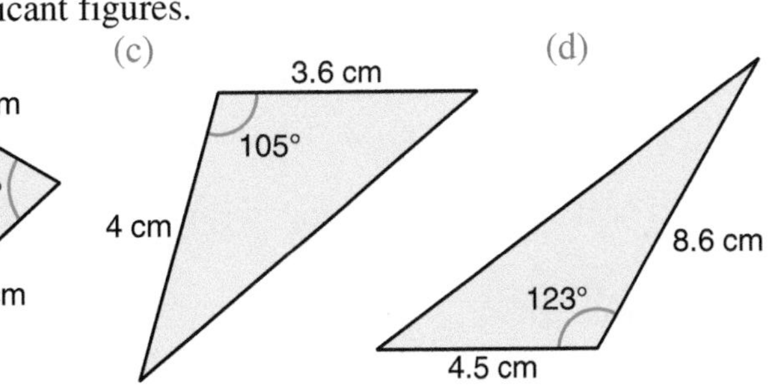

2. $PQRS$ is a parallelogram. $PQ = 9.2\ \text{cm}$, $PS = 11.4\ \text{cm}$ and angle $QPS = 54°$.
   Calculate the area of the parallelogram.

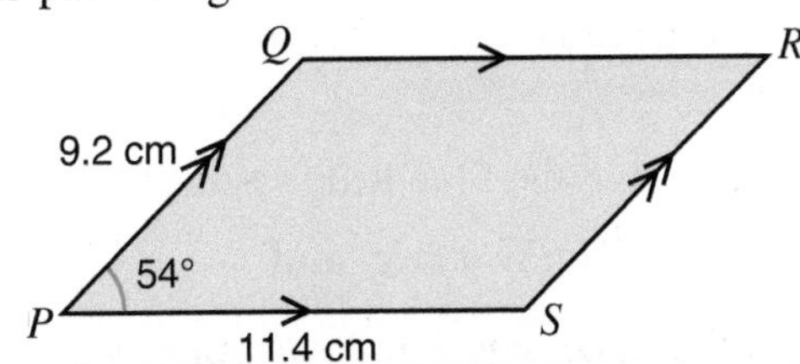

3. A regular hexagon, centre O, has sides of length 5 cm.
   Calculate the area of the hexagon.

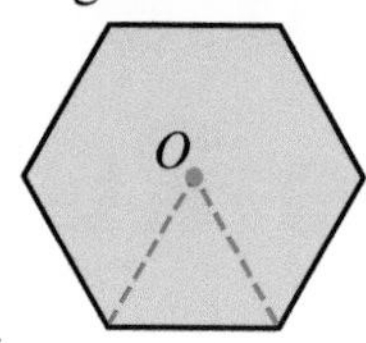

4. Triangle XYZ is isosceles, with $XY = YZ$.
   Angle $XYZ = 54°$.
   The area of triangle XYZ is $22\,\text{cm}^2$.
   Calculate the lengths of XY and YZ.

5. The area of triangle ABC is $31.7\,\text{cm}^2$.
   $AB = 9.6\,\text{cm}$ and angle $ABC = 48°$.
   Calculate the length of BC.
   Give your answer correct to one decimal place.

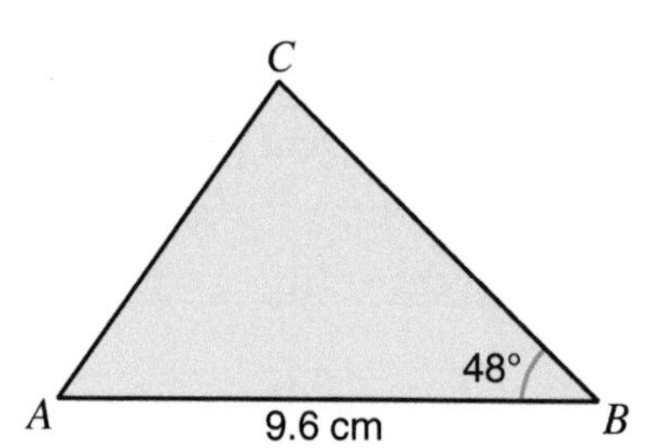

6. Triangle PQR has area $17.8\,\text{cm}^2$.
   $PQ = 5\,\text{cm}$ and $QR = 8\,\text{cm}$.
   Find the size of the acute angle PQR.

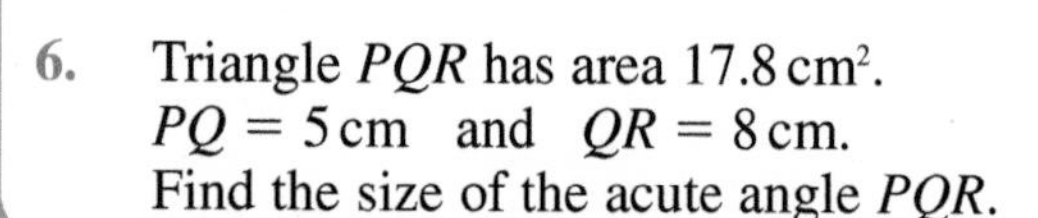

Triangles which are not right-angled

To solve problems involving right-angled triangles, you use the **trigonometric ratios** (sin, cos and tan) or **Pythagoras' Theorem**.
However, not all triangles are right-angled and to solve problems involving acute-angled and obtuse-angled triangles we need to know further rules.

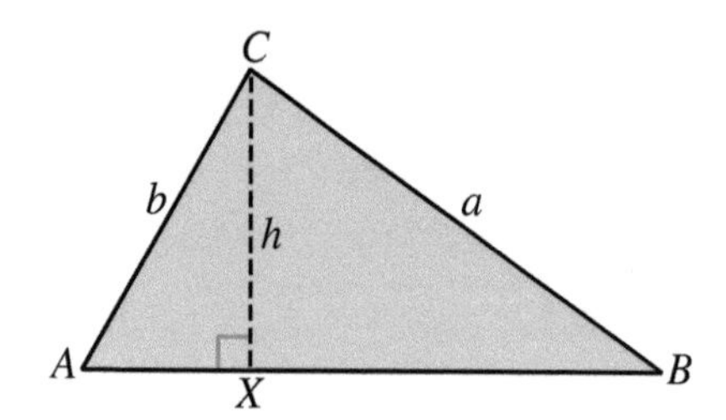

Angles are labelled with **capital letters** and the **sides** opposite the angles with **lower case letters.**

The **altitude**, or height of the triangle, is shown by the line CX.
The line CX divides triangle ABC into two right-angled triangles.

In triangle CAX.

$\sin A = \frac{h}{b}$

This can be rearranged to give: $h = b \sin A$.

In triangle CBX.

$\sin B = \frac{h}{a}$

This can be rearranged to give: $h = a \sin B$.

$h = b \sin A$ and $h = a \sin B$, so, $b \sin A = a \sin B$, which can be rearranged as: $\frac{a}{\sin A} = \frac{b}{\sin B}$.

The Sine Rule

In any triangle labelled ABC it can be proved that: $\frac{a}{\sin A} = \frac{b}{\sin B} = \frac{c}{\sin C}$.

The Sine Rule can also be written as: $\frac{\sin A}{a} = \frac{\sin B}{b} = \frac{\sin C}{c}$.

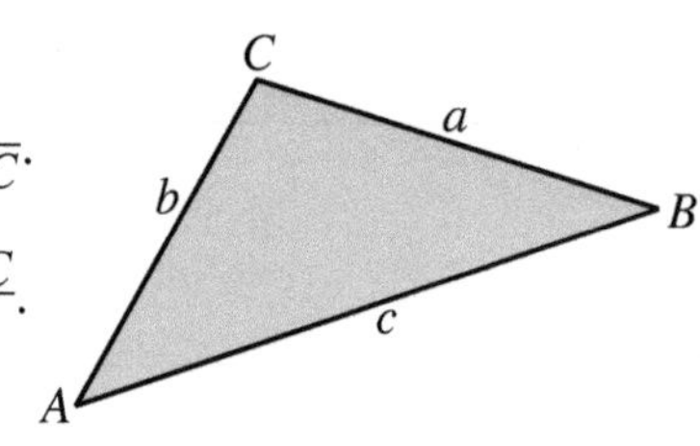

Using the Sine Rule to find sides

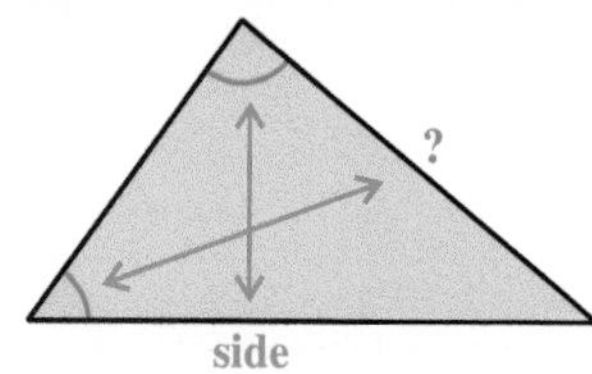

To find a side using the Sine Rule you need:

two angles of known size, **and**
the length of a side which is opposite one of the known angles.

Example 2

n triangle ABC, angle $A = 37°$, angle $C = 72°$ and $b = 12\,\text{cm}$.
alculate a, correct to 3 significant figures.

Jsing the Sine Rule:

$$\frac{a}{\sin A} = \frac{b}{\sin B} = \frac{c}{\sin C}$$

ubstitute known values.

$$\frac{a}{\sin 37°} = \frac{12}{\sin 71°} = \frac{c}{\sin 72°}$$

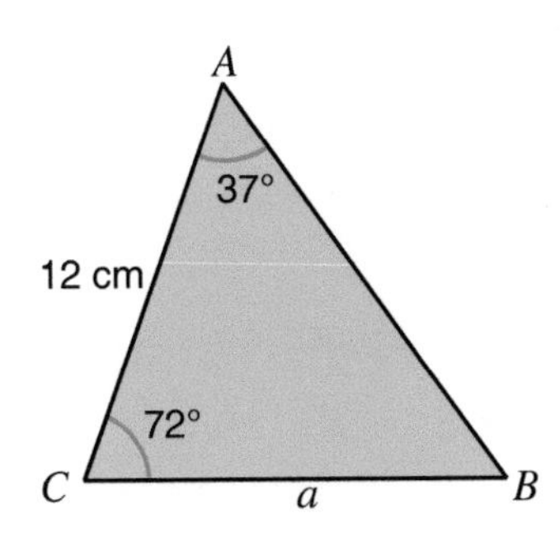

To find angle B:
$\angle A + \angle B + \angle C = 180°$
$37° + \angle B + 72° = 180°$
$\angle B = 71°$

Jsing:

$$\frac{a}{\sin 37°} = \frac{12}{\sin 71°}$$

Aultiply both sides by sin 37°.

$$a = \frac{12 \times \sin 37°}{\sin 71°}$$
$$= 7.637\ldots$$
$a = 7.64\,\text{cm}$, correct to 3 s.f.

Alternative method:
First rearrange the equation, then substitute known values.

Use $\frac{a}{\sin A} = \frac{b}{\sin B}$.

Rearrange to give $a = \frac{b \sin A}{\sin B}$.

Substitute known values $a = \frac{12 \times \sin 37°}{\sin 71°}$.

ractice Exercise 23.2

1. Find the side marked a in each of the following triangles.

(a) C, B, A; 75°, 34°; a; 10 cm

(b) C, A, B; 30°, 125°, 25°; 8 cm; a

(c) A, C, B; 12.5 cm; 47°, 115°; a

2. Find the marked side in each of these triangles.

(a) A, C, B; 83°, 38°; b; 16 cm

(b) C, A, B; 61°, 72°; 9 cm; c

(c) A, B, C; 3.4 cm; 56°, 19°; b

(d) R, Q, P; 17 cm; 46°, 116°; q

(e)

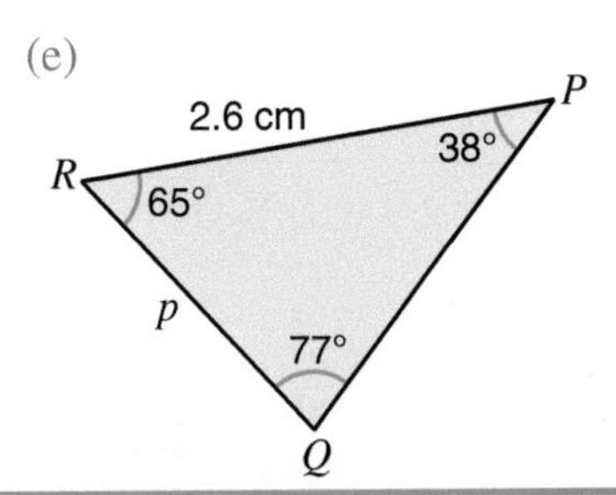

(f)

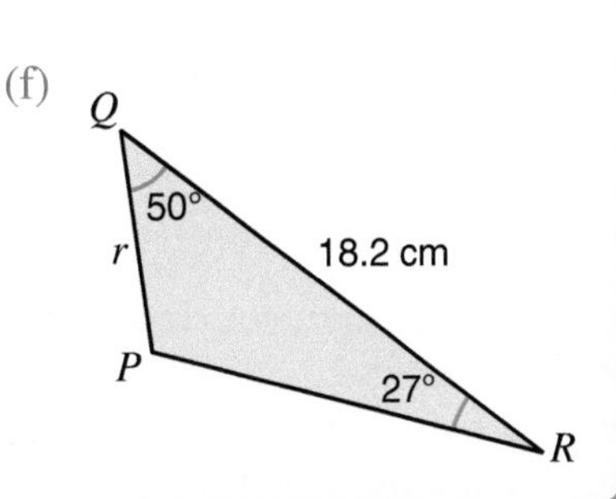

3. Find the remaining angles and sides of these triangles.

(a) ΔABC, when $\angle BAC = 57°$, $\angle ABC = 68°$ and $BC = 6.7$ cm.

(b) ΔLMN, when $\angle LMN = 33.2°$, $\angle LNM = 75.6°$ and $LN = 3.3$ cm.

(c) ΔPQR, when $\angle PQR = 62.8°$, $\angle PRQ = 47.4°$ and $PQ = 12.3$ cm.

(d) ΔSTU, when $\angle TSU = 94.9°$, $\angle STU = 53.3°$ and $SU = 19.4$ cm.

(e) ΔXYZ, when $\angle XYZ = 108.6°$, $\angle YXZ = 40.2°$ and $XZ = 13.8$ cm.

4. In ΔPQR, $\angle PQR = 110°$, $\angle QRP = 29°$ and $PQ = 5$ cm.
Calculate
(a) the length of PR,
(b) the length of QR.
Give your answers correct to one decimal place.

5. In ΔABC, $\angle BAC = 35°$ and $\angle ACB = 29°$.
$BC = 38$ m.
Calculate the perimeter of ΔABC.
Give your answer correct to 3 significant figures.

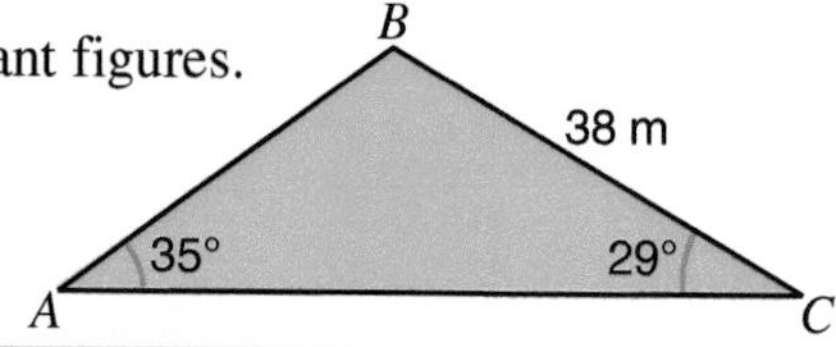

Using the Sine Rule to find angles

To find an angle using the Sine Rule you need:

the length of the side opposite the angle you are trying to find, **and**
the length of a side opposite an angle of known size.

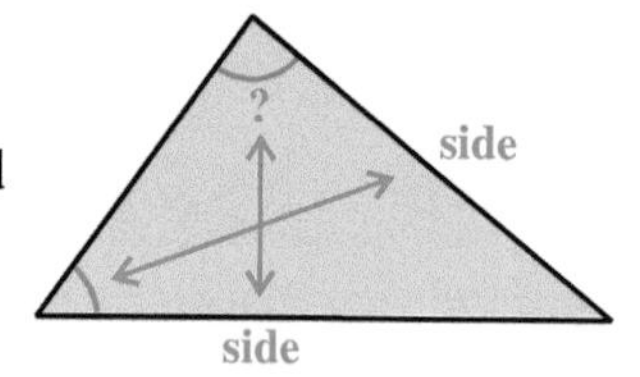

Example 3

Angle $A = 42°$, $a = 6$ cm and $c = 4$ cm.
Calculate angle C, correct to one decimal place.

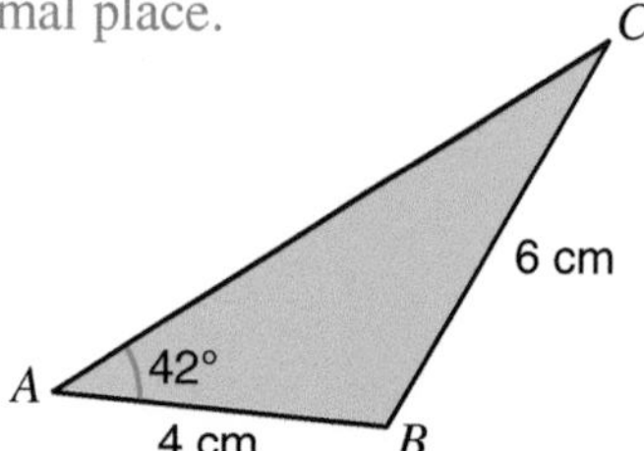

Using the Sine Rule:

$$\frac{\sin A}{a} = \frac{\sin B}{b} = \frac{\sin C}{c}$$

Substitute known values.

$$\frac{\sin 42°}{6} = \frac{\sin B}{b} = \frac{\sin C}{4}$$

Using:

$$\frac{\sin 42°}{6} = \frac{\sin C}{4}$$

Multiply both sides by 4.

$$\sin C = \frac{4 \times \sin 42°}{6}$$

$\sin C = 0.446...$

$C = \sin^{-1}(0.446...)$

$C = 26.49...$

$C = 26.5°$, correct to 1 d.p.

$C = \sin^{-1}(0.446...)$
gives possible angles of 26.5° and 153.5°.
$C \neq 153.5°$, since in an obtuse-angled triangle, the obtuse angle is always opposite the longest side of the triangle.

The ambiguous case

There are two angles between 0° and 180° which have the same sine.
When we use the Sine Rule to find an angle we must therefore look at the information to see if there are two possible values for the angle.

Example 4

In triangle PQR, $PQ = 8\text{ cm}$, $QR = 6\text{ cm}$ and angle $QPR = 38°$.
Find the size of angle QRP.

Using this information **two** triangles can be drawn.
Use a ruler and compasses to construct the triangles.

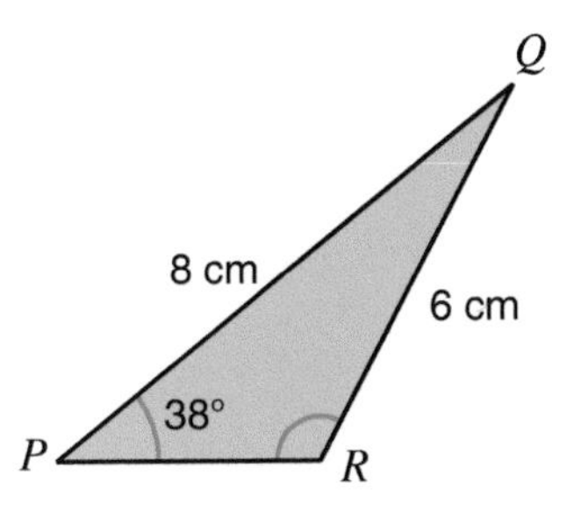

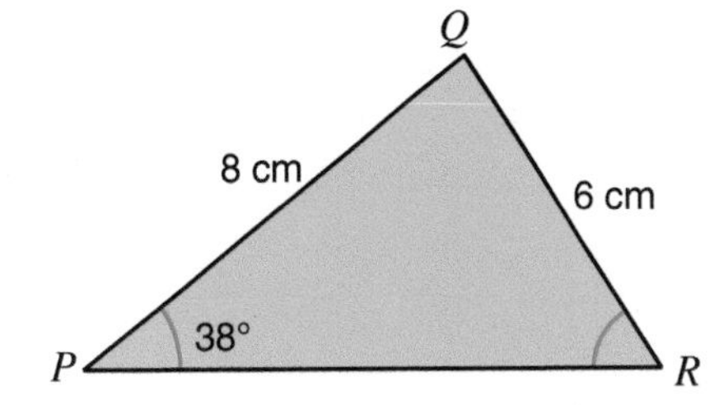

Using the Sine Rule:

$$\frac{\sin P}{p} = \frac{\sin Q}{q} = \frac{\sin R}{r}$$

Substitute known values.

$$\frac{\sin 38°}{6} = \frac{\sin Q}{q} = \frac{\sin R}{8}$$

Using:

$$\frac{\sin 38°}{6} = \frac{\sin R}{8}$$

Multiply both sides by 8.

$$\sin R = \frac{8 \times \sin 38°}{6}$$

$$\sin R = 0.820...$$

$$R = \sin^{-1}(0.820...)$$

$$R = 55.17...$$

$$R = 55.2°$$

Also, $R = 180° - 55.2° = 124.8°$.

$\angle QRP = 55.2°$ or $124.8°$, correct to 1 d.p.

Practice Exercise 23.3

1. Find angle A in each of these acute-angled triangles.

(a)

(b)

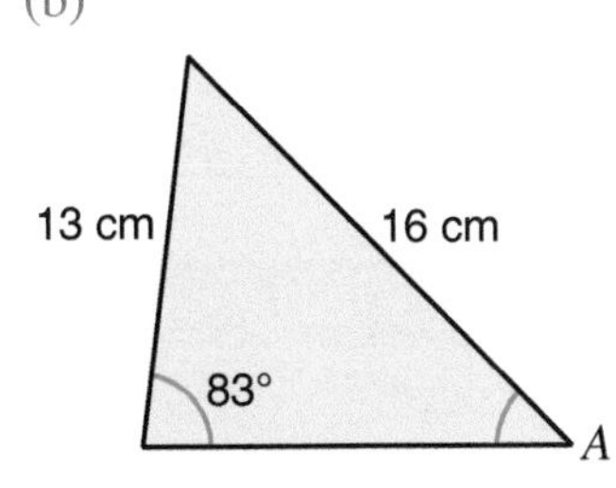

(c)

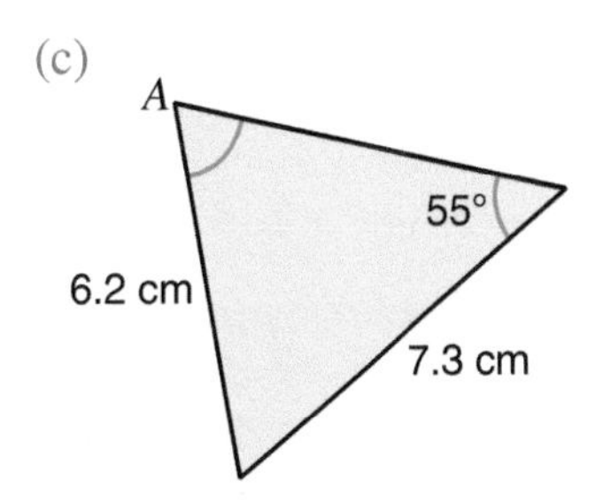

2. Find angle A in each of these obtuse-angled triangles.

(a)

(b)

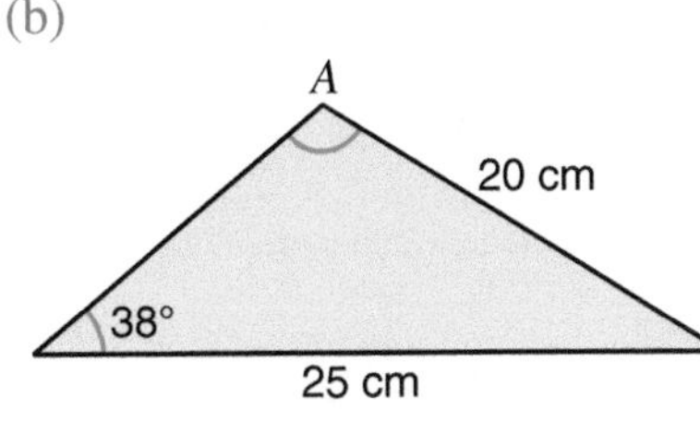

(c)

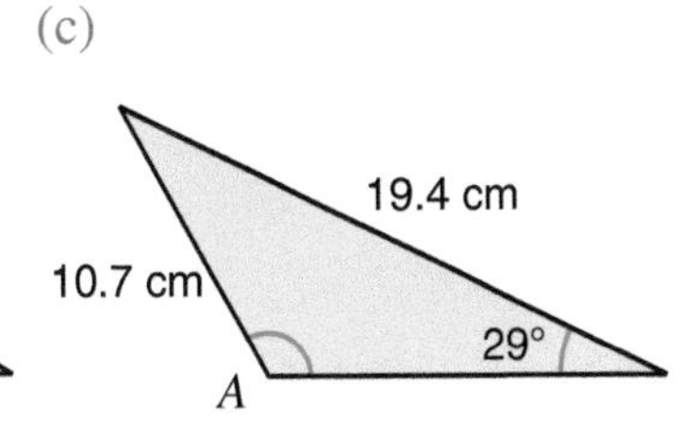

3. In triangle ABC, $AB = 6\text{ cm}$, $BC = 4\text{ cm}$ and angle $BAC = 30°$.
There are two possible triangles that can be constructed.
Calculate the two possible values of angle BCA.

4. In triangle PQR, $QR = 7.5\text{ cm}$, $RP = 7\text{ cm}$ and angle $PQR = 60°$.
There are two possible triangles that can be constructed.
Find the missing angles and sides of both triangles.

Deriving the Cosine Rule

$AB = c$
Let $AX = x$
So, $BX = c - x$

In triangle ACX.
Using Pythagoras' Theorem.

$$b^2 = x^2 + h^2$$

Using the cosine ratio.

$$\cos A = \frac{\text{adj}}{\text{hyp}}$$

$$\cos A = \frac{x}{b}$$

$$x = b \cos A$$

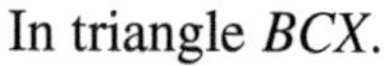

In triangle BCX.
Using Pythagoras' Theorem.

$$a^2 = (c - x)^2 + h^2$$

$$a^2 = c^2 - 2cx + x^2 + h^2$$

Replace: $x^2 + h^2$ by b^2 and x by $b \cos A$.

$$a^2 = c^2 - 2c(b \cos A) + b^2$$

This can be rearranged as: $a^2 = b^2 + c^2 - 2bc \cos A$, and is called the **Cosine Rule**.

The Cosine Rule

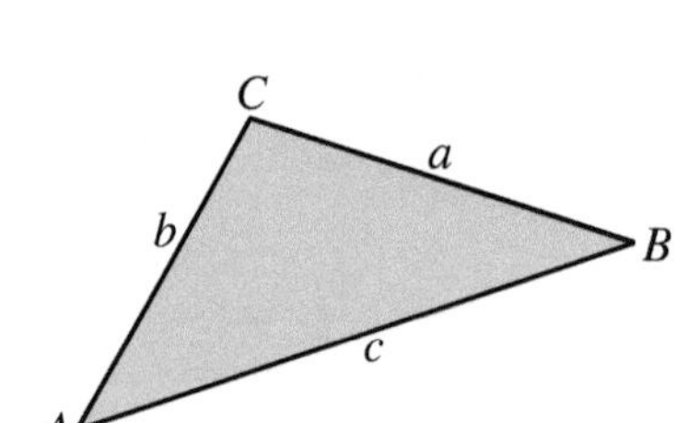

In any triangle labelled ABC it can be proved that:
$a^2 = b^2 + c^2 - 2bc \cos A$

In a similar way, we can find the Cosine Rule for b^2 and c^2.
$b^2 = a^2 + c^2 - 2ac \cos B$
$c^2 = a^2 + b^2 - 2ab \cos C$

When using the Cosine Rule to find the size of an angle it is sometimes easier to rearrange the above formulae as:

$$\cos A = \frac{b^2 + c^2 - a^2}{2bc} \qquad \cos B = \frac{a^2 + c^2 - b^2}{2ac} \qquad \cos C = \frac{a^2 + b^2 - c^2}{2ab}$$

Using the Cosine Rule to find sides

To find a side using the Cosine Rule you need:

two sides of known length, **and**
the size of the angle between the known sides.

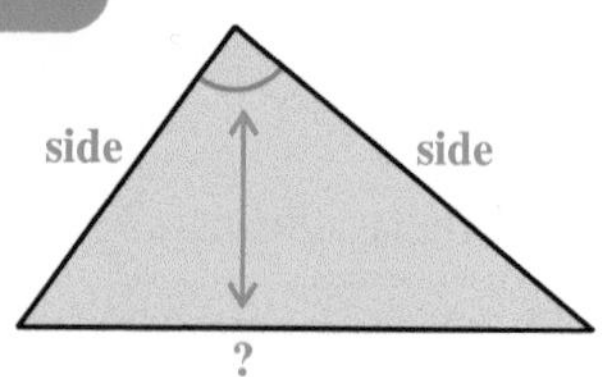

Example 5

Calculate the length of side a.

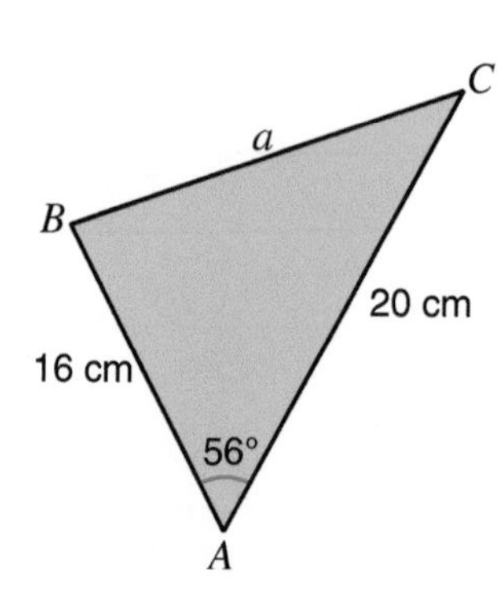

Using the Cosine Rule:
$a^2 = b^2 + c^2 - 2bc \cos A$
Substitute known values.
$a^2 = 20^2 + 16^2 - 2 \times 20 \times 16 \times \cos 56°$
$a^2 = 400 + 256 - 357.88...$
$a^2 = 298.116...$
Take the square root.
$a = 17.266...$
$a = 17.3\,\text{cm}$, correct to 3 s.f.

Practice Exercise 23.4

1. Calculate the lengths of the sides marked a.
   Give your answers correct to three significant figures.
   (a) (b) (c)

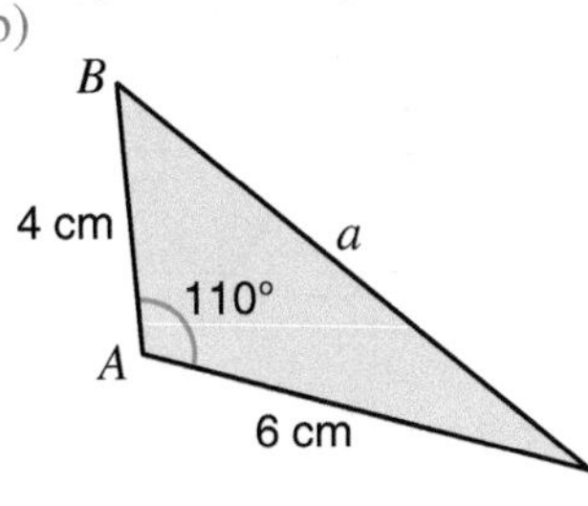

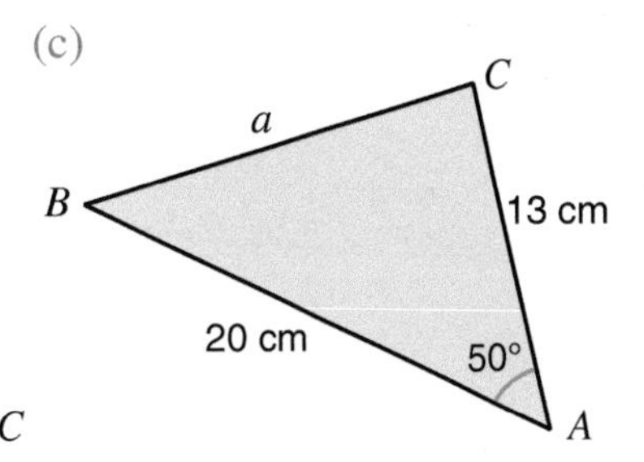

2. Calculate the lengths of the sides marked with letters.
   Give your answers correct to three significant figures.
   (a) (b) (c)

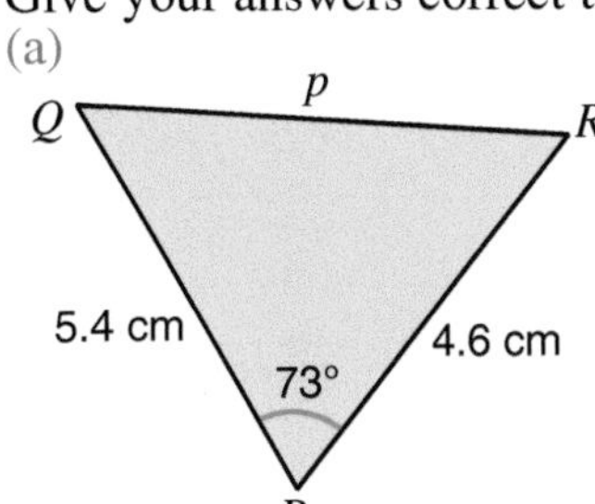

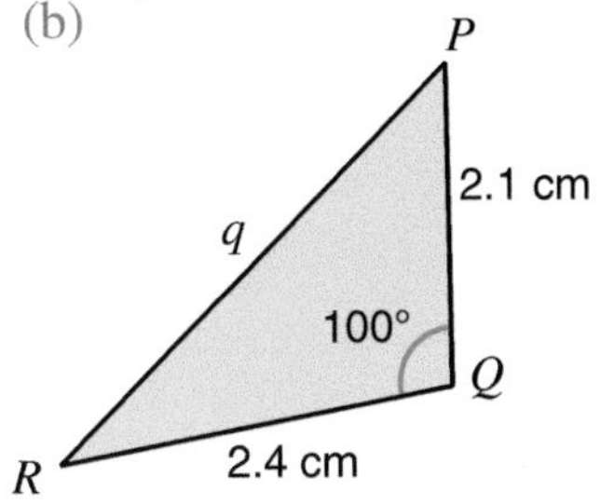

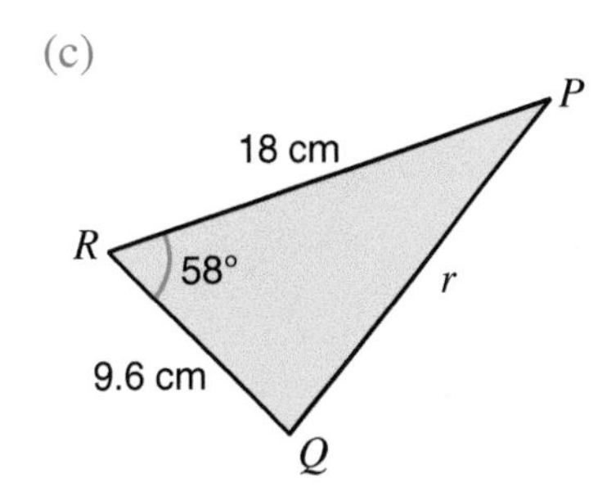

3. In triangle ABC, $AB = 8\,\text{cm}$, $AC = 13\,\text{cm}$ and $\angle BAC = 70°$.
   Calculate BC.

4. In triangle LMN, $LM = 16\,\text{cm}$, $MN = 9\,\text{cm}$ and $\angle LMN = 46.8°$.
   Calculate LN.

5. In triangle PQR, $PQ = 9\,\text{cm}$, $PR = 5.4\,\text{cm}$ and $\angle RPQ = 135°$.
   Calculate QR.

6. In triangle XYZ, $XY = 7.5\,\text{cm}$, $YZ = 13\,\text{cm}$ and $\angle XYZ = 120°$.
   Calculate XZ.

Using the Cosine Rule to find angles

To find an angle using the Cosine Rule you need:

three sides of known length.

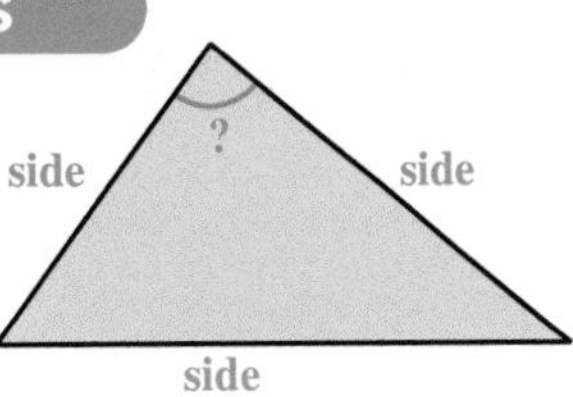

Example 6

Calculate the size of angle A.

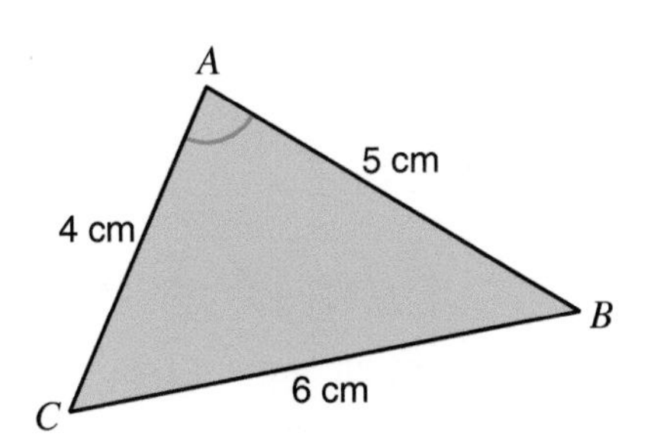

Using the Cosine Rule:

$$\cos A = \frac{b^2 + c^2 - a^2}{2bc}$$

Substitute known values.

$$\cos A = \frac{4^2 + 5^2 - 6^2}{2 \times 4 \times 5}$$

$\cos A = 0.125$

$A = \cos^{-1}(0.125)$

$A = 82.819...$

$A = 82.8°$, correct to 1 d.p.

Practice Exercise 23.5

1. Calculate angle A in each of the following.
Give your answers correct to one decimal place.

(a)

(b)

(c)

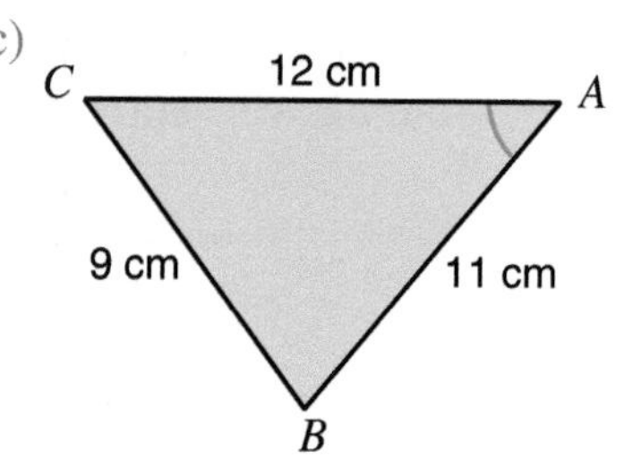

2. In this question give your answers correct to one decimal place.

(a) Calculate $\angle B$.

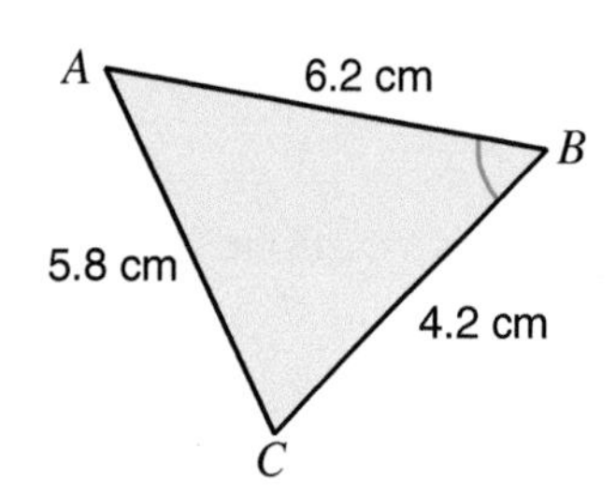

(b) Calculate $\angle C$.

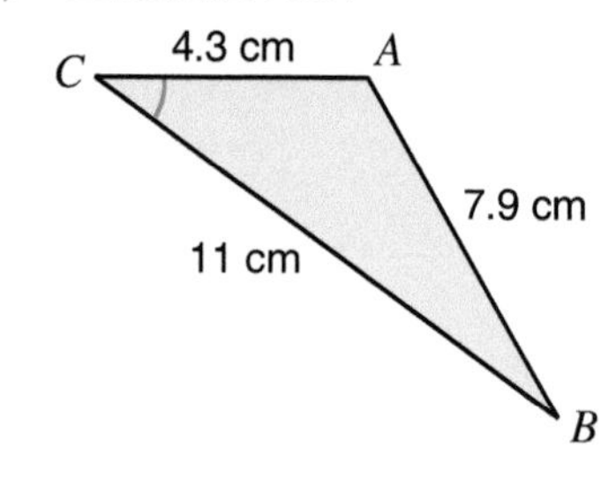

(c) Calculate $\angle P$.

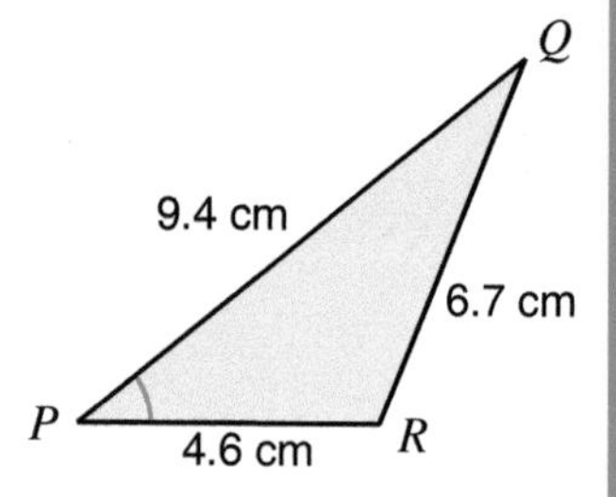

3. In triangle ABC, $AB = 18$ cm, $BC = 23$ cm and $CA = 24$ cm.
Calculate $\angle ACB$.

4. In triangle PQR, $PQ = 10$ cm, $QR = 12$ cm and $RP = 20$ cm.
Calculate $\angle PQR$.

5. In triangle LMN, $LM = 8$ cm, $MN = 6$ cm and $NL = 11$ cm.
Calculate $\angle LMN$.

6. In triangle XYZ, $XY = 9.6$ cm, $YZ = 13.4$ cm and $XZ = 20$ cm.
Calculate $\angle XYZ$.

Key Points

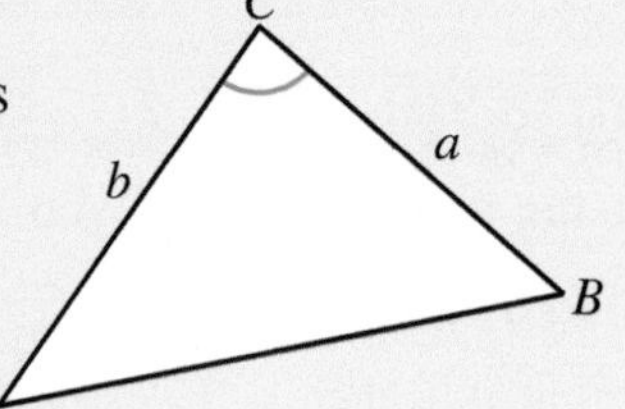

- To find the **area of a triangle** when you know two sides and the angle between the two sides use:
  Area of triangle $= \frac{1}{2}ab \sin C$.

- **The Sine Rule**
  $$\frac{a}{\sin A} = \frac{b}{\sin B} = \frac{c}{\sin C}$$
  This can also be written as: $\frac{\sin A}{a} = \frac{\sin B}{b} = \frac{\sin C}{c}$.

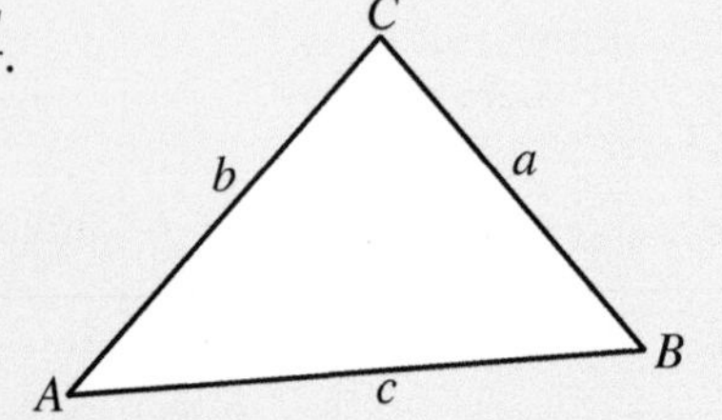

- **The Cosine Rule**
  $a^2 = b^2 + c^2 - 2bc \cos A$
  $b^2 = a^2 + c^2 - 2ac \cos B$
  $c^2 = a^2 + b^2 - 2ab \cos C$
  When using the Cosine Rule to find the size of an angle it is sometimes easier to rearrange the above formulae as:
  $$\cos A = \frac{b^2 + c^2 - a^2}{2bc} \qquad \cos B = \frac{a^2 + c^2 - b^2}{2ac} \qquad \cos C = \frac{a^2 + b^2 - c^2}{2ab}$$

Review Exercise 23

1. The diagram shows a triangular field.
   (a) Calculate the area of the field.
   (b) Calculate the perimeter of the field.
   Give your answers correct to 3 significant figures.

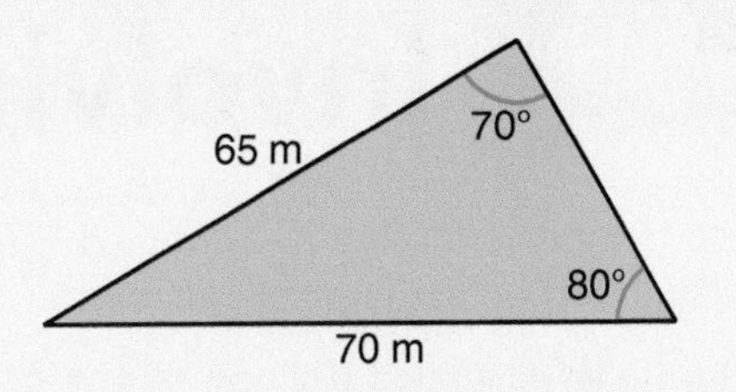

2. The area of triangle XYZ is $54.6\,\text{cm}^2$.
   $XY = 12.4\,\text{cm}$ and angle $XYZ = 35°$.
   Calculate the length of YZ.
   Give your answer correct to one decimal place.

3. $PQRS$ is a kite.
   $PQ = SP = 35\,\text{cm}$.
   $QR = RS = 24\,\text{cm}$.
   Angle $QPS = 48°$, angle $QRS = 72°$.
   Calculate the area of the kite.

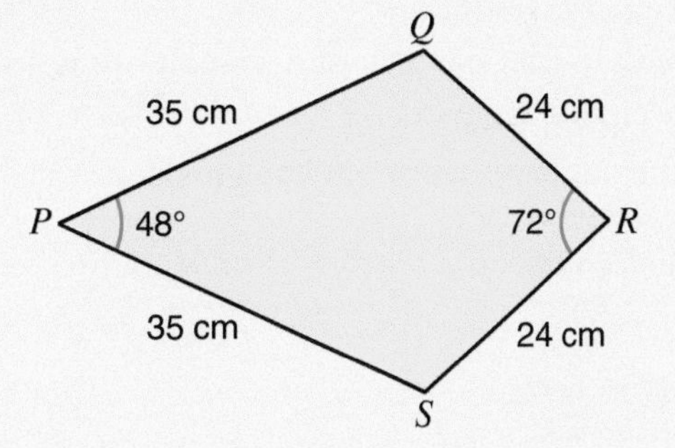

4. In triangle ABC, $AC = 3.8\,\text{cm}$, angle $CAB = 115°$ and angle $ABC = 28°$.
   (a) Calculate the length of BC.
   (b) Calculate the area of the triangle.

5. In the triangle PQR, angle $PQR = 108°$.
   $PQ = 12.3\,\text{cm}$ and $PR = 24.4\,\text{cm}$.
   (a) Calculate angle QRP.
   (b) Calculate the area of triangle PQR.

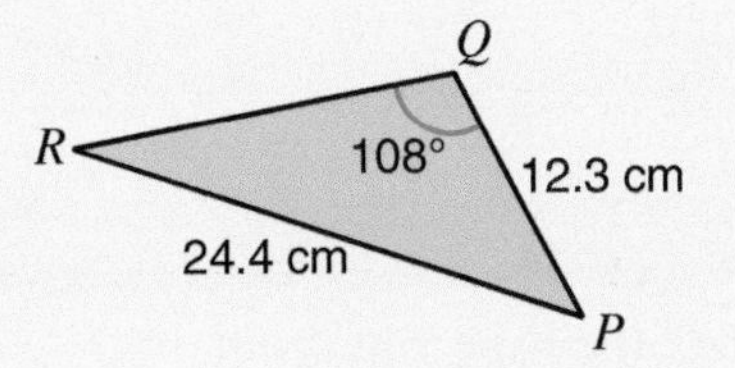

6. Triangle ABC has area $27.6\,\text{cm}^2$.
   $AB = 7.5\,\text{cm}$ and $BC = 9\,\text{cm}$.
   Find the size of the obtuse angle ABC.

7. The sides of a triangle are 5 cm, 7 cm and 10 cm.
   Find the size of the largest angle.

8. In ΔABC, $\angle ABC = 63°$, $AB = 3\,\text{cm}$ and $AC = 4\,\text{cm}$.
   (a) Calculate the size of $\angle ACB$.
   (b) Hence, or otherwise, find the size of $\angle BAC$.

9. The area of triangle ABC is $25.6\,\text{cm}^2$.
   $AB = 6.7\,\text{cm}$, $AC = 9.2\,\text{cm}$.
   Calculate the two possible sizes of angle A.

10. The diagram shows a circle with centre O and radius 6 cm.
    Angle $AOB = 110°$.
    Calculate the area of the blue shaded segment.

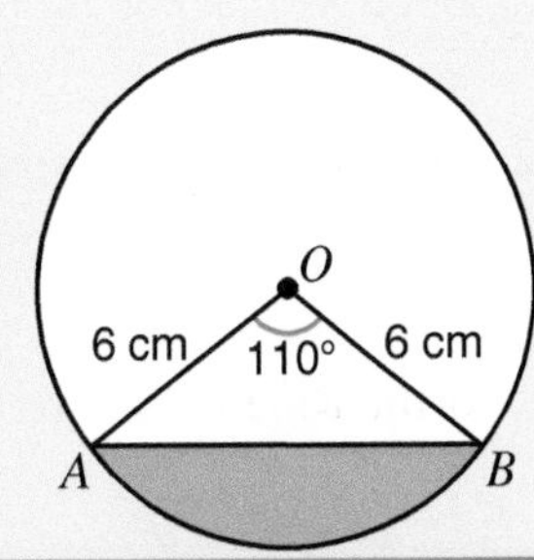

24 Solving Problems Involving Triangles

Solving problems involving right-angled triangles may involve using the trigonometric ratios (sin, cos and tan) or Pythagoras' Theorem.

The area of a right-angled triangle $= \frac{1}{2} \times$ base $\times$ perpendicular height.

In Chapter 23 you developed skills that can be applied to triangles which are not right-angled. Here is a reminder.

Area of a triangle

To use this formula to find the area of a triangle we need:

two sides of known length, **and**
the size of the angle between the known sides.

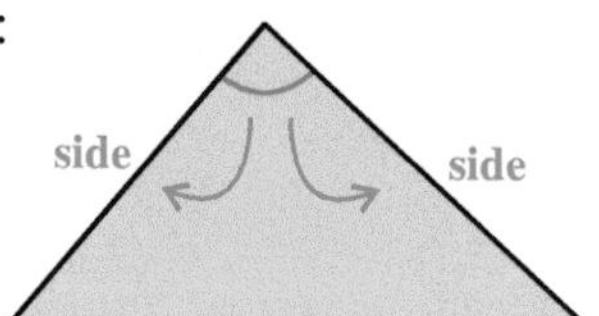

For any triangle ABC: Area $= \frac{1}{2} ab \sin C$

The Sine Rule

In any triangle labelled ABC it can be proved that:

$$\frac{a}{\sin A} = \frac{b}{\sin B} = \frac{c}{\sin C}.$$

The Sine Rule can also be written as:

$$\frac{\sin A}{a} = \frac{\sin B}{b} = \frac{\sin C}{c}.$$

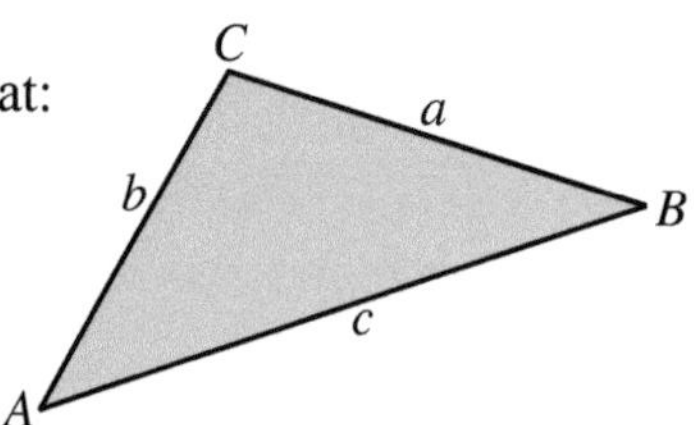

Using the Sine Rule to find sides

To find a side using the Sine Rule you need:

two angles of known size, **and**
the length of a side which is opposite one of the known angles.

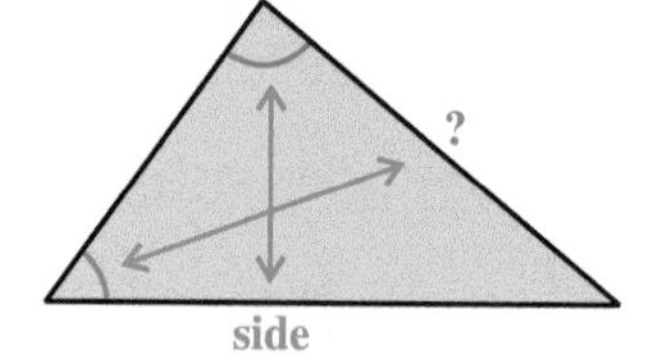

Using the Sine Rule to find angles

To find an angle using the Sine Rule you need:

the length of the side opposite the angle you are trying to find, **and**
the length of a side opposite an angle of known size.

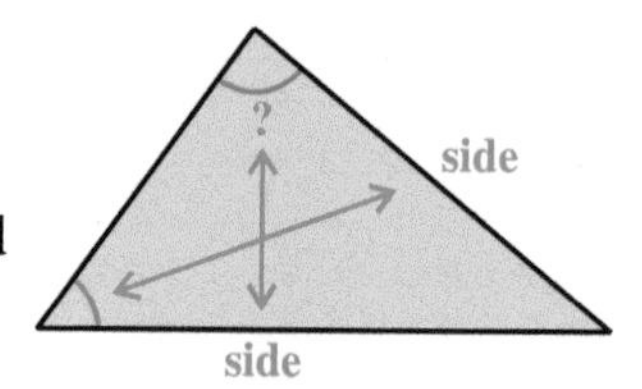

The Cosine Rule

In any triangle labelled ABC it can be proved that:

$a^2 = b^2 + c^2 - 2bc \cos A$

In a similar way, we can find the Cosine Rule for b^2 and c^2.

$b^2 = a^2 + c^2 - 2ac \cos B$

$c^2 = a^2 + b^2 - 2ab \cos C$

When using the Cosine Rule to find the size of an angle it is sometimes easier to rearrange the above formulae as:

$$\cos A = \frac{b^2 + c^2 - a^2}{2bc} \qquad \cos B = \frac{a^2 + c^2 - b^2}{2ac} \qquad \cos C = \frac{a^2 + b^2 - c^2}{2ab}$$

Using the Cosine Rule to find sides

To find a side using the Cosine Rule you need:

two sides of known length, **and**
the size of the angle between the known sides.

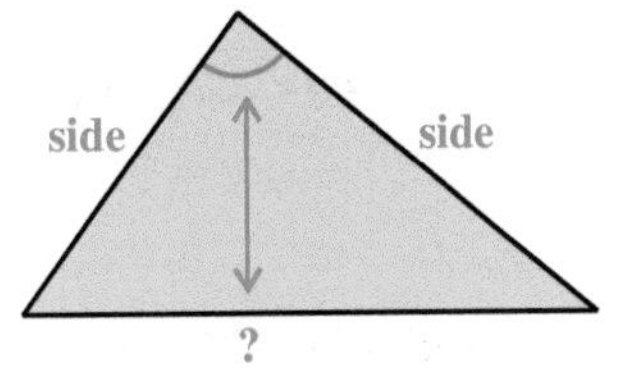

Using the Cosine Rule to find angles

To find an angle using the Cosine Rule you need:

three sides of known length.

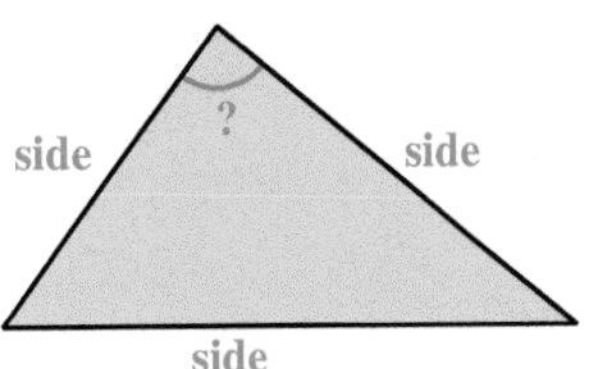

Solving problems involving triangles

Problems may involve using more than one skill.

Example 1

Find the unknown sides and angles in this triangle.

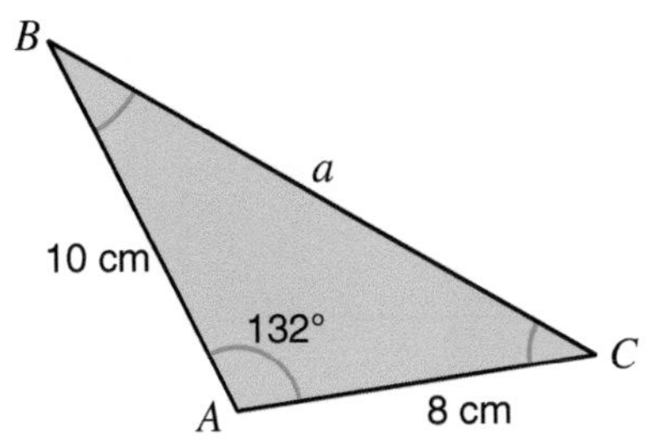

Using the Cosine Rule to find a.

$a^2 = b^2 + c^2 - 2bc \cos A$

$a^2 = 8^2 + 10^2 - 2 \times 8 \times 10 \times \cos 132°$

$a^2 = 271.06...$

$a = 16.46...$

Store this accurate value of a in the memory of your calculator.

$a = 16.5$ cm, correct to 3 s.f.

Now use the Sine Rule to find angle C.

$$\frac{\sin A}{a} = \frac{\sin B}{b} = \frac{\sin C}{c}$$

$$\frac{\sin 132°}{16.46...} = \frac{\sin B}{8} = \frac{\sin C}{10}$$

Using:

$$\frac{\sin 132°}{16.46...} = \frac{\sin C}{10}$$

$$\sin C = \frac{10 \times \sin 132°}{16.46...}$$

$\sin C = 0.451...$

$C = \sin^{-1}(0.451...)$

$C = 26.83...$

$C = 26.8°$, correct to 1 d.p.

$B = 180° - 26.8° - 132°$

$B = 21.2°$

Practice Exercise 24.1

1. (a) Find the unknown sides and angles in these triangles.
   (b) Calculate the area of each triangle.

(i)

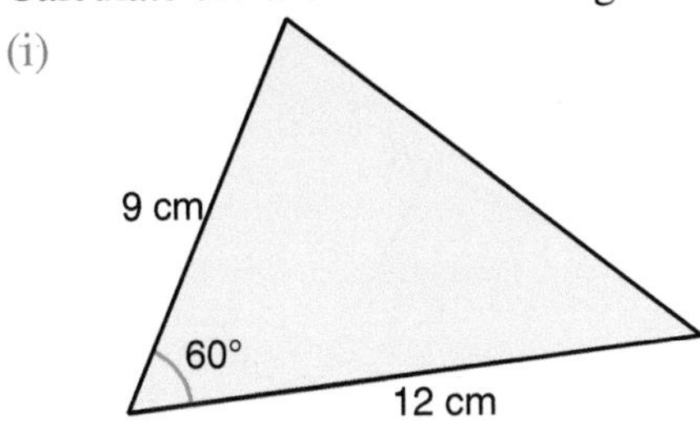

(ii)

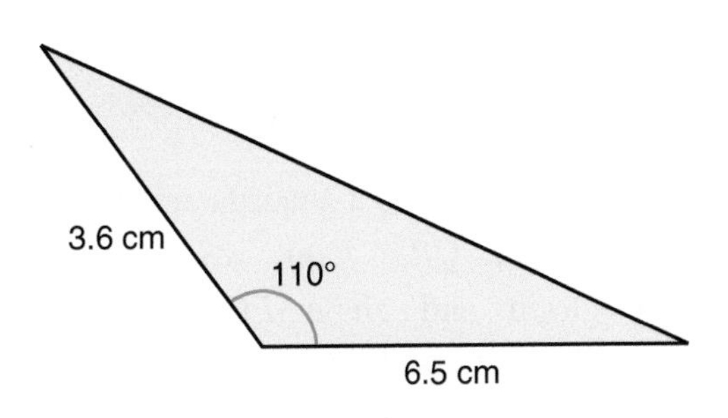

2. A gardener uses string and pegs to mark out a new flower bed. The string is 27 m long.
   (a) Calculate the size of angle QPR.
   (b) Calculate the area of the triangular flower bed.

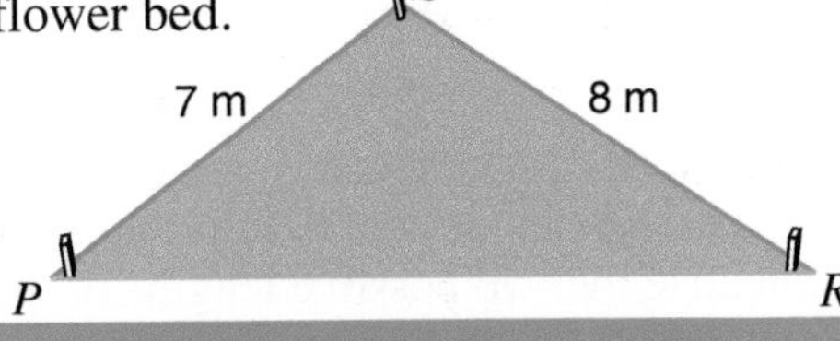

3. $ABCD$ is a quadrilateral with diagonal AC.
$AB = 38$ cm and $CD = 45$ cm.
Angle $BAC = 47°$, angle $ACB = 23°$ and angle $ACD = 78°$.
Calculate the length of AD.

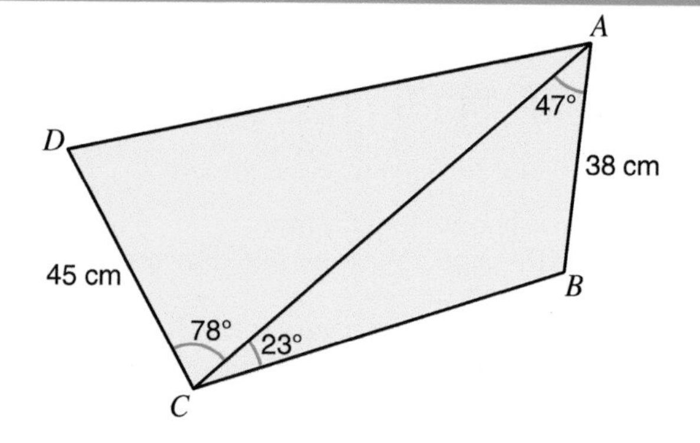

4. 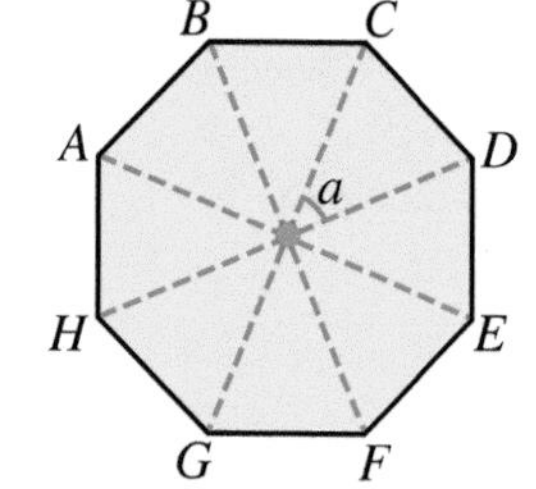

$ABCDEFGH$ is a regular octagon, centre O.
(a) Calculate the size of angle COD, marked a on the diagram.
(b) The length of each diagonal of the octagon is 20 cm.
Calculate the area of the octagon.

5. The diagram shows a tower.
At A the angle of elevation to the top of the tower is 43°.
At B the angle of elevation to the top of the tower is 74°.
The distance AB is 10 m.
Calculate the height of the tower.

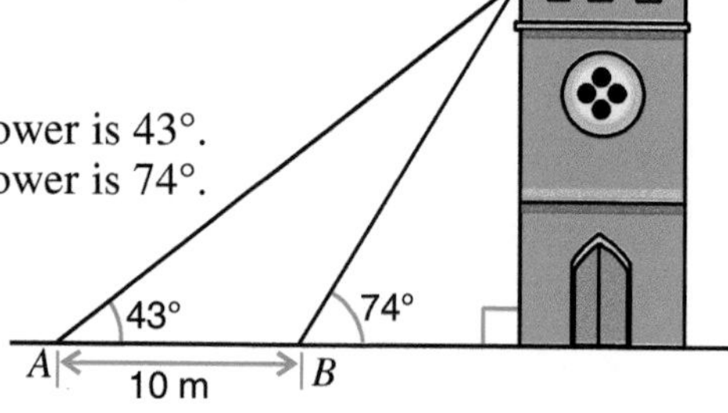

6. 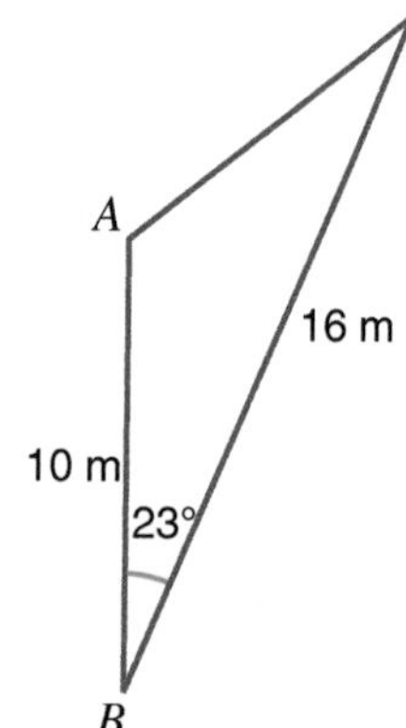

The diagram shows part of a steel framework, ABC.
B is on horizontal ground. A is 10 m vertically above B.
$BC = 16$ m. Angle $ABC = 23°$.
(a) Calculate the length of AC.
(b) Calculate angle BCA.
(c) Calculate the vertical height of C above B.

7. In triangle LMN, $MN = 7.5$ cm, angle $MLN = 104°$ and angle $LMN = 34°$.
(a) Calculate the length of LM.
(b) Calculate the area of the triangle.

8. The diagram shows a quadrilateral $PQRS$.
$PQ = 4$ cm, $QR = 5$ cm, $RS = 7$ cm, $QS = 8$ cm and angle $SPQ = 110°$.
(a) Calculate angle QSP.
(b) Calculate angle QSR.
(c) Calculate the area of $PQRS$.

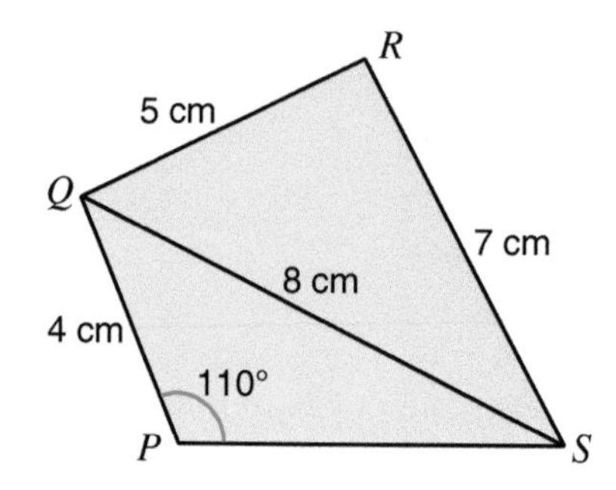

9. An isosceles triangle has an area of 25.6 cm^2.
The two equal sides are 8.4 cm long.
Calculate the two possible lengths of the third side.

Bearings

Bearings are used to describe the direction in which you must travel to get from one place to another.
They are measured from the North line in a clockwise direction.
A bearing can be any angle from 0° up to, but not including, 360° and is written as a three-figure number.

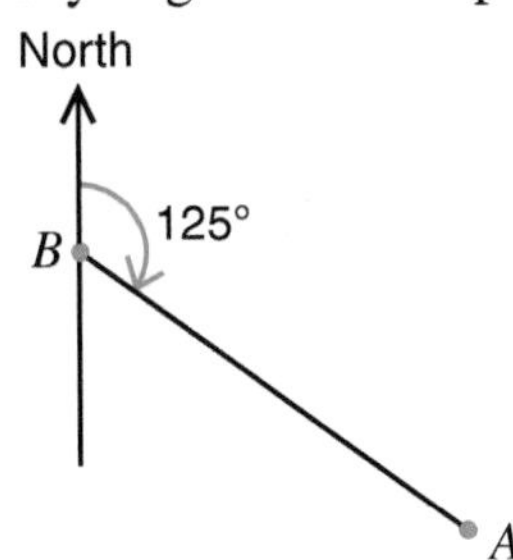

A is on a bearing of 125° from *B*.

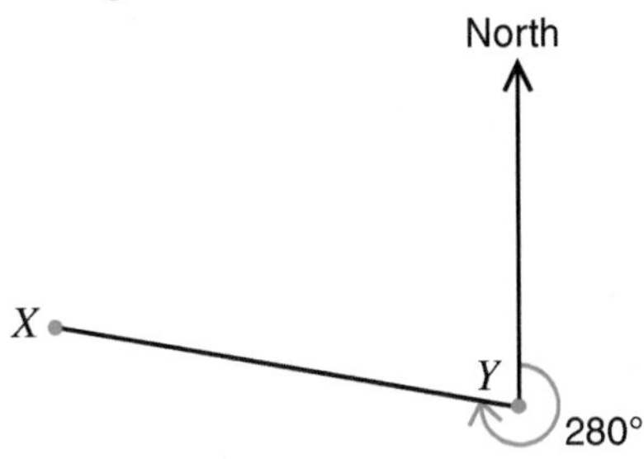

X is on a bearing of 280° from *Y*.

Example 2

A yacht sails on a bearing of 040° for 5000 m and then a further 3000 m on a bearing of 120°.
Find, by calculation, the distance of the yacht from its starting position.

When the yacht arrives at *B*, the line from *A* will be
180° − 40° = 140° anticlockwise from North.

Using the sum of angles at a point is 360°.

$\angle ABC = 360° - (140° + 120°)$

$\angle ABC = 360° - 260°$

$\angle ABC = 100°$

Using the Cosine Rule to find *AC*.

$AC^2 = 5000^2 + 3000^2 - 2 \times 5000 \times 3000 \times \cos 100°$

$AC^2 = 39\,209\,445.33...$

$AC = 6300$ m, correct to 2 s.f.

North
140°
120°
B
100°
3000 m
5000 m
C
North
040°
A

In a multi-step question, draw a sketch and add any information you find.

The distance of the yacht from its starting position is 6300 m, correct to 2 s.f.

Try to calculate the bearing on which it must sail to return directly to its starting position.

Practice Exercise 24.2

1. *L* is 19 km from *K* on a bearing of 068°.
   K is 13 km due North of *M*.
   Calculate the distance of *L* from *M*.

2. The diagram shows the positions of a tree, at *T*, and a pylon, at *P*, on opposite sides of a river.
   Two bridges cross the river at *X* and *Y*.

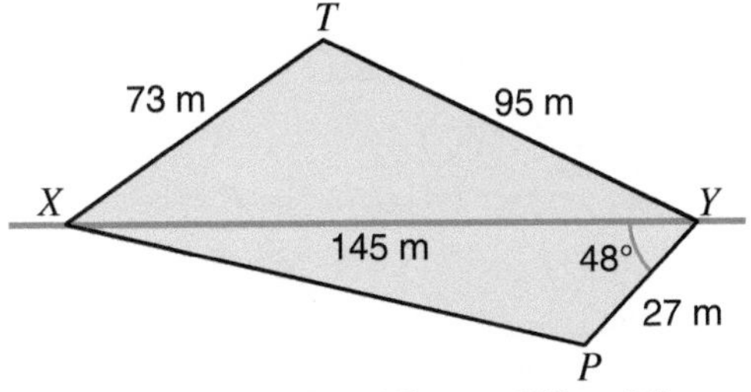

Remember:
The **longest side** is opposite the **largest angle**.
The **shortest side** is opposite the **smallest angle**.

$XY = 145$ m, $XT = 73$ m, $TY = 95$ m, $YP = 27$ m and $\angle XYP = 48°$.

(a) Calculate the angles of triangle *XTY*.
(b) Calculate *XP* and $\angle XPY$.

Bridge *X* is due west of bridge *Y*.

(c) What is the bearing of the tree, *T*, from the bridge, *X*?
(d) What is the bearing of the pylon, *P*, from the bridge, *Y*?

3. Two ships, P and Q, leave port at 2 pm.
P travels at 24 km/h on a bearing of 085°.
Q travels at 32 km/h on a bearing of 120°.
Calculate the distance between the ships at 2.30 pm.

4. A boat leaves port and and sails on a bearing of 144° for 4 km.
It then changes course and sails due East for 5 km to reach an island.
(a) Calculate the distance of the island from the port.
(b) Calculate the bearing of the island from the port.
(c) What is the bearing on which the boat must sail to return directly to the port?

5. An aircraft leaves an airport, at A, and flies on a bearing of 035° for 50 km and then on a bearing of 280° for a further 40 km before landing at an airport, at B.
(a) Calculate the distance between the airports.
(b) Find the bearing of B from A.
(c) Find the bearing of A from B.

6. 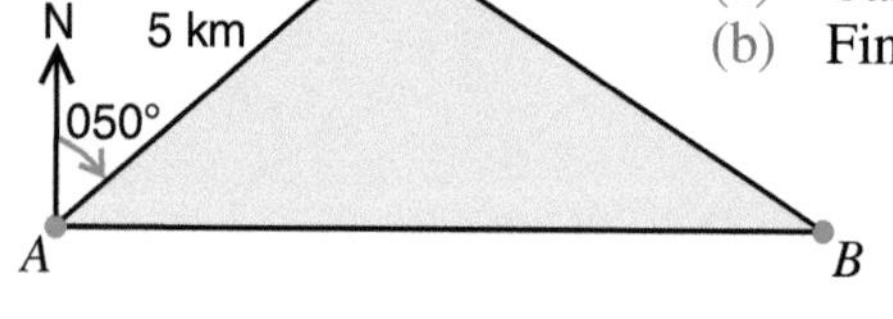

A fishing boat sails for 5 km on a bearing of 050° from port A.
It then turns and sails on a bearing of 120° to dock at port B.
Port B is due East of port A.
(a) Calculate the distance between ports A and B.
(b) Find how far the ship sailed altogether.

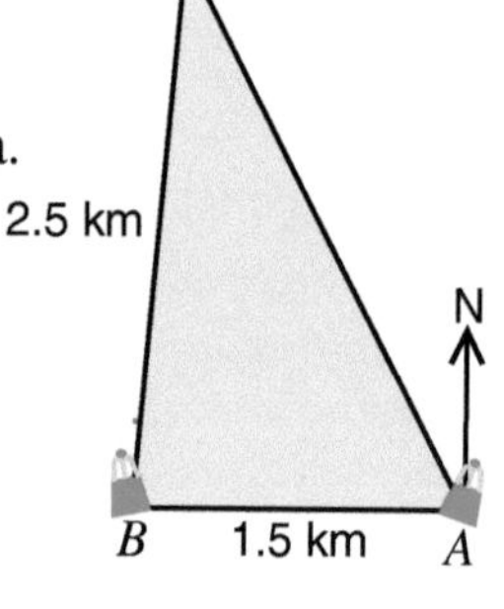

7. Buoys A, B and C are used to mark the course used in a sailing regatta.
B is 1.5 km due west of A.
C is on a bearing of 330° from A and is 2.5 km from B.
(a) Calculate the bearing of C from B.
(b) Calculate the shortest length of one circuit of the course.

8. A competitor in an orienteering contest travels 2000 m on a bearing of 330° and then 1000 m on a bearing of 260°.
(a) Draw a sketch of the route taken by the competitor.
(b) Find how far she has to return directly to her starting point and the bearing she must follow.

Key Points

- Solving problems involving triangles.

Right-angled triangles
Use the **trigonometric ratios** (sin, cos and tan) or **Pythagoras' Theorem.**

Triangles which are not right-angled

The Sine Rule

To find a **side** you need:

two angles of known size, **and**
the length of a side which is opposite one of the known angles.

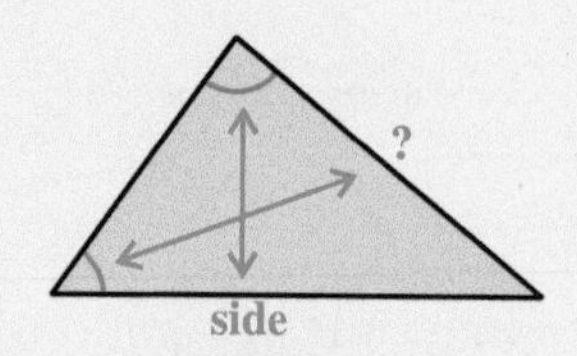

To find an **angle** you need:

the length of the side opposite the angle you are trying to find, **and**
the length of a side opposite an angle of known size.

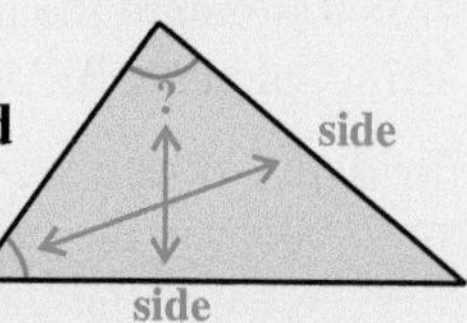

The Cosine Rule

To find a **side** you need:

two sides of known length, **and**
the size of the angle between the known sides.

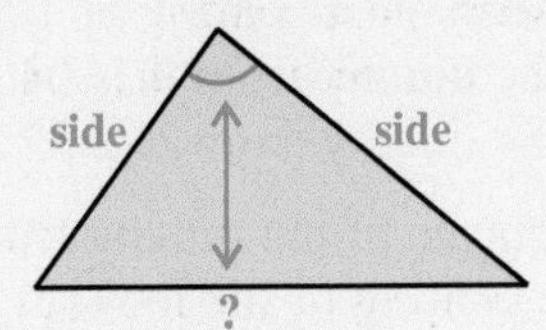

To find an **angle** you need:

three sides of known length.

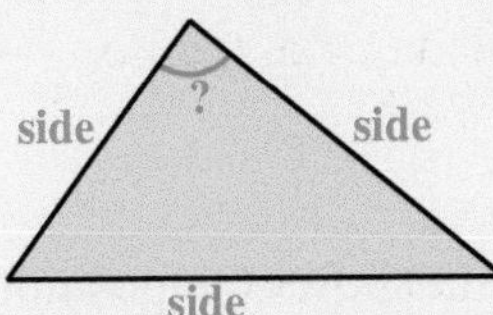

- **Compass points** and **three-figure bearings** are used to describe direction.
- A **bearing** is an angle measured from the North line in a clockwise direction.
- A bearing can be any angle from 0° up to, but not including, 360° and is written as a three-figure number.
- To find a bearing:

  measure angle a to find the bearing of Y from X,
  measure angle b to find the bearing of X from Y.

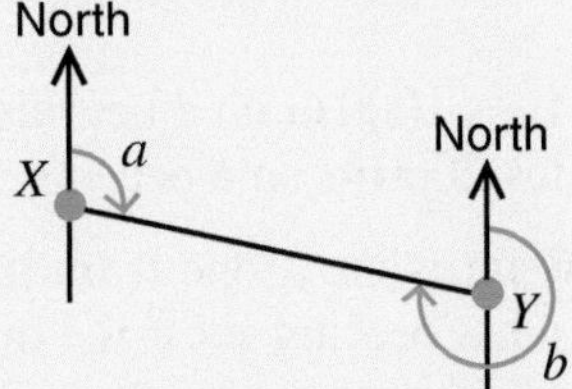

Review Exercise 24

1. (a) Find the unknown sides and angles in these triangles.
   (b) Calculate the area of each triangle.

   (i) 8 cm, 65°, 13 cm

   (ii)

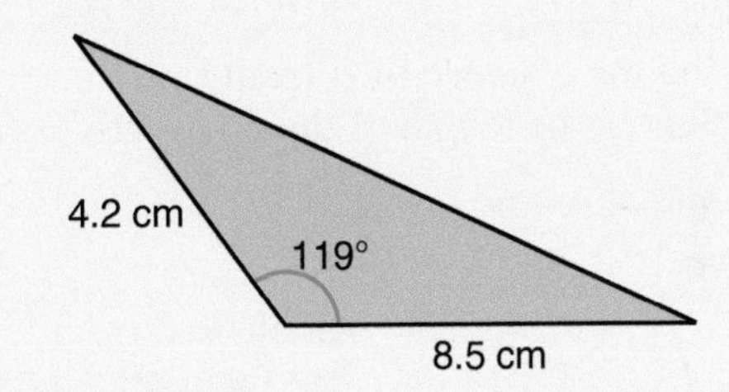

2. $ABCD$ is a tetrahedron.
   Triangle ABC is on a horizontal plane and BD is a vertical line.
   $BD = 12\,\text{cm}$, $AB = 16\,\text{cm}$ and $BC = 9\,\text{cm}$.
   Angle $ABC = 30°$.

   Calculate the length of:
   (a) AC, (b) AD, (c) CD.
   (d) Calculate the size of angle ADC.

   D, A, B, C

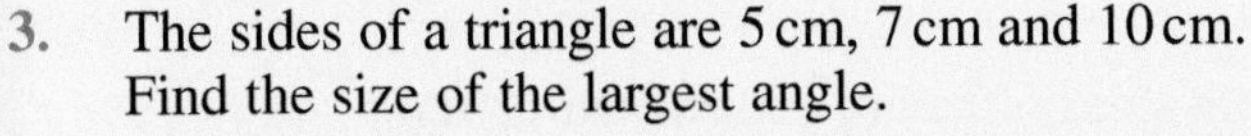

3. The sides of a triangle are 5 cm, 7 cm and 10 cm.
   Find the size of the largest angle.

4. In a sailing race the boats go round a triangular course, ABC, with $AB = 4\,\text{km}$, $BC = 5\,\text{km}$ and $CA = 6\,\text{km}$.
   B is due North of A.
   What is the bearing of C from B?

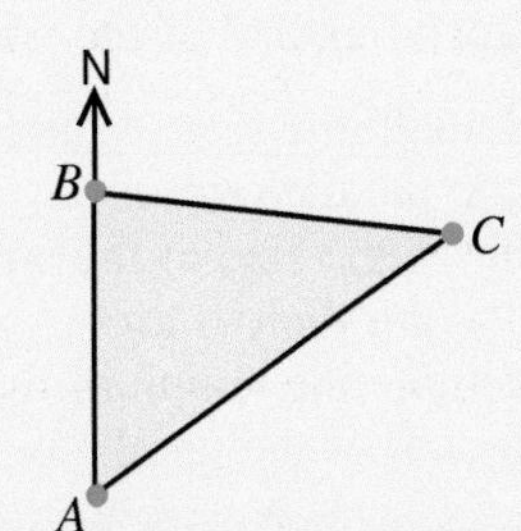

5. Two lookout stations, A and B, are 1500 m apart.
B is due East of A.
A dinghy, D, is seen out at sea.
The bearing of the dinghy from A is 042°.
The bearing of the dinghy from B is 325°.

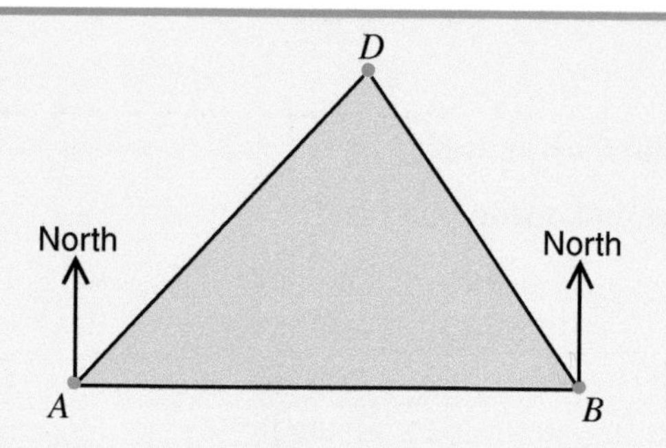

(a) Find the distance of the dinghy from A.
(b) What is the bearing of the lookout station A from the dinghy?

6. The diagram shows the positions of ships X, Y and Z.
X is 6.3 km due North of Z.
Y is 14.5 km from X on a bearing of 127°.
Calculate the distance and bearing of Z from Y.

North
127°
X
6.3 km
14.5 km
Z
Y

7. A drone flies 1000 m on a bearing of 070°.
It then flies 2000 m on a bearing of 150°.

(a) How far is the drone from its starting position?
(b) On what bearing must the drone fly to return to its starting position?

8. Two straight edges of a field meet at an angle of 80°.
A farmer wants to fence off a triangular section of the field which has an area of 10 000 m^2.
He makes a fence which starts at A, which is 125 m from C and finishes at B.

(a) Find the distance of B from C.
(b) What is the length of the fence, AB?

C
80°
B
125 m
A

9. An airfield is on level ground.
The main runway runs from East to West.
From E, the bearing of the control tower, C, is 320°.
From W, the bearing of the control tower is 020°.
The distance, EW, is 1800 m.

C
North
North
W
1800 m
E

(a) Calculate the distance of the control tower from E.
(b) Calculate the distance of the control tower from W.
(c) What is the shortest distance from the control tower to the runway?

10. A sign is in the shape of a parallelogram with sides of length 150 cm and 120 cm.
The longest diagonal of the parallelogram measures 240 cm.

(a) Calculate the size of the largest angle of the parallelogram.
(b) What is the size of the smallest angle of the parallelogram?
(c) Calculate the length of the shortest diagonal of the parallelogram.

25 Comparing Distributions

Statistics involves **data**.
Data is made up of a collection of **variables**.
Each variable can be described, numbered or measured.
Data can be displayed in a list, a table or a graph.
It can be **analysed** to make comparisons or to make decisions.

Analysing data

The **range** is a measure of **spread**. It measures how spread out the data is.

Range = Highest value − Lowest value

The range is influenced by extreme high or low values of data and can be misleading.
The **median** is the middle value (or mean of two middle values) when the data is arranged in order of size.

For n numbers, the rule for finding the median is:
Median = $\frac{1}{2}(n + 1)$th number.

A better way to measure spread is to find the range of the **middle 50%** of the data.
This is called the **interquartile range**.

Interquartile range = Upper quartile − Lower quartile
IQR = UQ − LQ

For n numbers the rules for finding the quartiles are:
Lower quartile = $\frac{1}{4}(n + 1)$th number.
Upper quartile = $\frac{3}{4}(n + 1)$th number.

The **semi-interquartile range** can also be used.

Semi-interquartile range = $\frac{1}{2}$ (Upper quartile − Lower quartile)
SIQR = $\frac{1}{2}$ (UQ − LQ)

Example 1

The number of posts made one day to social media by 8 friends is shown.

25 29 29 22 27 25 25 30

Find (a) the range, (b) the median.

(a) Range = Highest value − Lowest value.
= 30 − 22
Range = 8

(b) Arrange the data in order of size.

22 25 25 25 27 29 29 30

↑ Median

Where there is an even number of values, the median is the average of the middle two.

The middle value is $\frac{25 + 27}{2} = 26$
The median is 26.

Example 2

The number of text messages received by 7 students one day is shown.

4 6 3 4 7 8 5

Find (a) the median, (b) the interquartile range, (c) the semi-interquartile range.

(a) Arrange the data in order of size.

3 4 4 5 6 7 8

The median is the middle value.

Median = 5

(b)

3 4 4 5 6 7 8

LQ Median UQ

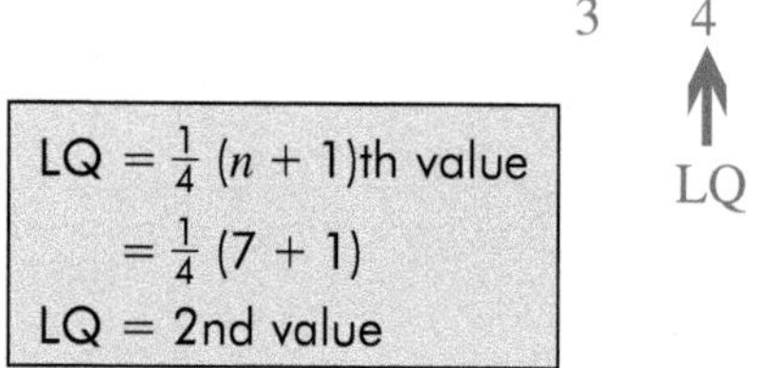
LQ = $\frac{1}{4}$ (n + 1)th value
= $\frac{1}{4}$ (7 + 1)
LQ = 2nd value

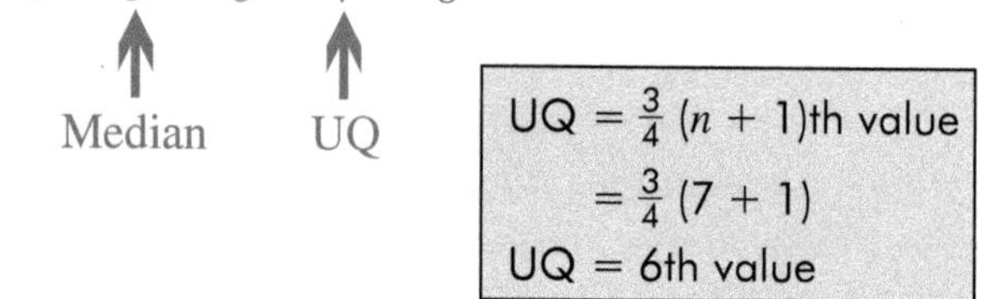
UQ = $\frac{3}{4}$ (n + 1)th value
= $\frac{3}{4}$ (7 + 1)
UQ = 6th value

Lower quartile = 4

Upper quartile = 7

Interquartile range = UQ − LQ

= 7 − 4

Interquartile range = 3

(c) Semi-interquartile range = $\frac{1}{2}$ (UQ − LQ)

SIQR = $\frac{1}{2}$ (7 − 4)

= $\frac{1}{2}$ (3)

= 1.5

Semi-interquartile range = 1.5

SIQR = $\frac{1}{2}$ IQR

Example 3

The table shows the boot sizes of players in a rugby team.

Boot size	8	9	10	11
Frequency	3	5	6	1

Find the semi-interquartile range.

LQ = 4th value.

The first 3 players wear size 8.

The next 5 wear boot size 9.

So, 4th player wears boot size 9.

UQ = 12th value.

The first 8 players wear sizes 8 and 9.

The next 6 players wear boot size 10.

So, 12th player wears boot size 10.

LQ = $\frac{1}{4}$ (n + 1)th value
= $\frac{1}{4}$ (15 + 1)
LQ = 4th value
UQ = $\frac{3}{4}$ (n + 1)th value
= $\frac{3}{4}$ (15 + 1)
UQ = 12th value

Semi-interquartile range = $\frac{1}{2}$ (UQ − LQ)

= $\frac{1}{2}$ (10 − 9)

= $\frac{1}{2}$ (1)

Semi-interquartile range = 0.5

Practice Exercise 25.1

1. Claire recorded the number of email messages she received each day last week.

   5 2 1 7 4 1 1

   Find the range and the median number of messages received each day.

2. Gail noted the number of stamps on parcels delivered to her office.

   3 2 4 5 7 3 4 3 5 3

   (a) Find the range of the number of stamps on a parcel.
   (b) Find the median number of stamps on a parcel.

3. The figures below show the length, in millimetres, of a sample of 7 leaves from an oak tree.

   142 132 139 125 143 128 151

   (a) Find the range of the data.
   (b) Find the median length of a leaf.
   (c) Find the lower quartile and upper quartile values.
   (d) Find the interquartile range of the data.
   (e) Find the semi-interquartile range of the data.

4. The prices paid for 11 different meals at a cafe are:

   £5 £7.50 £6 £8 £6.50 £9.50 £7 £10 £8.50 £12 £9

   Find the range, median and semi-interquartile range of the data.

5. A group of children were asked to estimate the weight of a bucket of water.
   Their estimates, in kilograms, are shown.

10	9	17.5	8	7.5
5	10	15	12.5	20
8	10	14	18	1

   (a) Find the range of the estimates.
   (b) Find the median estimate.
   (c) Find the semi-interquartile range of estimates.

6. Find the median and semi-interquartile range for the following sets of data.

   (a)

Number of letters delivered	1	2	3	4	5	6
Number of days	6	9	6	6	3	1

   (b)

Number of books read last month	0	1	2	3	4	5
Number of students	1	4	10	4	1	1

   (c)

Number of days absent in a year	0	1	2	3	4	5	6	7	8	9	10	11
Number of students	56	0	0	4	14	10	24	11	21	15	8	2

Comparing distributions

The **mean** is often used when comparing distributions.
The mean is found by finding the total of the values and dividing the total by the number of values.

$$\text{Mean} = \frac{\text{Total of all values}}{\text{Number of values}}$$

$$\text{Mean} = \frac{\text{Total of all values}}{\text{Number of values}} = \frac{\Sigma fx}{\Sigma f} \text{ for a frequency distribution.}$$

Example 4

The table shows the marks gained in a test.
Compare the marks obtained by the boys and the girls.

Mark (out of 10)	7	8	9	10
Number of boys	2	5	3	0
Number of girls	4	0	2	1

Use the range and the mean to compare the marks.

Boys: Range $= 9 - 7 = 2$
Girls: Range $= 10 - 7 = 3$

The girls had the higher range of marks.

Boys: Mean $= \frac{2 \times 7 + 5 \times 8 + 3 \times 9 + 0 \times 10}{10} = \frac{81}{10} = 8.1$

Girls: Mean $= \frac{4 \times 7 + 0 \times 8 + 2 \times 9 + 1 \times 10}{7} = \frac{56}{7} = 8$

The boys had the higher mean mark.

Overall the boys did better as: the girls' marks were more spread out with a lower average mark,
the boys' marks were closer together with a higher average mark.

Note: To compare the overall standard, the median could be used instead of the mean.

Practice Exercise 25.2

1. The weights, in grams, of a sample of 10 economy potatoes are shown.

70 76 83 86 95 98 113 117 122 130

(a) (i) What is the range of these weights?
(ii) Calculate the mean of these weights.

For a sample of 10 premium potatoes the range of their weights is 240 grams and the mean of their weights is 250 grams.

(b) Compare and comment on the weights of economy and premium potatoes.

2. The times, in minutes, taken by 7 boys to swim 50 metres are shown.

1.8 2.0 1.7 2.2 2.1 1.9 1.8

(a) (i) What is the range of these times?
(ii) What is the median time?
(iii) Find the semi-interquartile range of times.
(iv) Find the mean time.

The times, in minutes, taken by 7 girls to swim 50 m are shown.

2.1 1.9 1.8 2.3 1.6 2.0 2.6

(b) (i) What is the range of these times?
(ii) What is the median time?
(iii) Find the semi-interquartile range of times.
(iv) Find the mean time.

(c) Comment on the times taken by these boys and girls to swim 50 m.

3. Eleven people took part in a baking competition.
In Round 1 they had to bake a Victoria Sponge.
The judges awarded marks, out of 30, as follows.

24, 9, 23, 18, 15, 24, 15, 13, 19, 14, 18.

(a) Calculate the median mark awarded.
(b) Find the semi-interquartile range of the marks for Round 1.

In Round 2 the competitors had to bake a "Show Stopper".
For this round, the median mark was 23 and the semi-interquartile range was 2.

(c) Make **two** valid comparisons between the marks awarded by the judges in the two rounds.

4. Use the mean and the range to compare the number of visits to the cinema by these women and men.

Number of visits to the cinema last month	0	1	2	3	4	5	6	More than 6
Number of women	8	9	7	3	2	1	1	0
Number of men	0	12	7	1	0	0	0	0

5. Deepak thought that the girls in his class wore smaller shoes than the boys on average, but that the boys' shoe sizes were less varied than the girls'.
He did a survey to test his ideas. The table shows his results.

Shoe size	$4\frac{1}{2}$	5	$5\frac{1}{2}$	6	$6\frac{1}{2}$	7	$7\frac{1}{2}$	8	$8\frac{1}{2}$	9	$9\frac{1}{2}$
Number of boys	1	0	5	4	4	2	1	0	1	0	0
Number of girls	0	2	0	2	3	0	2	0	3	1	1

Was he correct? You must show your working.

Box plots

Box plots (or **box and whisker diagrams**) provide a useful way of representing the range, the median and the quartiles of a set of data.

They are also useful for comparing two (or more) distributions.

This box plot shows how the masses of fish are distributed.

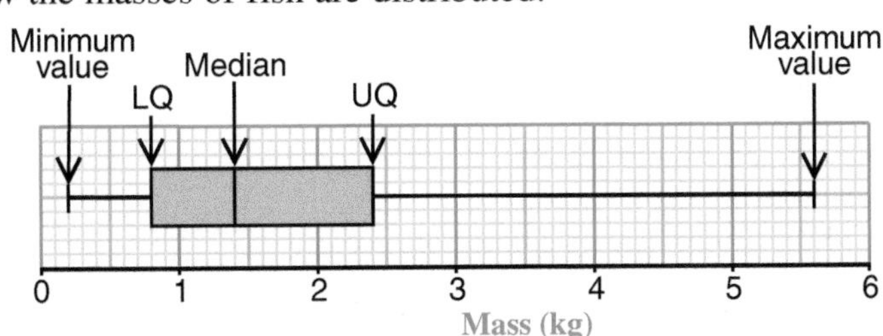

- the minimum mass is 0.2 kg,
- the maximum mass is 5.6 kg,
- the median mass is 1.4 kg,
- the lower quartile is 0.8 kg,
- the upper quartile is 2.4 kg,
- interquartile range is 2.4 kg − 0.8 kg = 1.6 kg,
- semi-interquartile range = $\frac{1}{2}$ (1.6 kg) = 0.8 kg.

The box plot shows how the masses of these fish are spread out and how the middle 50% are clustered.

Example 5

The number of points scored by the 11 players in a basketball team during a competition are shown.

13 12 5 34 8 10 11 46 25 23 4

Draw a box plot to illustrate the data.

Begin by putting the data in order and then locate the median, lower quartile and upper quartile.

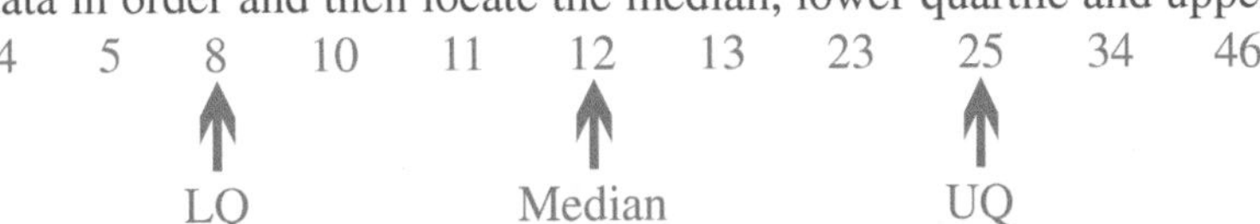

Then use these values to draw the box plot.

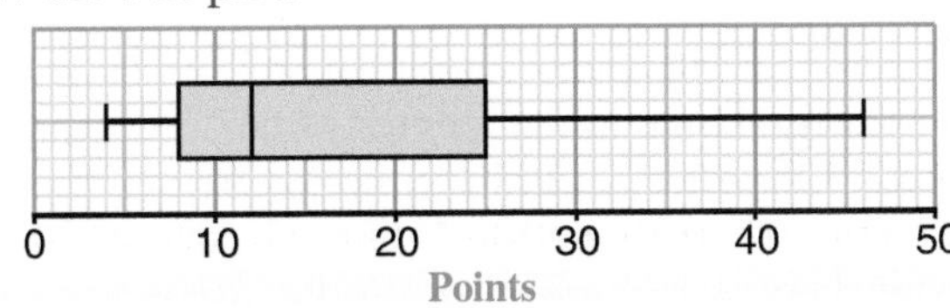

Practice Exercise 25.3

1. The midday temperatures, in °C, for 11 cities around the world are:

   9 12 20 24 25 28 28 30 31 32 35

   Draw a box plot to represent these temperatures.

2. The box plot illustrates the data obtained when a class of pupils were asked to guess the number of counters in a tin.

   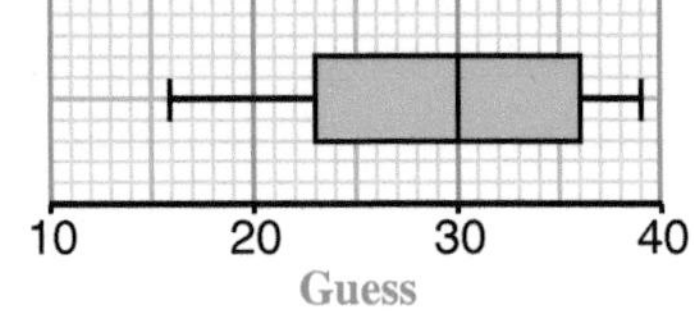

   Find: (a) the range,
   (b) the median,
   (c) the interquartile range.

3. 0 10 20 30 40
   Time (hundredths of a second)

   The box plot illustrates the reaction times of a group of people.
   (a) What was the minimum reaction time?
   (b) What is the value of the semi-interquartile range?

4. A group of students took examinations in Mathematics and English.
   The box plots illustrate the results.

   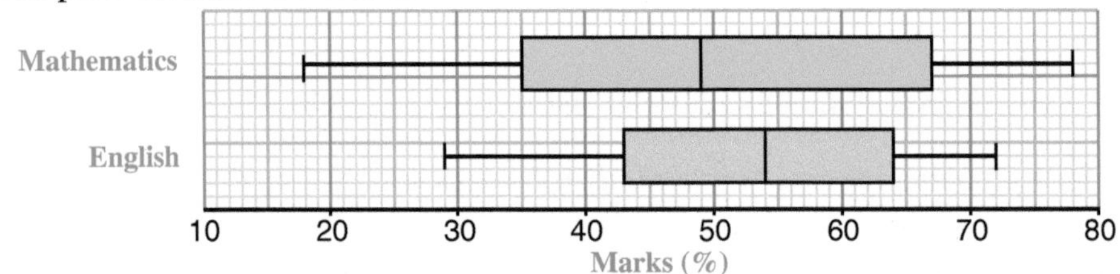

   (a) What was the highest mark scored in English?
   (b) What was the lowest mark scored in Mathematics?
   (c) Which subject has the higher median mark?
   (d) What is the value of the semi-interquartile range for English?
   (e) Comment on the results of these examinations.

5. Morag did a project to compare the times that males and females took to run 100 m.
   She summarised her data as follows.

   (a) Draw box plots to compare times for males and females.
   (b) Comment on the times for males and females.

	Males	Females
Slowest time	22 seconds	30 seconds
Fastest time	15 seconds	17 seconds
Lower quartile	18 seconds	23 seconds
Median	19 seconds	26 seconds
Upper quartile	20 seconds	27 seconds

6. Alan is comparing two varieties of runner beans that he is growing in his garden.
   He collected some runner beans from each variety, took measurements and then used his data to draw two box plots.

   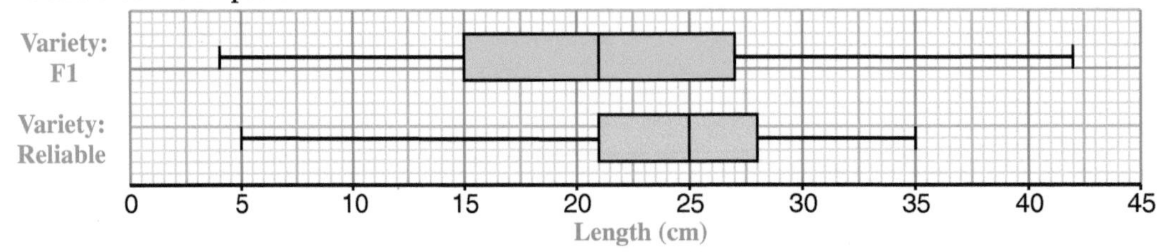

   (a) What variety of runner bean has the greater median?
   (b) Find the semi-interquartile range in the lengths of beans collected from each variety of plant.
   (c) Use your answer to compare the lengths of runner beans produced by each variety.

Standard deviation

You have already met three measures of spread.
The **range** measures the spread of all the data.
It is influenced by extreme high or low values of data and can be misleading.

Range = Highest value − Lowest value

The **interquartile range** measures the spread of the middle 50% of the data.

Interquartile range = Upper quartile − Lower quartile

Semi-interquartile range = $\frac{1}{2}$ (Upper quartile − Lower quartile)

A fourth measure of spread, called **standard deviation**, is more precise.
Standard deviation measures the average deviation (difference) of each piece of data from the mean.
Every piece of data is involved in the calculation of standard deviation.

> A **small** standard deviation from the mean indicates values are **clustered together**.
> A **large** standard deviation from the mean indicates values are **spread out**.

To calculate standard deviation, we use these formulae.

$$s = \sqrt{\frac{\Sigma(x - \bar{x})^2}{n - 1}} = \sqrt{\frac{\Sigma x^2 - (\Sigma x)^2/n}{n - 1}}$$

where n is the number of values in the sample,
$\bar{x}$ is the mean of the sample, $\bar{x} = \frac{\Sigma x}{n}$,
Σ means "the sum of",
e.g., Σx^2 means add together the squares of the data values.

Example 6

The number of texts sent each weekday was recorded.

10, 16, 11, 15, 13.

Find the mean and standard deviation for this data.

x	$x - \bar{x}$	$(x - \bar{x})^2$
10	−3	9
16	3	9
11	−2	4
15	2	4
13	0	0
$\Sigma x = 65$	$\Sigma(x - \bar{x}) = 0$	$\Sigma(x - \bar{x})^2 = 26$

> Set your working out in a table.
> **Notice that:**
> $\Sigma(x - \bar{x}) = 0$
> This is a useful check.

Mean $= \bar{x} = \frac{\Sigma x}{n} = \frac{65}{5} = 13$

$$s = \sqrt{\frac{\Sigma(x - \bar{x})^2}{n - 1}} = \sqrt{\frac{26}{5 - 1}} = \sqrt{\frac{26}{4}} = 2.549\ldots$$

Mean = 13
Standard deviation = 2.55, correct to 3 sig. figs.

Example 7

Find the mean and standard deviation of these two sets of results. Compare the groups.

Group A: 3, 7, 8, 5, 7.

For Group A: $n = 5$, $\Sigma x = 30$, $\Sigma x^2 = 196$

$$\bar{x} = \frac{\Sigma x}{n} = \frac{30}{5} = 6$$

$$s = \sqrt{\frac{\Sigma x^2 - (\Sigma x)^2/n}{n - 1}}$$

$$s = \sqrt{\frac{196 - 30^2/5}{5 - 1}}$$

$$s = 2$$

Group B: 10, 6, 9, 3.

For Group B: $n = 4$, $\Sigma x = 28$, $\Sigma x^2 = 226$

$$\bar{x} = \frac{\Sigma x}{n} = \frac{28}{4} = 7$$

$$s = \sqrt{\frac{\Sigma x^2 - (\Sigma x)^2/n}{n - 1}}$$

$$s = \sqrt{\frac{226 - 28^2/4}{4 - 1}}$$

$$s = 3.16$$

Group A: mean 6, standard deviation 2.
Group B: mean 7, standard deviation 3.16.
Group B did better overall (higher mean mark).
Group A were more consistent
(lower spread of marks).

Using a statistical calculator

A **statistical calculator** can be used to find the mean and standard deviation.

Example 8

Greg recorded the number of emails he received during a school week as:

5, 8, 11, 2, 4.

Use your calculator to find the mean and standard deviation.

Set calculator to SD mode. MODE 2

Clear statistical memory. SHIFT SCL =

Enter data. 5 DT 8 DT 1 1 DT 2 DT 4 DT

Mean SHIFT $\bar{x}$ = **Giving 6.**

Standard deviation SHIFT $x\sigma_{n-1}$ = **Giving 3.54.**

If your calculator works in a different way, refer to the instruction booklet supplied with the calculator or ask someone for help.

Practice Exercise 25.4

1. The exam marks of two groups of students were recorded as:

   Group 1: 44, 56, 53, 49. **Group 2**: 41, 72, 65, 28, 54.

   Use the mean and standard deviation to compare the marks of the two groups.

2. The temperatures, in °F, were recorded over 5 days.

   72, 68, 65, 60, 55.

   (a) Use the formula $s = \sqrt{\dfrac{\Sigma(x - \bar{x})^2}{n - 1}}$ to calculate the standard deviation.

   (b) Show that the formula $s = \sqrt{\dfrac{\Sigma x^2 - (\Sigma x)^2/n}{n - 1}}$ gives the same answer as part (a).

   (c) Use your calculator to find the standard deviation.

3. Calculate the mean and standard deviation of the following.
   (a) 3, 5, 6, 2.
   (b) 27, 38, 29, 31, 35.
   (c) 108 cm, 120 cm, 97 cm, 105 cm, 118 cm, 112 cm.
   (d) 1.2 kg, 3.2 kg, 3.9 kg, 1.7 kg.

4. The total amounts from sales of a product were recorded over two 5-week periods before and after Christmas.

	Week 1	Week 2	Week 3	Week 4	Week 5
Before Christmas	57	74	92	112	124
After Christmas	83	92	87	90	85

   Use the mean and standard deviation to compare sales in the two 5-week periods.

5. The students in two classes had a spelling test.
   Use the mean and standard deviation to compare the marks scored.

Class A	4	6	5	5	4	5	4	4	8	7
	4	5	7	5	8	7	4	8	4	5
	5	6	5	8	8	8	5	4	5	7
Class B	6	7	7	8	7	7	7	5	7	6
	7	7	7	9	7	7	7	8	7	7
	7	8	7	9	7					

6. A round of golf is made up of 18 holes.
The fewer the number of shots taken, the better the score.
The score cards of two golfers are shown.

Chris								
4	5	4	4	5	3	7	5	8
6	4	5	4	8	5	5	7	4

Robert								
3	4	5	3	4	2	4	3	5
3	3	4	4	4	3	4	3	4

Use the mean and standard deviation to compare the performance of the two golfers.

Key Points

- The **range** is a measure of **spread**.

Range = Highest value − Lowest value

- The **median** is the middle value (or the mean of the two middle values) when the values are arranged in order of size.

For n numbers, the rule for finding the median is:
Median $= \frac{1}{2}(n + 1)$th number.

- The **interquartile range** measures the spread of the middle 50% of the data.

Interquartile range = Upper quartile − Lower quartile
IQR = UQ − LQ

For n numbers the rules for finding the quartiles are:
Lower quartile $= \frac{1}{4}(n + 1)$th number.
Upper quartile $= \frac{3}{4}(n + 1)$th number.

- The **semi-interquartile range** can be used to measure the spread of data.

Semi-interquartile range $= \frac{1}{2}$ (Upper quartile − Lower quartile)
SIQR $= \frac{1}{2}$ (UQ − LQ)

- The **mean** can be used to compare distributions.

$$\text{Mean} = \frac{\text{Total of all values}}{\text{Number of values}}$$

$$\text{Mean} = \frac{\text{Total of all values}}{\text{Number of values}} = \frac{\Sigma fx}{\Sigma f} \quad \text{for a frequency distribution.}$$

- A **box plot** is used to represent the range, the median and the quartiles of a distribution.

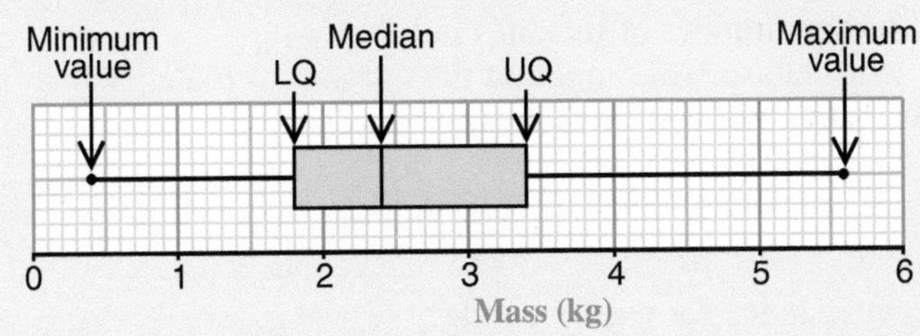

- A box plot shows how the data is spread out and how the middle 50% of data is clustered.
- Box plots can be used to compare two (or more) distributions.

- **Standard deviation** is the most widely used measure of spread in statistics. It takes into account all of the data values. The value of the standard deviation indicates the spread of the values from the mean. The values of the mean and standard deviation taken together can be used to compare two sets of data.

A **small** standard deviation from the mean indicates values are **clustered together**.
A **large** standard deviation from the mean indicates values are **spread out**.

$$s = \sqrt{\frac{\Sigma(x - \bar{x})^2}{n - 1}} = \sqrt{\frac{\Sigma x^2 - (\Sigma x)^2/n}{n - 1}}$$

where n is the number of values in the sample,

$\bar{x}$ is the mean of the sample, $\bar{x} = \frac{\Sigma x}{n}$,

Σ means "the sum of",
e.g., Σx^2 means add together the squares of the data values.

- A statistical calculator can be used to find the mean and standard deviation.

Review Exercise **25**

1. Frank counted the number of books on different shelves in a library. He recorded the following numbers.

   38 40 26 49 37 43 35

   (a) Find the median number of books on a shelf.
   (b) What is the range of the number of books on a shelf?
   (c) Calculate the mean number of books on a shelf. Give your answer correct to one decimal place.
   (d) Find the lower quartile and the upper quartile for the data.
   (e) Calculate the interquartile range.
   (f) Find the semi-interquartile range for the data.

2. The charges quoted by 7 companies for delivering and installing a cinema television system are shown below.

   £85 £45 £50 £110 £60 £60 £70

   (a) Find the median charge.
   (b) Find the upper quartile and lower quartile charges.
   (c) Find the interquartile range of charges.
   (d) Find the semi-interquartile range of charges.

3. Boxes of matches sold by a company claim to contain roughly 48 matches. A sample of 15 boxes was taken from the production line. The number of matches in each box was counted and recorded.

   47 48 49 46 45 48 49 52
   51 49 48 49 50 51 41

   (a) Calculate the median number of matches in a box.
   (b) Calculate the semi-interquartile range of the number of matches in a box.
   (c) Use your answers to comment on the claim that each box contains roughly 48 matches.

4. The attendances at meetings of a club for metal detectorists over a 5-week period were:

   10 11 13 13 8

   (a) Find the mean attendance for each meeting.
   (b) After the next meeting the club secretary reported the mean attendance for the last 6 meetings as 12. How many people attended the last meeting?

5. The box plot shows information about the price of a can of cola in a number of shops.

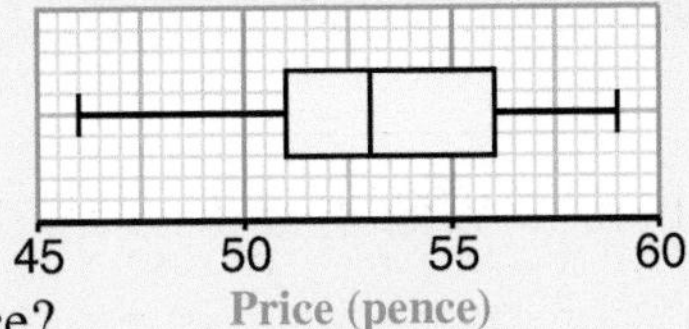

(a) What is the range in price?
(b) What is the median price?
(c) What is the value of the interquartile range?
(d) What is the value of the semi-interquartile range?

6. The times taken by students to access a website gave the following information.

Maximum time	5.1 minutes
Minimum time	1.5 minutes
Median time	3.8 minutes
Lower quartile	2.8 minutes
Upper quartile	4.2 minutes

Draw a box plot to represent this information.

7. A general knowledge test is marked out of 50.
The results for a group of people are shown.

30 25 38 45 36 40 27 43 39

Calculate the mean and standard deviation of these results.

8. The times, in seconds, taken by some boys and some girls to swim one length of a pool are shown.

Boys: 28.3 25.6 29.4 26.5 32.7 27.3 26.2 24.8
Girls: 33.3 29.7 32.5 29.4 30.6 33.2

Use the mean and standard deviation to compare the swimming times of the boys and girls.

9. Nesta notes the price of a loaf of bread sold in 5 different shops.

£1.20 £1.25 £1.40 £1.28 £1.35

Calculate the standard deviation of the prices.

10. Laura analysed this set of data.

2 5 6 9 10 16

She found the data had a mean of 8 and standard deviation of 4.9.
(a) Every number in the set of data had 6 added to it.
What is the mean and standard deviation of the new set of data?
(b) Every number in the original set of data was multiplied by 5.
What is the mean and standard deviation of the new set of data?

Craig analysed this set of data.

12 15 16 19 20 26

(c) Compare this set of data with Laura's original set of data.
State the mean and standard deviation for Craig's set of data.

11. (a) The heart rates, in beats per minute, of 6 athletes are:

61 45 48 52 53 49

Calculate the mean and standard deviation of this data.
(b) The heart rates, in beats per minute, of 6 sedentary adults have a mean of 65 beats per minute and a standard deviation of 7.6 beats per minute.
Make **two** comparisons between the heart rates of the athletes and the sedentary adults surveyed.

26 Using Scatter Graphs

When we investigate statistical information, we often find there are connections between sets of data, for example, height and weight.
In general, taller people weigh more than shorter people.

The table shows the weights and heights of 10 boys.

Weight (kg)	36.5	38.0	38.5	39.5	40.0	41.0	42.5	42.5	44.0	44.0
Height (cm)	123	124	127	130	136	136	135	140	142	146

To see if there is a connection between two sets of data we can plot a **scatter graph**.
The scatter graph below shows the data given in the table.

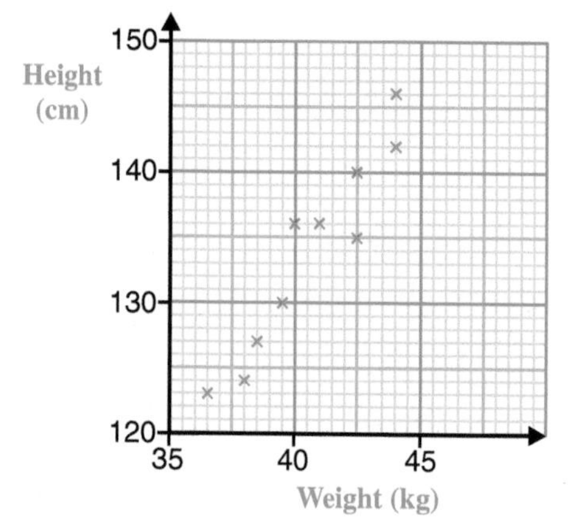

Each cross plotted on the graph represents the weight and height of one boy.

The diagram shows that taller boys generally weigh more than shorter boys.

Correlation

The relationship between two sets of data is called **correlation**.

In general the scatter graph of the heights and weights shows that as height increases, weight increases. This type of relationship shows there is a **positive correlation** between height and weight.

But if as the value of one variable increases the value of the other variable decreases, then there is a **negative correlation** between the variables.

When no linear relationship exists between two variables there is **zero correlation**.
This does not necessarily imply "no relationship", but merely "no linear relationship".

The following graphs show types of correlation.

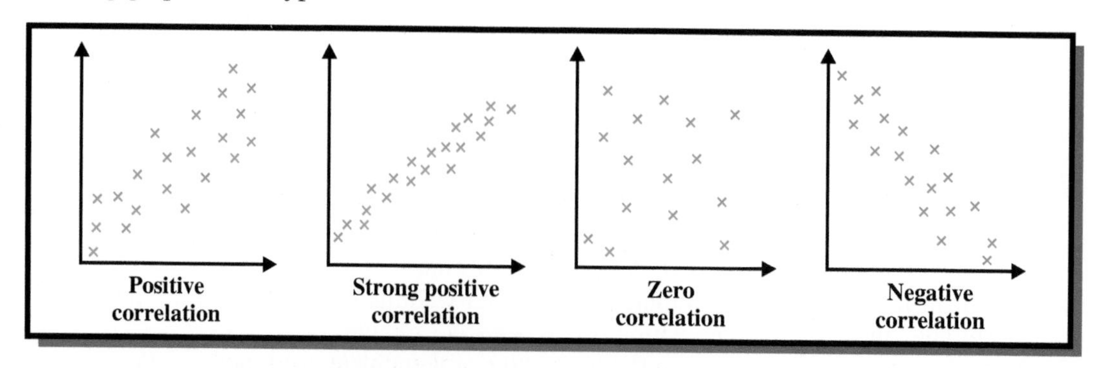

As points get closer to a straight line the stronger the correlation.
Perfect correlation is when all the points lie on a straight line.

Practice Exercise 26.1

1. Describe the type of correlation you would expect between:
   (a) the age of a car and its secondhand selling price,
   (b) the heights of children and their ages,
   (c) the shoe sizes of children and the distance they travel to school,
   (d) the number of cars on the road and the number of road accidents,
   (e) the engine size of a car and the number of kilometres it can travel on one litre of fuel.

2. (a) Which of these graphs shows the strongest positive correlation?
   (b) Which of these graphs shows perfect negative correlation?
   (c) Which of these graphs shows the weakest correlation?

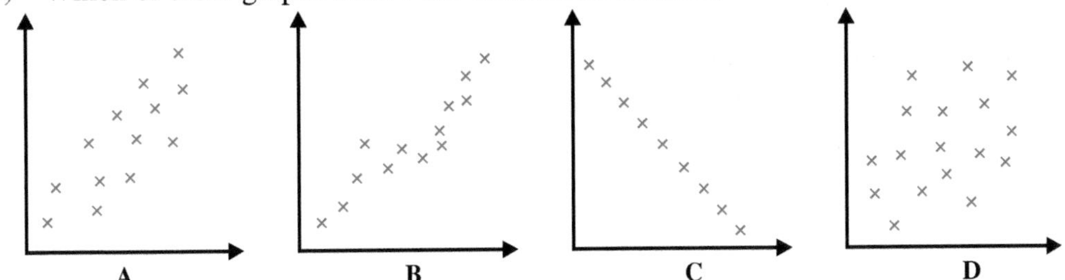

3. The scatter graph shows the shoe sizes and heights of a group of girls.

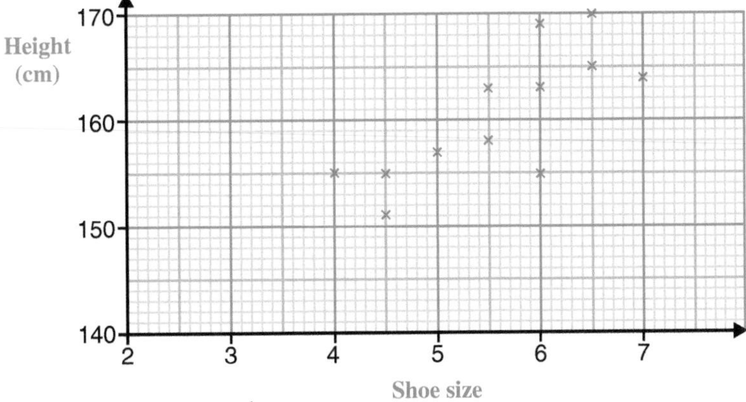

   (a) How many girls wear size $6\frac{1}{2}$ shoes?
   (b) How tall is the girl with the largest shoe size?
   (c) Does the shortest girl wear the smallest shoes?
   (d) What do you notice about the shoe sizes of taller girls compared to shorter girls?

4. The table shows the distance travelled and time taken by motorists on different journeys.

Distance travelled (km)	30	45	48	80	90	100	125
Time taken (hours)	0.6	0.9	1.2	1.2	1.3	2.0	1.5

   (a) Draw a scatter graph for the data.
   (b) What do you notice about distance travelled and time taken?
   (c) Give one reason why the distance travelled and the time taken are not perfectly correlated.

5. Tyres were collected from a number of different cars.
   The table shows the distance travelled and depth of tread for each tyre.

Distance travelled (1000 km)	4	5	9	10	12	15	18	25	30
Depth of tread (mm)	9.2	8.4	7.6	8	6.5	7.4	7	6.2	5

   (a) Draw a scatter graph for the data.
   (b) What do you notice about the distance travelled and the depth of tread?
   (c) Explain how you can tell that the relationship is quite strong.

Line of best fit

When there is a relationship between two sets of data, a **line of best fit** can be drawn on the scatter graph.
Then, using the methods introduced in Chapter 9, we can find the equation of the line of best fit.
The equation can then be used to estimate the value from one set of data when the corresponding value of the other set is given.

Example 1

The table shows the weights and heights of 10 boys.

Weight (kg)	36.5	38.0	38.5	39.5	40.0	41.0	42.5	42.5	44.0	44.0
Height (cm)	123	124	127	130	136	136	135	140	142	146

(a) Draw a scatter graph for the data.
(b) Draw a line of best fit.
(c) Find the equation of the line of best fit.
(d) Use your equation to estimate the height of a boy weighing 42 kg.

(a) (b)

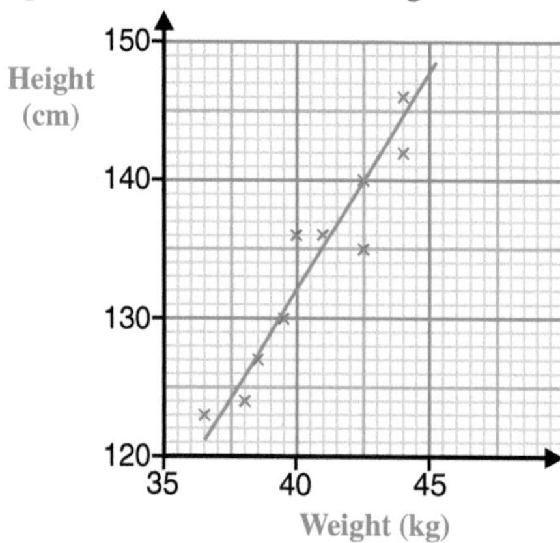

Draw the best line you can.
Place your ruler on the graph and try it in various positions.
Aim for a slope which matches the general slope of the points, where the points are balanced with some on both sides of the line.
The line of best fit does not have to go through the origin of the graph.
The **slope** of the line of best fit shows the **trend** of data.

(c) The equation is of the form $y - b = m(x - a)$.

First, find the gradient, m, of the line.

$$m = \frac{y_2 - y_1}{x_2 - x_1}$$

The line passes through the points (37.5, 124) and (42.5, 140).

$$m = \frac{140 - 124}{42.5 - 37.5} = \frac{16}{5} = 3.2$$

From the graph, the line passes through the point (40, 132).
Substitute $m = 3.2$, $a = 40$ and $b = 132$ into $y - b = m(x - a)$.

$y - 132 = 3.2(x - 40)$
Remove the brackets.
$y - 132 = 3.2x - 128$
Add 132 to both sides.
$y = 3.2x + 4$

If the line of best fit can be extended to the y axis, an equation of the form $y = mx + c$ can be used.
The value of c could be read from the y-intercept.

(d) When $x = 42$.
$y = 3.2(42) + 4$
$y = 134.4 + 4$
$y = 138.4$
An estimate of the height of a boy weighing 42 kg is 138 cm.

Practice Exercise 26.2

1. The scatter diagram shows the heights, h cm, of some plants, d days after germinating.

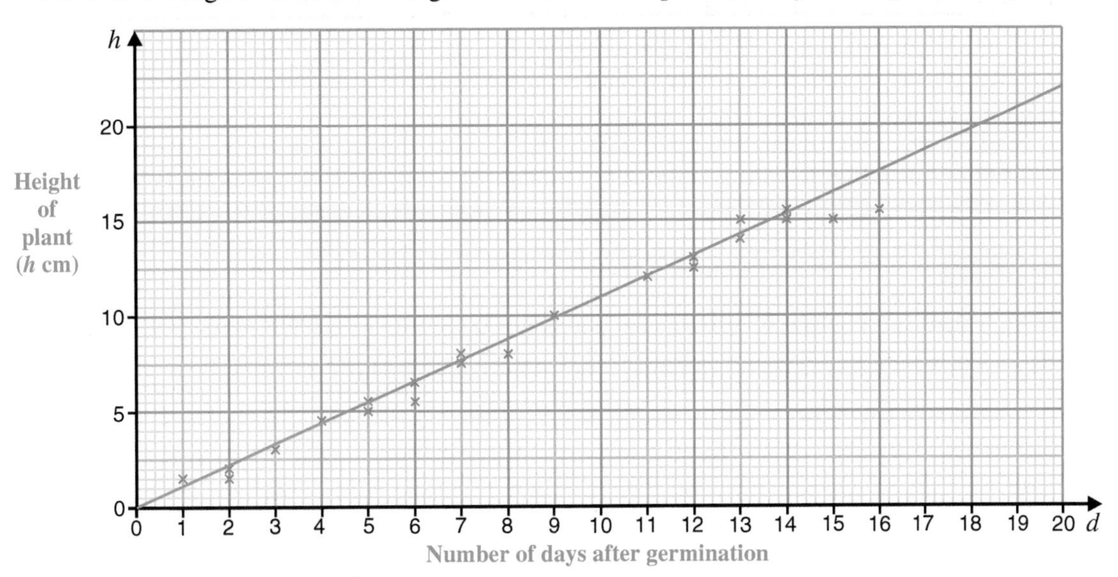

A line of best fit has been drawn on the diagram.

(a) Find the gradient of the line.

The equation of the line of best fit is in the form $h = md + c$.

(b) Explain why $c = 0$.
Write down the equation of the line of best fit.

(c) Use your equation to find the value of h when $d = 10$.

2. Two sets of related values, x and y, are recorded in a table.

x	10	20	30	40	50	60	70
y	2	3	12	25	33	36	47

(a) Draw a scatter graph to show the data.
Draw and label the x axis horizontally from 0 to 70.
Draw and label the y axis vertically from 0 to 50.

(b) Draw a line of best fit.

(c) Find an equation for your line of best fit.

(d) Find the value of y when $x = 45$.

(e) Find the value of x when $y = 47$.

3. The following table gives the marks obtained by some candidates taking examinations in French and German.

Mark in French	53	35	39	53	50	59	36	43
Mark in German	64	32	44	70	56	68	40	48

(a) (i) Use this information to draw a scatter graph.
(ii) Draw the line of best fit by eye.

(b) Find the equation of your line of best fit.

(c) Use your equation to estimate:
(i) the mark in German for a candidate who got 70 in French,
(ii) the mark in French for a candidate who got 58 in German.

(d) Which of the two estimates in (c) is likely to be more reliable?
Give a reason for your answer.

4. The table shows the weights and fitness factors for a number of women.
The higher the fitness factor the fitter a person is.

Weight (kg)	45	48	50	54	56	60	64	72	99	112
Fitness Factor	41	48	40	40	35	40	34	30	17	15

(a) Use this information to draw a scatter graph.
(b) What type of correlation is shown on the scatter graph?
(c) Draw a line of best fit.
(d) Find an equation for the line of best fit.
(e) Use your equation to estimate the fitness factor of a woman whose weight is 80 kg.

Key Points

- A **scatter graph** can be used to show the relationship between two sets of data.
- The relationship between two sets of data is referred to as **correlation.**
- You should be able to recognise **positive** and **negative** correlation.
  The correlation is stronger as points get closer to a straight line.

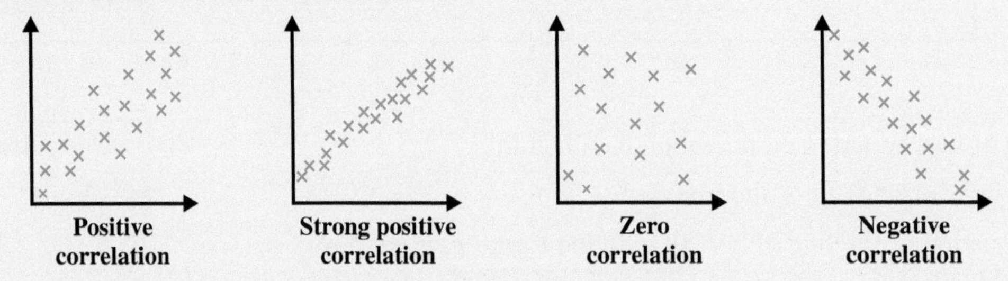

- **Perfect correlation** is when all the points lie on a straight line.
- When there is a relationship between two sets of data, a **line of best fit** can be drawn on the scatter graph.
  The **slope** of the line of best fit shows the **trend** of the data.
- The line of best fit can be used to **estimate** the value from one set of the data when the corresponding value of the other set is known.
- You should be able to find the equation of the line of best fit.
  The equation will generally be of the form $y - b = m(x - a)$.
  If the line cuts the y axis, it may be easier to use $y = mx + c$.

Review Exercise 26

1. The data from a survey of cars was used to plot these scatter graphs.

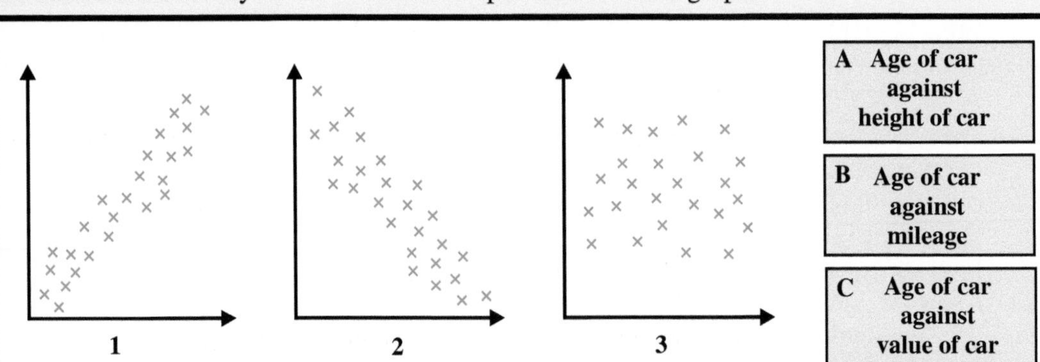

Match each scatter graph to the correct description.

2. In each of the following cases, decide if the possible correlation is positive, zero or negative.
(a) The outside temperature and the amount of electricity used in a house.
(b) The distances students live from school and their journey times to school.
(c) The marks obtained by pupils in a test and the house numbers where they live.

3. The scatter graph shows the marks obtained by 15 students in an exam that consisted of two papers. The line of best fit has been drawn.

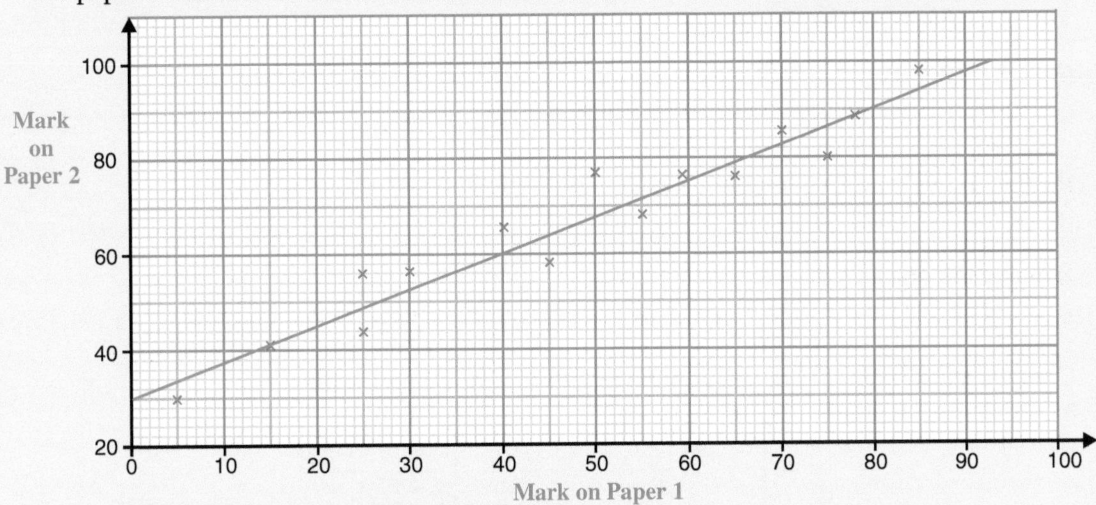

(a) Find the equation of the line of best fit.
(b) A student scored 60 on Paper 1 but was absent for Paper 2.
Use your equation to obtain an estimated mark for Paper 2.

4. The table shows the ages and weights of ten babies.

Age (weeks)	2	4	9	7	13	5	6	1	10	12
Weight (kg)	3.5	3.3	4.2	4.7	5	3.8	4	3	5	5.5

(a) Use this information to draw a scatter graph.
(b) What type of correlation is shown on the scatter graph?
(c) Draw a line of best fit.
(d) Find an equation for your line of best fit.
(e) Use your equation to estimate the weight of an 8-week-old baby.

5. The heights of 10 boys and their fathers are given in this table.

Height of father (cm)	167	168	169	171	172	172	174	175	176	182
Height of son (cm)	164	166	166	168	169	170	170	171	173	177

(a) Use this information to draw a scatter graph.
(b) Draw a line of best fit and find its equation.
(c) Use your equation to estimate the height of a boy whose father is 1.7 m tall.

6. A farmer carried out an experiment by treating seven fields with different amounts of fertiliser.

Units of fertiliser applied per square metre	0	1	2	3	4	5	6
Crop yield (kg per square metre)	1.4	1.6	1.7	2.0	2.1	2.3	2.3

(a) Draw a scatter graph to show the data.
(b) Draw a line of best fit and find its equation.
(c) Use your equation to estimate the crop yield if 2.5 units of fertiliser are applied per square metre.
(d) Explain why it might not be appropriate to use your equation to estimate the crop yield if 12 units of fertiliser are applied per square metre.

Revision Exercise 1 Non-calculator Paper

Do not use a calculator for this exercise.

1. Which is greater, 3^5 or 2^8?
   You must show your working.

2. Find the value of:
   (a) $2^2 \times 3$ (b) $(2 \times 3^2)^2$ (c) $2^4 \times 5^2$ (d) $4^3 \div 2^2$

3. If $p = \frac{1}{4}$, $q = \frac{1}{3}$, $r = \frac{2}{3}$ and $s = \frac{5}{6}$, calculate the value of:
   (a) $r + s$ (b) $q - p$ (c) qs (d) $s \div r$ (e) $pr + s$ (f) $\dfrac{r}{q + s}$

4. Remove the brackets.
   (a) $y^{\frac{1}{2}}\left(y^{\frac{1}{2}} + y\right)$ (b) $x^2(x^2 - x^{-2})$ (c) $x^{\frac{1}{3}}\left(x^{\frac{2}{3}} + x^{\frac{1}{3}}\right)$

5. Computer repair charges depend on the length of time taken for the repair, as shown on the graph.

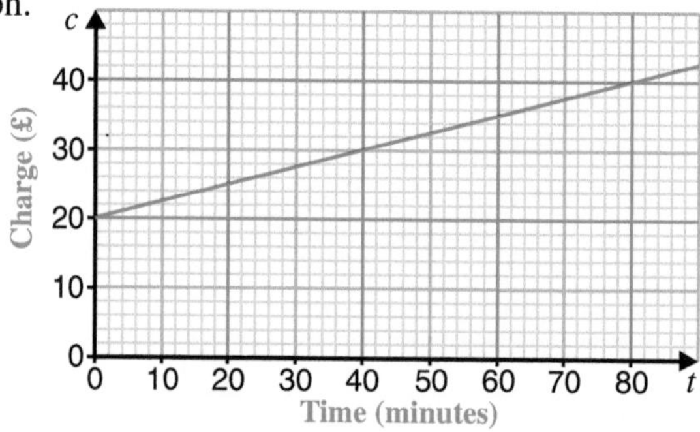

   The charge is made up of a fixed amount plus an extra amount which depends on the time.
   (a) What is the charge for a repair which takes 80 minutes?
   (b) (i) Calculate the gradient of the line.
       (ii) What does the gradient represent?
   (c) Write down the equation of the line.

6. Make a the subject of this formula $s = ut + \frac{1}{2}at^2$.

7. Draw and label axes for x and y from 0 to 10.
   (a) Draw the graphs of $y = x + 2$ and $y + 2x = 8$ on the same diagram.
   (b) What is the gradient of the straight line $y + 2x = 8$?
   (c) Show how your graph can be used to solve the simultaneous equations $y = x + 2$ and $y + 2x = 8$.

8. Triangle ABC is isosceles.
   $AB = AC$
   Angle $BAC = x$ and angle $ACB = 2x + 35$.
   Form an equation in x.
   Solve your equation and find the size of the angles in the triangle.

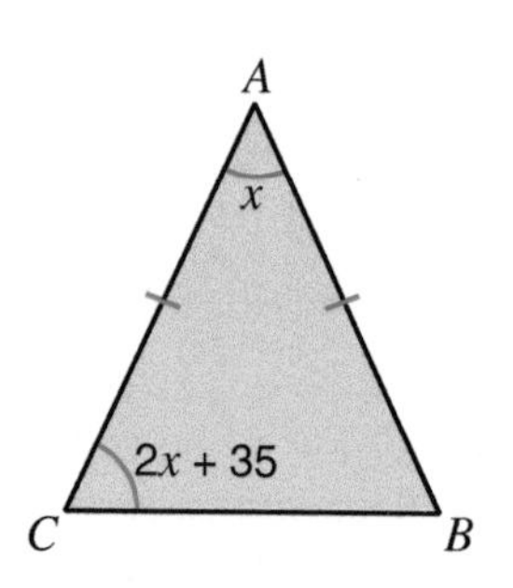

9. Solve the following equations.
   (a) $5x^2 + 3x = 0$ (b) $2x^2 + 5x - 3 = 0$
   (c) $x^2 - 2x = 15$ (d) $3x^2 = 12 - 5x$

Revision Exercise 2 Calculator Paper

You may use a calculator for this exercise.

1. (a) Expand and simplify $\sqrt{2}(3\sqrt{2} - 2)$.
   (b) Given that $150 = 5\sqrt{c}$, find the value of c.
   (c) Expand the brackets and simplify $(2 - \sqrt{5})(3 + \sqrt{5})$.

2. (a) Write the following numbers in scientific notation.
   (i) 140 000 000 (ii) 0.000 014
   (b) Calculate $(3.2 \times 10^{-4}) \times (2.5 \times 10^{2})$.
   Give your answer in scientific notation.
   (c) Calculate $(5.2 \times 10^{-3}) \times (1.4 \times 10^{-2})$.
   Give your answer as an ordinary number correct to two significant figures.

3. In a sale, all computers are sold with a discount of 35%.
   Amanda pays £273 for a computer in the sale.
   What was the price of the computer before the sale?

4. (a) Solve the equation $3p = -15$.
   (b) Solve the inequation $-2p < 8$.
   (c) Solve the equation $3(q - 4) = 12$.
   (d) Solve the inequation $2p + 3(p - 4) \geqslant p - 13$.

5. $ABCDE$ is a regular pentagon.
   F is a point such that CDF and AEF are straight lines.
   Calculate the size of:
   (a) angle DEF,
   (b) angle CDE,
   (c) angle DFE.

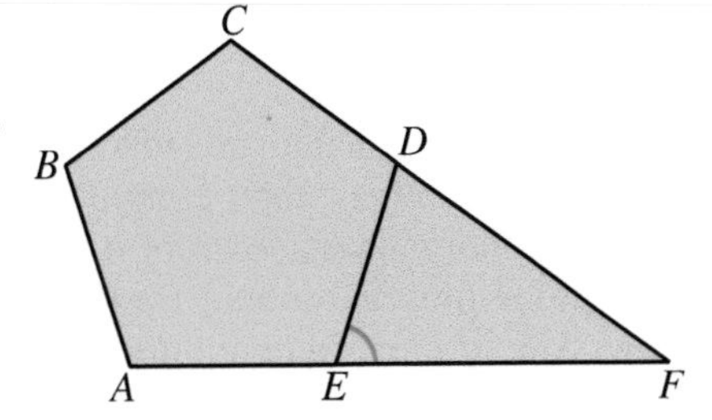

6. 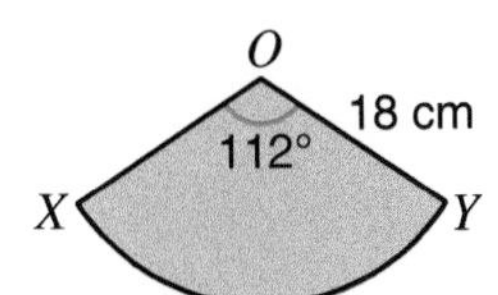

A sector of a circle, radius 18 cm, is cut out of card.
By joining OX to OY, a cone is formed.
(a) Calculate the length XY.
(b) Calculate the volume of the cone formed.

7. Find the value of v, if $v^3 = \dfrac{64P}{wA}$
   when $P = 1230$, $w = 65.4$ and $A = 0.0108$.
   Give your answer correct to 3 significant figures.

8. Trevor recorded his journey times, in minutes, to travel between home and school.

	Monday	Tuesday	Wednesday	Thursday	Friday
Home to school	35	37	37	38	35
School to home	38	31	41	43	36

   (a) Find the mean and standard deviation of the times, in minutes, for Trevor's journeys from home to school.
   (b) Find the mean and standard deviation of the times, in minutes, for Trevor's journeys from school to home.
   (c) Compare the times it takes Trevor to travel to and from school.

Revision Exercise 3 Non-calculator Paper

Do not use a calculator for this exercise.

1. (a) Write 0.000 45 in scientific notation.
(b) Write 984 000 in scientific notation.
(c) $1.2 \times 10^a \times b \times 10^5 = 4.8 \times 10^{-2}$
Find the values of a and b.

2. (a) Find the value of:
(i) $25^{-\frac{1}{2}}$ (ii) $16^{\frac{3}{4}}$ (iii) $8^{\frac{2}{3}}$
(b) Simplify:
(i) $2a^3b^2 \times 3a^2b^3$ (ii) $\dfrac{12a^2b^4}{3ab^3}$ (iii) $\dfrac{2a^2b^3 \times 5ab}{2(ab^2)^2}$

3. Work out.
(a) $\dfrac{x}{6} + \dfrac{2x}{3} - \dfrac{x}{2}$ (b) $\dfrac{2}{15x} + \dfrac{5}{12x}$ (c) $\dfrac{10x}{3} - \dfrac{2x}{9}$ (d) $\dfrac{2}{x} - \dfrac{3}{2x} + \dfrac{4}{3x}$

4.

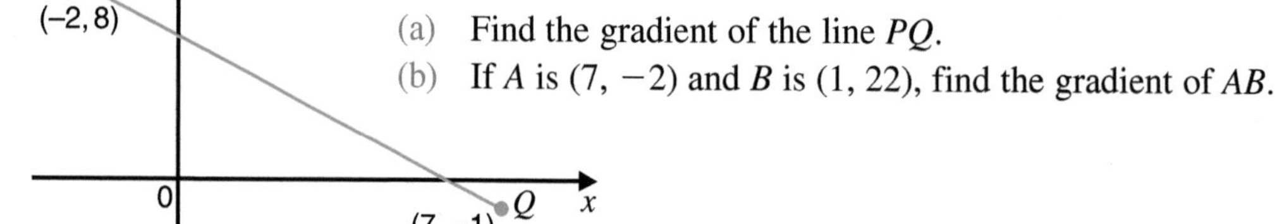

(a) Find the gradient of the line PQ.
(b) If A is $(7, -2)$ and B is $(1, 22)$, find the gradient of AB.

5. A computer technician uses this graph to show his customers how much he charges for home visits to repair their computers.
The total charge is made up of a call-out fee plus an amount for the time taken to repair a computer.

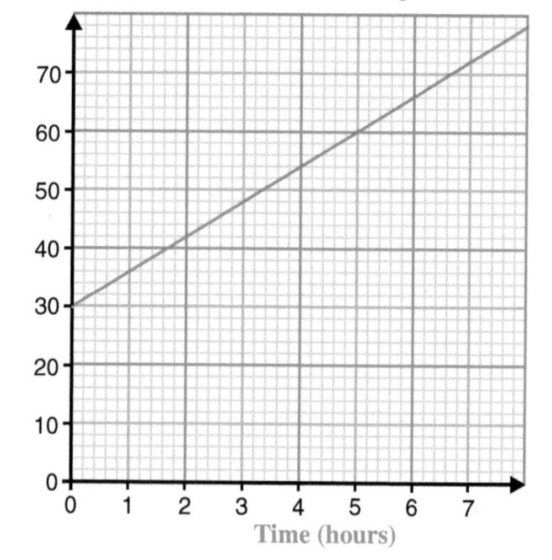

(a) How much is the call-out charge?
(b) How much would the technician charge if he takes 3 hours to repair a computer?
(c) Find the gradient of the line.
(d) Write a formula for the total charge, £C, for repairing a computer in t hours.
(e) Use your formula to find the number of hours required to repair a computer if the total charge was £78.

6. $\overrightarrow{OA} = \begin{pmatrix} -3 \\ 1 \end{pmatrix}$, $\overrightarrow{OB} = \begin{pmatrix} 4 \\ 2 \end{pmatrix}$, and $\overrightarrow{OC} = \begin{pmatrix} 1 \\ -4 \end{pmatrix}$.
Write the following vectors in column form.
(a) $\overrightarrow{AB}$ (b) $\overrightarrow{BC}$ (c) $\overrightarrow{AC}$

7. (a) On the same diagram, sketch the graphs of $y = \sin x$ and $y = 4 \sin x$ for values of x from 0° to 360°.
(b) Sin 30° = 0.5.
Solve the equation $2 \sin x + 1 = 0$ for values of x between 0° and 360°.

8. The formula for finding the total surface area of a cylinder is $A = 2\pi r(r + h)$.
Rearrange $A = 2\pi r^2 + 2\pi rh$ to make h the subject of the formula.

Revision Exercise 4 Calculator Paper

You may use a calculator for this exercise.

1. Calculate the value of: (a) $(\sqrt{7})^6$ (b) $7^{-\frac{3}{4}}$ (c) $\sqrt[4]{\dfrac{5.7}{(0.3)^2}}$ (d) $\left(\dfrac{2.4}{\sqrt{1.5}}\right)^5$

2. (a) Expand the brackets and simplify $(3x + 1)(x - 2)$.
   (b) Factorise the expression $6ab - 3b^2$.
   (c) Make p the subject of the formula $3p - 7q = r - p$.

3. Gregor is paid commission for every computer and printer he sells.
   One week he sold five computers and two printers.
   He was paid £116 commission.
   (a) Write down an equation to represent this information.

   The following week Gregor sold four computers and five printers and was paid £120 commission.
   (b) Write down an equation to represent this information.
   (c) Solve your equations simultaneously and use your answers to find the amount of commission Gregor would earn if he sold 6 computers and 7 printers in a week.

4. 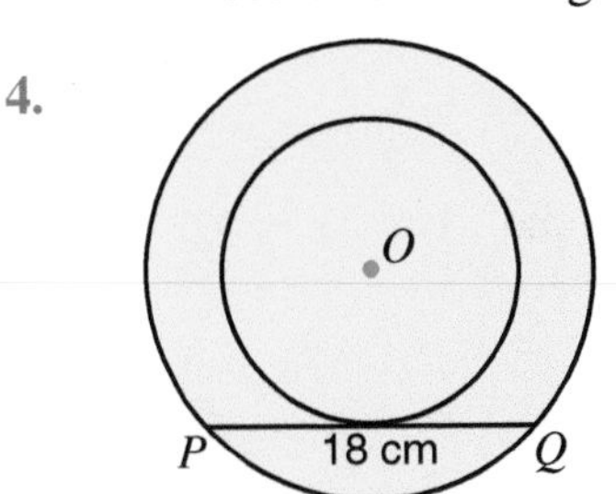

   Two circles have the same centre, O.
   The chord, PQ, in the larger circle is a tangent to the smaller circle.
   $PQ = 18\,\text{cm}$.
   The radius of the smaller circle is 12 cm.
   Find the radius of the larger circle.

5. 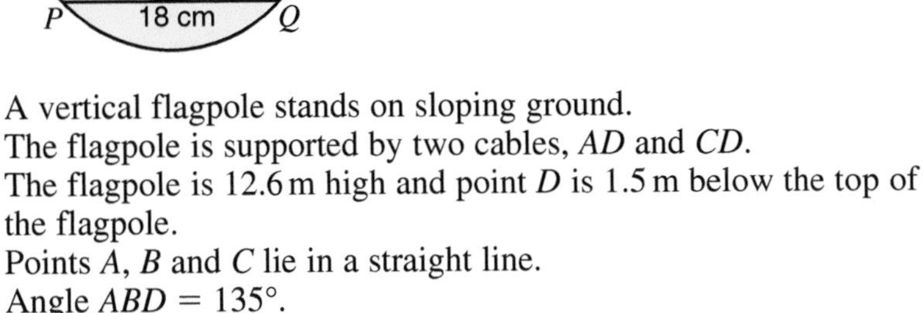

   A vertical flagpole stands on sloping ground.
   The flagpole is supported by two cables, AD and CD.
   The flagpole is 12.6 m high and point D is 1.5 m below the top of the flagpole.
   Points A, B and C lie in a straight line.
   Angle $ABD = 135°$.
   (a) Calculate the length of cable AD.
   (b) Calculate the size of angle BCD.

6. (a) Show that the equation $8 - \dfrac{3}{x} = x$
   can be written as $x^2 - 8x + 3 = 0$.
   (b) $x^2 - 8x + 3$ can be written in the form $(x + a)^2 + b$.
   Find the values of a and b.
   (c) Write down the coordinates of the turning point of the graph of $y = x^2 - 8x + 3$.
   (d) Use the quadratic formula to solve the equation $x^2 - 8x + 3 = 0$.
   Give your answer correct to 2 decimal places.

7. These two cones are similar.
   The larger cone is three times as tall as the smaller cone.
   (a) How many times bigger is the surface area of the bigger cone than the surface area of the smaller cone?
   (b) The volume of the smaller cone is $80\,\text{cm}^3$.
   What is the volume of the larger cone?

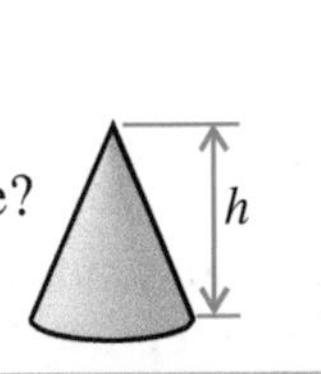

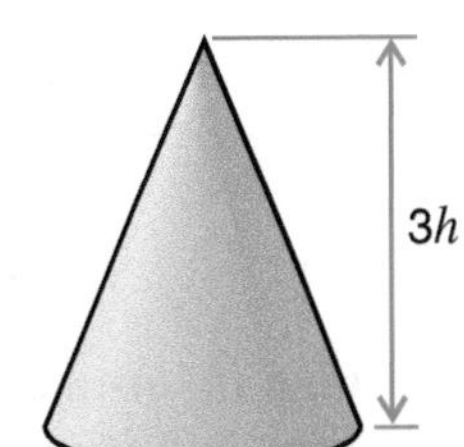

Revision Exercise 5 Non-calculator Paper

Do not use a calculator for this exercise.

1. Simplify.

(a) $\sqrt{20} \times \sqrt{5}$ (b) $\frac{20}{\sqrt{5}}$ (c) $\frac{\sqrt{20}}{\sqrt{5}}$

2. Work out.

(a) $\frac{2}{3} + \frac{1}{2}$ (b) $1\frac{1}{4} \times \frac{3}{5}$ (c) $6 \div 2\frac{2}{5}$

Give your answers as fractions in their simplest form.

3. (a) Expand $x^2(2x + 3)$.

(b) Expand and simplify $3x(x^2 + 2x - 1) - 2x^2(x + 3)$.

4. A line with gradient 4 passes through the points $C(2, 5)$ and $D(x, 17)$.
What is the value of x?

5. (a) The equation of a straight line is given by $y = 5x + 3$.

(i) Write down the gradient of the line.

(ii) Write down the coordinates of the point where the line $y = 5x + 3$ crosses the y axis.

(b) The equations of two straight lines are given by $2y = 3x - 5$ and $1 = 5x - 4y$.
Which line has the steeper gradient?
You must show your working.

6. A quadratic function is defined by $y = (x - 2)(x + 6)$.

(a) Find the equation of the line of symmetry of the graph.

(b) Write down the roots of the quadratic equation.

(c) Find the coordinates of the turning point of the graph.

7. In this cuboid, A is the point (1, 1, 1).
G is the point (13, 9, 4).

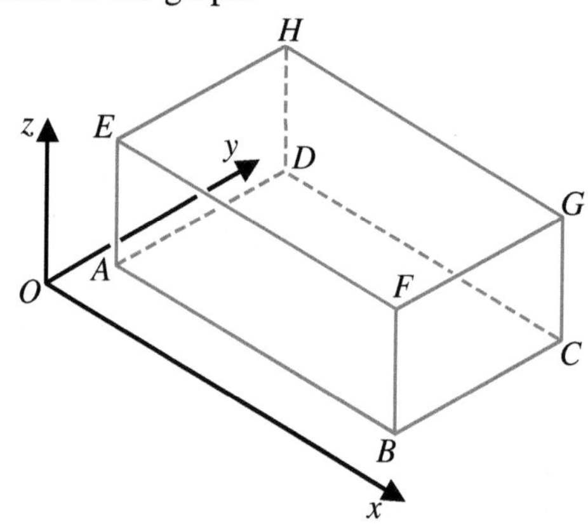

(a) Which point is (1, 9, 1)?

(b) Calculate the length AC.

(c) Calculate the length AG.

The position vector $\overrightarrow{OA} = \begin{pmatrix} 1 \\ 1 \\ 1 \end{pmatrix}$.

(d) Write down $\overrightarrow{OC}$.

(e) Find $\overrightarrow{BH}$ as a column vector.

8. A sample of 23 people were asked to record the amount they spent on food last week.
The amounts, in £s, are shown.

57	66	75	83	39	83	36	42
62	67	43	38	71	58	47	53
62	84	76	68	77	57	42	

(a) Find the median and quartiles of this distribution.

(b) Draw a box plot to represent the data.

Revision Exercise 6 Calculator Paper

You may use a calculator for this exercise.

1. In 2015 approximately 5.84×10^6 people visited Scotland.
They spent £1.154×10^9.

(a) Calculate the average spend per visit.
Give your answer as an ordinary number, to the nearest pound.

The total number of tourist visits all over the world in 2015 was roughly 800 times the number of people that visited Scotland.

(b) How many tourist visits were made all over the world in 2015?
Give your answer in scientific notation.

2. Greta invests £2500 for 3 years.
Compound Interest is paid at a rate of 4.5% per annum.

(a) Calculate the total value of Greta's investment after 3 years.

After 3 years, Greta is given the option of either withdrawing her money or continuing to invest it at the same rate of interest.

(b) After how many **more** years will Greta's investment be worth more than £3500?

3. (a) Make r the subject of the formula $F = \frac{mv^2}{r}$.

(b) Make c the subject of the formula $E = mc^2$.

4. (a) Find the volume of this square-based pyramid.

8 cm
6 cm
6 cm

(b) This cone has volume 377 cm³.
Find the radius of the cone.
Give your answer correct to 1 decimal place.

10 cm
r cm

(c)

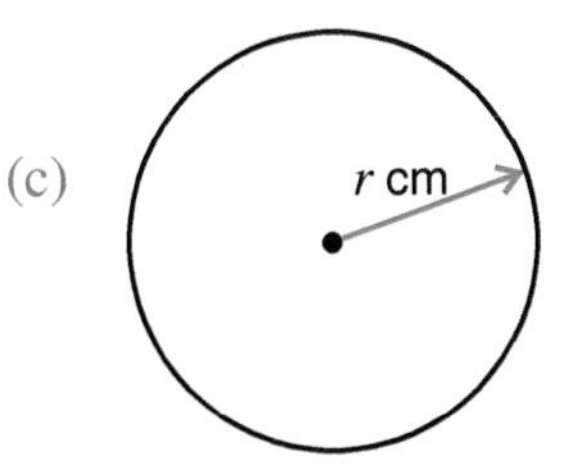

This sphere has volume 7100 cm³.
Show that the radius of the sphere is less than 12 cm.
You must show your working.

5. This diagram shows part of the graph of the function $y = 2 - \cos x°$.
What are the coordinates of the points A, B, C and D?

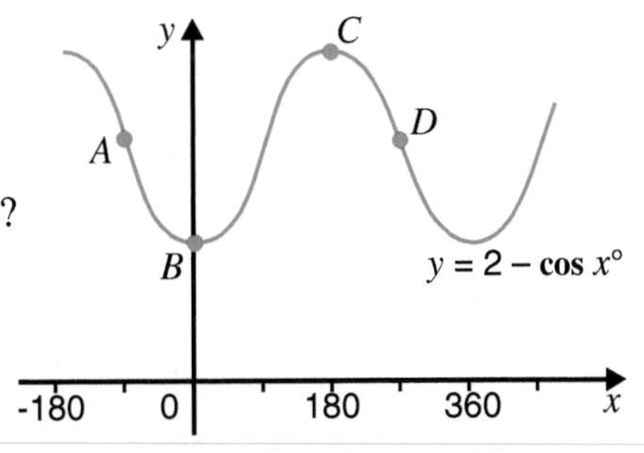

6. The times taken by a group of 6 people to complete a crossword were recorded in minutes.

40 24 15 35 62 32

Calculate the mean and standard deviation of these times.

Revision Exercise 7 Non-calculator Paper

Do not use a calculator for this exercise.

1. 9000 copies of the *Daily Dispatch* are sold each day.
The paper is printed 6 days a week for 52 weeks.
(a) How many copies of the *Daily Dispatch* are sold in a year?
Write your answer correct to 3 significant figures.
(b) Write your answer to part (a) using scientific notation.

2. Find the values of x in these equations.
(a) $3^x = \sqrt[3]{3}$ (b) $4^x = \frac{1}{64}$ (c) $5^x = \sqrt{125}$

3. If $a = \frac{2}{3}$, $b = \frac{5}{6}$ and $c = \frac{5}{8}$, find the value of $\frac{a+b}{b \div c}$.
Give your answer as a fraction in its simplest form.

4. (a) Factorise $x^2 + 8x$.
(b) Factorise $x^2 - 7x - 8$.
(c) Factorise $2x^2 - 18$.

5. Simplify.
(a) $8x^2 \div 4x$ (b) $3a^2 \times 5a^3$ (c) $(5t^3)^2$

6. The graph shows the taxi fare for short journeys.
Total fare = fixed charge + charge per kilometre.
(a) What is the fixed charge?
(b) What is the charge per kilometre?
(c) Calculate the total fare for a journey of 5 km.

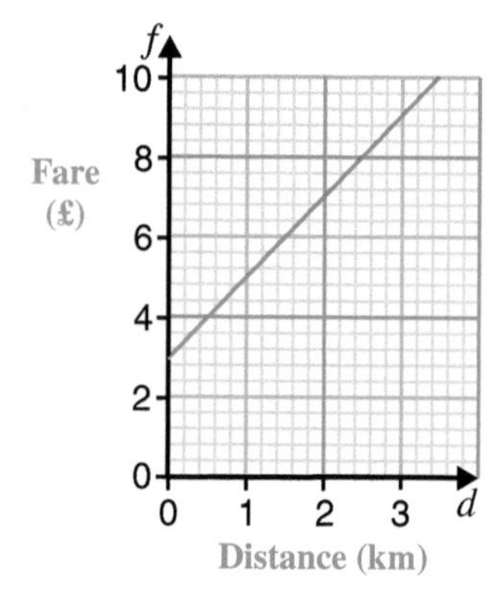

7. If $f(x) = \frac{9}{x} - x$ for what values of a is $f(a) = 0$?

8. If $\mathbf{a} = \begin{pmatrix} -3 \\ 2 \\ 1 \end{pmatrix}$, $\mathbf{b} = \begin{pmatrix} 4 \\ -1 \\ -2 \end{pmatrix}$ and $\mathbf{c} = \begin{pmatrix} -1 \\ -1 \\ 1 \end{pmatrix}$, find:
(a) $2\mathbf{a}$, (b) $\mathbf{a} + \mathbf{b}$, (c) $\mathbf{c} - \mathbf{b}$,
(d) $\mathbf{a} + 2\mathbf{c}$, (e) $3\mathbf{c} - 2\mathbf{b}$, (f) $\mathbf{a} + \mathbf{b} + \mathbf{c}$.

9. The monthly rainfall at a holiday resort was recorded in millimetres, to the nearest millimetre.

Month	J	F	M	A	M	J	J	A	S	O	N	D
Rainfall (mm)	26	18	17	13	14	22	18	24	19	26	18	19

(a) Find the range for the data.
(b) Find the median amount of rainfall in a month.
(c) Find the semi-interquartile range of rainfall.
You must show all your working.

10. A plane flies directly from A to B on a bearing of 055°.
On what bearing must the plane fly to travel directly back from B to A?

Revision Exercise 8 Calculator Paper

You may use a calculator for this exercise.

1. Calculate $\sqrt{\dfrac{523}{19.6}}$

2. Chris is on holiday and wants to hire a bike for the day. He wants to get the best deal for his money.

   (a) On graph paper, draw the graphs of $C = 3.5x + 25$ and $C = 6x + 10$.
   (b) Use your graphs to determine the number of hours for which the cost of hiring a bike from both shops is the same.
   (c) Show how an algebraic method can be used to answer part (b).
   (d) Describe how Chris should decide which shop to hire a bike from if he wants to get the best value for his money.

3. The volumes of these cuboids are equal.
   (a) Show that $3x^2 + 2x - 12 = 0$.
   (b) By solving the quadratic equation $3x^2 + 2x - 12 = 0$, find the value of x. Give your answer correct to 1 decimal place.

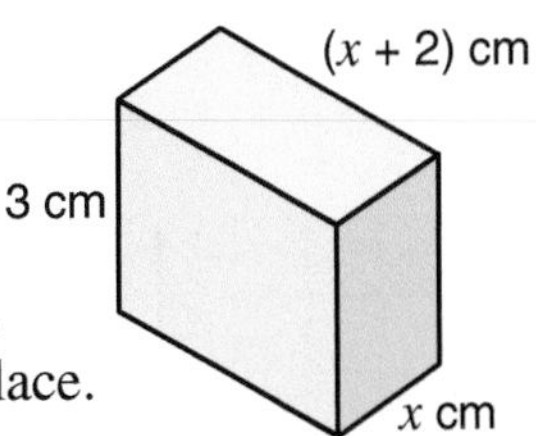

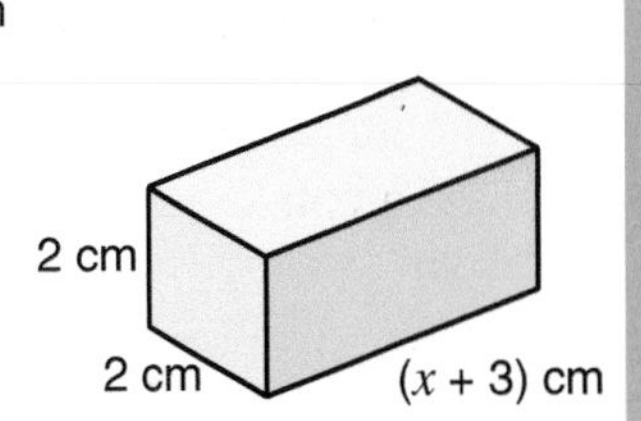

4. A box contains 2 balls of radius 2 cm, as shown. Find the volume of the space inside the box that could be filled with water.

   4 cm, 8 cm, 4 cm

5. $\mathbf{a} = \begin{pmatrix} -4 \\ 5 \end{pmatrix}$ and $\mathbf{p} = \begin{pmatrix} -1 \\ 6 \\ 2 \end{pmatrix}$.
   Which is the longer vector? You must show your working.

6. (a) Show that $2x^2 - 3x + 5 = 0$ has **no** solutions.
   (b) Show that $4x^2 - 4x + 1 = 0$ has **one** solution.
   (c) Use the quadratic formula to find **two** solutions for the equation $2x^2 - 3x - 5 = 0$. Give your answers correct to 2 decimal places.

7. If $AB = 8\,\text{cm}$, $BC = 5\,\text{cm}$ and $\angle ABC = 50°$, find:
   (a) the area of ΔABC, to the nearest cm^2,
   (b) the length of AC, to the nearest mm,
   (c) the size of $\angle BAC$, in degrees, to 1 decimal place.

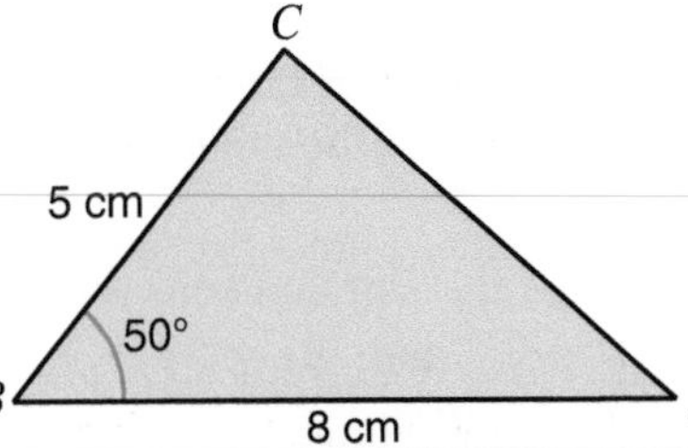

Revision Exercise 9 Non-calculator Paper

Do not use a calculator for this exercise.

1. Multiply out the brackets and collect like terms. $(x - 3)(x^2 - 2x - 3)$

2. Simplify by writing in the form $a\sqrt{b}$.
   (a) $\frac{1}{\sqrt{7}}$ (b) $\frac{3}{\sqrt{6}}$ (c) $\frac{16}{\sqrt{8}}$

3. A farmer decided to check and repair a wall which was 240 m long.
   On the first day, he did a length of 96 m, the second day he did $\frac{3}{4}$ of that distance, and on the third day he did $\frac{2}{3}$ of the distance of the second day.
   What fraction of the total length remained to be done?

4. Simplify.
   (a) $\frac{x^2 - 64}{x - 8}$ (b) $\frac{6x + x^2}{36 - x^2}$

5. Express $\frac{5}{x - 2} + \frac{3}{x + 7}$, where $x \neq 2$, $x \neq -7$, as a single fraction in its simplest form.

6. (a) Expand and simplify $(3x - 2)(x + 4) - 2x + 5$.
   (b) Solve the equation $16 - 3x = 5(x - 2)$.
   (c) Solve the inequality $4n - 5 < 2n + 11$.

7. The graph of the line $x + 3y = 7$ is shown in the diagram.

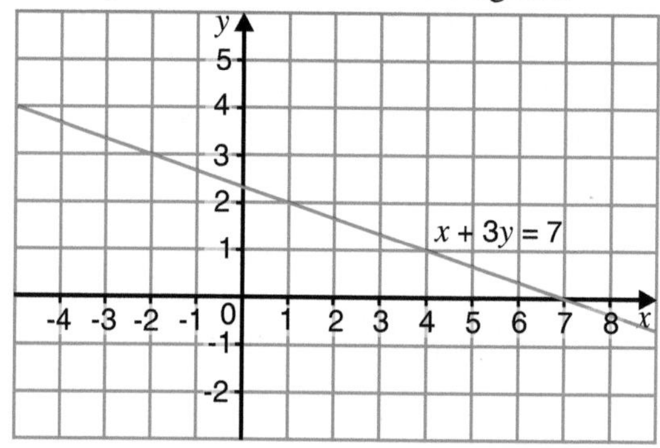

   (a) Copy the diagram and draw the graph of the line $2x + y = -1$.
   (b) Use your graphs to write down the solution to the simultaneous equations $x + 3y = 7$, $2x + y = -1$.

8. The amount of energy, c calories, needed to heat v ml of water from a°C to b°C is given by the formula $c = v(b - a)$.
   Rearrange this formula to make a the subject of the formula.

9. In the diagram, PT is a tangent to the circle with centre O.
   B, C and D are points on the circumference of the circle.
   AC is a diameter of the circle and $\angle ODC = 36°$.
   Find the sizes of the angles marked a, b, c, d and e.

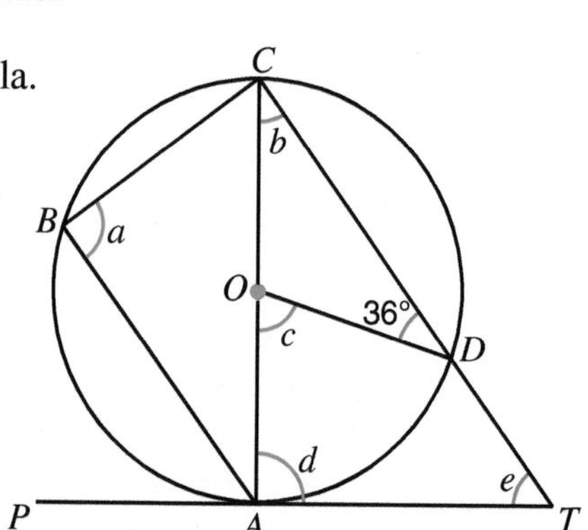

10. Show that the simultaneous equations
    $y - 1 = 3x$ and $9x = 3y + 7$
    have no solution.

Revision Exercise 10 Calculator Paper

You may use a calculator for this exercise.

1. An average person produces 2×10^{11} red blood cells per day.
   How many red blood cells are produced every minute?
   Give your answer in standard form, correct to 2 significant figures.

2. The diagram shows a circle with centre O.
   The area of sector AOB is $120\,\text{cm}^2$.
   Angle $AOB = 120°$.
   Calculate the radius of the circle,
   giving your answer correct to 1 decimal place.

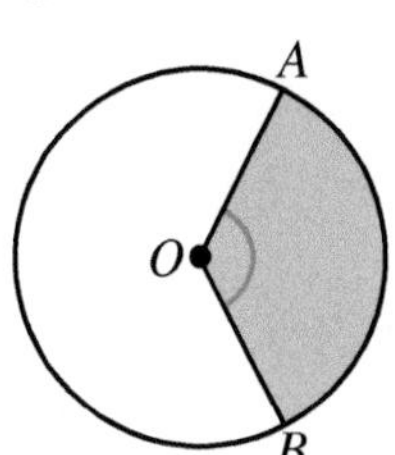

3. (a) The value of a second-hand car depreciates by 8% each year.
   What is the value of a second-hand car bought for £7500 after 3 years?
   Give your answer to a suitable degree of accuracy.
   (b) An antique vase was valued at £900 by an expert.
   The expert expects the vase to appreciate in value at a rate of 4% per year.
   What is the expected value of the vase after 4 years?

4. (a)

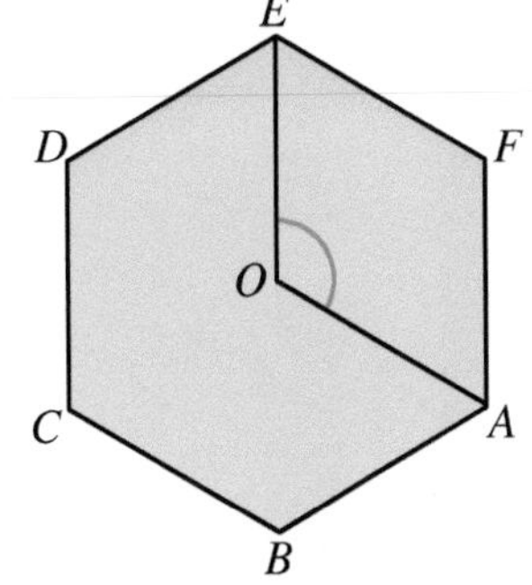

$ABCDEF$ is a regular hexagon.
Calculate the size of $\angle AOE$.

15°

(b) The diagram shows an exterior angle of a regular polygon.
How many sides has the regular polygon?

5. Triangle ABC is formed by points $A\,(1, 2)$, $B\,(-1, 5)$ and $C\,(7, 6)$.
   Show that triangle ABC is a right-angled triangle and name the right angle.

6. The area of a sector of a circle is $25\,\text{cm}^2$.
   The radius of the sector is 6 cm.
   Calculate the perimeter of the sector.

7. Peter travels at 80 km/h on a bearing of 050° from P, where he lives.
   Rachel lives at the point marked R.
   Rachel can intercept Peter one hour after leaving home,
   by travelling at 50 km/h from R.
   They both leave their homes at the same time.

   P is on a bearing of 250° from R.
   (a) Calculate the bearing on which Rachel should travel in order to intercept Peter after one hour.
   (b) How far does Peter live from Rachel's house?

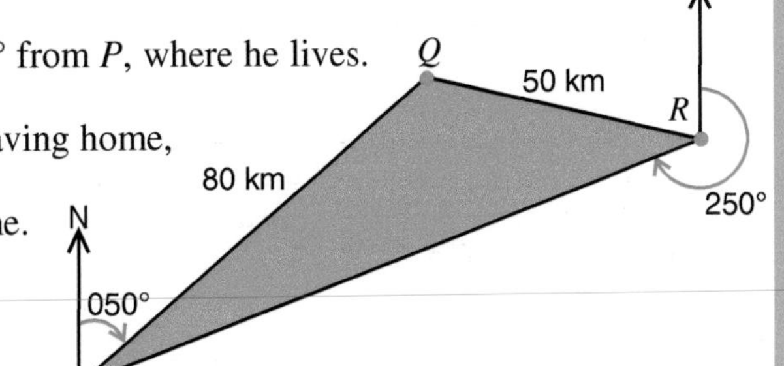

Revision Exercise 11 Non-calculator Paper

Do not use a calculator for this exercise.

1. (a) Find the values of a and b when $x^2 + 6x - 14 = (x + a)^2 + b$.

(b) Simplify $\frac{x^2 + 9x}{x^2 - 81}$.

2. Work out.

(a) $\frac{8a}{15} \times \frac{5}{4a}$ (b) $\frac{3}{8x} \div \frac{15}{16x}$ (c) $\frac{2x}{5} \div \frac{4x}{15}$ (d) $\frac{6}{5x} \times \frac{25x}{18}$

3. Show that $\frac{2}{x} - \frac{1}{x^2}$ can be written as $\frac{2x - 1}{x^2}$.

4. (a) Find the gradients of the lines labelled (**1**) to (**8**).

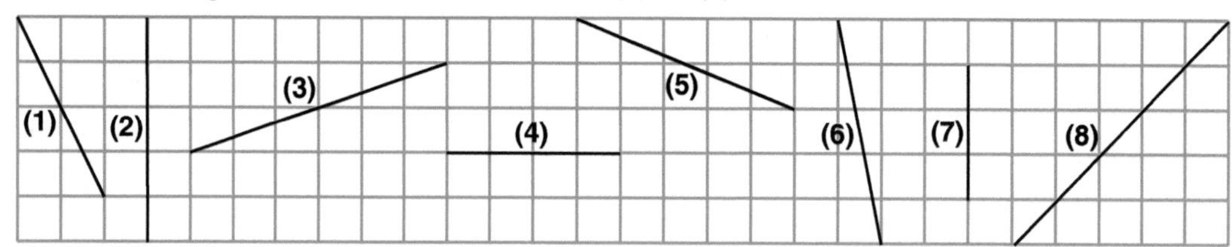

(b) Which two lines are parallel?

5. (a) Factorise $x^2 - 6x + 9$.

(b) Sketch the graph of $y = x^2 - 6x + 9$.

(c) Solve the equation $x^2 - 6x + 9 = 0$.

6. The cost, £C, of manufacturing n chairs is given by the formula $C = 350 + 5n$.

(a) Make n the subject of the formula $C = 350 + 5n$.

(b) Using your answer to part (a), or otherwise, find the number of chairs that can be made for £1000.

7. (a) Factorise $x^2 + 2x - 15$.

(b) Hence, or otherwise, solve the equation $x^2 + 2x - 15 = 0$.

(c) Find values of p and q such that $x^2 + 2x - 15 = (x + p)^2 + q$.

(d) Sketch the graph of $y = x^2 + 2x - 15$.
Show clearly where the graph cuts the x and y axes.

(e) What are the coordinates of the turning point of the graph of $y = x^2 + 2x - 15$?

8. (a) Draw the graph of $y = x^2 - 3x - 3$ for values of x from -2 to 3.

(b) $x^2 - 3x - 3$ can be written in the form $(x + a)^2 + b$.
Find the values of a and b.

(c) What is the minimum value of $y = x^2 - 3x - 3$?

(d) Show how the graph of $y = x^2 - 3x - 3$ can be obtained from the graph of $y = x^2$.

9. Show that $\sin^4 x - \cos^4 x = \sin^2 x - \cos^2 x$.

10. A function is defined as $f(x) = 5x - 3$.
Given that $f(p) = 17$, find the value of p.

Revision Exercise 12 Calculator Paper

You may use a calculator for this exercise.

1. Light travels at a speed of 186 000 miles per second.
   (a) How far does light travel in 1 minute? Give your answer in scientific notation.
   (b) The distance of the Earth from the Sun is approximately 9.6×10^7 miles.
   How long does it take light to travel to the Earth from the Sun?
   Give your answer in minutes, correct to 2 significant figures.

2. Simplify each of the following.
   (a) $3x^3y^4 \times 4x^2y$ (b) $8x^2y^5 \div 4xy^2$ (c) $(2x^3y^{-2})^3$
   (d) $\dfrac{(2x^2y)^2}{2xy^5}$ (e) $\dfrac{3x^2y \times 4x^{-2}y^2}{2x^{-3}y \times 6xy^4}$ (f) $\dfrac{(4x^{-2}y^3)^3}{(2x^3y^{-1})^4}$

3. The table shows by how much the value of a new car depreciates during the first three years.
   (a) Find the value after 3 years of a new car bought for £10 500.
   (b) Find the value after 3 years of a new car bought for £17 600.

	Depreciation		
Cost when new	**1st Year**	**2nd Year**	**3rd Year**
£10 001 - £14 000	17%	12%	9%
£14 001 - £18 000	24%	19%	14%

4.

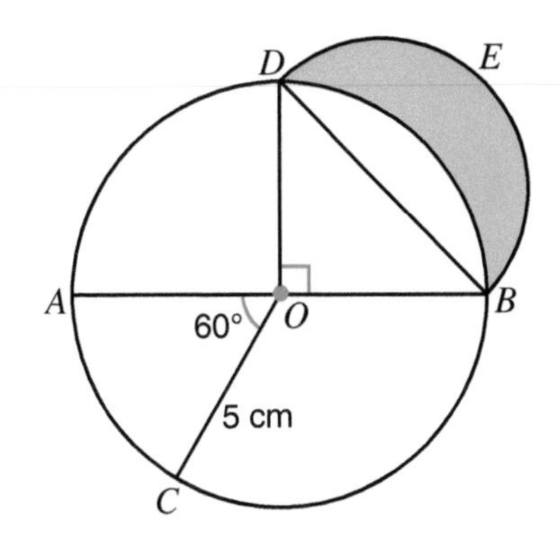

AB is a diameter of a circle with centre O.
The radius of the circle is 5 cm.
C and D are points on the circumference of the circle such that $\angle BOD = 90°$ and $\angle AOC = 60°$.
   (a) Calculate the length of the arc BC.
   (b) Calculate the area of the sector AOC.
   (c) The semicircle BED is drawn using BD as a diameter.
   Calculate the area of the shaded part in the diagram.

5. In this question, take $\pi = 3.14$.
A coffee dispenser is in the shape of a square-based pyramid.
To make one latte, a hemispherical scoop is completely filled.
The diameter of the scoop is 2.5 cm.

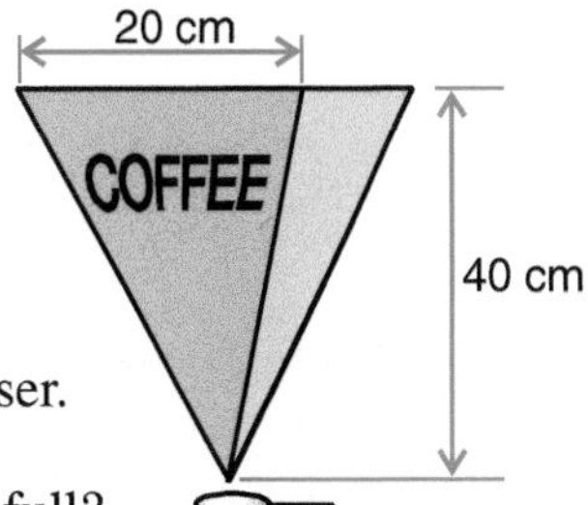

   (a) Calculate the volume of coffee that can be stored in the dispenser.
   (b) Calculate the volume of coffee used to make one latte.
   (c) How many cups of latte can be made if the coffee dispenser is full?

6.

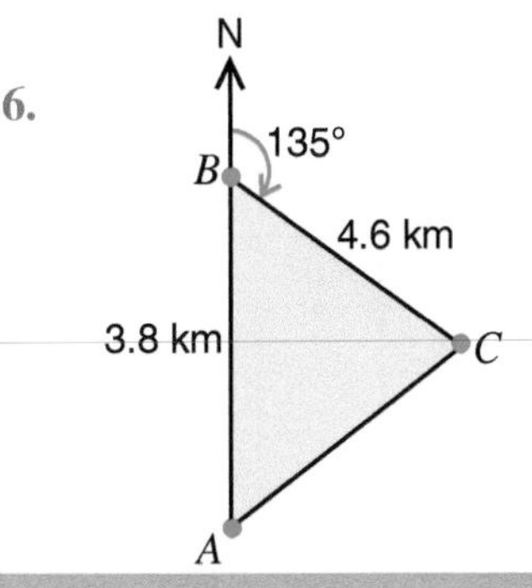

Three buoys, A, B and C, mark this course for a sailing race.
Competitors head 3.8 km due north from the start, A, to B.
At B, they head towards C, on a bearing of 135°.
The distance from B to C is 4.6 km.
At C, they head towards the finish at A.
   (a) What is the distance from C to A?
   (b) What is the total length of the race?
   (c) What is the bearing of A from C?

Exam Practice Non-calculator Paper

Do not use a calculator for this exercise.

1. (a) Simplify, leaving your answer where appropriate in surd form.

(i) $\sqrt{27} + \sqrt{147}$ (ii) $5\sqrt{3} + \sqrt{108}$ (iii) $\sqrt{18} \times \sqrt{12}$

(b) Simplify $\sqrt{\frac{2}{45}}$ leaving your answer with a rational denominator.

(c) Write $\frac{\sqrt{5}}{\sqrt{2}}$ in the form $\frac{\sqrt{a}}{b}$, where a and b are whole numbers.

2. Work out the value of $4^1 - 4^0 + 4^{-1}$.

3. Express each of the following as a fraction in its simplest form.

(a) 2^{-4} (b) $144^{-\frac{1}{2}}$ (c) $\left(\frac{27}{64}\right)^{\frac{2}{3}}$

4. (a) Solve the inequality $10 - 6x \geqslant 2(x - 3)$.

(b) Expand and simplify $3(2 - x) - 5(2x - 3)$.

5. (a) Remove the brackets and simplify.

(i) $m^{\frac{1}{2}}\left(m^{\frac{1}{2}} - m^2\right)$ (ii) $(3x - 1)^2 - 4(x^2 + 3)$

(b) Make a the subject of the formula $bc = ed + af$.

6. The area of a circle is exactly $\frac{36}{\pi}$ square centimetres.
Show that the circumference of the circle is 12 cm.

7. A relationship between P and Q is given by the formula $P = 4Q^3$.
When Q is doubled, what is the effect on P?

8. (a) Simplify. $\frac{a^2 + 2a}{5a + 10}$

(b) Factorise. $2x^2 - 18$

9. (a) $f(x) = x^2 + x$. Find the values of $f(-2)$, $f(-1)$, $f(0)$, $f(1)$ and $f(2)$.

(b) $f(a) = 2^a$. Find the value of a if $f(a) = 16$.

(c) $f(p) = 2^{-p}$. Find the value of $f(p)$ when $p = -3$.

10. (a) Multiply out and simplify $(x - 5)^2$.

(b) Remove the brackets and simplify $(x + 3)(4x^2 - 3x + 2)$.

(c) Factorise (i) $3m^2 - 6m$, (ii) $t^2 - t - 12$.

11. The diagram shows a square $ABCD$.

(a) Find the equation of the line parallel to AC which passes through D.

A line which is parallel to BD passes through point A.
The line cuts the line $y = -2$ at point P.

(b) Find the coordinates of P.

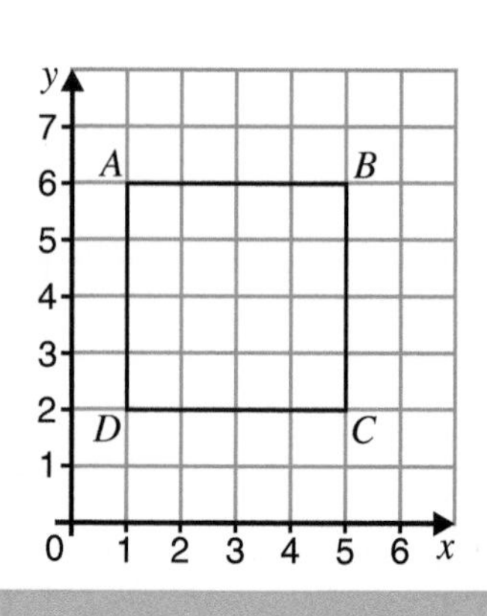

12. Express as a single fraction $\frac{1}{x} + \frac{3}{x + 4}$.

13. Solve the simultaneous equations $3x + y = 13$ and $2x - y = 7$.

14. (a) Solve the equation $\frac{m}{2} - \frac{m}{3} = 5$.
(b) Multiply out (i) $3y(x^2 - 2y)$, (ii) $(x - 2)(3x + 5)$.

15. (a) Remove the brackets and simplify $3(4x - 1) - 3(x - 5)$.
(b) Solve the equation $4(5 - x) = 3(2 - x)$.

16. You are give the formula $P = \left(\frac{m}{n}\right)^2$.
(a) Calculate the value of P when $m = -0.12$ and $n = \frac{2}{5}$.
(b) Rearrange the formula to give n in terms of P and m.

17. $x^2 + 2x - 2$ can be written in the form $(x + p)^2 + q$, where p and q are integers.
(a) Find the values of p and q.
(b) Sketch the graph of $y = x^2 + 2x - 2$.
(c) Write down the coordinates of the turning point of the graph of $y = x^2 + 2x - 2$.

18. In a sale, prices are reduced by 10%.
The sale price of a tennis racket is £36.
What was the original price of the tennis racket?

19. Work out. (a) $\frac{2}{3} + \frac{3}{4}$ (b) $\frac{1}{2} - \frac{2}{5}$ (c) $\frac{3}{4} \times \frac{2}{5}$

20. Solve the equations. (a) $y^2 + 5y = 0$ (b) $m^2 - 7m + 12 = 0$

21. The area of triangle $ABC = 27\,\text{cm}^2$.
$AB = 4\,\text{cm}$ and $\sin A = 0.75$.
Find the length of AC.

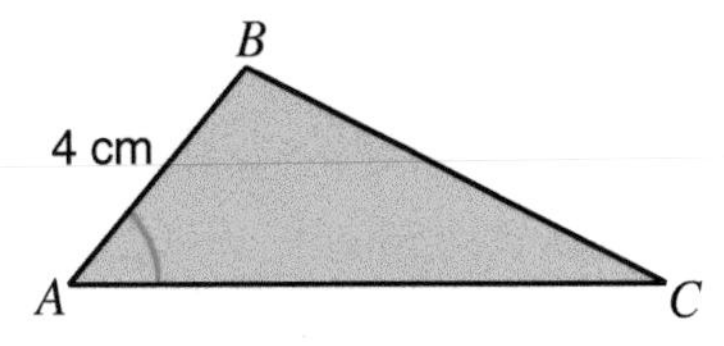

22. Work out. (a) $2\frac{2}{3} \times 3\frac{1}{2}$ (b) $3\frac{1}{2} \div 2\frac{4}{5}$

23. Express $\frac{3}{x - 4} - \frac{2}{x + 5}$ where $x \neq 4$, $x \neq -5$, as a single fraction in its simplest form.

24. A regular polygon has an exterior angle of 36°.
How many sides has the polygon?

25. A right-angled triangle has dimensions, as shown.
(a) Calculate the length of AB, leaving your answer as a surd in its **simplest form**.
(b) Find the value of $\cos A$, giving your answer in the form $\frac{\sqrt{b}}{a}$, where a and b are whole numbers.

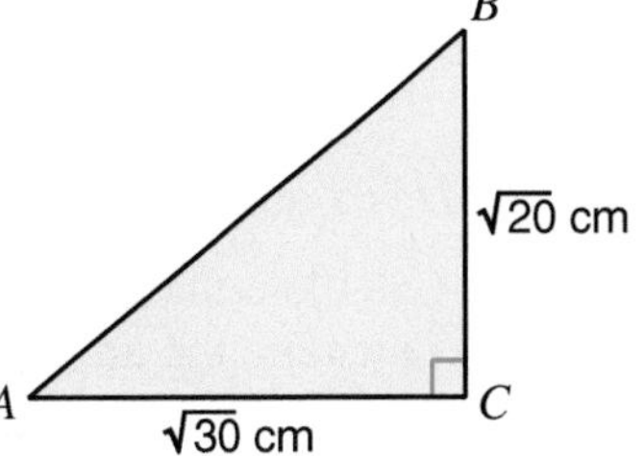

26. In triangle PQR, $PQ = 4\,\text{cm}$, $QR = 5\,\text{cm}$ and $PR = 7\,\text{cm}$.
Use the Cosine Rule to show that the angle PQR is greater than 90°.

27. Remove the brackets and simplify where possible.
(a) $(\sqrt{2} + 3)(\sqrt{2} - 1)$
(b) $(\sqrt{3} + 1)(1 - \sqrt{3})$
(c) $(3 + 2\sqrt{5})^2$

28. Given that $4x^2 - 2x - 1 = 0$, show that $x = \frac{1 \pm \sqrt{5}}{4}$.

29. Expand and simplify $(x - 4)(2x^2 - x + 1)$.

Exam Practice Calculator Paper

You may use a calculator for this exercise.

1. Work out the value of: (a) $8^{\frac{1}{3}} + 4^{\frac{3}{2}} - 16^{\frac{3}{4}}$ (b) $27^{-\frac{2}{3}}$

2. Simplify. (a) $x^{\frac{2}{3}} \times x^{\frac{1}{3}}$ (b) $x^{\frac{3}{4}} \div x^{\frac{1}{2}}$ (c) $(x^6)^{\frac{1}{2}}$

3. The area of the surface of the Earth is about 5.095×10^9 square miles.
Approximately 29.2% of this is land.
Use these figures to estimate the area of land surface on Earth.

4. (a) Factorise completely. $x^3 - 9x$
(b) Simplify. (i) $m^2 \times m^3$ (ii) $\frac{n^6}{n^3}$ (iii) $x^2y^3 \times \frac{x^3}{y}$ (iv) $(p^2)^3$
(c) Rearrange the formula $y = 5x + 10$ to make x the subject.
(d) Solve. (i) $\frac{2t + 3}{3} = 7$ (ii) $5(x - 3) = 3x + 1$

5. This formula is used in science. $f = \frac{uv}{u + v}$
Calculate the value of f when $u = 3.6 \times 10^5$ and $v = 8.9 \times 10^6$.
Give your answer in scientific notation.

6. AOB is a sector of a circle, centre O.
The radius of the circle is 12 cm.
Angle $AOB = 72°$.
Calculate the area of the shaded segment.

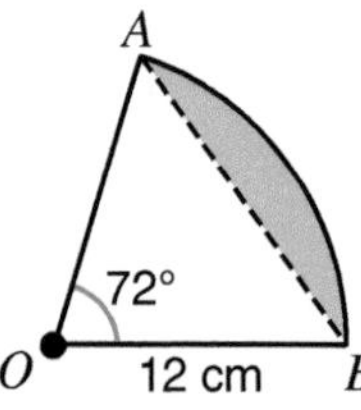

7. Solve the equation $\frac{3}{x + 1} + \frac{1}{x - 2} = 1$. Give your answers correct to 2 d.p.

8. Determine the nature of the roots of the equation $x^2 - 7x + 12 = 0$.

9. Solve algebraically the equation $4 \sin x° - 3 = 0$ for $0 \leqslant x \leqslant 360$.

10. Jim invests £10 000 at 5.2% per year Compound Interest.
What is the percentage increase in the value of his investment after 3 years?

11. $ABCD$ is a quadrilateral.
(a) Calculate the length of BD.
(b) Calculate the area of $ABCD$.

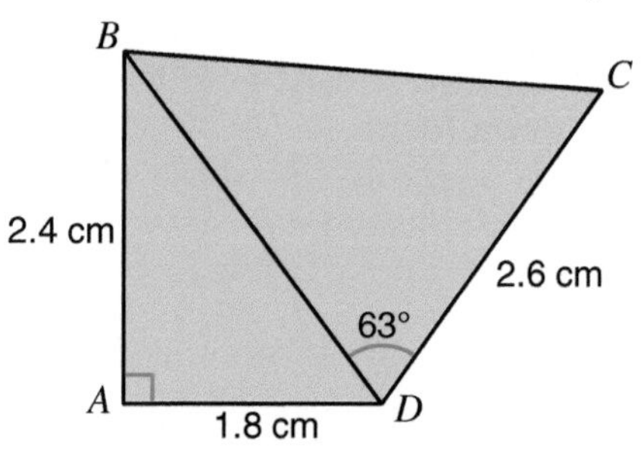

12. In the triangle ABC:
$AB = x$ cm, $AC = 8$ cm and $BC = 7$ cm.
Angle $BAC = 60°$.

(a) Using the Cosine Rule, form an equation in x and show that $x^2 - 8x + 15 = 0$.
(b) Solve the equation $x^2 - 8x + 15 = 0$ and find the **two** possible lengths of AB.

13. Solve the equation $2x^2 = 5x - 1$.
Give your answer correct to 2 decimal places.

14. Each shape is a regular polygon.
Work out the size of each lettered angle.

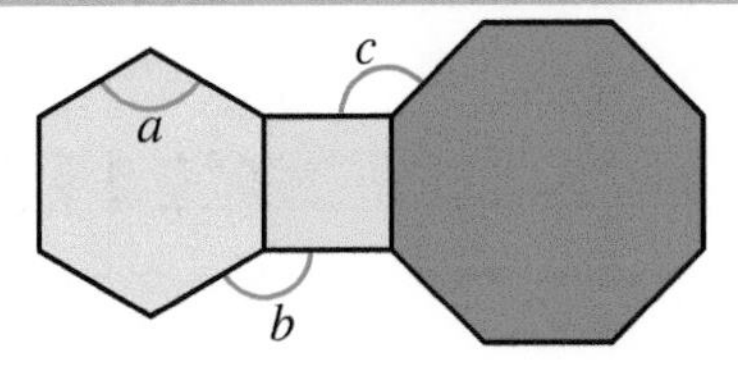

15. A triangle has sides of length 7 cm, 5 cm and 4 cm.
Work out the area of the triangle.

16. The lens cover on a security camera is in the shape of a circle with a segment removed at the bottom. The height of the lens cover is 8 cm and the radius of the circle, centre O, is 5 cm.

(a) Calculate the length of AB.

(b) Calculate the perimeter of the lens cover.

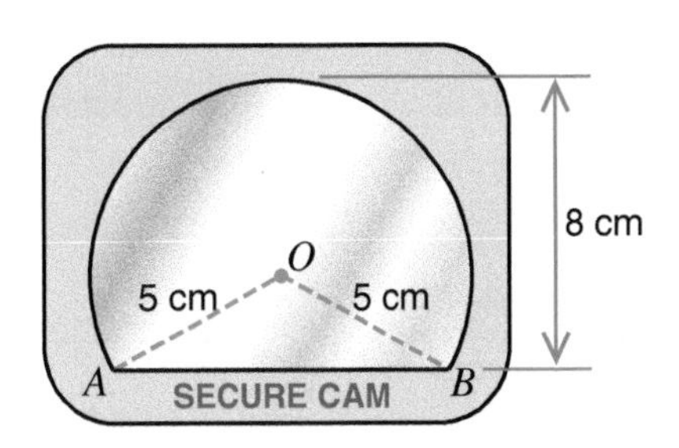

17. 10 cm 15 cm

These boxes are similar.

(a) The volume of the smaller box is 240 cm^3.
Work out the volume of the larger box.

(b) The surface area of the larger box is 558 cm^2.
Work out the surface area of the smaller box.

18. In the triangle ABC, B is the point (9, 8). $\overrightarrow{BA} = \begin{pmatrix} -6 \\ -4 \end{pmatrix}$ and $\overrightarrow{AC} = \begin{pmatrix} 10 \\ -2 \end{pmatrix}$

(a) Find the vector $\overrightarrow{BC}$.

(b) What type of triangle is ABC? Give a reason for your answer.

19. A golf ball is dropped from different heights (h cm) onto a concrete floor.
The height (b cm) the ball reaches after bouncing is recorded.

h cm	40	60	80	100	120	140
b cm	34	44	62	77	93	102

(a) (i) Use this information to draw a scatter graph.

(ii) Draw a line of best fit by eye and find its equation.

(b) Use your equation to estimate:

(i) the height a golf ball bounces when dropped from 50 cm,

(ii) the height of the golf ball after the **second** bounce when the ball is dropped from 110 cm above the ground.

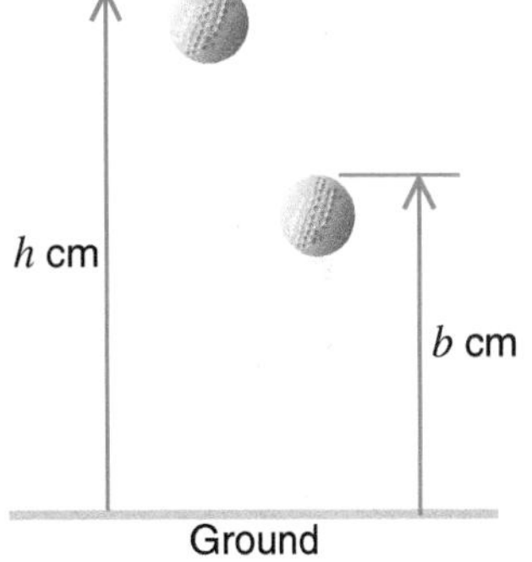

20.

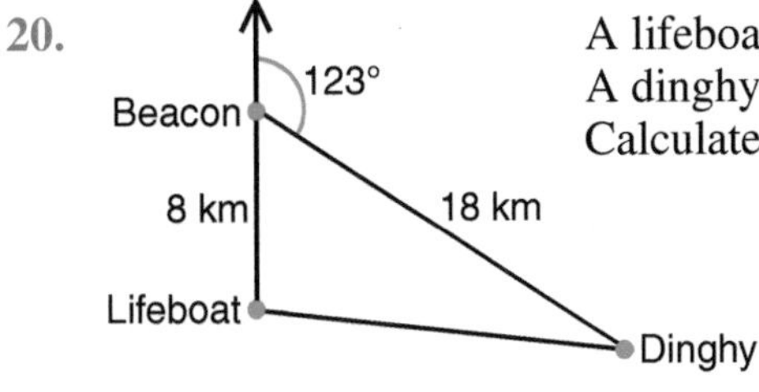

A lifeboat is 8 km due south of a beacon.
A dinghy is 18 km from the same beacon on a bearing of 123°.
Calculate the distance and bearing of the dinghy from the lifeboat.

21. (a) Sketch the graph of $y = \sin x$ for $-360° \leqslant x \leqslant 360°$.

(b) How many solutions are there to the equation $\sin x = -0.8$ between $-360°$ and $360°$?

(c) Solve the equation $\sin x = 0.6$ for $-360° \leqslant x \leqslant 360°$.
Give your answers to the nearest degree.

22. Triangle ABC has sides of length 5 cm, 12 cm and 13 cm.
Show that triangle ABC is a right-angled triangle.

23. Find $\overrightarrow{AC}$ if $\overrightarrow{AB}$ $\begin{pmatrix} 2 \\ 1 \\ 3 \end{pmatrix}$ and $\overrightarrow{BC} = \begin{pmatrix} -5 \\ 2 \\ -3 \end{pmatrix}$.

ANSWERS

Chapter 1 Approximation

Practice Exercise 1.1 — Page 1

1. (a) 3.962 (b) 3.96 (c) 4.0
2. 4.86
3. 68.8 kg
4. Missing entries are:
   0.96, 0.97, 15.281, 0.06, 4.99, 5.00.
5. (a) (i) 46.1 (ii) 59.7 (iii) 569.4 (iv) 17.1 (v) 0.7
   (b) (i) 46.14 (ii) 59.70 (iii) 569.43 (iv) 17.06 (v) 0.66
   (c) (i) 46.145 (ii) 59.697 (iii) 569.434 (iv) 17.059 (v) 0.662
6. (a) 40.9 litres, nearest tenth of a litre.
   (b) £4.30, nearest penny.
   (c) 35.7 cm, nearest millimetre.
   (d) £1.33, nearest penny.
   (e) £20.89, nearest penny.

Practice Exercise 1.2 — Page 2

1. (a) 20 (b) 500 (c) 400
   (d) 2000 (e) 20 (f) 0.08
   (g) 0.09 (h) 0.009 (i) 0.01
2. Missing entries are:
   450 000, 7 980 000, 0.00057, 0.094, 0.0937.
3. 490
4. (a) (i) 80 000 (ii) 80 (iii) 1000 (iv) 0.007 (v) 0.002
   (b) (i) 83 000 (ii) 83 (iii) 1000 (iv) 0.0073 (v) 0.0019
   (c) (i) 82 700 (ii) 82.7 (iii) 1000 (iv) 0.00728 (v) 0.00190
5. 472 m^2 (3 s.f.).
6. (a) 157 cm^2 (3 s.f.). (b) 6100 m^2 (2 s.f.).
   (c) 1.23 m (nearest cm).
   (d) 15.9 m^2 (round up 1 d.p.).

Review Exercise 1 — Page 3

1. (a) 28.71 (b) 6.91 (c) 12.40
   (d) 0.04 (e) 0.01
2. (a) 2310 (b) 23.6 (c) 37.0
   (d) 504 (e) 0.000565
3. (a) 25.57 (b) 25.6 (c) 30
4. (a) 5.7 (b) 5.73 (c) 5.728
5. (a) 1 (b) 1.5
6. £3900
7. (a) 19 m^2
   (b) Round up to the nearest square metre for enough carpet to cover whole floor.
8. £13.17

Chapter 2 Working with Surds

Practice Exercise 2.1 — Page 4

1. Rational: (b), (d), (e), (f), (h), (i).
   Irrational: (a), (c), (g), (j).
2. (a) E.g. $\sqrt{2} \times \sqrt{18} = \sqrt{36} = 6$
   (b) E.g. $\sqrt{3} \times \sqrt{2} = \sqrt{6}$
   (c) E.g. $\sqrt{8} \div \sqrt{2} = \sqrt{4} = 2$
   (d) E.g. $\sqrt{3} \div \sqrt{2} = \sqrt{\frac{3}{2}}$

Practice Exercise 2.2 — Page 6

1. (a), (d), (e), (i), (k), (m), (o).
2. (a) $2\sqrt{3}$ (b) $3\sqrt{3}$ (c) $3\sqrt{5}$
   (d) $4\sqrt{3}$ (e) $4\sqrt{2}$ (f) $5\sqrt{2}$
   (g) $3\sqrt{6}$ (h) $2\sqrt{6}$ (i) $7\sqrt{2}$
   (j) $4\sqrt{5}$
3. (a) $2\sqrt{11}$ (b) $5\sqrt{3}$ (c) $8\sqrt{2}$
   (d) $6\sqrt{2}$ (e) $10\sqrt{2}$
4. (a) $\frac{3}{4}$ (b) $\frac{7}{8}$ (c) $\frac{3}{2}$
   (d) 2 (e) 4
5. (a) $2\sqrt{2}$ (b) $\sqrt{5}$ (c) $7\sqrt{3}$
   (d) $2\sqrt{2}$ (e) $5\sqrt{5}$ (f) $5\sqrt{3}$
   (g) $5\sqrt{2}$ (h) $\sqrt{2}$ (i) $7\sqrt{5}$
   (j) $3\sqrt{3}$ (k) $6\sqrt{3}$ (l) $6\sqrt{2}$
   (m) $12\sqrt{5}$ (n) $14\sqrt{3}$ (o) $5\sqrt{5}$
   (p) $10\sqrt{2}$ (q) $5\sqrt{3}$ (r) $6\sqrt{2}$
6. (a) 3 (b) 6 (c) 30
   (d) 4 (e) 6 (f) $5\sqrt{2}$
   (g) $6\sqrt{2}$ (h) $10\sqrt{2}$ (i) 12
   (j) $6\sqrt{6}$ (k) $8\sqrt{6}$ (l) $12\sqrt{6}$
7. (a) $2 + \sqrt{2}$ (b) $3\sqrt{2} - 3$
   (c) $2\sqrt{3} + 2$ (d) $5\sqrt{2} - 5$

Practice Exercise 2.3 — Page 7

1. (a) $\frac{\sqrt{3}}{3}$ (b) $\frac{\sqrt{5}}{5}$ (c) $\frac{\sqrt{7}}{7}$ (d) $\sqrt{2}$
   (e) $\sqrt{5}$ (f) $2\sqrt{2}$ (g) $2\sqrt{3}$ (h) $2\sqrt{7}$
   (i) $\frac{\sqrt{6}}{2}$ (j) $3\sqrt{5}$ (k) $3\sqrt{3}$ (l) $\frac{\sqrt{15}}{3}$
   (m) $3\sqrt{6}$ (n) $7\sqrt{5}$ (o) $\frac{\sqrt{21}}{3}$ (p) $\frac{\sqrt{22}}{2}$
   (q) $\frac{\sqrt{30}}{3}$ (r) $\frac{3\sqrt{14}}{2}$ (s) $\frac{2\sqrt{35}}{5}$ (t) $\frac{3\sqrt{10}}{2}$

2. (a) $\frac{3\sqrt{2}}{2}$ (b) $\sqrt{3}$ (c) $\frac{\sqrt{6}}{2}$ (d) $\sqrt{2}$
(e) $\frac{3\sqrt{2}}{2}$ (f) $\frac{\sqrt{2}}{2}$ (g) $\sqrt{3}$ (h) 2
(i) 2 (j) 3 (k) $\frac{1}{2}$ (l) 4
(m) $\frac{\sqrt{3}}{2}$ (n) 1 (o) $\sqrt{3}$ (p) $2\sqrt{2}$
(q) 1 (r) $\sqrt{5}$ (s) $\frac{\sqrt{2}}{2}$ (t) $\frac{\sqrt{3}}{3}$

Review Exercise 2 — Page 8

1. $n = 15$
2. $\pi + 2$ (irrational), 5.1416 (rational), $5\frac{1}{7}$ (rational), 2.268^2 (rational), $\sqrt{27}$ (irrational).
3. Rational: (b), (c), (d). Irrational: (a).
4. (b)
5. (a) $2\sqrt{5}$ (b) $3\sqrt{2}$ (c) $5\sqrt{3}$ (d) $10\sqrt{2}$ (e) $4\sqrt{7}$
6. (a) $8\sqrt{2}$ (b) $7\sqrt{3}$ (c) $2\sqrt{7}$ (d) 15 (e) $12\sqrt{2}$ (f) 5
7. $7\sqrt{3}$
8. (a) $\frac{5\sqrt{6}}{6}$ (b) $4\sqrt{2}$ (c) $\frac{\sqrt{10}}{4}$ (d) $\frac{\sqrt{6}}{6}$
9. (a) $12 - 3\sqrt{3}$ (b) $k = 5$ (c) $\frac{\sqrt{6}}{2}$
10. (a) $\frac{5\sqrt{3}}{3}$ (b) $x = 175$
11. $\sqrt{ab} = \sqrt{4 \times 50} = \sqrt{4} \times \sqrt{50} = 2 \times 5\sqrt{2} = 10\sqrt{2}$ (irrational)
12. (a) $\frac{5}{3}$ (b) $6 + 2\sqrt{2} + 3\sqrt{3} + \sqrt{6}$ (c) $\frac{\sqrt{10}}{4}$

Chapter 3 Using Indices

Practice Exercise 3.1 — Page 10

1. (a) 2^5 (b) 3^7 (c) 5^8 (d) 7^4 (e) 9^5
2. (a) 2 (b) 3^3 (c) 5^4 (d) 7^2 (e) 9^5
3. (a) y^3 (b) t^5 (c) g^{10} (d) m^6 (e) y^2 (f) a (g) 1 (h) g^{-1}
4. (a) 3^6 (b) 10^2 (c) 4^5 (d) 5^3 (e) 2^3 (f) $5^0 = 1$ (g) 7^{-2} (h) 3^3
5. (a) 2^6 (b) 3^{10} (c) 5^6 (d) 7^9 (e) $9^0 = 1$
6. (a) t^6 (b) y^6 (c) x^2y^2 (d) m^3n^6
7. (a) 8 (b) 7 (c) 11 (d) 3 (e) 1 (f) 4 (g) 6 (h) 20 (i) 2 (j) 8 (k) 6 (l) 3

Practice Exercise 3.2 — Page 11

1. (a) 9^0 (b) 2^{-2} (c) 5^{-2} (d) 8^{-5} (e) 2^{-4} (f) 5^{12} (g) 11^{-5} (h) 7^{-1}
2. (a) $6d^5$ (b) $8x^5$ (c) $6t^5$ (d) $24r^6$ (e) $6b^2$ (f) $5m$ (g) $4t$ (h) $3h^{-1}$ (i) $9a^2$ (j) $8h^3$ (k) $2m^6$ (l) $4m^6$
3. (a) 8^2 (b) 7^{-5} (c) 2.5^{-1} (d) 4^0 (e) 10^{-1} (f) 6^{-4} (g) 0.1^{-12} (h) 5^{-15} (i) 4^5 (j) $4^2 \times 8^7$ (k) 4×5^{-2} (l) $2^{-2} \times 5^5$ (m) 3 (n) 5^3 (o) 2^{-6}
4. (a) $\frac{1}{3^2}$ (b) $\frac{1}{2^3}$ (c) $\frac{1}{3}$ (d) $\frac{1}{5^3}$ (e) 3^2 (f) 5^2 (g) 2^4 (h) $\frac{1}{3^2}$ (i) $\frac{1}{5^6}$ (j) 3^6
5. (a) t^2 (b) x^2 (c) x^{10} (d) $3x^4$ (e) $\frac{1}{a^6}$ (f) $\frac{n^4}{9}$
6. (a) t (b) g^{-1} (c) m^2 (d) y
7. (a) 1 (b) $\frac{1}{3}$ (c) $\frac{1}{8}$ (d) $\frac{1}{9}$ (e) 8 (f) 27
8. (a) $11\frac{1}{10}$ (b) $2\frac{7}{8}$ (c) $1\frac{6}{25}$ (d) $\frac{5}{6}$ (e) $\frac{29}{100}$ (f) $\frac{1}{100}$ (g) $\frac{4}{25}$ (h) $1\frac{7}{30}$
9. (a) −4 (b) −2 (c) −2 (d) −3 (e) −1 (f) −2 (g) −3 (h) −1
10. 2^{14}

Practice Exercise 3.3 — Page 13

1. (a) 3×10^{11} (b) 8×10^7 (c) 7×10^8 (d) 2×10^9 (e) 4.2×10^7 (f) 2.1×10^{10} (g) 3.7×10^9 (h) 6.3×10^2
2. (a) 600 000 (b) 2000 (c) 50 000 000 (d) 900 000 000 (e) 3 700 000 000 (f) 28 (g) 99 000 000 000 (h) 71 000
3. (a) (i) 4.5×10^3 (ii) 7.8×10^7 (iii) 5.3×10^5 (iv) 3.25×10^4
(b) (i) 4500 (ii) 78 000 000 (iii) 530 000 (iv) 32 500
4. (a) 7×10^{-3} (b) 4×10^{-2} (c) 5×10^{-9} (d) 8×10^{-4} (e) 2.3×10^{-9} (f) 4.5×10^{-8} (g) 2.34×10^{-2} (h) 2.34×10^{-9} (i) 6.7×10^{-3}
5. (a) 0.35 (b) 0.0005 (c) 0.000 072 (d) 0.0061 (e) 0.000 000 000 117 (f) 0.000 000 813 5 (g) 0.064 62 (h) 0.000 000 004 001

6. (a) (i) 3.4×10^{-3} (ii) 5.65×10^{-5}
(iii) 7.2×10^{-4} (iv) 9.13×10^{-1}
(b) (i) 0.0034 (ii) 0.000 056 5
(iii) 0.000 72 (iv) 0.913

7. (a) (i) Brazil
(ii) 250 000 000
(b) 1.288×10^{9}

8. 7.5×10^{-4} cm
2×10^{-7} mm
2.25×10^{-4} mm.

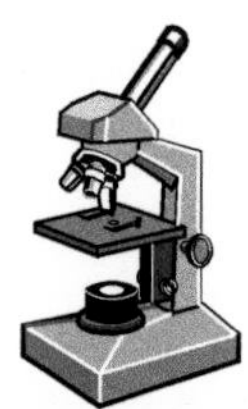

Practice Exercise 3.4 Page 14

1. (a) 262 500
(b) 105 000 000 000 000
(c) 72 250 000 000 000
(d) 0.000 000 125
(e) 30 000 000 000
(f) 263 160 000 000 000 000 000

2. (a) 9.38×10^{19} (b) 1.6×10^{1}
(c) 2.25×10^{26} (d) 1.25×10^{-37}
(e) 2.4×10^{20} (f) 5×10^{-2}

3. (a) 2.088×10^{9}
(b) $1.525\,965 \times 10^{16}$ km

4. (a) Pluto, Saturn, Jupiter
(b) 1.397×10^{5} km, 139 700 km

5. (a) 9.273×10^{6} km^2
(b) 4.93×10^{5} km^2

6. 2.52×10^{12}

7. 2.2×10^{-5}

8. 1.49×10^{9} square miles

9. 30%

10. 9×10^{-28} grams

11. (a) 4.55×10^{9} years
(b) About 6.6×10^{3} times

Practice Exercise 3.5 Page 16

1. (a) 5.2×10^{3} (b) 8.4×10^{6}
(c) 4×10^{4} (d) 5×10^{4}

2. (a) 2.8×10^{3} (b) 9×10^{5}
(c) 3×10^{7} (d) 9.5×10^{8}

3. (a) 8×10^{7} (b) 6×10^{7}
(c) 2.4×10^{8} (d) 2.7×10^{15}

4. (a) 3×10^{3} (b) 3×10^{3}
(c) 5×10^{1} (d) 2×10^{11}
(e) 4×10^{5} (f) 3×10^{-3}

5. (a) 1.5×10^{0} (b) 2.7×10^{13}
(c) 4.5×10^{4} (d) 1.25×10^{-13}
(e) 6×10^{8}

Practice Exercise 3.6 Page 17

1. (a) 20 (b) 3 (c) 10 (d) 2
(e) 4 (f) 2.5

2. (a) 8 (b) 2 (c) 3 (d) 2
(e) 5 (f) 6

3. (a) $\frac{1}{10}$ (b) $\frac{1}{7}$ (c) $\frac{1}{2}$ (d) $\frac{1}{5}$
(e) $\frac{1}{4}$ (f) $\frac{1}{3}$

4. (a) 100 (b) 27 (c) 8 (d) 4
(e) 32 (f) 243 (g) 25 (h) 32
(i) 81 (j) 216

5. (a) $\frac{1}{100}$ (b) $\frac{1}{64}$ (c) $\frac{1}{4}$ (d) $\frac{1}{8}$
(e) $\frac{1}{8}$ (f) $\frac{1}{100\,000}$ (g) $\frac{1}{125}$ (h) $\frac{1}{8}$
(i) $\frac{1}{32}$ (j) $\frac{1}{25}$

6. (a) (i) $\frac{1}{2}$ (ii) $\frac{3}{7}$
(b) (i) 25 (ii) 9

7. (a) 3 (b) 4 (c) 5 (d) $\frac{1}{2}$
(e) $\frac{2}{3}$ (f) $\frac{1}{4}$ (g) $\frac{3}{4}$ (h) 256
(i) 125 (j) 1.5

8. (a) $\frac{3}{4}$ (b) $\frac{1}{2}$ (c) 3 (d) $\frac{15}{56}$
(e) $6\frac{3}{4}$

10. (a) 19.300 259 07… (b) 291.866 535 1…
(c) 3.169 674 155… (d) 3.024 773 331…
(e) 8.556 307 843… (f) 6.695 526 189…

Review Exercise 3 Page 19

1. (a) 16 (b) 343 (c) 10 000
(d) 8 (e) 3 (f) 108
(g) 300 (h) 90 000

2. (a) 3^8 (b) 7^4 (c) 4^9
(d) 2^{12} (e) 2^{-2} (f) 5^{-4}
(g) 6^4 (h) $4^{\frac{2}{3}}$ (i) 3^6

3. (a) $6y^5$ (b) $2t^3$ (c) m^3n^2 (d) a^4b^2
(e) a^{-1} (f) y (g) m^2n^{-1} (h) $3rs^2$
(i) $9d^6$ (j) $9p^4q^2$ (k) x^5 (l) x^7
(m) y^2 (n) m^{-1} (o) $2t^2$ (p) 3

4. (a) $\frac{1}{49}$ (b) 1 (c) $\frac{1}{16}$ (d) 4
(e) $\frac{8}{6} = \frac{4}{3}$ (f) 9 (g) 100 (h) $\frac{1}{2}$
(i) 27 (j) $\frac{1}{125}$ (k) $\frac{1}{4}$ (l) 16

5. (a) 2^0 (b) 2^6

6. (a) $x = 5$ (b) $x = 1$
(c) $x = 6$ (d) $x = 4$

7. (a) 5.6×10^{4} (b) 3×10^{8}
(c) 1.2×10^{-1}

8. (a) 3.45×10^{10} (b) 5.43×10^{-7}
(c) 1.125×10^{2}

9. 8×10^{4}

10. 6×10^{8}

11. (a) 4.746×10^{7} (b) 2.67×10^{8}

12. £1.81×10^{4}

13. 1.35×10^{9} km^3

14. 1.5×10^{8} km

Chapter 4 Using Percentages

Practice Exercise 4.1 Page 20

1. 600 ml
2. £600
3. £300
4. 1.60 m
5. £312 500
6. 67.5 mg
7. £1620
8. £8000
9. School A 425, School B 450
10. £400
11. £150

Practice Exercise 4.2 Page 22

1. £10
2. (a) £30 (b) £15
3. £225
4. £48
5. £225
6. £242
7. (a) earns 78p more interest
8. (a) £96 (b) £101.92 (c) £5.92
9. £11 910.16
10. (a) £2187.91 (b) £17 910.78
11. (a) £10 534.47 (b) £2534.47
12. (a) £6151.15 (b) £1151.15
13. **Option 2**
    Option 1: 9042.59
    Option 2: 9093.69

Practice Exercise 4.3 Page 24

1. £1715
2. (a) 2860 (b) 12 days
3. £47 200
4. (a) (i) £8270 (ii) 63.6%
   (b) (i) £12 700 (ii) 63.6%
   (c) Same
5. (a) £1156 (b) 5 years
6. £6220
7. (a) 39 400 (b) 9 years
8. At the end of: Year 1: £7500, Year 2: £6000, Year 3: £5400, Year 4: £5130, Year 5: £4870.
9. £208 000
10. (a) £10 600 (b) £9390 (c) £9680 (d) 14 years

Review Exercise 4 Page 25

1. 270
2. £42
3. £700
4. £4
5. £17 190
6. 780
7. 9 years
8. (a) £140 (b) £147.52
9. £892.67
10. 11 m
11. 81 m
12. (a) 6900
    (b) Yes. $6000 \times 1.15^4 = 10\,500$
13. (a) 49 days (b) Never
14. 3.75%
15. (a) £10 160 (b) No. Bond £10 130

Chapter 5 Working with Fractions

Practice Exercise 5.1 Page 28

1. E.g. $\frac{15}{24}$, $\frac{30}{48}$, $\frac{45}{72}$, …, $\frac{10}{16}$, $\frac{20}{32}$, …
   Simplest form $\frac{5}{8}$

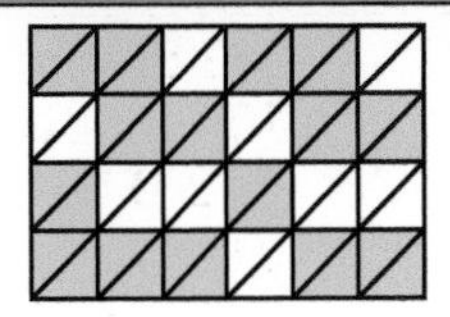

2. $\frac{8}{15}$, $\frac{3}{5}$, $\frac{2}{3}$, $\frac{7}{10}$
3. $\frac{4}{5}$
4. (a) $\frac{4}{5}$ (b) $\frac{2}{3}$ (c) $\frac{2}{3}$ (d) $\frac{2}{5}$
5. (a) $1\frac{1}{2}$ (b) $2\frac{1}{8}$ (c) $3\frac{3}{4}$ (d) $4\frac{3}{5}$
6. (a) $\frac{27}{10}$ (b) $\frac{8}{5}$ (c) $\frac{35}{6}$
7. $\frac{3}{5}$
8. $\frac{30}{50} = \frac{3}{5}$
9. (a) 3 (b) 4 (c) 3 (d) 8 (e) 8 (f) 9 (g) 12 (h) 20 (i) 40 (j) 12
10. (a) 6 (b) 24
11. £5
12. £148.40
13. £3187.50
14. (a) 9 squares (b) 10 squares (c) $\frac{5}{24}$

Practice Exercise 5.2 Page 29

1. (a) $\frac{3}{8}$ (b) $\frac{7}{12}$ (c) $\frac{7}{10}$ (d) $\frac{8}{15}$ (e) $\frac{9}{14}$
2. (a) $\frac{1}{8}$ (b) $\frac{1}{12}$ (c) $\frac{3}{10}$ (d) $\frac{2}{15}$ (e) $\frac{5}{14}$
3. (a) $1\frac{1}{4}$ (b) $1\frac{1}{2}$ (c) $1\frac{11}{20}$ (d) $1\frac{8}{21}$ (e) $1\frac{5}{24}$
4. (a) $\frac{1}{8}$ (b) $\frac{8}{15}$ (c) $\frac{5}{8}$ (d) $\frac{1}{15}$ (e) $\frac{1}{3}$
5. (a) $4\frac{1}{4}$ (b) $3\frac{5}{6}$ (c) $4\frac{3}{8}$ (d) $5\frac{17}{20}$ (e) $6\frac{13}{30}$
6. (a) $1\frac{1}{10}$ (b) $\frac{5}{12}$ (c) $1\frac{3}{8}$ (d) $3\frac{3}{10}$ (e) $2\frac{1}{4}$

7. $\frac{4}{15}$

8. $\frac{13}{60}$

9. (a) $\frac{5}{12}$ (b) $\frac{11}{12}$

10. (a) $\frac{1}{4}$ (b) Billy

Practice Exercise 5.3 — Page 30

1. (a) $3\frac{1}{2}$ (b) $1\frac{4}{5}$ (c) $6\frac{1}{4}$ (d) 9 (e) $10\frac{1}{2}$
2. (a) $\frac{1}{20}$ (b) $\frac{1}{10}$ (c) $\frac{3}{8}$ (d) $\frac{1}{16}$ (e) $\frac{1}{20}$
3. (a) $7\frac{1}{2}$ (b) $7\frac{1}{2}$ (c) $4\frac{4}{5}$ (d) $7\frac{1}{3}$ (e) $7\frac{4}{5}$
   (f) $2\frac{1}{4}$ (g) $\frac{3}{4}$ (h) $\frac{4}{5}$ (i) $\frac{2}{3}$ (j) $\frac{5}{8}$
4. (a) $\frac{1}{2}$ (b) $\frac{3}{10}$ (c) $\frac{1}{3}$ (d) $\frac{2}{7}$ (e) $\frac{1}{4}$
5. (a) $1\frac{1}{8}$ (b) 2 (c) $3\frac{3}{4}$ (d) $3\frac{17}{20}$ (e) $4\frac{7}{8}$
   (f) $1\frac{13}{15}$ (g) 14 (h) $13\frac{1}{2}$ (i) $2\frac{7}{10}$ (j) $8\frac{3}{4}$
6. (a) $\frac{4}{15}$ (b) $\frac{2}{5}$
7. (a) $\frac{4}{15}$ (b) 15
8. $\frac{1}{10}$

Practice Exercise 5.4 — Page 31

1. (a) $\frac{1}{10}$ (b) $\frac{3}{8}$ (c) $\frac{1}{3}$ (d) $\frac{1}{10}$ (e) $\frac{2}{7}$
2. (a) $\frac{3}{4}$ (b) $\frac{1}{3}$ (c) $\frac{4}{7}$ (d) $\frac{4}{9}$ (e) $\frac{2}{3}$
3. (a) 2 (b) $\frac{2}{5}$ (c) $2\frac{5}{8}$ (d) $3\frac{1}{3}$ (e) 6
4. (a) $\frac{5}{6}$ (b) $\frac{9}{16}$ (c) $\frac{4}{5}$ (d) $1\frac{1}{3}$ (e) $\frac{2}{3}$
   (f) $1\frac{1}{2}$ (g) $\frac{2}{3}$ (h) $1\frac{1}{6}$ (i) $1\frac{1}{2}$ (j) $1\frac{4}{5}$
5. (a) 14 (b) 5 (c) 2 (d) 6 (e) 6
   (f) $\frac{4}{5}$ (g) $1\frac{1}{2}$ (h) $1\frac{1}{4}$ (i) $1\frac{1}{7}$ (j) $\frac{21}{25}$
6. $\frac{1}{10}$
7. $\frac{1}{9}$
8. 23

Practice Exercise 5.5 — Page 32

1. £3.20
2. 70 km
3. £3.20
4. £8
5. 96 cm

Practice Exercise 5.6 — Page 32

1. (a) $3\frac{19}{24}$ (b) $1\frac{7}{12}$ (c) $13\frac{11}{36}$ (d) $\frac{11}{20}$
   (e) $6\frac{1}{12}$ (f) $\frac{1}{9}$ (g) $6\frac{19}{24}$ (h) $2\frac{17}{30}$
2. (a) $\frac{1}{4}$ (b) 3 (c) $4\frac{1}{5}$ (d) $12\frac{3}{4}$
   (e) $\frac{20}{21}$ (f) $1\frac{1}{2}$ (g) 4 (h) $5\frac{1}{7}$
3. (a) $5\frac{1}{12}$ (b) $\frac{2}{3}$ (c) $7\frac{1}{3}$ (d) $4\frac{5}{12}$
   (e) $2\frac{3}{5}$ (f) $3\frac{1}{2}$ (g) 2 (h) $1\frac{1}{7}$
4. (a) $8\frac{1}{16}$ (b) $6\frac{77}{80}$ (c) $2\frac{1}{8}$
5. (a) 1 (b) $3\frac{2}{3}$ (c) 3 (d) $\frac{5}{16}$
   (e) $1\frac{13}{20}$ (f) $\frac{1}{4}$

Review Exercise 5 — Page 33

1. (a) $\frac{3}{11}$ (b) $\frac{2}{5}$ (c) $\frac{3}{8}$ (d) $\frac{5}{8}$ (e) $\frac{1}{10}$
2. (a) 64 (b) $\frac{7}{16}$
3. $\frac{7}{10}$
4. £80
5. (a) $1\frac{1}{12}$ (b) $4\frac{3}{10}$ (c) $3\frac{7}{24}$ (d) $2\frac{3}{5}$
   (e) $\frac{2}{3}$ (f) $4\frac{33}{40}$
6. (a) $\frac{15}{28}$ (b) $1\frac{1}{2}$ (c) $3\frac{3}{4}$ (d) $\frac{9}{35}$
   (e) $1\frac{9}{16}$ (f) $\frac{10}{27}$
7. (a) $4\frac{10}{63}$ (b) $2\frac{19}{30}$ (c) 10 (d) $1\frac{5}{7}$
8. (a) $25\frac{1}{6}$ (b) $2\frac{1}{5}$ (c) $1\frac{3}{4}$ (d) $7\frac{1}{20}$
9. (a) £4.17 (b) £2.94 (c) £0.39
   (d) £1.44 (e) £20.13
10. (a) 140 (b) $\frac{1}{4}$
11. (a)

$$\frac{5}{6}-\frac{4}{5}=\frac{5\times5-6\times4}{6\times5}=\frac{1}{30}$$
$$\frac{6}{7}-\frac{5}{6}=\frac{6\times6-7\times5}{7\times6}=\frac{1}{42}$$
$$\frac{7}{8}-\frac{6}{7}=\frac{7\times7-8\times6}{8\times7}=\frac{1}{56}$$
$$\frac{8}{9}-\frac{7}{8}=\frac{8\times8-9\times7}{9\times8}=\frac{1}{72}$$
$$\frac{9}{10}-\frac{8}{9}=\frac{9\times9-10\times8}{10\times9}=\frac{1}{90}$$

   (b) $\frac{1}{380}$

12. P(left-handed female) $= \frac{1}{10}\times\frac{2}{5}=\frac{1}{25}$
    P(left-handed male) $= \frac{1}{10}\times\frac{3}{5}=\frac{3}{50}$
    P(right-handed female) $= \frac{9}{10}\times\frac{2}{5}=\frac{9}{25}$
    P(right-handed male) $= \frac{9}{10}\times\frac{3}{5}=\frac{27}{50}$
13. $\frac{9}{20}$
14. £300
15. Yes
16. (a) $14\frac{5}{24}$ (b) $2\frac{3}{4}$ (c) $81\frac{3}{8}$

Chapter 6

Algebraic Expressions

Practice Exercise 6.1 Page 35

1. (a) $3n$ (b) $5y$ (c) $10g$ (d) $4y$ (e) $8m$ (f) $-4x$ (g) $-14a$
2. (a) $8x + 4$ (b) $3 + x$ (c) $x + 4$ (d) $3x + 3$ (e) $5 - 5x$ (f) $-3x - 7$ (g) $7x + 4$ (h) $-x - 1$ (i) $2 + x$
3. (a) $2x^2$ (b) $2x^2 + 7x$ (c) $5x + 2x^2$ (d) $2x^2 - 3x$ (e) $-x^2 - 4$ (f) $3 + x^2$
4. (a) $-3x$ (b) $-5x$ (c) $2x$ (d) $-6x$ (e) $-x^2$ (f) $-2x^2$ (g) $-10x^2$ (h) $10x^2$

Practice Exercise 6.2 Page 36

1. (a) $2x + 10$ (b) $3a + 18$ (c) $4y + 12$
2. (a) $3x + 6$ (b) $2y + 10$ (c) $4x + 2$ (d) $3p + 3q$
3. (a) $a^2 + a$ (b) $2d + d^2$ (c) $2x^2 + x$
4. (a) $2x + 8$ (b) $3t - 6$ (c) $6a + 2b$ (d) $15 - 10d$ (e) $x^2 + 3x$ (f) $t^2 - 3t$ (g) $2g^2 + 3g$ (h) $2m - 3m^2$
5. (a) $2p^2 + 6p$ (b) $6d - 9d^2$ (c) $m^2 - mn$ (d) $5x^2 + 10xy$
6. (a) $2x + 5$ (b) $3a + 11$ (c) $6w - 17$ (d) $10 + 2p$ (e) $3q$ (f) $7 - 3t$ (g) $5z + 8$ (h) $8t + 15$ (i) $2c - 6$ (j) $5a - 9$ (k) $3y - 10$ (l) $2x + 6$ (m) $8a + 23$ (n) $10x - 12$ (o) $2p - 11$ (p) $5a + 2b$ (q) $3x + y$ (r) $2p - 5q$ (s) $5x - x^2$ (t) $a^2 - 2a$ (u) $2y$
7. (a) $5x + 8$ (b) $5a + 13$ (c) $9y + 23$ (d) $9a + 5$ (e) $26t + 30$ (f) $5z + 13$ (g) $12q + 16$ (h) $11x - 3$ (i) $20e - 16$ (j) $12d + 6$ (k) $3m^2 - 3m$ (l) $5a^2 - 4a$
8. (a) $-3x - 6$ (b) $-3x + 6$ (c) $-2y + 10$ (d) $-6 + 2x$ (e) $-15 + 3y$ (f) $-4 - 4a$ (g) $3 - 2a$ (h) $2d + 6$ (i) $2b - 6$ (j) $-6p - 9$ (k) $m - 6$ (l) $5d + 1$ (m) $-a^2 + 2a$ (n) $d - d^2$ (o) $2x^2$ (p) $-6g^2 - 9g$ (q) $7t^2 - 6t$ (r) $8m - 2m^2$
9. (a) $3a - 1$ (b) $y - 5$ (c) $5m - 1$ (d) $3x - 4$ (e) $1 - d$ (f) $t - 10$ (g) $4m + 9$ (h) $-x - 18$ (i) $23 - 6a$

Practice Exercise 6.3 Page 38

1. $x^2 + 7x + 12$
2. $x^2 + 6x + 5$
3. $x^2 - 3x - 10$
4. $2x^2 - 3x - 2$
5. $3x^2 - 20x + 12$
6. $6x^2 + 7x + 2$
7. $x^2 + 6x - 16$
8. $x^2 + 3x - 10$
9. $x^2 + 2x - 3$
10. $x^2 - 5x + 6$
11. $x^2 - 5x + 4$
12. $x^2 - 5x - 14$
13. $2x^2 + 7x + 3$
14. $3x^2 + 13x - 10$
15. $5x^2 - 8x + 3$
16. $4x^2 + 4x - 3$
17. $6x^2 + 13x - 5$
18. $12x^2 + 14x - 10$
19. $x^2 - 9$
20. $x^2 - 25$
21. $x^2 - 49$
22. $x^2 + 10x + 25$
23. $x^2 - 14x + 49$
24. $4x^2 - 12x + 9$
25. (a) $x^3 + 5x^2 + 10x + 12$ (b) $2x^3 - 3x^2 - 5x + 6$ (c) $6x^3 - 5x^2 + 9x - 4$ (d) $-3x^4 + 2x^3 + x^2 - 4x + 10$

Practice Exercise 6.4 Page 39

1. (a) $2(x + y)$ (b) $3(a - 2b)$ (c) $2(3m + 4n)$ (d) $x(x - 2)$ (e) $a(b + 1)$ (f) $x(2 - y)$ (g) $2a(b - 2)$ (h) $2x(2x + 3)$ (i) $dg(1 - g)$
2. (a) $2(a + b)$ (b) $5(x - y)$ (c) $3(d + 2e)$ (d) $2(2m - n)$ (e) $3(2a + 3b)$ (f) $2(3a - 4b)$ (g) $4(2t + 3)$ (h) $5(a - 2)$ (i) $2(2d - 1)$ (j) $3(1 - 3g)$ (k) $5(1 - 4m)$ (l) $4(k + 1)$
3. (a) $x(y - z)$ (b) $g(f + h)$ (c) $b(a - 2)$ (d) $q(3 + p)$ (e) $a(1 + b)$ (f) $g(h - 1)$ (g) $a(a + 3)$ (h) $t(5 - t)$ (i) $d(1 - d)$ (j) $m(m + 1)$ (k) $r(5r - 3)$ (l) $x(3x + 2)$
4. (a) $3(y + 2 - 3x)$ (b) $t(t^2 - t + 1)$ (c) $2d(d + 2)$ (d) $3m(1 - 2n)$ (e) $2g(f + 2g)$ (f) $4q(p - 2)$ (g) $3y(2 - 5y)$ (h) $2x(3x + 2y)$ (i) $2n(3n - 1)$ (j) $2b(2a + 3)$ (k) $\frac{1}{2}a(1 - a)$ (l) $4x(5 + y)$ (m) $a^2(a + a^3 + 1)$ (n) $\pi r(2 + r)$ (o) $4ab(5a + 3b)$ (p) $3pq(1 - 3p)$

Practice Exercise 6.5 Page 40

1. (a) $x(x + 5)$ (b) $x(x - 7)$ (c) $y(y - 6)$ (d) $2y(y - 6)$ (e) $t(5 - t)$ (f) $y(8 + y)$ (g) $x(x - 20)$ (h) $3x(x - 20)$
2. (a) $(x + 3)(x - 3)$ (b) $(x + 9)(x - 9)$ (c) $(y + 5)(y - 5)$ (d) $(y + 1)(y - 1)$ (e) $(x + 8)(x - 8)$ (f) $(10 + x)(10 - x)$ (g) $(6 + x)(6 - x)$ (h) $(x + a)(x - a)$ (i) $(10x + 3)(10x - 3)$ (j) $(5 + 4x)(5 - 4x)$ (k) $(5y + 7)(5y - 7)$ (l) $(3x + 5y)(3x - 5y)$
3. (a) $(x + 5)(x + 1)$ (b) $(x + 7)(x + 2)$ (c) $(x + 2)(x + 4)$ (d) $(x + 3)(x + 6)$ (e) $(x - 5)(x - 1)$ (f) $(x - 5)(x - 2)$ (g) $(x - 4)(x - 3)$ (h) $(x + 4)(x - 1)$ (i) $(x + 7)(x - 2)$ (j) $(x - 5)(x + 1)$

4. (a) $(x+1)(x+2)$ (b) $(x+1)(x+7)$
(c) $(x+3)(x+5)$ (d) $(x+2)(x+6)$
(e) $(x+1)(x+11)$ (f) $(x+4)(x+5)$
(g) $(x+4)(x+6)$ (h) $(x+4)(x+9)$
(i) $(x+1)(x+14)$ (j) $(x+2)(x+8)$

5. (a) $(x-3)^2$ (b) $(x-4)(x-2)$
(c) $(x-10)(x-1)$ (d) $(x-15)(x-1)$
(e) $(x-5)(x-3)$ (f) $(x-8)(x-2)$
(g) $(x-10)(x-2)$ (h) $(x-8)(x-3)$
(i) $(x-1)(x-12)$ (j) $(x-2)(x-6)$

6. (a) $(x-3)(x+2)$ (b) $(x-6)(x+1)$
(c) $(x-4)(x+6)$ (d) $(x-3)(x+8)$
(e) $(x-5)(x+3)$ (f) $(x-3)(x+6)$
(g) $(x-8)(x+5)$ (h) $(x-6)(x+2)$
(i) $(x-2)(x+5)$ (j) $(x+4)(x-5)$

7. (a) $(x-2)^2$ (b) $(x+5)(x+6)$
(c) $(x-2)(x+4)$ (d) $(x-7)(x+3)$
(e) $(x-4)(x+5)$ (f) $(x+3)(x+4)$
(g) $(x+4)^2$ (h) $(x-1)^2$
(i) $(x+7)(x-7)$ (j) $t(t+12)$
(k) $(x-2)(x-7)$ (l) $(x-6)(x-1)$
(m) $(x+2)(x+9)$ (n) $(x+3)(x+8)$
(o) $(x+1)(x+18)$ (p) $(x+y)(x-y)$
(q) $(x+3)(x-2)$ (r) $4(x+3)(x-3)$
(s) $(y-5)^2$ (t) $(x-6)^2$

Practice Exercise 6.6 Page 42

1. (a) $(2x+6)(x+3)$ (b) $(2y+3)(y-6)$
(c) $(2a+9)(a-2)$ (d) $(2m-2)(m-9)$
(e) $(4x-1)(x-3)$ (f) $(2d-1)(2d+3)$

2. (a) $(2x+1)(x+3)$ (b) $(2x+3)(x+4)$
(c) $(2a+1)(a+12)$ (d) $(2x+1)^2$
(e) $(3y+2)(2y+1)$ (f) $(3x+4)(2x+3)$
(g) $(11m+1)(m+1)$ (h) $(3x+2)(x+5)$
(i) $(5k+2)(k+2)$

3. (a) $(2a-3)(a-1)$ (b) $(2y-7)(y-1)$
(c) $(2x-1)^2$ (d) $(3x-2)(x-1)$
(e) $(3d-1)(2d-3)$ (f) $(2x-1)(x-1)$
(g) $(2x-5)(x-3)$ (h) $(3y-2)^2$
(i) $(2t-3)^2$

4. (a) $(2x-3)(x+2)$ (b) $(2a-3)(a+1)$
(c) $(3x+1)(x-5)$ (d) $(3t-1)(t+5)$
(e) $(2y-7)(y+1)$ (f) $(4y-1)(y+2)$
(g) $(3x-1)(3x+2)$ (h) $(10x-3)(x+3)$
(i) $(5m+3)(m-1)$

Practice Exercise 6.7 Page 42

1. (a) $3(x+4y)$ (b) $(t+4)(t-4)$
(c) $(x+1)(x+3)$ (d) $y(y-1)$
(e) $2(d^2-3)$ (f) $(p+q)(p-q)$
(g) $a(a-2)$ (h) $(x-1)^2$
(i) $2y(y-4)$ (j) $(a-3)^2$
(k) $3(m+2)(m-2)$ (l) $(v-3)(v+2)$
(m) $a(x+y)(x-y)$ (n) $-2(4+x)$
(o) $(2-k)^2$ (p) $2(3+x)(3-x)$
(q) $2(5+x)(5-x)$ (r) $3(2x+3y)(2x-3y)$
(s) $3a(2a-1)$ (t) $(x-7)(x-8)$
(u) $(x-3)(x+5)$

2. (a) $(3a+4)(a-2)$ (b) $(3x+1)(x+10)$
(c) $(8y+7)(8y-7)$ (d) $(3x+5)(x+2)$
(e) $(7x-5)(x-2)$ (f) $(6x+5)(x-5)$
(g) $(5m-3)(m-3)$ (h) $(2x+3)(4x-5)$
(i) $(5-2x)(x-4)$ (j) $2(y+4)(2y+3)$
(k) $3(x+2)(2x-3)$ (l) $2(3n-1)(n+5)$
(m) $2(2x+3)(x+4)$ (n) $3(x-2)^2$
(o) $x(x+1)(x-1)$

Practice Exercise 6.8 Page 43

1. (a) 1 (b) 4 (c) 9 (d) 16
(e) 1 (f) 2.25 (g) 25 (h) 6.25

2. (a) $x^2+6x+9=(x+3)^2$
(b) $a^2-4a+4=(a-2)^2$
(c) $b^2+2b+1=(b+1)^2$
(d) $m^2-8m+16=(m-4)^2$
(e) $n^2-n+\frac{1}{4}=\left(n-\frac{1}{2}\right)^2$
(f) $x^2+5x+6.25=(x+2.5)^2$

3. (a) $(x+3)^2+11$ (b) $(x+3)^2-4$
(c) $(x+5)^2-29$ (d) $(x-2)^2+1$
(e) $(x-2)^2-2$ (f) $(x-2)^2-8$
(g) $(x-3)^2-5$ (h) $(x-4)^2-16$
(i) $(x+6)^2-36$

4. (a) $a=3,\ b=6$
(b) $a=5,\ b=-20$
(c) $a=-3,\ b=-14$
(d) $a=-4,\ b=-12$
(e) $a=6,\ b=-32$
(f) $a=3,\ b=0$

5. (a) $a=-2,\ b=-5$ (b) -5

6. $a=8,\ b=-3$

7. 1

Review Exercise 6 Page 45

1. (a) y^2-4y (b) $7y+4$

2. (a) $3x-6y$ (b) $2x^2+6x$
(c) x^3-3x^2

3. (a) $2x^2-x$ (b) $1-3x$

4. (a) (i) $3(2x-5)$ (ii) $y(y+7)$
(b) $y+12$

5. (a) x^2-x-6 (b) $2x^2-9x+4$
(c) $x-x^2$ (d) $x+1$

6. $x^2-3x-10$

7. (a) $2x^2+9x+4$ (b) $2x(4x-3)$

8. (a) $2xy(y+2)$ (b) $(m-1)(m-8)$

9. (a) $x^2-10x+21$ (b) $x^2+8x+16$
(c) $6x^2+7x-5$ (d) $16x^2-9$
(e) $4x^2-4x+1$ (f) $2x^3-x^2-7x-4$

10. (a) $3(x^2-2)$ (b) $(3x+5)(3x-5)$
(c) $x(x+1)(x-1)$ (d) $(x+1)(x+10)$
(e) $(x-1)^2$ (f) $(x-3)(x+11)$
(g) $(2x+7)(x+1)$ (h) $(2x-5)(x+1)$
(i) $4(x+2)(x-2)$

11. (a) $64,\ (x+8)^2$ (b) $144,\ (x-12)^2$

12. $3x^2+x-2=(3x-2)(x+1)$
$3x^2-5x+2=(3x-2)(x-1)$
Common factor is $(3x-2)$.

13. $p=4,\ q=4$

14. $(x+3)$ cm
15. (a) 27.2 (b) 50 (c) 79.74
16. $2x^2 - 8y^2 = 2(x+2y)(x-2y)$

Chapter 7 Algebraic Fractions

Practice Exercise 7.1 Page 46

1. (a) $2d+3$ (b) $3x+2$ (c) $4a+5b$
   (d) $3m-2n$ (e) $4x-2y$ (f) $a+b$
   (g) $x-1$ (h) x^2-2
   (i) $\frac{4}{x-3}$ (j) $\frac{x+2}{3}$ (k) $\frac{x+2y+3z}{2x-3y+z}$
   (l) $\frac{x+2}{3x+2}$ (m) $\frac{x}{3x-2}$ (n) $\frac{x}{2x^2-1}$
   (o) $\frac{1}{2x-1}$ (p) $\frac{1}{3x}$ (q) $\frac{3}{2}$
   (r) $\frac{m}{3}$ (s) $\frac{x-3}{x+2}$ (t) $-\frac{5}{3}$
2. (a) **E** (b) **C** (c) **B** (d) **A** (e) **D**
3. (a) $\frac{x+5y}{3+4xy}$ (b) $\frac{x+y+2z}{2x-3y+z}$
   (c) $\frac{ab-3}{b+a}$ (d) $\frac{x+b+y}{ab-2y}$
4. (a) $\frac{x}{x+1}$ (b) $\frac{x+2}{x+3}$ (c) $\frac{x}{x+3}$
   (d) $\frac{x-5}{x+3}$ (e) $\frac{x-2}{x-3}$ (f) $\frac{x-5}{x+4}$
   (g) $x-1$ (h) $\frac{1}{x+2}$ (i) $\frac{x+1}{2}$
   (j) $\frac{x+2}{x+3}$ (k) $\frac{x-5}{x-4}$ (l) $\frac{2x-5}{3x-1}$

Practice Exercise 7.2 Page 49

1. (a) $\frac{8}{7}$ (b) $\frac{x^2}{4}$ (c) $\frac{3}{4}$ (d) $\frac{3}{4}$
2. (a) $\frac{20}{21}$ (b) $\frac{5}{8}$ (c) $\frac{3}{4}$ (d) 6
3. (a) $\frac{2x}{15}$ (b) $\frac{8}{3y}$ (c) $\frac{2x}{3}$ (d) $\frac{10x^2}{3y}$
4. (a) $\frac{3y}{10}$ (b) $\frac{2}{3y}$ (c) $\frac{3x}{4}$ (d) $\frac{5}{18}$
5. (a) $\frac{2x^2}{3y}$ (b) $\frac{45y}{4}$ (c) $\frac{21y}{2}$ (d) $4xy$

Practice Exercise 7.3 Page 50

1. (a) $\frac{7x}{12}$ (b) $\frac{7x}{10}$ (c) $\frac{25x}{24}$ (d) $\frac{29x}{14}$
2. $\frac{2x}{3} + \frac{4x}{9} = \frac{18x+12x}{27} = \frac{30x}{27} = \frac{10x}{9}$
3. (a) $\frac{35+x^2}{5x}$ (b) $\frac{12}{x}$ (c) $\frac{10x^2+7}{35x}$ (d) $\frac{7}{3x}$
4. (a) $\frac{x}{12}$ (b) $\frac{7x}{30}$ (c) $\frac{x}{5}$ (d) $\frac{17x}{40}$
5. $\frac{3x}{4} - \frac{5x}{12} = \frac{9x-5x}{12} = \frac{4x}{12} = \frac{x}{3}$
6. (a) $\frac{12-x^2}{4x}$ (b) $\frac{7}{2x}$ (c) $\frac{15x^2-8}{20x}$ (d) $\frac{23}{60x}$

Practice Exercise 7.4 Page 52

1. (a) $\frac{5x+7}{6}$ (b) $\frac{7x+5}{12}$ (c) $\frac{7x+11}{10}$
2. (a) $\frac{1-x}{6}$ (b) $\frac{3x+12}{10}$ (c) $\frac{-x-5}{12}$
3. (a) $\frac{5x+5}{(x+3)(x-2)}$ (b) $\frac{x-11}{(x+4)(x-1)}$
   (c) $\frac{2x+14}{(x-3)(x+1)}$ (d) $\frac{4x-11}{(x-5)(x+4)}$
   (e) $\frac{x+3}{(x-1)(x+1)}$ (f) $\frac{7x+13}{(x+4)(x-1)}$
4. (a) $\frac{5x+2}{x(x+1)}$ (b) $\frac{3x+2}{x(x-2)}$ (c) $\frac{5x-9}{x(x-3)}$
5. (a) $\frac{2x+4}{x(x+1)}$ (b) $\frac{3y+2}{y(y+2)}$ (c) $\frac{2p-6}{p(p-2)}$

Review Exercise 7 Page 52

1. (a) (i) $(x+2)(x-2)$ (b) $\frac{x(x-2)}{3x-4}$
   (ii) $(3x-4)(x+2)$
2. $\frac{6}{x-2}$
3. (a) (i) $(x+3)(x-3)$ (b) $\frac{x+3}{2x+1}$
   (ii) $(2x+1)(x-3)$
4. (a) $\frac{a}{a-d}$ (b) $\frac{p-2}{p+3}$ (c) $\frac{c+2}{c+4}$
5. (a) $\frac{2}{3}$ (b) $\frac{d}{15}$ (c) $\frac{5s}{3q}$
   (d) $\frac{2}{5}$ (e) $\frac{3g}{2h}$ (f) $3q$
6. (a) $\frac{13x}{24}$ (b) $\frac{28x}{9}$ (c) $\frac{11x}{10}$
7. (a) $\frac{3x+4}{4}$ (b) $\frac{x+13}{(x-3)(x+5)}$
   (c) $\frac{8x+5}{x(x+1)}$

Chapter 8 Gradient of a Straight Line Graph

Practice Exercise 8.1 Page 53

1. (a) **(1)** 1 **(2)** $-\frac{3}{4}$ **(3)** 0
   (4) undefined **(5)** -3, **(6)** 0
   (7) $\frac{4}{5}$ **(8)** -3, **(9)** 1
   (10) undefined **(11)** -4, **(12)** -1
   (13) 2 **(14)** 0
   (b) **(9)** (c) **(5)**
2. (a) **(1)** undefined **(2)** $-\frac{1}{4}$ **(3)** $\frac{1}{2}$
   (b) **(4)** $\frac{1}{5}$ **(5)** 0 **(6)** $-\frac{2}{5}$

Practice Exercise 8.2 Page 54

1. $\frac{6}{5} = 1.2$
2. (a) $\frac{2}{3}$ (b) -2 (c) 4
   (d) undefined (e) 5 (f) 0
3. (a) $x = 4$ (b) $y = -4$
4. (a) $m_{AB} = 3$ (b) $a = 9$
5. $b = 9$

6. $a = 2$
7. $a = 11$
8. $b = 2$
9. (a) $m_{AD} = 0$, $m_{BC} = 0$.
   (b) $m_{AB} = -1.5$, $m_{CD} = -1.5$.
   (c) Opposite sides are parallel.

Review Exercise 8 — Page 55

1. $-\frac{3}{5}$
2. $\frac{1}{4}$
3. -2
4. (a) $m_{AB} = 1$, $m_{BC} = -\frac{2}{3}$, $m_{AD} = -\frac{2}{3}$, $m_{DC} = 1$.
   (b) Opposite sides are parallel.
5. $a = 14$
6. $c = 1$

Chapter 9 Straight Line Graphs

Practice Exercise 9.1 — Page 57

1. (b) Same slope, parallel. y-intercept is different.
2. (a) gradient 3, y-intercept -1
3. $y = 3x$, $y = 3x + 2$
4. gradients: 3, 2, -2, $\frac{1}{2}$, 2, 0
   y-intercepts: 5, -3, 4, 3, 0, 3
5. (a) $y = 5x - 4$ (b) $y = -\frac{1}{2}x + 6$
6. (a) **(1)** $y = 4$ **(2)** $x = -3$ **(3)** $x = 1$ **(4)** $y = -1$ **(5)** $y = x$
   (b) $y = 3x$
7. **(1) C** **(2) D** **(3) B** **(4) A**
8. (a) $y = x - 2$ (b) $y = 2x - 2$
   (c) $y = -2x - 2$
9. (0, 4)

Practice Exercise 9.2 — Page 59

1. (a) $y = x + 5$ (b) $y = 2x - 11$
   (c) $y = -5x - 18$ (d) $y = -3x - 20$
2. $y = 2x + 1$
3. $y = mx + c$ and $m = 3.5$, $c = -2$
   $y = 3.5x - 2$
   Multiply both sides by 2.
   $2y = 7x - 4$
4. (a) $y = -x + 8$ (b) $y = 5x - 4$
   (c) $m_{BC} = \frac{1}{2}$, $(a, b) = (1, 1)$
   $y - b = m(x - a)$
   $y - 1 = \frac{1}{2}(x - 1)$
   $y - 1 = \frac{1}{2}x - \frac{1}{2}$
   Add 1 to both sides.
   $y = \frac{1}{2}x + \frac{1}{2}$
5. $y = -2x - 1$

Practice Exercise 9.3 — Page 60

1. $y = \frac{3x}{2} + 3$
2. (a) $y = -\frac{1}{2}x + 2$ (b) $y = -\frac{4}{5}x + 4$
   (c) $y = -\frac{2}{3}x + \frac{4}{3}$ (d) $y = \frac{2}{7}x - 2$
3. (a) $\frac{1}{2}$ (b) 1 (c) 2 (d) $-\frac{1}{3}$ (e) $\frac{3}{4}$ (f) $-\frac{2}{5}$
4. (a) $a = 0$ (b) $a = \frac{1}{2}$ (c) $a = -2$
   (d) $a = 2\frac{1}{2}$ (e) $a = 2$ (f) $a = \frac{2}{5}$
5. (a) $y = x +$ (any number)
   (b) $y = 2x +$ (any number)
   (c) $y = \frac{1}{2}x +$ (any number)
6. $y = -\frac{3}{2}x +$ (any number)
7. (a) $A(0, 2)$, $B(-4, 0)$
   (b) $\frac{1}{2}$
   (c) (ii) $y = \frac{1}{2}x +$ (any number)
8. (a) $P(0, 3)$, $Q(6, 0)$
   (b) $-\frac{1}{2}$
   (c) (ii) $y = -\frac{1}{2}x +$ (any number)

Practice Exercise 9.4 — Page 61

1. **A:** $f(x) = 2 - 2x$, **B:** $f(x) = x + 2$,
   C: $f(x) = 2 - x$, **D:** $f(x) = 2$,
   E: $f(x) = 2x$, **F:** $f(x) = 2x + 2$.
2. (a) $f(0) = 5$ (b) (0, 5)
   (c) $x = -5$ (d) $(-5, 0)$
3. (a) (0, 6) (b) $(-12, 0)$

Review Exercise 9 — Page 62

1. (a) 4, $(0, -1)$ (b) -5, (0, 2)
   (c) 0, (0, 2) (d) -1, (0, 3)
   (e) $\frac{1}{3}$, (0, 1) (f) $\frac{1}{2}$, (0, 7)
   (g) Undefined, line does not cut the y axis.
   (h) 0, (0, 0)
2. $A(0, 5)$
3. (a) $A(2, 0)$
   (b) $y = 4x - 8$
4. (a) $y = -x + 8$
   (b) $y = 2x - 9$
   (c) $y = \frac{1}{3}x + \frac{1}{3}$
5. (a) $P(0, -12)$, $Q(4, 0)$
   (b) $q = 5$
6. $y = 4x + 3$
7. $y = -\frac{3}{5}x + \frac{22}{5}$
8. (a) $y = \frac{1}{2}x$ (b) $D(5, 5)$ (c) $y = 5$
9. $y = -\frac{1}{2}x + \frac{3}{2}$
10. $f(-3) = -15$, $f(-2) = 0$, $f(-1) = 3$, $f(0) = 0$, $f(1) = -3$, $f(2) = 0$, $f(3) = 15$
11. (a) $f(4) = 81$
    (b) $x = -1$
12. Gradient 3.1, $c = 3.1d$ ($c = \pi d$)

Chapter 10 Equations and Inequalities

Practice Exercise 10.1 Page 64

1. 5
2. 14
3. 6
4. 11
5. 4
6. 3
7. 4
8. 7
9. (a) 2 (b) $2(x + 3) = 2x + 6$
10. (a) 9 (b) $3(x - 2) = 3x - 6$

Practice Exercise 10.2 Page 66

1. (a) $y = 3$ (b) $n = 16$ (c) $x = 2$ (d) $y = 19$ (e) $b = 7$ (f) $x = 29$ (g) $m = 4$ (h) $k = 5$ (i) $y = 7$
2. (a) $c = 4$ (b) $a = 4$ (c) $f = 3$ (d) $q = 7$ (e) $x = 5$ (f) $y = 1$
3. (a) $p = 4$ (b) $t = 3$ (c) $h = 7$ (d) $b = 2$ (e) $d = 10$ (f) $x = 6$ (g) $c = 5$ (h) $n = 3$ (i) $x = 2$

Practice Exercise 10.3 Page 66

1. (a) $k = \frac{1}{2}$ (b) $a = -3$ (c) $d = -4$ (d) $n = -\frac{1}{2}$ (e) $t = -5$ (f) $n = 1$ (g) $m = 1\frac{1}{2}$ (h) $x = 2\frac{1}{3}$ (i) $y = -\frac{1}{2}$
2. (a) $x = -2$ (b) $y = -3$ (c) $t = -2$ (d) $a = -2$ (e) $d = -3$ (f) $g = -3$ (g) $t = \frac{1}{2}$ (h) $x = 7\frac{1}{2}$ (i) $d = 1\frac{2}{5}$ (j) $a = 1\frac{1}{2}$ (k) $g = \frac{1}{5}$ (l) $b = 4\frac{1}{2}$
3. (a) $x = -2$ (b) $n = -\frac{1}{2}$ (c) $a = -1$ (d) $y = -3$ (e) $x = -1$ (f) $d = -3$ (g) $x = -1\frac{1}{2}$ (h) $x = -4$ (i) $x = -1\frac{1}{2}$ (j) $n = 1\frac{1}{2}$ (k) $z = -9$ (l) $m = -3$ (m) $n = 4$ (n) $p = -1\frac{1}{2}$ (o) $y = -9$

Practice Exercise 10.4 Page 67

1. (a) $x = 3$ (b) $a = 2$ (c) $t = 2$ (d) $p = 5$ (e) $c = 6$ (f) $x = 3$ (g) $d = 9$ (h) $e = 5$ (i) $f = 4$
2. (a) $w = 2$ (b) $s = 3$ (c) $t = 3$ (d) $y = 1$ (e) $x = 3$ (f) $y = 5$
3. (a) $p = -1$ (b) $d = -2$ (c) $g = -2$ (d) $x = 8\frac{1}{2}$ (e) $y = \frac{2}{5}$ (f) $t = \frac{1}{2}$ (g) $t = 1\frac{3}{4}$ (h) $a = 2\frac{1}{2}$ (i) $m = 2\frac{3}{5}$
4. (a) $x = 4$ (b) $a = 2$ (c) $m = -\frac{1}{2}$ (d) $y = 2\frac{1}{2}$ (e) $w = -3$ (f) $e = 2$ (g) $a = 1\frac{1}{2}$ (h) $t = -11$ (i) $x = 1$
5. (a) $x = 3$ (b) $x = -2$ (c) $x = 6$ (d) $x = -7$ (e) $x = -1\frac{4}{7}$ (f) $x = \frac{1}{3}$

Practice Exercise 10.5 Page 68

1. (a) $x = 5$ (b) $q = 2$ (c) $t = 3$ (d) $e = 3$ (e) $g = 4$ (f) $y = 1$ (g) $x = 2$ (h) $k = 1$ (i) $a = 4$ (j) $p = 6$ (k) $m = 2$ (l) $d = 5$ (m) $y = 5$ (n) $u = 3$ (o) $q = 0$
2. (a) $d = 8$ (b) $q = 3$ (c) $c = 2$ (d) $t = 3$ (e) $w = 2$ (f) $e = 3$ (g) $g = 5$ (h) $z = 4$ (i) $m = 6$ (j) $a = 5$ (k) $x = 4$ (l) $y = 3$
3. (a) $m = -4$ (b) $t = -2$ (c) $p = -2$ (d) $x = 3\frac{1}{2}$ (e) $a = \frac{1}{2}$ (f) $b = \frac{4}{5}$ (g) $y = \frac{4}{5}$ (h) $d = \frac{3}{4}$ (i) $f = -3\frac{1}{2}$
4. (a) $x = 4$ (b) $a = 1\frac{2}{3}$ (c) $m = \frac{1}{2}$ (d) $a = -5$ (e) $y = 5\frac{1}{2}$ (f) $n = -1\frac{1}{2}$ (g) $d = -4\frac{1}{2}$ (h) $k = -11$ (i) $t = -4$ (j) $q = 2$ (k) $x = \frac{2}{3}$ (l) $a = 2\frac{1}{2}$
5. (a) $h = 2$ (b) $x = -3$ (c) $w = 3$ (d) $y = 1\frac{1}{2}$ (e) $v = -8$ (f) $c = -1\frac{1}{2}$ (g) $x = -18$ (h) $x = -7$

Practice Exercise 10.6 Page 69

1. (a) $x = 6$ (b) $d = -10$ (c) $w = 8$ (d) $n = 24$ (e) $a = 6$ (f) $m = 6$ (g) $a = \frac{15}{32}$ (h) $p = \frac{16}{21}$ (i) $t = \frac{4}{9}$ (j) $b = \frac{5}{6}$
2. (a) $h = 11$ (b) $x = 8$ (c) $a = -3$ (d) $d = -3$ (e) $a = \frac{1}{10}$ (f) $h = 1\frac{1}{2}$ (g) $x = \frac{7}{9}$ (h) $a = 5$ (i) $x = 1\frac{1}{4}$
3. (a) $x = 1\frac{1}{3}$ (b) $x = 12$ (c) $x = -1\frac{3}{5}$ (d) $x = -4$ (e) $x = 1$ (f) $x = 19$ (g) $x = -1$ (h) $x = -1$ (i) $x = 3\frac{1}{2}$

Practice Exercise 10.7 Page 70

1. (a) $6k$ kg (b) $2\frac{1}{2}$ kg

2. (a) $(n + 7)$ years old (b) Dominic is 18 years old
   Marcie is 25 years old
3. (a) $(4y - 2)$ cm (b) 19 cm
4. (a) (i) £$(p + 4)$ (ii) £$(p - 3)$ (iii) £$(3p + 1)$
   (b) Aimee £12, Grace £8, Lydia £5
5. (a) $(x + 10)$ pence (b) $(3x + 20)$ pence (c) 15 pence
6. $3x + 2(x - 4) = 77$, $x = 17$
   Cream biscuit 13 pence.
7. (a) $(7x - 4)$ cm (b) $7x - 4 = 59$, $x = 9$
   (c) 13 cm, 17 cm, 29 cm

ALGEBRA

8. (a) $(x - 4)$ cm (b) $(4x - 8)$ cm
(c) $x = 7$

9. (a) $18x = 540°$ (b) largest angle = 150°

10. Cake costs 37 pence

11. (a) $(x + 7)(x + 2) = (x + 4)^2$, $x = 2$
(b) Rectangle (26 cm) has greater perimeter than square (24 cm).

12. $(x + 10)^2 = 12^2 + (x + 7)^2$, $x = 15.5$

Practice Exercise 10.8 Page 72

1. (a) $n > 2$ (b) $x < -2$ (c) $a < 4$
(d) $a < 4$ (e) $d \leqslant 3$ (f) $t < -3$
(g) $g > -2$ (h) $y \geqslant 0$

2. (a) $a < 4$ (b) $x \geqslant -2$ (c) $y < -3$
(d) $c > 5$ (e) $d < -3$ (f) $b \geqslant 1$
(g) $b \leqslant 1$ (h) $c \leqslant 3$ (i) $d > 4$
(j) $f < -2$ (k) $g \leqslant \frac{1}{2}$ (l) $h < 2$
(m) $x < -3$ (n) $j \geqslant 2\frac{1}{2}$ (o) $k > -4$
(p) $m \leqslant 1\frac{2}{5}$

3. (a) $x \geqslant 5$ (b) $p > 8$ (c) $b \leqslant 1\frac{1}{7}$
(d) $m > -6\frac{1}{2}$ (e) $n \leqslant 2$ (f) $t > 13$
(g) $a > 2\frac{1}{3}$ (h) $x \leqslant -\frac{1}{3}$ (i) $x > \frac{4}{5}$

Practice Exercise 10.9 Page 73

1. $a < -2$	**6.** $f < 1$	**11.** $m < -3$
2. $b \geqslant 3$	**7.** $g > -4$	**12.** $n > -4$
3. $c \leqslant -4$	**8.** $h \geqslant 2$	**13.** $p \geqslant \frac{5}{9}$
4. $d > -1$	**9.** $j \leqslant 1$	**14.** $q > -2\frac{1}{5}$
5. $e \geqslant 2$	**10.** $k > -\frac{1}{3}$	**15.** $n < 9$

Review Exercise 10 Page 73

1. (a) 3 (b) 23
(c) Answer should always be a whole number.

2. (a) $x = 12$ (b) $y = 4$
(c) $y = 3$ (d) $y = -1$

3. (a) 34 years old (b) 19 years old

4. (a) $6x + 3$ (b) 7

5. 2.5 litres

6. (a) $(x - 15)$ pence (b) 42 pence

7. 105 cm^2

8. (a) $2(x + 5)$ (b) 11

9. Triangle A: $8x + 84 = 180$; $x = 12$;
angles: 60°, 60°, 60°; equilateral.
Triangle B: $3x + 72 = 180$; $x = 36$;
angles: 72°, 72°, 36°; isosceles.
Triangle C: $6x + 12 = 180$; $x = 28$;
angles: 32°, 58°, 90°; right-angled.

10. (a) $x + y > 200$ (b) $y > 5x$
(c) $x \geqslant 70$

11. (a) $a = -5$ (b) $x = 0.8$
(c) $x = -0.5$

12. (a) $(6x + 16)$ cm (b) $x = 2$
(c) 28 cm

13. (a) $x = -14$ (b) $y = 22$
(c) $x = 4$

14. (a) Mr Archer: £$(5x + 400)$.
Mr Barton: £$13x$.
(b) $13x < 5x + 400$
(c) $8x < 400$, so, $x < 50$.
Cheaper to employ Mr Barton if the job takes less than 50 hours.
(d) Mr Archer (£700).

15. (a) $x \leqslant 8$ (b) $x < -6$ (c) $x \leqslant \frac{1}{2}$
(d) $x \leqslant \frac{3}{4}$ (e) $x < -1$ (f) $x > 0$

16. (a) $2x$
(b) Jack: $x - 6$, Rachel: $2x - 6$
(c) Rachel: 24, Jack 12

Chapter 11 Simultaneous Equations

Practice Exercise 11.1 Page 77

1. $x = 4$, $y = 2$	**5.** $x = 3$, $y = 1$
2. $x = 3$, $y = 5$	**6.** $x = 4$, $y = 3$
3. $x = 2$, $y = 3$	**7.** $x = -2$, $y = 3$
4. $x = 2$, $y = 3$	**8.** $x = 5$, $y = 0.5$

Practice Exercise 11.2 Page 78

1. (a) Both lines have gradient -1.
(b) Both lines have gradient 4.
(c) Both lines are the same.
(d) Both lines have gradient 0.4.

2. (a) $y = -2x + 6$, $y = -2x + 3$
(b) $y = 2x + 3.5$, $y = 2x + 2$
(c) $y = 2.5x - 4$, $y = 2.5x + 1.75$
(d) $y = -3x + 1.25$, $y = -3x + 0.5$

3. (b) and (d) have no solution.
(a) $x = 0.4$, $y = 4$
(c) $x = -0.125$, $y = 1.875$

Practice Exercise 11.3 Page 80

1. $x = 1$, $y = 2$	**13.** $x = 2$, $y = 3$
2. $x = 3$, $y = 4$	**14.** $x = 4$, $y = -2$
3. $x = 2$, $y = 1$	**15.** $x = 2$, $y = -1.5$
4. $x = 1$, $y = 7$	**16.** $x = -1$, $y = 3$
5. $x = 3$, $y = 1$	**17.** $x = -5$, $y = 6$
6. $x = 2$, $y = 3$	**18.** $x = 2.5$, $y = -1$
7. $x = 5$, $y = 2$	**19.** $x = -6$, $y = 2.5$
8. $x = 4$, $y = 2$	**20.** $x = 3$, $y = -1.5$
9. $x = 5$, $y = 4$	**21.** $x = -1$, $y = -3$
10. $x = 3$, $y = 2$	**22.** $x = 3.5$, $y = -1$
11. $x = 4$, $y = 1$	**23.** $x = 1.5$, $y = -0.5$
12. $x = 4$, $y = 2$	**24.** $x = -0.5$, $y = -2.5$

Practice Exercise 11.4 — Page 81

1. $x = 2,\ y = 1$
2. $x = 2,\ y = 3$
3. $x = 3,\ y = 1$
4. $x = 4,\ y = 2$
5. $x = 2,\ y = -1$
6. $x = 4,\ y = -3$
7. $x = 4,\ y = 0.5$
8. $x = -3,\ y = 0.5$
9. $x = 4,\ y = 1$
10. $x = 2.2,\ y = 5.6$
11. $x = 1.5,\ y = 2$
12. $x = 1,\ y = 2$
13. $x = 5,\ y = 2$
14. $x = 1,\ y = -2$
15. $x = -1,\ y = 2$
16. $x = 2,\ y = 1$
17. $x = -1,\ y = 1$
18. $x = -2,\ y = 3$
19. $x = -1,\ y = 7$
20. $x = 4,\ y = -1$
21. $x = 2.5,\ y = 3$
22. $x = 2,\ y = -1$
23. $x = 1,\ y = -2$
24. $x = -2,\ y = 0.5$

Practice Exercise 11.5 — Page 82

1. $x = 2,\ y = 6$
2. $x = 9,\ y = 18$
3. $x = 8,\ y = 2$
4. $x = 3,\ y = 6$
5. $x = 2,\ y = 13$
6. $x = 0,\ y = 2$
7. $x = 2,\ y = 4$
8. $x = 7,\ y = 6$
9. $x = 4,\ y = 8$
10. $x = 10.5,\ y = 0.5$
11. $x = -68,\ y = -122$
12. $x = 8,\ y = 6$

Practice Exercise 11.6 — Page 83

1. (a) $6x + 3y = 93,\ \ 2x + 5y = 91$
   (b) Pencil 8p, pen 15p
2. (a) $5x + 30y = 900,\ \ 10x + 15y = 1260$
   (b) Apples £1.08 per kg, oranges 12p each
3. $x = 95,\ y = 82$
4. 32 children, 4 adults
5. $x = 100,\ y = 70$
6. Coffee 90p, Tea 80p
7. $x = 160,\ y = 120$
8. $x = \boxed{13},\ y = \boxed{9}$
9. $p = 5,\ q = -2$
10. Ticket £12.50, CD £4.50

Review Exercise 11 — Page 84

1. (a) $x = 3,\ y = 7$ (b) $x = 4.5,\ y = 1.25$
2. $x = -1,\ y = -2$
3. (a) Same gradient, 2.
   $y = 2x - 2,\ \ y = 2x + \frac{3}{2}$
   (b) Same gradient, $\frac{1}{4}$.
   $y = \frac{1}{4}x + \frac{1}{4},\ \ y = \frac{1}{4}x + \frac{3}{8}$
   (c) $a = 3$, b = any number, except 2.
   (d) Any numbers such that $\frac{q}{p} = -3,\ \ \frac{r}{p} \neq 2$
4. $x = 3,\ y = 2$
5. (a) $x = 1,\ y = 3$ (b) $x = 2,\ y = -2$
   (c) $x = -1,\ y = 2$ (d) $x = 2,\ y = 3\frac{1}{2}$
6. $x = 6,\ y = 2$
7. (a) $4x + y = 58,\ \ 6x + 2y = 92$
   (b) $x = 12,\ y = 10$
8. (a) Possible equations:
   $x - y = 15,\ \ 7x - 5y = 155,$
   $3x - 2y = 70$
   (b) Using any pair of equations:
   $x = 40,\ y = 25$
   (c) $\angle ABD = 70°$
   $\angle CBD = 110°$
   $\angle BCE = 70°$
9. (a) 1 rose costs 25p, 1 carnation costs 20p.
   (b) £2.95
10. (a) Possible equations:
    $x - y = 1,\ \ 2x - 3y = -1,\ \ 3x - 4y = 0$
    (b) Using any pair of equations:
    $x = 4,\ y = 3$
    (c) 39 cm
11. (a) $x + y = 15$ (b) $5x + 2y = 57$ (c) 9
12. (a) $x + y = 1850$
    (b) $12.5x - 5.5y = 22\,225$
    Multiply both sides by 2.
    $25x - 11y = 44\,450$
    (c) 1800 miles

Chapter 12 Working with Formulae

Practice Exercise 12.1 — Page 87

1. (a) $P = 4g$
   (b) $P = 4y + 4$
   (c) $P = 3x - 1$
   (d) $P = 2a + 2b$
2. $C = 25d$
3. (a) £80
   (b) $C = 12x + 8$
4. 17 points
5. $T = 97$
6. (a) $M = -7$
   (b) $n = 4\frac{1}{2}$
7. (a) $H = 2.5$
   (b) $g = 6$
8. $F = 75$
9. $V = 2$
10. $P = 12$
11. 96 m
12. 86°F
13. 138 minutes
14. 25°C
15. 240 volts
16. $F = 4100$
17. £45.14
18. 0.3
19. $S = 4$
20. (a) $R = 15$
    (b) $R = 3$
21. $K = 36$
22. (a) $S = 18$
    (b) $S = 18$
23. (a) $S = 36$
    (b) $S = 36$
24. $T = \frac{3}{4}$
25. (a) $L = 10$
    (b) $L = 0.5$
26. $F = 144$
27. $R = 2.25$
28. (a) $v = 6.3$
    (b) $v = 7.4$

Practice Exercise 12.2 — Page 89

1. (a) $m = a - 5$ (b) $m = a - x$
   (c) $m = a + 2$ (d) $m = a + b$
2. (a) $x = \frac{y}{4}$ (b) $x = \frac{y}{a}$ (c) $x = 2y$
   (d) $x = ay$ (e) $x = \frac{5y}{3}$

ALGEBRA

3. (a) $p = \frac{1}{2}y - 3$ (b) $p = \frac{t - q}{5}$
(c) $p = \frac{m + 2}{3}$ (d) $p = \frac{q + r}{4}$
4. $n = \frac{C - 35}{24}$
5. $R = \frac{V}{I}$
6. (a) $x = \frac{3a}{2}$ (b) $x = \frac{3b - 2a}{5}$
7. (a) $a = \frac{3x}{2}$ (b) $a = \frac{b}{1 - x}$
(c) $a = \frac{b + 2}{1 - x}$
8. (a) $p = \frac{2q - r}{q + r}$ (b) $b = \frac{ax - a}{x + 1}$

Practice Exercise 12.3 Page 90

1. (a) $c = \pm\sqrt{y}$ (b) $c = y^2$
(c) $c = \pm\sqrt{\frac{y}{d}}$ (d) $c = 9y^2$
(e) $c = \pm\sqrt{y - x}$ (f) $c = (y - x)^2$
(g) $c = \pm\sqrt{dy - dx}$ (h) $c = (ax + ay)^2$
2. (a) $a = b - c^2$ (b) $a = \pm\sqrt{b}$
(c) $a = \frac{p - d}{m}$ (d) $a = \pm\sqrt{\frac{F}{m}}$
3. (a) $t = \pm\sqrt{\frac{x}{3}}$ (b) $t = \pm\sqrt{\frac{V}{a}}$
(c) $t = \pm\sqrt{\frac{b - a}{2}}$ (d) $t = \pm\sqrt{a - b}$
(e) $t = \pm\sqrt{\frac{c - a}{b}}$ (f) $t = \pm\sqrt{\frac{a - c}{b}}$
4. (a) $x = a^2 - 3$ (b) $x = 9a^2 - 2$
(c) $x = 8b^2 + \frac{1}{2}a$
5. (a) $a = \frac{3x}{2}$ (b) $a = \frac{b}{1 - x}$
(c) $a = \frac{b + 2}{1 - x}$ (d) $a = \frac{3x - 2}{1 - x}$
(e) $a = \frac{5y + 3}{1 + y}$ (f) $a = \frac{2b + x}{x - b}$
(g) $a = \frac{6 + y}{y + 3}$ (h) $a = \frac{5x}{3}$
(i) $a = \frac{4x^3}{1 + 4x^2}$

Practice Exercise 12.4 Page 90

1. (a) $d = \frac{P}{4}$ (b) $d = 0.7\,\text{cm}$
2. (a) $l = \frac{A}{b}$ (b) $l = 6\,\text{cm}$
3. (a) (i) $D = ST$ (ii) $D = 96\,\text{km}$
(b) (i) $T = \frac{D}{S}$ (ii) $T = 2.5$ hours
4. (a) $b = \frac{1}{2}P - l$ (b) $b = 4.2\,\text{cm}$
5. (a) $x = \frac{y - c}{m}$ (b) $x = 3$
6. (a) $b = \frac{2A}{h}$ (b) $b = 6.4$
7. (a) $q = \pm\sqrt{p^2 - r^2}$ (b) $q = \pm 6.8$

8. (a) $r = \sqrt{\frac{A}{4\pi}} = \frac{1}{2}\sqrt{\frac{A}{\pi}}$ (b) $r = 5.3\,\text{cm}$
9. (a) $h = \frac{3V}{\pi r^2}$ (b) $r = \sqrt{\frac{3V}{\pi h}}$

Review Exercise 12 Page 91

1. (a) £17 (b) $C = 0.15n + 5$ (c) 480
2. (a) 115 (b) 175 (c) 45
(d)

n	$n+1$	$n+2$
	$n+11$	
	$n+21$	

(e) $\mathbf{S}_n = 5n + 35$
3. $y = x + 3,\ x = y - 3$
$y = 3x + 1,\ x = \frac{1}{3}(y - 1)$
$y = \frac{1}{3}x,\ x = 3y$
$y = 3x,\ x = \frac{1}{3}y$
$y = 3x - 1,\ x = \frac{1}{3}y + \frac{1}{3}$
4. (a) £107 (b) $n = \frac{C - 35}{24}$; 9 days
5. $t = \frac{v - u}{5}$
6. (a) $a = -2.5$ (b) $d = \frac{3a}{b + c}$
7. $C = \frac{F - 32}{1.8}$
8. $s = -8$
9. $x = (3a - 2)^2$
10. (a) $V = 6$ (b) $h = 18.75$ (c) $x = \sqrt{\frac{6V}{h}}$
11. (a) $h = \frac{A}{2\pi r}$ (b) $h = 2.9\,\text{cm}$
(c) $r = \sqrt{\frac{V}{\pi h}}$ (d) $r = 2.0\,\text{cm}$

Chapter 13 Graphs of Quadratic Functions

Practice Exercise 13.1 Page 94

1. (a)

x	−3	−2	−1	0	1	2	3
y	7	2	−1	−2	−1	2	7

(c) $y = 0.25$
2. (b) $y = 7.25$
(c) $x = -1.7$ and $x = 1.7$
(d) (0, 1)
3. (a)

x	−3	−2	−1	0	1	2	3
y	−3	2	5	6	5	2	−3

(c) (−2.4, 0), (2.4, 0)
(d) (0, 6)
4.

x	−2	−1	0	1	2
y	8	2	0	2	8

5.

x	−2	−1	0	1	2	3
y	8	4	2	2	4	8

Practice Exercise 13.2 — Page 96

1. **a**: $y = x^2 + 1$ **b**: $y = x^2 - 4$
   c: $y = (x - 1)^2$ **d**: $y = (x + 1)^2$
2. (a) (b) (c) (d)
3. (a) $y = (x - 3)^2$ (b) $y = (x + 5)^2$
   (c) $y = x^2 - 2$ (d) $y = x^2 + 6$
   (e) $y = 2x^2$
4. **a**: $y = \frac{1}{2}x^2$ **b**: $y = 16x^2$ **c**: $y = -2x^2$

Practice Exercise 13.3 — Page 98

1. (a) $y = (x + 2)^2 - 5$; $(-2, -5)$
   (b) $y = (x + 3)^2 - 4$; $(-3, -4)$
   (c) $y = (x - 1)^2 + 2$; $(1, 2)$
2. (a) 1 (b) $x = 1$
3. $b = -2$, $c = -8$
4. (a) $y = x^2 - 10x + 16$ (b) $x = 5$ (c) -9
5. (a) $(-2, 0)$ and $(3, 0)$
   (b)

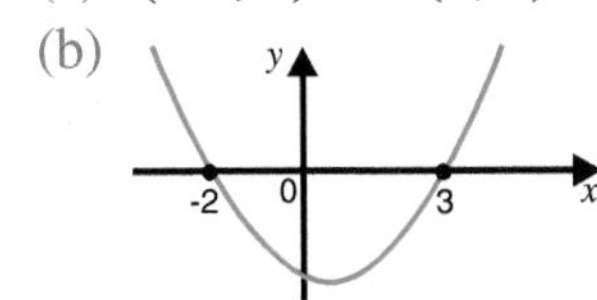

6. (a) $y = (x - 2)^2 - 5$
   (b) Translate $y = x^2$ by 2 units to right to get $y = (x - 2)^2$.
   Translate $y = (x - 2)^2$ by 5 units down to get $y = (x - 2)^2 - 5$.
7. (a)

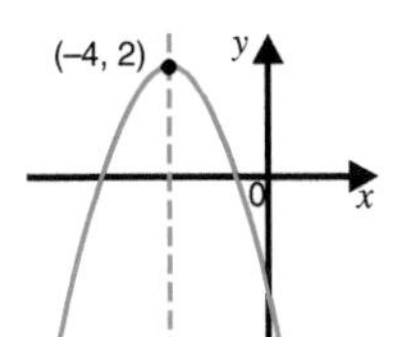

   (b) Reflect $y = x^2$ in x axis to get $y = -x^2$.
   Translate $y = -x^2$ by 4 units left to get $y = -(x + 4)^2$.
   Translate $y = -(x + 4)^2$ by 2 units up to get $y = -(x + 4)^2 + 2$.
8. (a) $y = (x - 4)^2 - 4$; $x = 4$
   (b) $y = (x - 1)^2 - 16$; $x = 1$
   (c) $y = (x + 3)^2 - 9$; $x = -3$
9. (a) $y = 5 - (x - 4)^2$; $x = 4$
   (b) $y = 10 - (x - 4)^2$; $x = 4$
   (c) $y = 25 - (x - 5)^2$; $x = 5$

Review Exercise 13 — Page 99

1. (a) (b) (c)

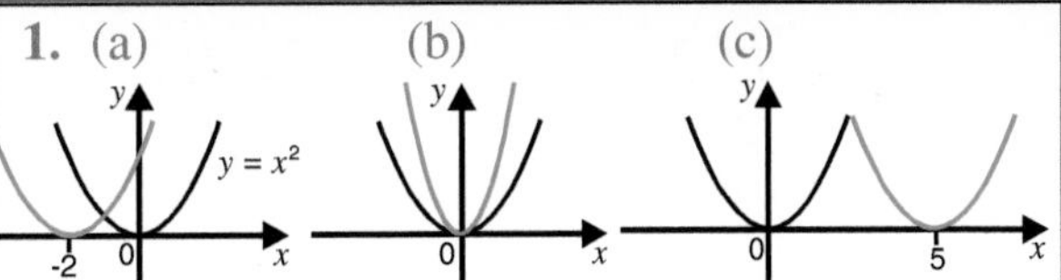

2. (a) $y = (x - 4)^2 - 14$
   (b) Translate $y = x^2$ by 4 units right to get $y = (x + 4)^2$.
   Translate $y = (x - 4)^2$ by 14 units down to get $y = (x - 4)^2 - 14$.
3. **A**: $y = x^2 - 4x + 4$, **B**: $y = x^2 + 4$
   C: $y = 4x^2$, **D**: $y = x^2 - 4$, **E**: $y = 4 - x^2$
4. $b = -1$, $c = -2$
5. (a) $(-3, 0)$ and $(4, 0)$
   (b) $(0, -12)$ (c) $x = \frac{1}{2}$ (d) $\left(\frac{1}{2}, -12\frac{1}{4}\right)$
6. (a) $x = -1$ and $x = 9$ (b) $x = 4$
   (c) $(4, 25)$ (d) $(0, 9)$
7. $y = (x + 1)^2 + 4 = x^2 + 2x + 5$

Chapter 14 Quadratic Equations

Practice Exercise 14.1 — Page 101

1. $x = -3.6$ or $x = 0.6$
2. (b) $x = -1$ or $x = 0$ (c) $(-0.5, -0.25)$
3. (a) Entries are: 9, 4, 1, 0, 1, 4
   (c) $x = 1$
4. (b) $x = \pm 3.2$ (c) $(0, 10)$
5. (a) $x = \pm 3.2$ (b) $x = \pm 2.2$
   (c) $x = 1$ or $x = 2$ (d) $x = \pm 2.4$

Practice Exercise 14.2 — Page 103

1. (a) $x = 2$ or 3 (b) $x = -4$ or -6
   (c) $x = 3$ or -1 (d) $x = 5$ or -2
   (e) $x = 0$ or 4 (f) $x = 0$ or -2
2. (a) $x = 1$ or 2 (b) $y = -3$ or -4
   (c) $m = 4$ or -2 (d) $a = 3$ or -4
   (e) $n = 9$ or -4 (f) $z = 6$ or 3
   (g) $k = -3$ or -5 (h) $c = -7$ or -8
   (i) $b = 4$ or -5 (j) $v = 12$ or -5
   (k) $w = 4$ or -12 (l) $p = 9$ or -8
3. (a) $x = 0$ or 5 (b) $y = 0$ or -1
   (c) $p = 0$ or -3 (d) $a = 0$ or 4
   (e) $t = 0$ or 6 (f) $g = 0$ or 4
4. (a) $x = 2$ or -2 (b) $y = 12$ or -12
   (c) $a = 3$ or -3 (d) $d = 4$ or -4
   (e) $x = 10$ or -10 (f) $x = 6$ or -6
   (g) $x = 2.5$ or -2.5 (h) $x = 2.25$ or -2.25
   (i) $x = \frac{5}{12}$ or $-\frac{5}{12}$
5. (a) $x = -6$ or $x = 0$ (b) $x = -1$ or $x = -4$
   (c) $x = -8$ or $x = 8$ (d) $x = 1$ or $x = 3$
   (e) $x = 0$ or $x = 2$ (f) $x = -2$ or $x = 3$
   (g) $x = -5$ or $x = 3$ (h) $x = -2$ or $x = 2$
   (i) $x = 7$ or $x = 8$

6. (a) $x = -2\frac{1}{2}$ or -1 (b) $x = \frac{1}{2}$ or 5
(c) $x = 7$ or $-1\frac{1}{3}$ (d) $y = -\frac{1}{2}$ or $-\frac{2}{3}$
(e) $x = -3$ or $\frac{2}{3}$ (f) $z = -\frac{1}{3}$ or 2
(g) $m = \frac{3}{5}$ or 1 (h) $a = -3\frac{1}{2}$ or 3
(i) $y = 2$ or $-1\frac{1}{4}$

7. (a) $y = 5$ or -1 (b) $x = 0$ or 1
(c) $x = 4$ (d) $x = 5$ or -3
(e) $n = 12$ or -2 (f) $m = 7$ or 1
(g) $a = 8$ or -3 (h) $x = 1$ or 3
(i) $x = \frac{1}{2}$ or $-\frac{1}{3}$ (j) $m = -5$ or $1\frac{1}{2}$
(k) $a = \frac{1}{2}$ or $-\frac{2}{3}$ (l) $x = \frac{1}{2}$ or $-\frac{3}{4}$

Practice Exercise 14.3 Page 104

1. (a) $x = -1$ or $x = -3$
(b) $x = 1$ or $x = 2$
(c) $x = -1$ or $x = 3$
2. (a) $x = -0.59$ or -3.41
(b) $x = -6.19$ or -0.81
(c) $x = -1.62$ or 0.62
(d) $x = -0.56$ or 3.56
(e) $x = -0.54$ or 5.54
(f) $x = -4.30$ or -0.70
(g) $x = -5.74$ or 1.74
(h) $x = 0.38$ or 2.62
(i) $x = -3.56$ or 0.56
3. (a) $x = -1.5$ or $x = 1$
(b) $x = 0.5$ or $x = 1$
(c) $x = -1$ or $x = 0.6$
4. (a) $x = 0.23$ or 1.43
(b) $x = 0.72$ or 2.78
(c) $x = -7.30$ or -0.37
(d) $x = -0.79$ or 2.12
(e) $z = -2.37$ or -0.63
(f) $x = -2.27$ or 2.94

Practice Exercise 14.4 Page 105

1. (a) 36 (b) 0 (c) -7
2. (a) 13, two irrational roots.
(b) 0, one real root.
(c) 25, two rational roots.
(d) -4, no real solutions.
(e) 16, two rational roots.
(f) -156, no real solutions.
(g) 172, two irrational roots.
(h) 49, two rational roots.
(i) 121, two rational roots.
(j) 81, two rational roots.
(k) 56, two irrational roots.
(l) 0, one real root.
3. $p < 16$
4. $q = 10$ or $q = -10$
5. $p > 1$
6. $a = 4$ (note $a \neq 0$)
7. $p > \sqrt{12}$ or $p < -\sqrt{12}$
8. $p < 4$

9. $a = 0$ or $a = 16$
10. (a) $x = -3$ or 1
(b) $(x + 3)(x - 1)$, $x = -3$ or $x = 1$
(c) $x = -3$ or 1
11. (a) $x = 0.55$ or 5.45
(b) $x = -1.14$ or 2.64
(c) $x = -0.85$ or 2.35
(d) $x = -0.74$ or 0.54
(e) $x = -0.65$ or 4.65
(f) $x = -3.81$ or 1.31
(g) $x = 0.15$ or 4.52
(h) $x = -0.58$ or 2.58
(i) $x = -1.16$ or 2.16

Practice Exercise 14.5 Page 107

1. (a) $x(x - 4) = 21$
(b) $x = 7$ (x cannot equal -3)
2. (a) $(x + 3)(x + 1) - \frac{1}{2}x^2 = 7.5$
$x^2 + 4x + 3 - \frac{1}{2}x^2 = 7.5$
$\frac{1}{2}x^2 + 4x - 4.5 = 0$
$x^2 + 8x - 9 = 0$
(b) $x = 1$ (x cannot equal -9)
3. (a) $V = 2(x - 4)^2$
$V = 2(x^2 - 8x + 16)$
$V = 2x^2 - 16x + 32$
(b) $x = 9.5$
4. (a) 72 litres (b) 32 litres (c) 6 minutes
5. (a) 5 m (b) 9 m (c) 3.2 seconds

Review Exercise 14 Page 108

1. (b) $x = -2.7$ or 0.7
2. $x = -1$ or 2
3. (a) $(x - 5)(x + 3)$ (b) $x = -3$ or 5
4. (a) (i) $t(t - 4)$
(ii) $t = 0$ or $t = 4$
(b) (i) $(y + 1)(y + 2)$
(ii) $y = -1$ or $y = -2$
5. $y = -0.46$ or 6.46
6. (a) $x = -3$ or 3 (b) $x = 0$ or 9
(c) $x = -1$ or 9 (d) $x = 1$ or 8
(e) $x = -0.8$ or 9.8
7. $x = -1.22$ or 0.55
8. (a) $2x(x + 1) = 3(2x + 2)$
$2x^2 + 2x = 6x + 6$
$2x^2 - 4x - 6 = 0$
(b) $x = 3$, $BC = 4$ cm
9. $a = 3$, $b = 2$
10. 2 m
11. $k = -12$
12. (a) $p > -4$
(b) $q = 6$ or $q = -6$
13. (a) $(x + 2)(x + 1) = 6$
$x^2 + x + 2x + 2 = 6$
$x^2 + 3x - 4 = 0$
(b) 25, two rational roots.
(c) $x = 1$; 3 cm × 2 cm

Chapter 15 Working with Arcs and Sectors

Practice Exercise 15.1 Page 109

1. (a) 3.14 cm (b) 2.09 cm (c) 6.28 cm (d) 10.5 cm (e) 15.6 cm (f) 26.4 cm
2. (a) 1.88 cm (b) 22.7 cm (c) 2.91 cm (d) 26.6 cm (e) 11.8 cm
3. (a) $\frac{1}{12}$ (b) $\frac{1}{8}$ (c) $\frac{2}{5}$ (d) 135° (e) 240°
4. 53.5°
5. 6.62 cm
6. (a) 2.23 cm (b) 73.9°

Practice Exercise 15.2 Page 111

1. (a) 7.07 cm² (b) 6.98 cm² (c) 28.3 cm² (d) 67.5 cm² (e) 34.2 cm² (f) 68.4 cm²
2. (a) $\frac{3}{8}$ (b) $\frac{1}{5}$ (c) $\frac{7}{12}$ (d) 160° (e) 300° (f) 108°
3. (a) 17.7 cm² (b) 11.5 cm² (c) 55.9 cm² (d) 52.9 cm² (e) 471 cm²
4. $a = 19.1°$
5. 6.91 cm

Practice Exercise 15.3 Page 112

1. (a) 170 cm² (b) 54 cm
2. 14.0 cm²
3. (a) 764 cm² (b) 954 cm² (c) 30.5 cm
4. (a) 9.5 cm (b) 14.1 cm
5. 14.3 cm²

Review Exercise 15 Page 113

1. (a) 3.14 cm (b) 11.8 cm²
2. (a) 78.5 m (b) 3540 m²
3. 3180 cm²
4. (a) 26.6 cm² (b) 24.1 cm
5. (a) 18.0 cm (b) 36.2 cm (c) 21.1 cm
6. 43.4°
7. (a) 54° (b) 4.8 cm

Chapter 16 Volume of Solids

Practice Exercise 16.1 Page 115

1. (a) 101 cm³ (b) 188 cm³ (c) 37.3 cm³ (d) 36.8 cm³ (e) 2140 cm³ (f) 8180 cm³
2. Hemispherical bowl has greater volume. Cone 393 cm³, hemisphere 452 cm³.
3. 37.5 cm
4. 405
5. 0.26 cm
6. 3.89 cm
7. 4.73 cm
8. 2.40 cm
9. (a) 18.8 cm (b) 3 cm (c) 11.6 cm (d) 110 cm³

Practice Exercise 16.2 Page 116

1. (a) 170 cm³ (b) 33.5 cm³ (c) 91.6 cm³
2. 23 600 cm³
3. 1470 cm³
4. 8860 cm³
5. 94.2 cm³

Review Exercise 16 Page 117

1. (a) 209 cm³ (b) 905 cm³ (c) 108 cm³
2. 644 000 cm³
3. (a) 377 cm³ (b) 374 cm³
4. 1040 cm³
5. $x = 3.7$ cm
6. 1890 cm³
7. 11.2 cm
8. (a) 2.5 cm (b) 63.4 cm³
9. 175 cm³
10. 347 cm³

Chapter 17 Pythagoras' Theorem

Practice Exercise 17.1 Page 120

1. (a) $a = 25$ cm (b) $b = 12.8$ cm (c) $c = 2$ cm
2. (a) $\sqrt{52}$ cm (b) $\sqrt{20}$ cm
3. (a) 5 (b) 9.22 (c) 13 (d) 8.06 (e) 7.21
4. (a) $R(9, 7)$ (b) $X(2, 3)$ (c) $Y(4, 2)$ (d) 2.24
5. 339 m
6. 36 cm²
7. 3.6 cm
8. (a) Right-angled. (b) Not right-angled. (c) Right-angled.
9. $AB = \sqrt{8}$, $AC = \sqrt{40}$, $BC = \sqrt{32}$
   $AC^2 = AB^2 + BC^2$, $\angle ABC = 90°$
10. $AB = \sqrt{20}$ cm, $BC = \sqrt{5}$ cm
    $AC^2 = AB^2 + BC^2$
11. $PQ = \sqrt{34}$, $PR = \sqrt{50}$, $QR = \sqrt{32}$
    $PR^2 \neq PQ^2 + PR^2$. ΔPQR is not right-angled.

Practice Exercise 17.2 Page 122

1. (a) 2.9 cm (b) 5.7 cm (c) 2.1 cm (d) 2.0 cm
2. 17 cm
3. 10 cm
4. 8.5 cm
5. 10.6 cm
6. 15 cm
7. 6.9 cm
8. 74.3 cm

Practice Exercise 17.3 Page 123

1. 13 cm
2. 24 cm
3. 10.4 cm
4. 20.7 cm
5. 11.9 cm
6. 19.5 cm
7. (a) 11.2 cm (b) 14.1 cm (c) 9.49 cm
8. 4.95 cm
9. (a) 2 units (b) 4 units (c) 4.5 units (d) 3.6 units (e) 5 units (f) 4.5 units

Review Exercise 17 Page 125

1. $PQ = 15$ cm
2. $AB = 8.6$ units
3. $25^2 = 24^2 + 7^2$. ΔPQR is right-angled at R.
4. 361 m
5. $40^2 > 35^2 + 15^2$, so, $\angle PQR > 90°$
6. 17.3 cm
7. $AC = 10$ units, $AE = \sqrt{80}$ units
   $AF = \sqrt{116}$ units

8. 9.75 cm 9. 5.66 cm 10. 5.4 units

11. $EF = \sqrt{50}$, $CF = \sqrt{72}$, $CE = \sqrt{122}$.
In ΔCFE, $CE^2 = EF^2 + CF^2$,
so, $\angle CFE = 90°$.

Chapter 18 Properties of Shapes

Practice Exercise 18.1 Page 128

1. (a) Yes (b) Yes (c) No
(d) No (e) Yes (f) No
2. (a) Yes, obtuse-angled (b) No
(c) Yes, acute-angled (d) Yes, right-angled
(e) Yes, obtuse-angled
3. (a) $a = 45°$ (b) $b = 148°$, $c = 32°$
(c) $d = 52°$, $e = 64°$ (d) $f = 63°$
4. (a) $a = 18°$, $b = 144°$
(b) $c = 26.5°$, $d = 153.5°$
5. (a) Isosceles (b) 74° (c) 46°
6. (a) $\angle BCD = 120°$
(b) $\angle PRQ = 80°$, $\angle QRS = 160°$
(c) $\angle MNX = 50°$

Practice Exercise 18.2 Page 130

1. $S(3, 1)$
2. $Y(6, 4)$
3. $A(1, 3)$
4. (3, 1), (3, 5)
5. (a) $a = 62°$ (b) $b = 54°$, $c = 36°$
(c) $d = 62°$ (d) $e = 116°$, $f = 86°$
(e) $g = 124°$ (f) $h = 75°$
(g) $i = 38°$, $j = 42°$
(h) $k = 55°$, $l = 45°$
(i) $m = 65°$, $n = 90°$, $o = 25°$
(j) $p = 117°$ (k) $q = 70°$
(l) $r = 85°$ (m) $s = 28°$, $t = 80°$
(n) $u = 70°$
6. $\angle SQR = 50°$
7. $a = 22\frac{1}{2}°$, $b = 112\frac{1}{2}°$, $c = 22\frac{1}{2}°$

Practice Exercise 18.3 Page 132

1. (a) $x + 130° = 180°$ (supp. $\angle$'s)
$x = 180° - 130° = 50°$
(b) $y + 130° + 70° + 100° = 360°$
(sum of $\angle$'s in a quad. $= 360°$)
$y = 360° - 300° = 60°$
2. (a) $a = 62°$, $b = 55°$
(b) $c = 62°$ (c) $d = 76°$
3. (a) 900° (b) 1080° (c) 1260°
4. (a) $a = 130°$ (b) $b = 250°$ (c) $c = 85°$
5. 720°

Practice Exercise 18.4 Page 133

1. (a) (i) 120° (ii) 90° (iii) 60° (iv) 45°
(b) (i) 60° (ii) 90° (iii) 120° (iv) 135°
2. 20
3. 8
4. (a) 72° (b) 108°
(c) 540°
5. (a) $a = 90°$, $b = 60°$, $c = 210°$
(b) $d = 90°$, $e = 120°$, $f = 150°$
(c) $g = 90°$, $h = 135°$, $i = 135°$
(d) $j = 105°$
6. 153°
7. 150°
8. At any vertex, sum of angles cannot equal 360°.

Practice Exercise 18.5 Page 135

1. (a) $a = 90°$, $b = 50°$ (b) $c = 45°$
(c) $d = 40°$, $e = 54°$ (d) $f = 20°$
2. (a) $f = 50°$, $g = 65°$
(b) $h = 130°$, $i = 48°$
(c) $j = 30°$, $k = 39°$
3. (a) $a = 90°$, $b = 50°$
(b) $c = 35°$, $d = 35°$
(c) $e = 90°$, $f = 49°$, $g = 49°$
(d) $h = 63°$ (e) $j = 67°$
(f) $k = 50°$ (g) $l = 65°$
(h) $m = 146°$
4. 8 cm
5. (a) $\angle AXO = 90°$
(b) Radius OX meets tangent AB at 90°.
OX is perpendicular bisector of AB.
So, $AX = XB$.
(c) $AB = 19.2$ cm
6. 22.4 cm 7. 14 cm 8. 32 cm 9. 1.66 m

Review Exercise 18 Page 138

1. $x = 20°$, $\angle PSR = \angle PRS = 40°$
2. $\angle BDC = 112°$
3. (a) 1 (b) 54°
4. $a = 115°$, $b = 44°$
5. (a) $q = 12°$ (b) 30 sides
6. (a) $x = 108°$ (b) $y = 36°$
7. $PQ = 24$ cm
8. $x = 20$
9. (a) 90° (b) 32° (c) 116° (d) 26°
10. (a) 72° (b) 36° (c) $\frac{1}{3}$
11. $x = 18$
12. 7 cm
13. (a) $\angle AED = 54°$ (b) $\angle AEI = 36°$
(c) $\angle BAJ = 180° - \frac{360°}{10} = 144°$
$\angle BAE = \frac{1}{2}(540° - 3 \times 144°) = 54°$
$\angle EAJ = 144° - 54° = 90°$

Chapter 19 Similar Figures

Practice Exercise 19.1 Page 140

1. Corresponding lengths not in same ratio.
2. (a) Two circles (d) Two squares
3. $a = 4$ cm, $b = 24$ cm
4. (a) Scale factor $= \frac{3}{2} = 1.5$
(b) $x = 1.8$ cm (c) $a = 120°$
5. 30 cm

6. (a) $x = 1.5$ cm, $y = 2.4$ cm, $a = 70°$
(b) $x = 5$ cm, $y = 1.5$ cm, $a = 53°$
(c) $x = 30$ cm, $y = 17.5$ cm, $z = 10$ cm
7. 15 cm 9. 18°
8. 2.8 cm 10. 5 cm 11. $x = 16$, $y = 48$

Practice Exercise 19.2 Page 142

1. (a) $x = 10$ cm, $y = 27$ cm
(b) $x = 6$ cm, $y = 10$ cm
(c) $x = 12$ cm, $y = 12$ cm
(d) $x = 12$ cm, $y = 5$ cm
2. (a) $AQ = 2.5$ cm, $BC = 7$ cm
(b) $AQ = 4\frac{2}{3}$ cm, $BP = 16\frac{2}{3}$ cm
3. (a) $x = 58°$ (b) $x = 56°$
4. (a) (i) 61° (ii) 29°
(b) (i) 13 cm, 5.2 cm (ii) 36 cm, 27 cm
Same
5. (b) $AB = 1.8$ cm (c) 12.5 cm
6. 13.5 cm^2

Practice Exercise 19.3 Page 145

1. 30 m
2. (a) 1.5 (b) 9 cm
3. (a) 1 : 50 (b) 2.8 cm
4. Original area multiplied by $2^2 = 4$.
5. 40 kg
6. 100 times larger.
7. 2.4 cm
8. Area is quartered.
9. 90 cm^2
10. (a) 8 cm (b) 96 cm^2
11. 2.4 m
12. (a) 1.5 km (b) 5 km^2
13. (a) 14 cm (b) 750 000 m^2 (c) 1.92 cm^2
14. 8 times original volume.
15. 4000 ml
16. (a) 880 cm^2 (b) 55 cm^2, 25 cm^3
17. 5400 litres
18. 12 kg
19. $\left(\frac{1}{2}\right)^3 = \frac{1}{8}$ of original volume.
20. 25 cm^3
21. 405 cm^3
22. (a) 10 cm (b) 16 : 25
23. 36 cm
24. 391 ml
25. (a) 328 m^2 (b) 45 cm^3
26. (a) 1 : 300 (b) 7.41 cm^3 (c) 2880 m^2
27. 1.41
28. (a) 33.5 cm^3, 524 cm^3
(b) Ratio of surface areas = 4 : 25.
As a percentage, $\frac{4}{25} \times 100 = 16\%$.
29. 15 cm
30. 0.81 kg

Review Exercise 19 Page 147

1. (a) $x = 14.4$ cm, $y = 12.5$ cm
(b) $p = 44$ cm, $q = 10$ cm
2. (a) 2.5 (b) 6.25 (c) $\angle DFE$
3. 80 m
4. (a) 1 : 50 000 (b) 4.2 km
5. 8 cm by 5.4 cm
6. (a) 6 : 7
(b) 11.9 cm
(c) 36 : 49
(d) 216 : 343
7. Height = 8 m
Floor area = 200 m^2 Volume = 1600 m^3
8. 15 cm
9. (a) 6 cm (b) 8 : 27
10. (a) 3125 cm^2 (b) 12 000 m^3
11. (a) 3.1 km (b) 0.2 km^2
12. (a) 3 : 4 (b) 9 : 16 (c) 27 : 64
13. (a) 3 : 5 (b) 9 : 25
14. 12.8 m

Chapter 20 Vectors and 3D Coordinates

Practice Exercise 20.1 Page 150

3. $\mathbf{a} = \begin{pmatrix}3\\5\end{pmatrix}$ $\mathbf{b} = \begin{pmatrix}3\\-1\end{pmatrix}$ $\mathbf{c} = \begin{pmatrix}-2\\4\end{pmatrix}$
$\mathbf{d} = \begin{pmatrix}-6\\-1\end{pmatrix}$ $\mathbf{e} = \begin{pmatrix}4\\3\end{pmatrix}$ $\mathbf{f} = \begin{pmatrix}2\\-5\end{pmatrix}$

Practice Exercise 20.2 Page 151

2. (a) $\mathbf{a} = \begin{pmatrix}3\\6\end{pmatrix}$ $\mathbf{b} = \begin{pmatrix}4\\2\end{pmatrix}$ $\mathbf{c} = \begin{pmatrix}0\\4\end{pmatrix}$ $\mathbf{d} = \begin{pmatrix}-4\\2\end{pmatrix}$
$\mathbf{e} = \begin{pmatrix}6\\-3\end{pmatrix}$ $\mathbf{f} = \begin{pmatrix}-4\\2\end{pmatrix}$ $\mathbf{g} = \begin{pmatrix}8\\4\end{pmatrix}$
(b) **d** and **f** (c) **e** (d) **g**
3. (a) $\mathbf{a} = 2\mathbf{p}$, $\mathbf{e} = -\mathbf{p}$
(b) $\mathbf{b} = 2\mathbf{q}$, $\mathbf{c} = -\mathbf{q}$, $\mathbf{h} = -3\mathbf{q}$
(c) $\mathbf{d} = \frac{1}{2}\mathbf{r}$, $\mathbf{f} = 2\mathbf{r}$, $\mathbf{g} = -\frac{1}{2}\mathbf{r}$
4. (a) $\begin{pmatrix}4\\-2\end{pmatrix}$ (b) $\begin{pmatrix}-2\\8\end{pmatrix}$ (c) $\begin{pmatrix}6\\-3\end{pmatrix}$ (d) $\begin{pmatrix}-2\\1\end{pmatrix}$
(e) $\begin{pmatrix}-6\\-2\end{pmatrix}$ (f) $\begin{pmatrix}2\\-4\end{pmatrix}$ (g) $\begin{pmatrix}4\\-16\end{pmatrix}$ (h) $\begin{pmatrix}-5\\10\end{pmatrix}$
5. $|\mathbf{a}| = 13$, $|\mathbf{b}| = 5$, $|\mathbf{c}| = 17$
6. **e**, **f** and **h**
7. (a) $\sqrt{65}$ (b) $\sqrt{34}$

Practice Exercise 20.3 Page 153

1. (a) $\overrightarrow{AB} = \begin{pmatrix}4\\2\end{pmatrix}$, $\overrightarrow{BC} = \begin{pmatrix}1\\-4\end{pmatrix}$ (b) $\overrightarrow{AC} = \begin{pmatrix}5\\-2\end{pmatrix}$
2. (b) (i) $\begin{pmatrix}6\\4\end{pmatrix}$ (ii) $\begin{pmatrix}6\\1\end{pmatrix}$ (iii) $\begin{pmatrix}4\\1\end{pmatrix}$ (iv) $\begin{pmatrix}3\\2\end{pmatrix}$
3. (b) (i) $\begin{pmatrix}5\\0\end{pmatrix} + \begin{pmatrix}3\\3\end{pmatrix} = \begin{pmatrix}8\\3\end{pmatrix}$
(ii) $\begin{pmatrix}-1\\-2\end{pmatrix} + \begin{pmatrix}4\\-4\end{pmatrix} = \begin{pmatrix}3\\-6\end{pmatrix}$
(iii) $\begin{pmatrix}-1\\2\end{pmatrix} + \begin{pmatrix}4\\-2\end{pmatrix} = \begin{pmatrix}3\\0\end{pmatrix}$
(iv) $\begin{pmatrix}-4\\0\end{pmatrix} + \begin{pmatrix}0\\-2\end{pmatrix} = \begin{pmatrix}-4\\-2\end{pmatrix}$

4. (b) (i) $\begin{pmatrix}4\\1\end{pmatrix} - \begin{pmatrix}2\\3\end{pmatrix} = \begin{pmatrix}2\\-2\end{pmatrix}$

(ii) $\begin{pmatrix}1\\3\end{pmatrix} - \begin{pmatrix}5\\-2\end{pmatrix} = \begin{pmatrix}-4\\5\end{pmatrix}$

(iii) $\begin{pmatrix}4\\-2\end{pmatrix} - \begin{pmatrix}0\\3\end{pmatrix} = \begin{pmatrix}4\\-5\end{pmatrix}$

(iv) $\begin{pmatrix}5\\0\end{pmatrix} - \begin{pmatrix}-2\\2\end{pmatrix} = \begin{pmatrix}7\\-2\end{pmatrix}$

5. (b) (i) $\begin{pmatrix}5\\0\end{pmatrix} - \begin{pmatrix}3\\3\end{pmatrix} = \begin{pmatrix}2\\-3\end{pmatrix}$

(ii) $\begin{pmatrix}-1\\-2\end{pmatrix} - \begin{pmatrix}4\\-4\end{pmatrix} = \begin{pmatrix}-5\\2\end{pmatrix}$

(iii) $\begin{pmatrix}-1\\2\end{pmatrix} - \begin{pmatrix}4\\-2\end{pmatrix} = \begin{pmatrix}-5\\4\end{pmatrix}$

(iv) $\begin{pmatrix}-4\\0\end{pmatrix} - \begin{pmatrix}0\\-2\end{pmatrix} = \begin{pmatrix}-4\\2\end{pmatrix}$

6. (a) $\begin{pmatrix}3\\7\end{pmatrix}$ (b) $\begin{pmatrix}1\\-2\end{pmatrix}$ (c) $\begin{pmatrix}8\\8\end{pmatrix}$ (d) $\begin{pmatrix}5\\-1\end{pmatrix}$ (e) $\begin{pmatrix}-2\\8\end{pmatrix}$

7. (a) **r** (b) **s** (c) **t** (d) **q**
(e) **v** (f) **u** (g) **w** (h) **p**

8. (a) $\begin{pmatrix}5\\3\end{pmatrix}$ (b) $\begin{pmatrix}1\\5\end{pmatrix}$ (c) $\begin{pmatrix}9\\12\end{pmatrix}$ (d) $\begin{pmatrix}11\\0\end{pmatrix}$ (e) $\begin{pmatrix}0\\11\end{pmatrix}$

9. (a) $\overrightarrow{AD} = \begin{pmatrix}-5\\-2\end{pmatrix}$, $\overrightarrow{BC} = \begin{pmatrix}-5\\-2\end{pmatrix}$
(b) Parallelogram

10. (a) $\overrightarrow{AO} = -\mathbf{a}$ (b) $\overrightarrow{AB} = -\mathbf{a} + \mathbf{b}$
(c) $\overrightarrow{MB} = \frac{1}{2}\overrightarrow{AB} = -\frac{1}{2}\mathbf{a} + \frac{1}{2}\mathbf{b}$
(d) $\overrightarrow{MO} = -\frac{1}{2}\mathbf{a} - \frac{1}{2}\mathbf{b}$

11. (a) $\overrightarrow{BD} = -\mathbf{u} + \mathbf{v}$ (b) $\overrightarrow{CA} = -\mathbf{u} - \mathbf{v}$

12. (a) $\overrightarrow{AD} = 3\mathbf{a} - 2\mathbf{b}$ $\overrightarrow{BC} = 6\mathbf{a} - 4\mathbf{b}$
(b) *ABCD* is a trapezium.

Practice Exercise 20.4 Page 156

1. $C(8, 4, 0)$, $F(8, 1, 6)$, $G(8, 4, 6)$, $H(2, 4, 6)$.
2. $B(12, 4, 5)$, $C(12, 9, 5)$, $D(1, 9, 5)$, $E(1, 4, 9)$, $F(12, 4, 9)$, $H(1, 9, 9)$.

3. (a) $\overrightarrow{OC} = \begin{pmatrix}9\\4\\4\end{pmatrix}$, $\overrightarrow{OD} = \begin{pmatrix}2\\4\\4\end{pmatrix}$, $\overrightarrow{OE} = \begin{pmatrix}2\\3\\7\end{pmatrix}$

$\overrightarrow{OF} = \begin{pmatrix}9\\3\\7\end{pmatrix}$, $\overrightarrow{OG} = \begin{pmatrix}9\\4\\7\end{pmatrix}$, $\overrightarrow{OH} = \begin{pmatrix}2\\4\\7\end{pmatrix}$

(b) $\overrightarrow{BF} = \begin{pmatrix}0\\0\\3\end{pmatrix}$ (c) $\overrightarrow{AG} = \begin{pmatrix}7\\1\\3\end{pmatrix}$

4. (a) $\begin{pmatrix}4\\2\\6\end{pmatrix}$ (b) $\begin{pmatrix}-6\\12\\-3\end{pmatrix}$ (c) $\begin{pmatrix}-4\\-2\\-6\end{pmatrix}$

(d) $\begin{pmatrix}2\\-4\\1\end{pmatrix}$ (e) $\begin{pmatrix}0\\5\\2\end{pmatrix}$

5. (a) $\begin{pmatrix}-1\\5\\1\end{pmatrix}$ (b) $\begin{pmatrix}6\\-1\\6\end{pmatrix}$ (c) $\begin{pmatrix}-4\\4\\-1\end{pmatrix}$ (d) $\begin{pmatrix}-2\\13\\0\end{pmatrix}$

(e) $\begin{pmatrix}7\\-6\\5\end{pmatrix}$ (f) $\begin{pmatrix}-9\\18\\-20\end{pmatrix}$ (g) $\begin{pmatrix}-11\\10\\-6\end{pmatrix}$ (h) $\begin{pmatrix}1\\-7\\16\end{pmatrix}$

6. (a) $\overrightarrow{AC} = \begin{pmatrix}-1\\3\\3\end{pmatrix}$

7. (a) $\overrightarrow{OA} = \begin{pmatrix}-8\\-4\\-3\end{pmatrix}$, $\overrightarrow{OB} = \begin{pmatrix}7\\-4\\-3\end{pmatrix}$, $\overrightarrow{OC} = \begin{pmatrix}7\\9\\-3\end{pmatrix}$

$\overrightarrow{OD} = \begin{pmatrix}-8\\9\\-3\end{pmatrix}$, $\overrightarrow{OE} = \begin{pmatrix}-8\\-4\\4\end{pmatrix}$, $\overrightarrow{OF} = \begin{pmatrix}7\\-4\\4\end{pmatrix}$

$\overrightarrow{OG} = \begin{pmatrix}7\\9\\4\end{pmatrix}$, $\overrightarrow{OH} = \begin{pmatrix}-8\\9\\4\end{pmatrix}$

(b) $\overrightarrow{AG} = \begin{pmatrix}15\\13\\7\end{pmatrix}$

(c) $\sqrt{443}$

(d) $\overrightarrow{BH} = \begin{pmatrix}-15\\13\\7\end{pmatrix}$, $\sqrt{443}$

(e) Diagonals of a cuboid are equal in length.

Review Exercise 20 Page 158

2. (a) $\begin{pmatrix}2\\7\end{pmatrix}$ (b) $\begin{pmatrix}-8\\1\end{pmatrix}$ (c) $\begin{pmatrix}-3\\0\end{pmatrix}$ (d) $\begin{pmatrix}-5\\-7\end{pmatrix}$

(e) $\begin{pmatrix}-6\\8\end{pmatrix}$ (f) $\begin{pmatrix}17\\-3\end{pmatrix}$ (g) $\begin{pmatrix}-6\\4\end{pmatrix}$ (h) $\begin{pmatrix}-3\\-15\end{pmatrix}$

(i) $\begin{pmatrix}36\\0\end{pmatrix}$ (j) $\begin{pmatrix}0\\1\end{pmatrix}$

(k) $|\mathbf{a}| = 5$
$|\mathbf{b}| = \sqrt{34}$
$|\mathbf{c}| = \sqrt{45} = 3\sqrt{5}$
$|\mathbf{d}| = \sqrt{73}$

3. (a) $\begin{pmatrix}4\\12\end{pmatrix}$ (b) $\begin{pmatrix}-5\\-1\end{pmatrix}$ (c) $\begin{pmatrix}1\\16\end{pmatrix}$

4. (a) $\mathbf{a} + \mathbf{b}$
(b) $\mathbf{a} + \mathbf{b} + \mathbf{c}$
(c) $-\mathbf{c} - \mathbf{b}$

5. (a) $\overrightarrow{BC} = \mathbf{v} - \mathbf{u}$ $\overrightarrow{PQ} = \frac{1}{2}\mathbf{v} - \frac{1}{2}\mathbf{u}$
(b) $\overrightarrow{PQ}$ is parallel to $\overrightarrow{BC}$.
(c) $\overrightarrow{PQ}$ is half the length of $\overrightarrow{BC}$.

6. (a) $\overrightarrow{OA} = \begin{pmatrix}-2\\-3\\-1\end{pmatrix}$, $\overrightarrow{OB} = \begin{pmatrix}8\\-3\\-1\end{pmatrix}$, $\overrightarrow{OC} = \begin{pmatrix}8\\6\\-1\end{pmatrix}$

$\overrightarrow{OD} = \begin{pmatrix}-2\\6\\-1\end{pmatrix}$, $\overrightarrow{OE} = \begin{pmatrix}-2\\-3\\5\end{pmatrix}$, $\overrightarrow{OF} = \begin{pmatrix}8\\-3\\5\end{pmatrix}$

$\overrightarrow{OG} = \begin{pmatrix}8\\6\\5\end{pmatrix}$, $\overrightarrow{OH} = \begin{pmatrix}-2\\6\\5\end{pmatrix}$

(b) $\overrightarrow{AF} = \begin{pmatrix}10\\0\\6\end{pmatrix}$

(c) $\sqrt{136} = 2\sqrt{34}$

(d) $\sqrt{217}$

7. (a) $\begin{pmatrix}9\\-6\\12\end{pmatrix}$ (b) $\begin{pmatrix}1\\-1\\9\end{pmatrix}$ (c) $\begin{pmatrix}-5\\3\\1\end{pmatrix}$

(d) $\begin{pmatrix}5\\-4\\22\end{pmatrix}$ (e) $\begin{pmatrix}11\\-7\\7\end{pmatrix}$

8. $\overrightarrow{AC} = \begin{pmatrix}-2\\7\\0\end{pmatrix}$

Chapter 21 Graphs of Trigonometric Functions

Practice Exercise 21.1 Page 162

1. (a)

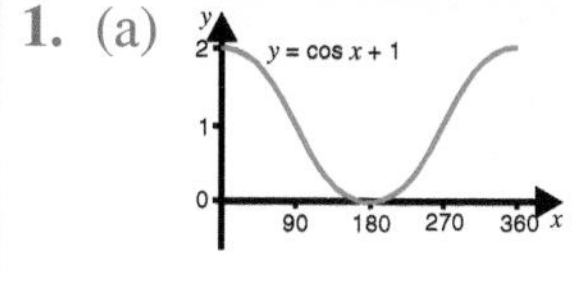

(b)

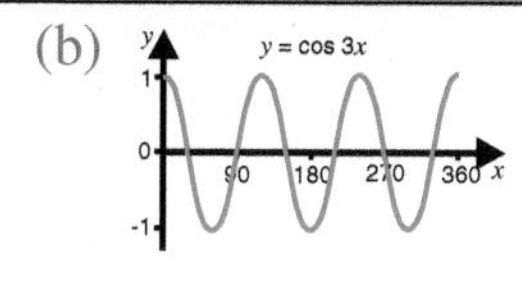

(c)

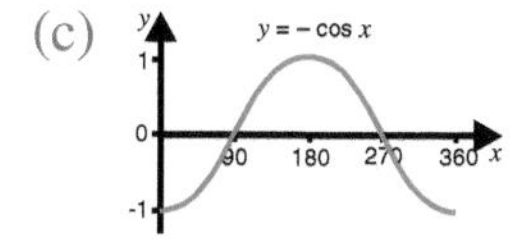

2. (a) Amplitude = 2, period = 120°.
(b) Amplitude = 5, period = 60°.
(c) Amplitude = 4, period = 720°.
(d) Amplitude = 3, period = 1440°.

3. Amplitude = undefined, period = 60°.

4. (a) (i) Reflection in x axis.
(ii) Stretch, from y axis, parallel to x axis, scale factor $\frac{1}{2}$.
(b) Reflection in x axis.
(c) Reflection in x axis, followed by, stretch, from y axis, parallel to x axis, scale factor $\frac{1}{2}$. (Transformations in either order.)
(d) **a**: $y = -\sin x$
b: $y = \sin 2x$
c: $y = -\sin 2x$

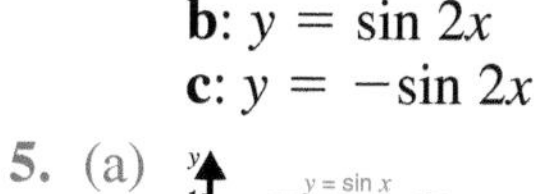

5. (a)

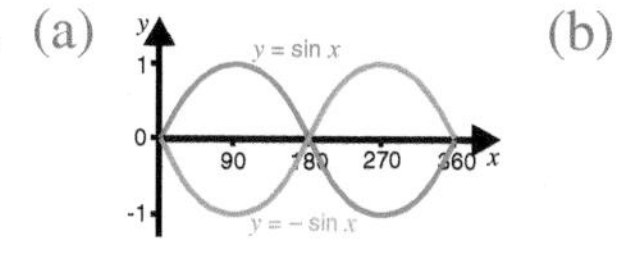

(b)

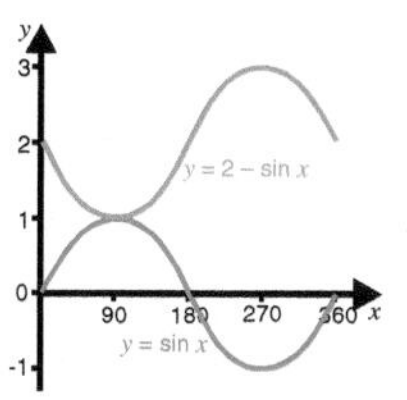

6. (a) $p = 45°$
(b)

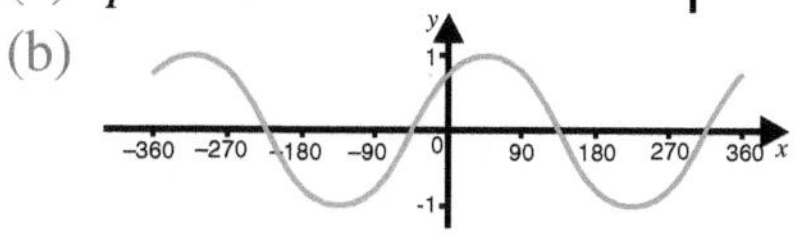

7.

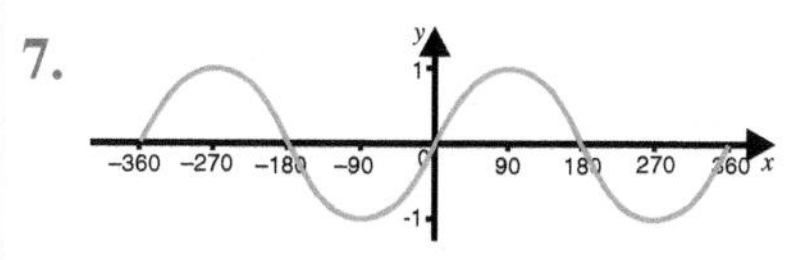

8. (a) $y = \cos(x + 45°)$
(b) $y = \cos x - 1$

Practice Exercise 21.2 Page 165

1. (a) positive (b) positive (c) positive
(d) positive (e) negative (f) negative
(g) negative (h) negative (i) positive
(j) positive (k) negative (l) positive
(m) negative (n) positive (o) negative
(p) positive (q) negative (r) negative
(s) negative (t) positive (u) negative
(v) positive (w) positive (x) negative
(y) negative (z) positive

2. (a) $-\tan 80°$ (b) $-\cos 30°$ (c) $-\sin 20°$
(d) $\tan 70°$ (e) $\cos 60°$ (f) $\sin 60°$
(g) $-\tan 10°$ (h) $-\cos 30°$ (i) $-\sin 70°$
(j) $-\tan 30°$ (k) $\cos 10°$ (l) $-\sin 80°$
(m) $\tan 10°$ (n) $\sin 60°$ (o) $-\cos 60°$
(p) $-\sin 50°$ (q) $-\cos 80°$ (r) $\tan 30°$

Practice Exercise 21.3 Page 166

1. (a) $p = 60°, 300°$ (b) $p = 210°, 330°$
(c) $p = 45°, 225°$ (d) $p = 30°, 150°$
(e) $p = 135°, 315°$ (f) $p = 120°, 240°$
(g) $p = 230°, 310°$ (h) $p = 40°, 320°$
(i) $p = 60°, 120°$ (j) $p = 116°, 296°$
(k) $p = 10.9°, 190.9°$ (l) $p = 55.6°, 304.4°$
(m) $p = 20°, 160°$ (n) $p = 150°, 210°$
(o) $p = 35°, 215°$

2. $x = 228.6°,\ 311.4°,\ 588.6°,\ 671.4°$

3. $x = -245°,\ -115°,\ 245°$

4. $x = -307°,\ -233°,\ 53°,\ 127°$

Review Exercise 21 Page 167

1. (a)

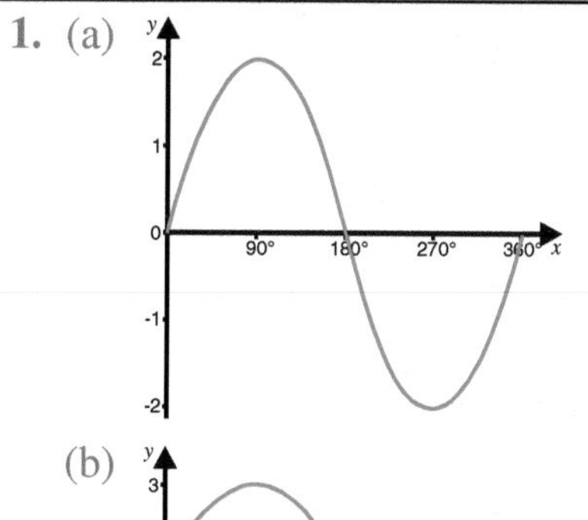

(b)

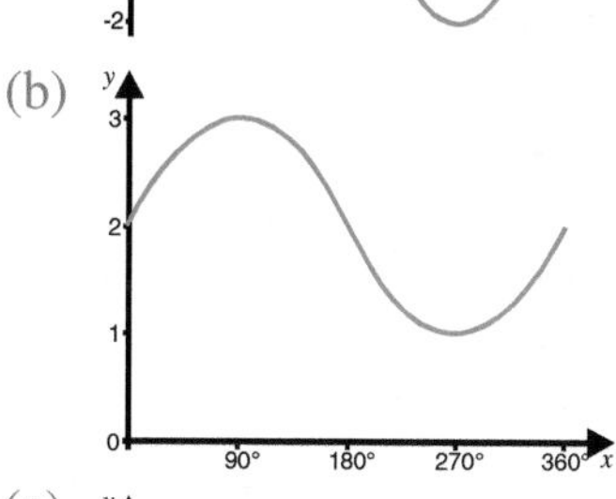

(c)

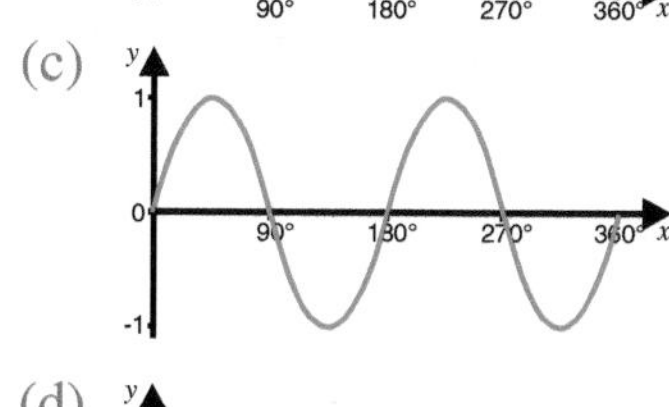

(d)

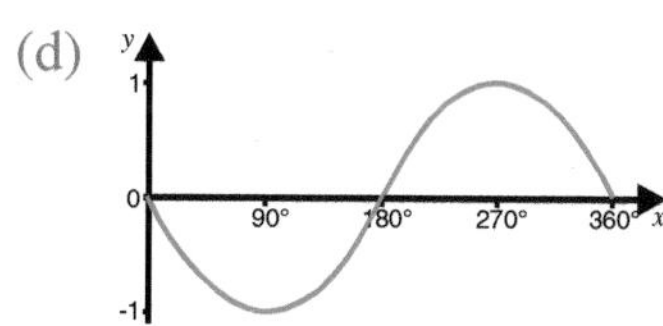

2.

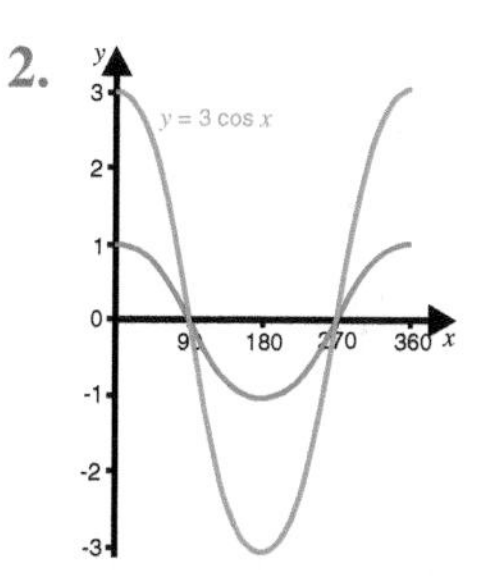

3. (a) Amplitude = 3, period = 180°.
(b) Amplitude = 4, period = 720°.
(c) Amplitude = undefined, period = 90°.
(d) Amplitude = 5, period = 1440°.

4. (a) $x = 30°$ and $x = 150°$
(b) $p = 120°$ and $p = 240°$

5. (b) $x = 143°$

6. $x = 138.6$ and $x = 221.4$

7. (a) $x = 45$ and $x = 225$
(b) $x = 0$, $x = 180$, $x = 360$
(c) $x = 135$ and $x = 315$

8. (a)

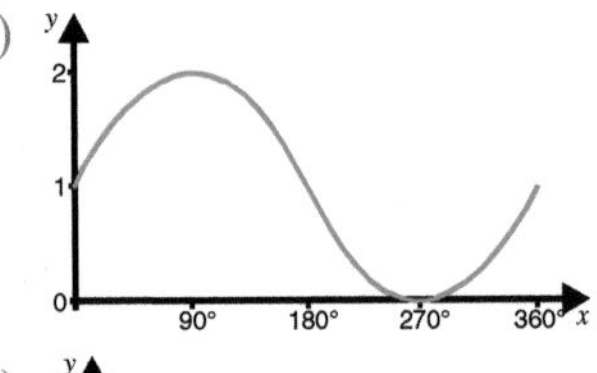

(b)

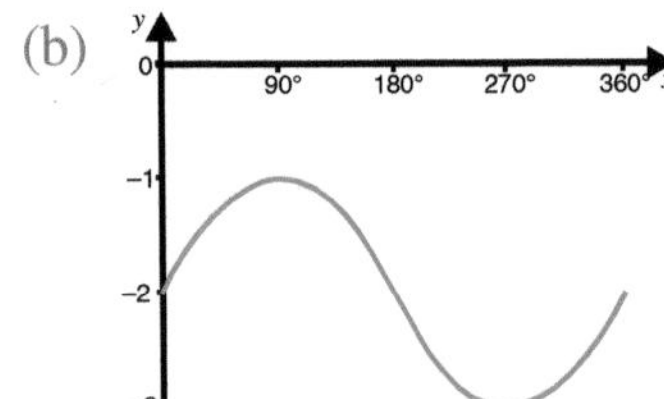

9. (a) 0.5 (b) $x = 210°$ and 330°

10. (a) $-\frac{\sqrt{3}}{2}$ (b) $x = 330°$
(c) $y = 150°$ and 210°

Chapter 22 Working with Trigonometric Relationships

Practice Exercise 22.1 Page 168

1. $x = 135°$, 315°
2. $x = 53.1°$, 306.9°
3. $x = 60°$, 300°
4. $x = 45°$, 225°
5. $x = 143.1°$, 216.9°
6. $x = 131.8°$, 228.2°
7. $x = 66.6°$, 246.6°
8. $x = 210°$, 330°

9. (a) $A(45, 2)$, $B(225, 2)$
(b) Graph of $\tan x$ has period 180°.
x-coordinate of $B = 225$.
x-coordinate of $C = 225 + 180 = 405$.
(c) $P(116.6, -1)$, $Q(296.6, -1)$

10. (a)

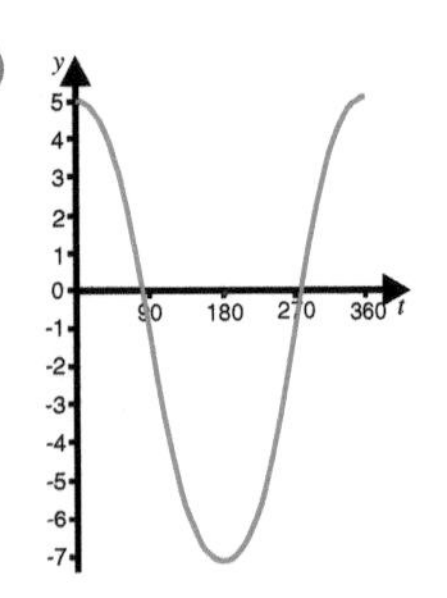

(b) $t = 80.4$ seconds, 279.6 seconds.
Particle passes point O after 80 seconds and 280 seconds.
(c) 7 metres, after 180 seconds (3 minutes).

Review Exercise 22 Page 170

1. (a) $x = 60°$ and 300°
(b) $x = 109.5°$ and 250.5°

2. (a) $x = 82.9°$ and 262.9°
(b) $x = 23.6°$ and 156.4°
(c) $x = 80.4°$ and 279.6°

3. $x = 126.9°$, 233.1°

4. $x = 48.6°$, 131.4°

5. $\cos x \tan x = \sin x$

$$\text{LHS} = \cos x \tan x$$

Substitute $\tan x = \frac{\sin x}{\cos x}$

$$= \cos x \frac{\sin x}{\cos x}$$
$$= \sin x$$
$$= \text{RHS}$$

So, $\cos x \tan x = \sin x$

6. $a = 4$

7. $\frac{\cos^2 x}{1 - \sin x} = 1 + \sin x$

$$\text{LHS} = \frac{\cos^2 x}{1 - \sin x}$$

Substitute $\cos^2 x = 1 - \sin^2 x$.

$$\text{LHS} = \frac{1 - \sin^2 x}{1 - \sin x}$$
$$= \frac{(1 - \sin x)(1 + \sin x)}{(1 - \sin x)}$$
$$= 1 + \sin x$$
$$= \text{RHS}$$

So, $\frac{\cos^2 x}{1 - \sin x} = 1 + \sin x$.

8. (a) 15 m (b) 1 m
(c) 3 am (d) noon

Chapter 23 Sine Rule *and* Cosine Rule

Practice Exercise 23.1 Page 171

1. (a) 6.83 cm² (b) 14.2 cm²
(c) 6.95 cm² (d) 16.2 cm²

2. 84.8 cm²

3. 65 cm²

4. $XY = YZ = 7.37$ cm

5. 8.9 cm

6. 62.9°

Practice Exercise 23.2 Page 173

1. (a) $a = 5.79$ cm
(b) $a = 13.1$ cm
(c) $a = 4.26$ cm

2. (a) $b = 9.92$ cm (b) $c = 8.28$ cm
(c) $b = 10.1$ cm (d) $q = 13.6$ cm
(e) $p = 1.64$ cm (f) $r = 8.48$ cm

3. (a) $\angle ACB = 55°$, $AC = 7.41$ cm, $AB = 6.54$ cm
(b) $\angle MLN = 71.2°$, $LM = 5.84$ cm, $MN = 5.71$ cm
(c) $\angle RPQ = 69.8°$, $QR = 15.7$ cm, $PR = 14.9$ cm
(d) $\angle SUT = 31.8°$, $ST = 12.8$ cm, $UT = 24.1$ cm
(e) $\angle XZY = 31.2°$, $XY = 7.54$ cm, $ZY = 9.4$ cm
4. (a) $PR = 9.7$ cm (b) $QR = 6.8$ cm
5. 130 m

Practice Exercise 23.3 Page 175

1. (a) 61.7° (b) 53.7° (c) 74.7°
2. (a) 137.3° (b) 129.7° (c) 118.5°
3. 48.6°, 131.4°
4. $\angle QPR = 68.1°$, $\angle PRQ = 51.9°$, $PQ = 6.36$ cm
$\angle QPR = 111.9°$, $\angle PRQ = 8.1°$, $PQ = 1.14$ cm

Practice Exercise 23.4 Page 177

1. (a) $a = 10.8$ cm
(b) $a = 8.27$ cm
(c) $a = 15.3$ cm
2. (a) $p = 5.98$ cm
(b) $q = 3.45$ cm
(c) $r = 15.3$ cm
3. 12.7 cm
4. 11.8 cm
5. 13.4 cm
6. 18 cm

Practice Exercise 23.5 Page 178

1. (a) $A = 130.5°$
(b) $A = 32.2°$
(c) $A = 45.8°$
2. (a) $B = 64.5°$
(b) $C = 35.4°$
(c) $P = 41.6°$
3. 45°
4. 130.5°
5. 102.6°
6. 119.9°

Review Exercise 23 Page 179

1. (a) 1140 m² (b) 168 m
2. $YZ = 15.4$ cm
3. 729 cm²
4. (a) $BC = 7.3$ cm (b) 8.4 cm²
5. (a) 28.6° (b) 103 cm²
6. 125.1°
7. 111.8°
8. (a) 41.9° (b) 75.1°
9. 56.2°, 123.8°
10. 17.6 cm²

Chapter 24 Solving Problems Involving Triangles

Practice Exercise 24.1 Page 181

1. (a) (i) 10.8 cm, 73.9°, 46.1°
(ii) 8.44 cm, 46.4°, 23.6°
(b) (i) 46.8 cm²
(ii) 11 cm²
2. (a) $\angle QPR = 39.8°$
(b) 26.9 m²
3. $AD = 93.1$ cm
4. (a) $\angle COD = 45°$
(b) 283 cm²
5. 12.7 m
6. (a) 7.84 m (b) 29.9° (c) 14.7 m
7. (a) 5.17 cm
(b) 10.8 cm²
8. (a) 28° (b) 38.2° (c) 28 cm²
9. 6.63 cm, 15.4 cm

Practice Exercise 24.2 Page 183

1. 26.7 km
2. (a) 118.7°, 35.1°, 26.2°
(b) $XP = 128.5$ m, $\angle XPY = 123°$
(c) 055°
(d) 222°
3. 9.24 km
4. (a) 8 km (b) 114° (c) 294°
5. (a) 49.1 km (b) 347° (c) 167°
6. (a) 9.4 km (b) 11.4 km
7. (a) 001.3° (001°) (b) 6.9 km
8. (b) 2520 m, 128°

Review Exercise 24 Page 185

1. (a) (i) 12.0 cm, 78.0°, 37.0°
(ii) 11.2 cm, 41.8°, 19.2°
(b) (i) 47.1 cm²
(ii) 15.6 cm²
2. (a) 9.4 cm (b) 20 cm
(c) 15 cm (d) 26.4°
3. 111.8°
4. 097.2° (097°)
5. (a) 1260 m
(b) 222°
6. 11.8 km, 282°
7. (a) 2390 m
(b) 306°
8. (a) 162 m
(b) 187 m
9. (a) 1950 m (b) 1590 m (c) 1490 m
10. (a) 125.1° (b) 54.9° (c) 127 cm

Chapter 25

Comparing Distributions

Practice Exercise 25.1 Page 189

1. Range = 6, median = 2.
2. (a) 5
   (b) 3.5
3. (a) 26 mm
   (b) 139 mm
   (c) LQ = 128 mm, UQ = 143 mm
   (d) 15 mm
   (e) 7.5 mm
4. Range = £7
   Median = £8
   SIQR = £1.50
5. (a) 19 kg
   (b) 10 kg
   (c) 3.5 kg
6. (a) Median = 3 letters,
   SIQR = 1 letter.
   (b) Median = 2 books,
   SIQR =1 book.
   (c) Median = 5 days,
   SIQR = 4 days.

Practice Exercise 25.2 Page 190

1. (a) (i) 60 g (ii) 99 g
   (b) Premium potatoes have a greater range and a much higher mean weight.
2. (a) (i) 0.5 mins (ii) 1.9 mins
   (iii) 0.15 mins (iv) 1.93 mins
   (b) (i) 1 min (ii) 2.0 mins
   (iii) 0.25 mins (iv) 2.04 mins
   (c) Girls a little slower on average and more variation in times.
3. (a) 18
   (b) 4.5
   (c) Higher marks awarded in Round 2, and also less variation in marks awarded.
4. Women: mean 1.6, range 6
   Men: mean 1.5, range 2
   Women made more visits to the cinema, though the number of visits is more spread.
5.

Average:	Range:
Boys 6.2	Boys 4
Girls 7.2	Girls $4\frac{1}{2}$

No.
Girls' average greater than boys'.
Correct about variation.

Practice Exercise 25.3 Page 192

1.

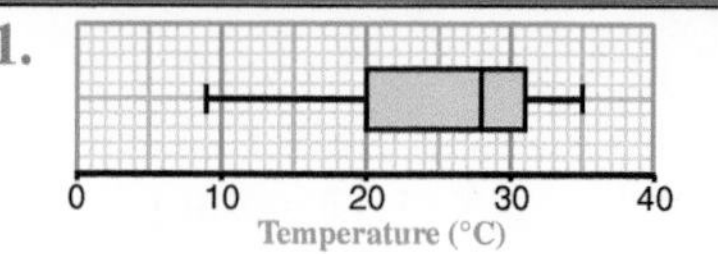

2. (a) 23 (b) 30 (c) 13
3. (a) 0.06 s
   (b) 0.04 s
4. (a) 72%
   (b) 18%
   (c) English
   (d) 10.5%
   (e) English mark has a higher median, Maths marks more spread.
5. (a)

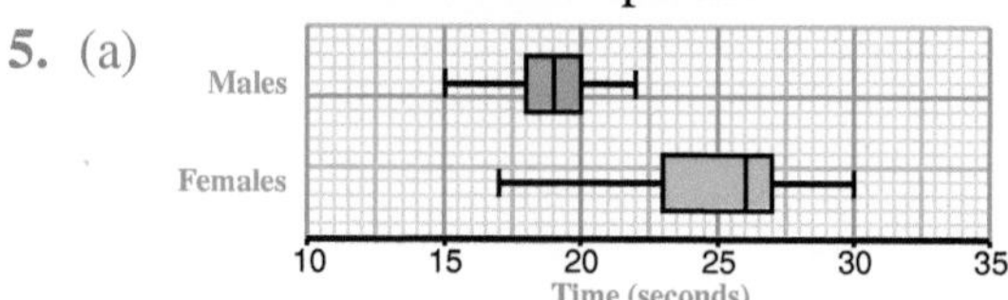

   (b) Males have lower median time and not spread out.
   Females have higher median time and much greater variation.
6. (a) **Reliable**.
   (b) **F1**: 6, **Reliable**: 3.5.
   (c) **F1** variety have lower median length and lengths are spread out.
   Reliable variety have higher median and the lengths are more clustered.

Practice Exercise 25.4 Page 194

1. **Group 1**: mean = 50.5,
   standard deviation = 5.2.
   Group 2: mean = 52,
   standard deviation = 17.8.
   Means are similar, but marks of **Group 2** are very spread out.
2. (a) 6.67 (b) 6.67 (c) 6.67
3. (a) Mean = 4,
   SD = 1.83.
   (b) Mean = 32,
   SD = 4.47.
   (c) Mean = 110 cm,
   SD = 8.56.
   (d) Mean = 2.5 kg,
   SD = 1.26 kg.
4.

Before Christmas:	After Christmas:
Mean = £91.80	Mean = £87.40
SD = £27.26	SD = £3.65

Mean sales per week slightly higher before Christmas.
Values of sales tightly clustered after Christmas, more consistent.

5. Class A:
   Mean = 5.67, SD = 1.52.
   Class B:
   Mean = 7.12, SD = 0.83.
   Class B had a higher mean mark, and marks were more clustered than Class A.

6. Chris:
   Mean = 5.17, SD = 1.47.
   Robert:
   Mean = 3.61, SD = 0.78.
   Robert had a lower mean, better golfer, and lower standard deviation, more consistent scores.

Review Exercise 25 Page 196

1. (a) 38 (b) 23
   (c) 38.3 (d) LQ = 35, UQ = 43
   (e) 8 (f) 4

2. (a) £60
   (b) UQ = £85, LQ = £50
   (c) £35
   (d) £17.50

3. (a) 49
   (b) 1.5
   (c) It is a fair claim.
   Low semi-interquartile range shows that data collected is tightly clustered.

4. (a) 11 (b) 17

5. (a) 13p (b) 53p
   (c) 5p (d) 2.5p

6.

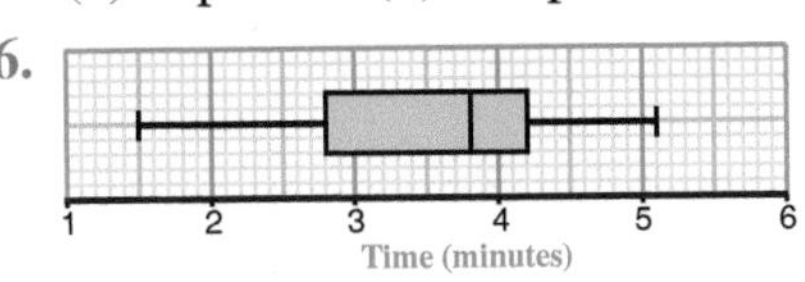

7. Mean = 35.9, SD = 7.04.

8. Boys:
   Mean = 27.6 s, SD = 2.53 s.
   Girls:
   Mean = 31.5 s, SD = 1.76 s.
   Boys have lower mean, faster on average. Times for girls are more tightly clustered, lower standard deviation.

9. £0.08

10. (a) Mean = 14, SD = 4.9
    (b) Mean = 40,
    SD = 5 × 4.9 = 24.5 (or 24.3)
    (c) Craig's data has a higher mean, but the same spread, same SD.
    Mean = 18, SD = 4.9

11. (a) Mean = 51.3 bpm, SD = 5.53 bpm.
    (b) Athletes have slower heart rate, lower mean.
    Heart rates of sedentary people are more spread out, higher SD.

Chapter 26 Using Scatter Graphs

Practice Exercise 26.1 Page 199

1. (a) Negative
   (b) Positive
   (c) Zero
   (d) Positive
   (e) Negative

2. (a) **B** (b) **C** (c) **D**

3. (a) 2
   (b) 164 cm
   (c) No.
   (d) Taller girls usually have larger shoe sizes than shorter girls.

4. (b) Positive correlation.
   (c) Different conditions, types of road, etc.

5. (b) Negative correlation.
   (c) Points are close to a straight line.

Practice Exercise 26.2 Page 201

1. (a) 1.1
   (b) The line passes through the point (0, 0), so, $c = 0$. $h = 1.1d$
   (c) $h = 11$

2. (c) $y = 0.8x - 9$
   (d) $y = 27$
   (e) $x = 70$

3. (b) $y = 1.5x - 16$
   (c) (i) 89 (ii) 49
   (d) Estimate (ii), as estimated value is within the range of known values.

4. (b) Negative correlation.
   (d) $y = 65 - 0.5x$
   (e) 25

Review Exercise 26 Page 202

1. **A**: 3, **B**: 1, **C**: 2

2. (a) Negative (b) Positive (c) Zero

3. (a) Mark Paper 2 = 0.75 (Mark Paper 1) + 30
   (b) 75

4. (b) Positive
   (d) $y = 0.2x + 3$
   (e) 4.6 kg

5. (b) $y = 0.8x + 24$
   (c) 1.6 m

6. (b) $y = 0.2x + 1.4$
   (c) 1.9 kg per square metre.
   (d) Required value is outside range of known values.

Revision Exercise 1 Non-calculator Paper Page 204

1. 2^8. $3^5 = 243$ and $2^8 = 256$
2. (a) 12 (b) 324
   (c) 400 (d) 16
3. (a) $1\frac{1}{2}$ (b) $\frac{1}{12}$ (c) $\frac{5}{18}$
   (d) $1\frac{1}{4}$ (e) 1 (f) $\frac{4}{7}$
4. (a) $y + y^{\frac{3}{2}}$ (b) $x^4 - 1$ (c) $x + x^{\frac{2}{3}}$
5. (a) £40
   (b) (i) 0.25
   (ii) £0.25 per minute charge.
   (c) $c = 0.25t + 20$
6. $a = \dfrac{2(s - ut)}{t^2}$
7. (b) -2
   (c) $x = 2$ and $y = 4$
8. $\angle BAC = 22°$, $\angle ACB = \angle ABC = 79°$
9. (a) $x = 0$ or $x = -\frac{3}{5}$
   (b) $x = \frac{1}{2}$ or $x = -3$
   (c) $x = 5$ or $x = -3$
   (d) $x = 1\frac{1}{3}$ or $x = -3$

Revision Exercise 2 Calculator Paper Page 205

1. (a) $6 - 2\sqrt{2}$ (b) $c = 900$ (c) $1 - \sqrt{5}$
2. (a) (i) 1.4×10^8 (ii) 1.4×10^{-5}
   (b) 8×10^{-2}
   (c) 0.000073
3. £420
4. (a) $p = -5$ (b) $p > -4$
   (c) $q = 8$ (d) $p \geqslant -\frac{1}{4}$
5. (a) 72° (b) 108° (c) 36°
6. (a) 35.2 cm (b) 562 cm³
7. $v = 48.1$
8. (a) Mean = 36.4 mins, SD = 1.34 mins.
   (b) Mean = 37.8 mins, SD = 4.66 mins.
   (c) Trevor takes less time to travel to school, lower mean. Greater variation in times to travel home, higher standard deviation.

Revision Exercise 3 Non-calculator Paper Page 206

1. (a) 4.5×10^{-4}
   (b) 9.84×10^5
   (c) $a = -7$, $b = 4$
2. (a) (i) $\frac{1}{5}$ (ii) 8 (iii) 4
   (b) (i) $6a^5b^5$ (ii) $4ab$ (iii) $5a$
3. (a) $\dfrac{x}{3}$ (b) $\dfrac{11}{20x}$
   (c) $\dfrac{28x}{9}$ (d) $\dfrac{11}{6x}$
4. (a) -1 (b) -4
5. (a) £30 (b) £48
   (c) 6 (d) $C = 6t + 30$
   (e) 8 hours
6. (a) $\begin{pmatrix} 7 \\ 1 \end{pmatrix}$ (b) $\begin{pmatrix} -3 \\ -6 \end{pmatrix}$ (c) $\begin{pmatrix} 4 \\ -5 \end{pmatrix}$
7. (a)

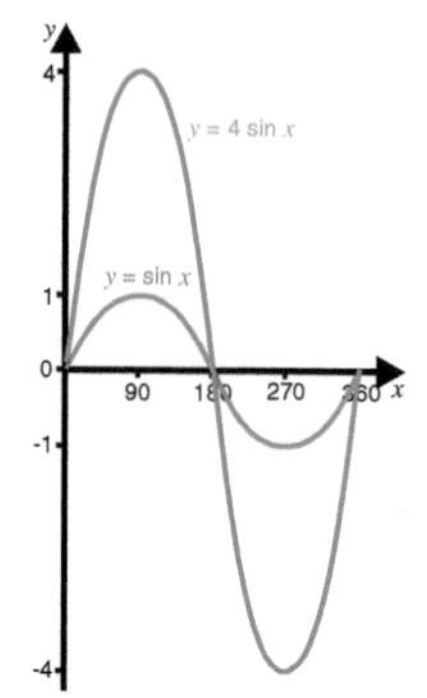

   (b) $x = 210°$ and 330°
8. $h = \dfrac{A - 2\pi r^2}{2\pi r}$

Revision Exercise 4 Calculator Paper Page 207

1. (a) 343 (b) 0.232
   (c) 2.82 (d) 28.9
2. (a) $3x^2 - 5x - 2$
   (b) $3b(2a - b)$
   (c) $p = \dfrac{r + 7q}{4}$
3. (a) $5c + 2p = 116$
   (b) $4c + 5p = 120$
   (c) $c = 20$, $p = 8$; £176
4. 15 cm
5. (a) 14.4 m (b) 57.6°
6. (a) $8 - \dfrac{3}{x} = x$
   Multiply both sides by x.
   $8x - 3 = x^2$
   Rearranging gives
   $x^2 - 8x + 3 = 0$
   (b) $a = -4$, $b = -13$
   (c) $(4, -13)$
   (d) $x = 0.39$ or $x = 7.61$
7. (a) 9 times bigger.
   (b) 2160 cm³

Revision Exercise 5 Non-calculator Paper Page 208

1. (a) 10 (b) $4\sqrt{5}$ (c) 2
2. (a) $1\frac{1}{6}$ (b) $\frac{3}{4}$ (c) $2\frac{1}{2}$

3. (a) $2x^3 + 3x^2$ (b) $x^3 - 3x$

4. $x = 5$

5. (a) (i) 5 (ii) (0, 3)

(b) $2y = 3x - 5$, $\left(y = \frac{3}{2}x - \frac{5}{2}\right)$, gradient $\frac{3}{2}$

$1 = 5x - 4y$ $\left(y = \frac{5}{4}x - \frac{1}{4}\right)$, gradient $\frac{5}{4}$

$2y = 3x - 5$ has steeper gradient.

6. (a) $x = -2$ (b) 2 and -6

(c) $(-2, -16)$

7. (a) D

(b) $\sqrt{208} = 14.4$ (c) $\sqrt{217} = 14.7$

(d) $\begin{pmatrix} 13 \\ 9 \\ 1 \end{pmatrix}$ (e) $\begin{pmatrix} -12 \\ 8 \\ 3 \end{pmatrix}$

8. (a) Median = £62

LQ = £43, UQ = £75

Revision Exercise 6 Calculator Paper Page 209

1. (a) £198 (b) 4.7×10^9

2. (a) £2852.92 (b) 5 more years

3. (a) $r = \frac{mv^2}{F}$ (b) $c = \pm\sqrt{\frac{E}{m}}$

4. (a) $96\,\text{cm}^3$ (b) 6.0 cm

(c) $\frac{4}{3} \times \pi \times r^3 = 7100$

$$r^3 = \frac{7100 \times 3}{4\pi}$$

$$r = \sqrt[3]{\frac{7100 \times 3}{4\pi}}$$

$$r = 11.92\ldots$$

So, radius of sphere is less than 12 cm.

5. $A(-90, 2)$, $B(0, 1)$, $C(180, 3)$, $D(270, 2)$

6. Mean = 34.7 minutes, SD = 16.0 minutes.

Revision Exercise 7 Non-calculator Paper Page 210

1. (a) 2 810 000 (b) 2.81×10^6

2. (a) $x = \frac{1}{3}$ (b) $x = -3$ (c) $x = \frac{3}{2}$

3. $1\frac{1}{8}$

4. (a) $x(x + 8)$ (b) $(x - 8)(x + 1)$

(c) $2(x - 3)(x + 3)$

5. (a) $2x$ (b) $15a^5$ (c) $25t^6$

6. (a) £3 (b) £2 (c) £13

7. $x = 3$ or $x = -3$

8. (a) $\begin{pmatrix} -6 \\ 4 \\ 2 \end{pmatrix}$ (b) $\begin{pmatrix} 1 \\ 1 \\ -1 \end{pmatrix}$ (c) $\begin{pmatrix} -5 \\ 0 \\ 3 \end{pmatrix}$

(d) $\begin{pmatrix} -5 \\ 0 \\ 3 \end{pmatrix}$ (e) $\begin{pmatrix} -11 \\ -1 \\ 7 \end{pmatrix}$ (f) $\begin{pmatrix} 0 \\ 0 \\ 0 \end{pmatrix}$

9. (a) 13 mm

(b) 18.5 mm

(c) UQ = 23 mm

LQ = 17.5 mm

SIQR = 2.75 mm

10. 235°

Revision Exercise 8 Calculator Paper Page 211

1. 5.17

2. (b) 6 hours

(c) $C = 3.5x + 25$ and $C = 6x + 10$

$3.5x + 25 = 6x + 10$

$x = 6$

(d) Less than 6 hours:

MOUNTAIN BIKE HIRE

More than 6 hours:

HEALTHY CYCLE HIRE

3. (a) $3x(x + 2) = 4(x + 3)$

$3x^2 + 6x = 4x + 12$

$3x^2 + 2x - 12 = 0$

(b) $x = 1.7$. Note: $x \neq -2.4$.

4. $61.0\,\text{cm}^3$

5. Same length.

$|\mathbf{a}| = \sqrt{(-4)^2 + 5^2} = \sqrt{41}$

$|\mathbf{b}| = \sqrt{(-1)^2 + 6^2 + 2^2} = \sqrt{41}$

6. (a) Substitute

$a = 2$, $b = -3$, $c = 5$ into $b^2 - 4ac$

$b^2 - 4ac = -31$

$b^2 - 4ac < 0$, so, no solutions.

(b) Substitute

$a = 4$, $b = -4$, $c = 1$ into $b^2 - 4ac$

$b^2 - 4ac = 0$

$b^2 - 4ac = 0$, so, one solution.

(c) $x = -1.00$ or $x = 2.50$

7. (a) $15\,\text{cm}^2$ (b) 6.1 cm (c) 38.7°

Revision Exercise 9 Non-calculator Paper Page 212

1. $x^3 - 5x^2 + 3x + 9$

2. (a) $\frac{\sqrt{7}}{7}$ (b) $\frac{\sqrt{6}}{2}$ (c) $4\sqrt{2}$

3 $\frac{1}{10}$

4. (a) $x + 8$ (b) $\frac{x}{6 - x}$

5. $\frac{8x + 29}{(x - 2)(x + 7)}$

6. (a) $3x^2 + 8x - 3$

(b) $x = 3\frac{1}{4}$

(c) $n < 8$

7. (a)

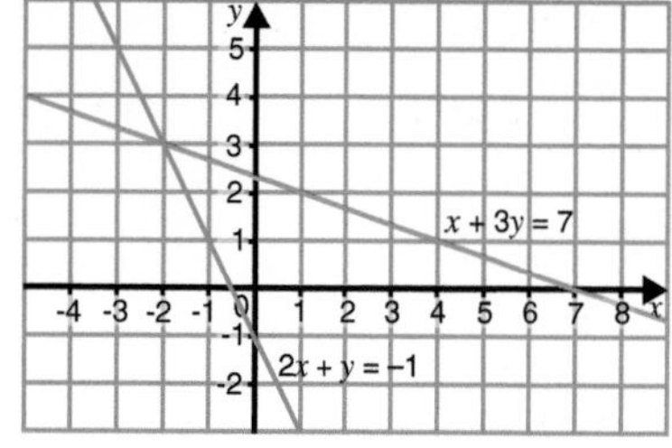

(b) $x = -2, \quad y = 3$

8. $a = \dfrac{bv - c}{v}$

9. $a = 90°, \quad b = 36°, \quad c = 72°,$
$d = 90°, \quad e = 54°$

10. $y - 1 = 3x$ rearranged is $y = 3x + 1$.
$9x = 3y + 7$ rearranged is $3y = 9x - 7$,
which can be written as $y = 3x - \frac{7}{3}$.
$y - 1 = 3x$ and $9x = 3y + 7$
have same gradient 3, are parallel,
never intersect, so, no solution.

Revision Exercise 10 Calculator Paper Page 213

1. 1.4×10^8
2. 10.7 cm
3. (a) £5800 (b) £1050
4. (a) $\angle AOE = 120°$ (b) 24 sides
5. $AB = \sqrt{13}, \quad AC = \sqrt{52}, \quad BC = \sqrt{65}$
$BC^2 = AB^2 + AC^2$,
so, ΔABC is right-angled.
$\angle BAC = 90°$
6. 20.3 cm
7. (a) 283° (b) 117 km

Revision Exercise 11 Non-calculator Paper Page 214

1. (a) $a = 3, \quad b = -23$
(b) $\dfrac{x}{x - 9}$
2. (a) $\frac{2}{3}$ (b) $\frac{2}{5}$ (c) $\frac{3}{2}$ (d) $\frac{5}{3}$
3. $\dfrac{2}{x} - \dfrac{1}{x^2} = \dfrac{2x}{x^2} - \dfrac{1}{x^2} = \dfrac{2x - 1}{x^2}$
4. (a) **(1)** -2
(2) undefined
(3) $\frac{1}{3}$
(4) 0
(5) $-\frac{2}{5}$
(6) -5
(7) undefined
(8) 1
(b) **(2)** and **(7)**

5. (a) $(x - 3)^2$
(b)
(c) $x = 3$
6. (a) $n = \dfrac{C - 350}{5} = \left(\dfrac{C}{5} - 70\right)$
(b) 130
7. (a) $(x + 5)(x - 3)$
(b) $x = 3$ or $x = -5$
(c) $p = 1, \quad q = -16$
(d)

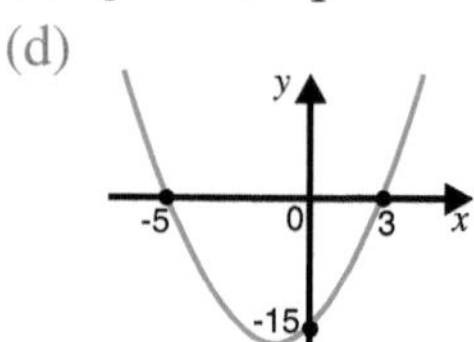

(e) $(-1, -16)$
8. (b) $a = -1\frac{1}{2}, \quad b = -5\frac{1}{4}$
(c) $-5\frac{1}{4}$
(d) Translate $y = x^2$
by $1\frac{1}{2}$ units right to get
$y = \left(x - 1\frac{1}{2}\right)^2$.
Translate $y = \left(x - 1\frac{1}{2}\right)^2$
by $5\frac{1}{4}$ units down to get
$y = x^2 - 3x - 3$.
9. $\sin^4 x - \cos^4 x = \sin^2 x - \cos^2 x$
LHS $= \sin^4 x - \cos^4 x$
$= (\sin^2 x - \cos^2 x)(\sin^2 x + \cos^2 x)$
$\sin^2 x + \cos^2 x = 1$
LHS $= \sin^2 x - \cos^2 x$
= RHS
So, $\sin^4 x - \cos^4 x = \sin^2 x - \cos^2 x$.
10. $p = 4$

Revision Exercise 12 Calculator Paper Page 215

1. (a) 1.116×10^7 miles
(b) 8.6 minutes
2. (a) $12x^5y^5$ (b) $2xy^3$
(c) $8x^9y^{-6}$ (d) $2x^3y^{-3}$
(e) x^2y^{-2} (f) $4x^{-18}y^{13}$
3. (a) £6980 (b) £9320
4. (a) 10.5 cm (b) 13.1 cm²
(c) 12.5 cm²
5. (a) 5330 cm³ (b) 4.1 cm³
(c) 1300
6. (a) 3.3 km (b) 11.7 km
(c) 260°

Exam Practice — Non-calculator Paper — Page 216

1. (a) (i) $10\sqrt{3}$ (ii) $11\sqrt{3}$ (iii) $6\sqrt{6}$
   (b) $\frac{\sqrt{10}}{15}$ (c) $\frac{\sqrt{10}}{2}$
2. $3\frac{1}{4}$
3. (a) $\frac{1}{16}$ (b) $\frac{1}{12}$ (c) $\frac{9}{16}$
4. (a) $x \leqslant 2$ (b) $21 - 13x$
5. (a) (i) $m - m^{\frac{5}{2}}$ (ii) $5x^2 - 6x - 11$
   (b) $a = \frac{bc - ed}{f}$
6. $\pi r^2 = \frac{36}{\pi}$, $r^2 = \frac{36}{\pi^2}$, $r = \frac{6}{\pi}$
   Using $C = 2\pi r$.
   $C = 2 \times \pi \times \frac{6}{\pi}$, $C = 12\,\text{cm}$
7. Value of P is multiplied by 8.
8. (a) $\frac{a}{5}$ (b) $2(x - 3)(x + 3)$
9. (a) $f(-2) = 2$, $f(-1) = 0$, $f(0) = 0$, $f(1) = 2$, $f(2) = 6$
   (b) $a = 4$ (c) $f(p) = 8$
10. (a) $x^2 - 10x + 25$ (b) $4x^3 + 9x^2 - 7x + 6$
    (c) (i) $3m(m - 2)$ (ii) $(t - 4)(t + 3)$
11. (a) $y = 3 - x$ (b) $(-7, -2)$
12. $\frac{4x + 4}{x(x + 4)}$
13. $x = 4$, $y = 1$
14. (a) $m = 30$
    (b) (i) $3x^2y - 6y^2$ (ii) $3x^2 - x - 10$
15. (a) $9x + 12$ (b) $x = 14$
16. (a) $P = 0.09$ (b) $n = \frac{m}{\sqrt{P}}$
17. (a) $p = 1$, $q = -3$ (b)
    (c) $(-1, -3)$
18. £40
19. (a) $1\frac{5}{12}$ (b) $\frac{1}{10}$ (c) $\frac{3}{10}$
20. (a) $y = 0$ or $y = -5$
    (b) $m = 3$ or $m = 4$
21. 18 cm
22. (a) $9\frac{1}{3}$ (b) $1\frac{1}{4}$
23. $\frac{x + 23}{(x - 4)(x + 5)}$
24. 10 sides.
25. (a) $AB = 5\sqrt{2}\,\text{cm}$ (b) $\cos A = \frac{\sqrt{15}}{5}$
26. $\cos PQR = \frac{4^2 + 5^2 - 7^2}{2 \times 4 \times 5} = \frac{16 + 25 - 49}{40} = -0.2$
    As $\cos PQR$ is negative, angle $PQR > 90°$.
27. (a) $2\sqrt{2} - 1$ (b) -2 (c) $29 + 12\sqrt{5}$
28. Substitute $a = 4$, $b = -2$ and $c = -1$ into the quadratic formula and simplify.
29. $2x^3 - 9x^2 + 5x - 4$

Exam Practice — Calculator Paper — Page 218

1. (a) 2 (b) $\frac{1}{9}$
2. (a) x (b) $x^{\frac{1}{4}}$ (c) x^3
3. 1.49×10^9 square miles.
4. (a) $x(x - 3)(x + 3)$
   (b) (i) m^5 (ii) n^3 (iii) x^5y^2 (iv) p^6
   (c) $x = \frac{y - 10}{5}$
   (d) (i) $t = 9$ (ii) $x = 8$
5. 3.46×10^5
6. $22.0\,\text{cm}^2$
7. $x = 0.70$ or $x = 4.30$
8. Real, unequal, rational.
9. $x = 48.6, 131.4$
10. 16.4%
11. (a) 3 cm (b) $5.63\,\text{cm}^2$
12. (a) $7^2 = x^2 + 8^2 - 2 \times x \times 8 \times \cos 60°$
    $7^2 = x^2 + 8^2 - 2 \times x \times 8 \times 0.5$
    $49 = x^2 + 64 - 8x$
    $x^2 - 8x + 15 = 0$
    (b) $AB = 3\,\text{cm}$ or $5\,\text{cm}$
13. $x = 0.22$ or $x = 2.28$
14. $a = 120°$, $b = 150°$, $c = 135°$
15. $9.8\,\text{cm}^2$
16. (a) 8 cm (b) 30.1 cm
17. (a) $810\,\text{cm}^3$ (b) $248\,\text{cm}^2$
18. (a) $\begin{pmatrix} 4 \\ -6 \end{pmatrix}$
    (b) Right-angled isosceles triangle.
    $|\overrightarrow{AB}| = \sqrt{52}$
    $|\overrightarrow{BC}| = \sqrt{52}$
    $|\overrightarrow{AC}| = \sqrt{104}$
    $AB^2 + BC^2 = AC^2$
    So, ΔABC is right-angled at B, converse of Pythagoras' Theorem.
    $AB = BC$, so, ΔABC is also isosceles.
19. (a) (ii) $b = 0.7h + 4$
    (b) (i) 39 cm (ii) 61 cm
20. Distance 15.2 km, bearing 097° (**not** 083°).
21. (a)

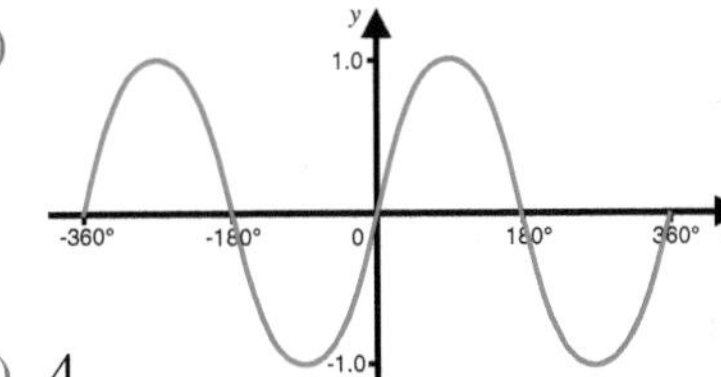

    (b) 4
    (c) $x = 37°, 143°, -217°, -323°$
22. By converse of Pythagoras' Theorem
    $13^2 = 5^2 + 12^2$
    So, ABC is a right-angled triangle.
23. $\overrightarrow{AC} = \begin{pmatrix} -3 \\ 3 \\ 0 \end{pmatrix}$

INDEX

H

I

K

L

M

N

P

Q

R

S

T

U

V

W

X

Y

Z